S0-EFM-657

Advances in Physical Organic Chemistry

In Memoriam

Professor Victor Gold, FRS
29 June 1922 – 29 September 1985
Founding Editor of *Advances in Physical Organic Chemistry*

Advances in Physical Organic Chemistry

Volume 22

Edited by

V. GOLD

Department of Chemistry
King's College London
Strand, London WC2R 2LS

and

D. BETHELL

The Robert Robinson Laboratories
University of Liverpool
P.O. Box 147, Liverpool L69 3BX

ACADEMIC PRESS 1986
Harcourt Brace Jovanovich, Publishers

London Orlando
San Diego New York Austin
Boston Montreal Sydney Tokyo

ACADEMIC PRESS INC. (LONDON) LTD
24/28 Oval Road
London NW1 7DX

United States Edition published by
ACADEMIC PRESS INC.
Orlando, Florida 32887

ISBN 0-12-033522-0
ISSN 0065-3160

TYPESET BY BATH TYPESETTING LTD., BATH, U.K.
AND PRINTED IN GREAT BRITAIN BY
ST. EDMUNDSBURY PRESS, BURY ST. EDMUNDS

Contents

Contributors to Volume 22 vii

Intramolecular Reactions of Chain Molecules 1

LUIGI MANDOLINI

1 Introduction 2
2 General background. Basic concepts and definitions 4
3 Hypothetical cyclisation reactions 12
4 Cyclisation reactions in solution 30
5 EM and transition state structure 84
6 EM and the synthesis of ring compounds 102

Mechanisms of Proton Transfer between Oxygen and Nitrogen Acids and Bases in Aqueous Solution 113

FRANK HIBBERT

1 Introduction 113
2 Simple proton transfers of oxygen and nitrogen acids 115
3 Proton transfer along hydrogen bonds 127
4 Proton removal from intramolecular hydrogen bonds 148
5 Hindered proton transfer from molecular cavities 184
6 Multiple proton transfers 190
7 Future work 205

Organic Reactivity in Aqueous Micelles and Similar Assemblies 213

CLIFFORD A. BUNTON and GIANFRANCO SAVELLI

1 Introduction 214
2 Micellar structure and ion binding 219
3 Quantitative treatments of rates and equilibria 222

4 Spontaneous, unimolecular and water-catalysed reactions 244
5 The source of micellar rate enhancements 251
6 Functional micelles and comicelles 259
7 Micellar effects on acid-base equilibria 265
8 Reactions in non-micellar aggregates 268
9 Stereochemical effects 277
10 Applications in synthetic and analytical chemistry 279
Appendix 282

Structure and Reactivity of Carbenes having Aryl Substituents **311**

GARY B. SCHUSTER

1 Introduction 311
2 Orbitals and energetics of carbenes 312
3 The menagerie of aromatic carbenes 316
4 Experimental investigation of chemical and physical properties of carbenes 320
5 The structure and reactivity of aromatic carbenes 331
6 Understanding the properties of aryl-substituted carbenes 352
7 Conclusions 357

Author Index **363**

Cumulative Index of Authors **379**

Cumulative Index of Titles **381**

Contributors to Volume 22

Clifford A. Bunton Department of Chemistry, University of California, Santa Barbara, California 93106, U.S.A.

Frank Hibbert Department of Chemistry, King's College London, Strand, London WC2R 2LS, U.K.

Luigi Mandolini Dipartimento di Chimica, Università di Roma "La Sapienza", Piazzale Aldo Moro 2, 00185 Rome, Italy

Gianfranco Savelli Dipartimento di Chimica, Università di Perugia, Via Elce di Sotto 8, 06100 Perugia, Italy

Gary B. Schuster School of Chemical Science, University of Illinois at Urbana-Champaign, Box 58, Roger Adams Laboratory, 1209 W. California Street, Urbana, Illinois 61801, U.S.A.

Intramolecular Reactions of Chain Molecules

LUIGI MANDOLINI

Dipartimento di Chimica and Centro di Studio CNR sui Meccanismi di Reazione, Università "La Sapienza", Roma, Italy

1 Introduction 2
2 General background. Basic concepts and definitions 4
Kinetics of cyclisation. The cyclisation constant C 4
The effective concentration C_{eff} and the effective molarity EM 6
Equilibria in ring formation. Macrocyclisation equilibria. The chelate effect 9
3 Hypothetical cyclisation reactions 12
Thermodynamic properties of alkane chains 13
Thermodynamic properties of ring compounds 15
Enthalpy and entropy effects on ring closure 21
The *gem*-dimethyl effect 27
4 Cyclisation reactions in solution 30
The experimental approach to the kinetics of cyclisation 30
Early studies 31
Cyclisation of polymethylene and polyoxyethylene chains 35
Cyclisation of polymer chains. Theory and experiment 64
Entropy changes for cyclisation reactions in solution 74
5 EM and transition state structure 84
Extrathermodynamic calculations of EM's 86
The question of the smallest rings in S_N2 reactions 89
Miscellaneous reactions 95
The tight and loose transition state hypothesis. Intramolecular proton transfer 99
6 EM and the synthesis of ring compounds 102
Ring closure under batchwise conditions 103
Ring closure under influxion conditions 103
Acknowledgements 106
Note added in proof 106
References 107

ADVANCES IN PHYSICAL ORGANIC CHEMISTRY
VOLUME 22 ISBN 0 12 033522 0

1 Introduction

This chapter is concerned with reaction rates, equilibria, and mechanisms of cyclisation reactions of chain molecules. A detailed analysis of the historical development of experimental approaches and theories concerning the intramolecular interactions of chain molecules and the processes of ring closure is outside the scope of this chapter. It must be borne in mind, however, that the present state of the art in the field is the result of investigations which have been approached with a variety of lines of thought, methods, and objectives.

About a century ago, a brilliant young organic chemist named Perkin synthesised for the first time derivatives of trimethylene and tetramethylene – now better known as cyclopropane and cyclobutane, respectively – by treatment with base of the 2-bromoethyl and 3-bromopropyl derivatives of malonic ester and acetoacetic ester. An account of these pioneering investigations was presented by Perkin himself in the first Pedler lecture (Perkin, 1929). Since that time, countless small and common ring[1] compounds have been synthesised via the intramolecular counterparts of known intermolecular reactions. A significant impetus to the field was given by the development of synthetic procedures leading to macrocycles, which were pioneered by Ruzicka, Stoll, and Ziegler. A comprehensive review on earlier methods of macrocycle synthesis is that of Ziegler (1955). That this is still a research area attracting the attention of numerous investigators is shown in a very recent and comprehensive review by Rossa and Vögtle (1983).

Parallel to the search for synthetic methods to make rings of all sizes, kinetic studies have appeared with the aim of providing insight into physical aspects of ring closure. Earlier kinetic work has been reviewed by Salomon (1936a) and Bennett (1941), and more recent work by Illuminati and Mandolini (1981) and Winnik (1981a).

A significant portion of the evidence available on the kinetics of intramolecular reactions falls in the domain of neighbouring group participation (Capon and McManus, 1976). In principle, the term is sufficiently wide to encompass "all intramolecular reactions and all reactions which involve nonelectrostatic through-space interactions between groups in the same molecule" (Capon and McManus, 1976). However, the most common types of neighbouring group participation involve nucleophilic participation either with or without anchimeric assistance, and intramolecular acid and base catalysis. Available examples in this field essentially refer to reactions involving cyclic structures with 3–6 ring atoms. In practice, therefore, the term neighbouring group participation is restricted to the shorter chains, namely,

[1]The usual distinction between small (3- and 4-membered), common (5- to 7-membered), medium (8- to 11-membered), and large (12-membered and higher) rings will be adopted throughout this chapter.

the ones which experience the highest ease of ring closure. Intramolecular catalysis phenomena have attracted much attention, both from the experimental and theoretical points of view, not only on account of their inherent mechanistic interest, but also because they are believed to model the extraordinary efficiency of enzyme catalysis (Page, 1973; Jencks, 1975; Fife, 1975; Guthrie, 1976; Gandour, 1978; Fersht and Kirby, 1980; Kirby, 1980).

Finally, there is a large body of experimental and theoretical contributions from investigators who are mainly interested in the dynamic and conformational properties of chain molecules. The basic idea is that the cyclisation probability of a chain is related to the mean separation of the chain ends (Morawetz, 1975). Up to date comprehensive review articles are available on the subject (Semlyen, 1976; Winnik, 1977, 1981a; Imanishi, 1979). Rates and equilibria of the chemical reactions occurring between functional groups attached to the ends or to the interior of a flexible chain molecule are believed to provide a convenient testing ground for theories of chain conformations and chain dynamics in solution.

In the last two decades, there has been a large accumulation of experimental evidence as well as of theoretical interpretations of intramolecular reactions. One notes, however, that attention has been focused on the phenomena of immediate interest to the various specialists. As a consequence of the fact that specialisation implies intensification of knowledge on the one hand but limitation on the other, there has still been insufficient communication and cross-fertilisation between the different schools. This situation is well exemplified by the two most extensive reviews on intramolecular phenomena, namely, that of Kirby (1980), entitled "Effective Molarities for Intramolecular Reactions", and that of Winnik (1981a), entitled "Cyclisation and the Conformation of Hydrocarbon Chains", which present different approaches and apparently unrelated facts and theories.

This author is perfectly aware that he could add very little to the work done by these workers if an attempt was made to focus on intramolecular catalysis phenomena or on the relevance to cyclisation of available models of chain conformation and chain dynamics: instead, the aim will be the presentation of a general treatment of the subject, namely, one that includes the cyclisation of very short chains as well as that of very long chains of, say, 100 atoms or more. With a subject as vast as this, an encyclopaedic review would be a hopeless task. Therefore, the subject will be treated in a systematic and critical way, with more concentration on reaction series with regular and wide variations in structure, rather than on scattered examples. The aim will be to show that the field of intramolecular reactions is a mature area in which the merging of concepts from both physical organic chemistry and polymer chemistry leads to a unified treatment of cyclisation rates and equilibria in terms of a few simple generalisations and theories.

General background. Basic concepts and definitions

KINETICS OF CYCLISATION. THE CYCLISATION CONSTANT *C*

The cyclisation of a bifunctional molecule X∼∼Y is made complex by the competition of the polymerisation reaction through head-to-tail condensation. A complete reaction scheme would include countless consecutive as well as parallel condensation reactions. Although the system may be a great deal simplified by the usual assumption of functional group reactivity independent of the length of the chain (Billmeyer, 1971), a significant complication arises from the cyclisation of the linear x-meric species to the x-meric ring. Exact solution of the complex system of differential equations derived therefrom is a hopeless task. Since cyclisation is first order, whereas polymerisation is second order, it is always possible in principle to carry out a given cyclisation reaction at a concentration low enough to suppress polymerisation (Ruggli, 1912). It is instructive, however, to consider the general case where the α,ω-bifunctional reactant undergoes simultaneous cyclisation and polycondensation.

In their pioneering work on the lactonisation of an extensive series of ω-hydroxycarboxylic acids, Stoll *et al.* (1934) proposed a procedure for determining the specific rate for cyclisation from the yield of the monomeric lactone at the end of the reaction. To this end they considered the simplified

$$\text{monomer} \xrightarrow{k_{\text{intra}}} \text{ring product} \tag{1a}$$

$$2\ \text{monomer} \xrightarrow{k_{\text{dim}}} \text{polymer} \tag{1b}$$

reaction scheme (1a, b) which led to the integrated equation (2) relating the weight ratio Y of monomeric ring product to polymer to the quantity X defined as $k_{\text{intra}}/k_{\text{dim}}\ [\text{M}]_0$, $[\text{M}]_0$ being the initial monomer concentration.

$$Y/(1+Y) = X \ln(1+X^{-1}) \tag{2}$$

Thus, within the approximations on which scheme (1) is based, the yield $Y/(1+Y)$ of ring product is solely determined by the value of the initial monomer concentration relative to the $k_{\text{intra}}/k_{\text{dim}}$ ratio, which Stoll *et al.* termed the cyclisation constant C. C has units of mol l^{-1} and represents the monomer concentration at which intra- and intermolecular processes occur at the same rate. That Stoll *et al.* disregarded the fact that two monomer units are used up in the intermolecular condensation, as well as that the monomer disappears by reaction with functional groups at the ends of polymer chains was pointed out by Morawetz and Goodman (1970), who proposed an alternative approximate solution to the problem, as based on

$$\text{monomer} \xrightarrow{k_{\text{intra}}} \text{ring product} \tag{3a}$$

$$2\ \text{monomer} \xrightarrow{k_{\text{pol}}} \text{polymer} \tag{3b}$$

$$\text{monomer} + \text{polymer} \xrightarrow{k_{\text{pol}}} \text{polymer} \tag{3c}$$

$$\text{polymer} + \text{polymer} \xrightarrow{k_{\text{pol}}} \text{polymer} \tag{3d}$$

the kinetic scheme (3a–d) in which the possible formation of polymeric ring products was neglected. Introduction of the dimensionless parameters $m = [M]/[M]_0$ and $p = [P]/[M]_0$ led to expressions (4) and (5). Numerical solution of (4) and (5) showed that the cyclisation constants estimated by Stoll *et al.* (1934) and Stoll and Rouvé (1935) from experimental Y-values are too low by a factor of from 5 to 2.5 when Y varies from 0.01 to 2.

$$dp/dm = (p^2 - m^2)/m(X + 2m + p) \tag{4}$$

$$Y/(1 + Y) = X\int_1^0 (2m + p + X)^{-1}\, dm \tag{5}$$

Neglect of the formation of polymeric rings, however, is sometimes too crude an approximation. It may happen that the cyclisation constant C_2 of the linear dimer is larger than the cyclisation constant C_1 of the monomer. It may also happen that the concentration of the monomer is comparable to C_1 but smaller than C_2. When this is the case, the open chain dimer, once formed, will show a higher tendency to cyclise than to react with the monomer to give the linear trimer. Under the above conditions the stepwise polymerisation is truncated after the first step, and the system is described to a useful approximation by scheme (6).

$$\text{monomer} \xrightarrow{k_{\text{intra}}} \text{monomeric ring} \tag{6a}$$

$$2\ \text{monomer} \xrightarrow{k_{\text{dim}}} \text{dimer} \tag{6b}$$

$$\text{dimer} \xrightarrow{k'_{\text{intra}}} \text{dimeric ring} \tag{6c}$$

A clear-cut example of this kind of behaviour is found in Freundlich's early work on the cyclisation (7) of ω-bromoalkylamines quoted by Salomon

$$Br(CH_2)_2NH_2 \longrightarrow \text{aziridine (NH)} + \text{piperazine (HN, NH)} \tag{7}$$

(1936a). The high ease of formation of the 6-membered ring prevents the linear dimer $Br(CH_2)_2NH(CH_2)_2NH_2$ from further intermolecular condensation to trimer. Many additional examples can be found in the synthetic literature where major or exclusive formation of a dimeric 6-membered ring occurs on attempted formation of a 3-membered ring. Additional important examples concern the cyclisation of chains in the neighbourhood of 10 skeletal atoms. Here the dimeric rings are quite often the major products, little or no monomeric ring being formed (Ziegler, 1955).

For a system which is well-described by scheme (6) the rate equation is (8) where C is Stoll's cyclisation constant (Galli *et al.*, 1973). Equation (8) is a

$$\ln \frac{[\mathrm{M}]_0}{[\mathrm{M}]_t} + \ln \frac{[\mathrm{M}]_t + C/2}{[\mathrm{M}]_0 + C/2} = k_{\mathrm{intra}} t \tag{8}$$

mixed first- and second-order rate equation which reduces to the familiar first-order equation when $[\mathrm{M}]_0 \ll C$. Product composition under the conditions of scheme (6) is predicted by (9) (Galli and Mandolini, 1975). This

$$\text{Yield} = \frac{1}{2\alpha} \ln (1 + 2\alpha) \tag{9}$$

equation relates the (normalised) yield of monomeric ring product to the dimensionless parameter α, which is defined as $[\mathrm{M}]_0/C$ and may be viewed as a reduced initial concentration. Equation (9) is equivalent to (2), apart from the presence of the coefficients 2. Figure 1 shows plots of calculated yield of monomeric ring product as a function of α, as based on (9) and (5) (curves A and B, respectively). The two curves are practically superposable when α is small, but significant differences exist in the region of large α's, where polymerisation predominates. Yields are overestimated by (9) and underestimated by (5). Thus, any actual cyclisation reaction where polymerisation is significant is represented by a point lying somewhere in the region between the two curves, depending on the particular system at hand.

THE EFFECTIVE CONCENTRATION C_{eff} AND THE EFFECTIVE MOLARITY EM

In the same year in which Stoll *et al.* (1934) defined the cyclisation constant C, Kuhn (1934) laid the foundations of the theoretical approaches to the conformational statistics of hydrocarbon chains and considered the cyclisation probability of the chain as a fundamental, chain length dependent phenomenon related to chain shape. He proposed to view the specific rate k_{intra} of an intramolecular reaction between a pair of reactive groups attached to the ends of a chain molecule as the product of the effective concentration C_{eff} of

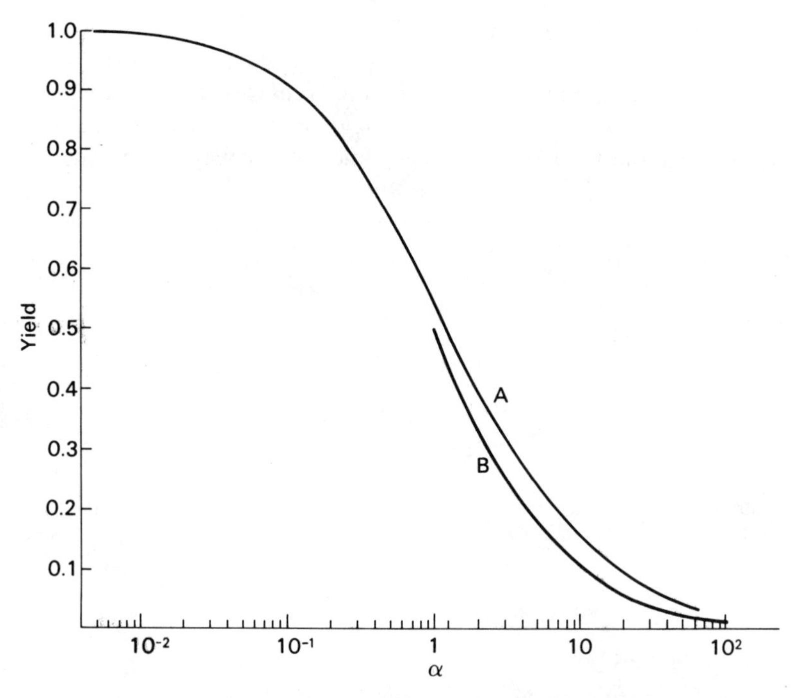

FIG. 1 Yield of monomeric ring product from a bifunctional precursor as a function of the dimensionless parameter α. Plot A from (9) and Plot B from (5)

one chain end in the neighbourhood of the other multiplied by the specific rate k_{inter} for the reaction between the chain ends on different molecules (10).

$$k_{intra} = C_{eff} \cdot k_{inter} \tag{10}$$

In this model, C_{eff} is viewed as a "physically real" concentration. The intramolecular reaction between A and B can be visualised (Fig. 2) as occurring within a sphere centred at A, with radius l equal to the length of the fully extended chain, from which B cannot escape. If the distribution of B in the sphere were homogeneous, as it would be if the motion restraints due to the interconnecting chain were absent, then C_{eff} would be easily calculated as the concentration of one molecule or $1/N_A$ moles in a volume of $(4/3)\pi l^3$. But this is far from being the case. In any realistic model, an appropriate form of the distribution function $W(r)$ of the end-to-end chain displacement r must be taken into account. The simplest solution to this problem was presented by Kuhn (1934) in terms of the random walk model, which leads to a Gaussian distribution in the limit of a large number of links. Kuhn showed that when r approaches zero, $W(r)$ approaches the value given by (11), where $\langle r_0^2 \rangle$ is

$$W(0) = 4\pi r^2 [3/(2\pi \langle r_0^2 \rangle)]^{3/2} \qquad (11)$$

the mean square end-to-end length. W(0) is the cyclisation probability of the chain and the ratio $W(0)/4\pi r^2$. In the limit of $r \to 0$, represents the local concentration C_{eff} of one chain end in the immediate vicinity of the other.

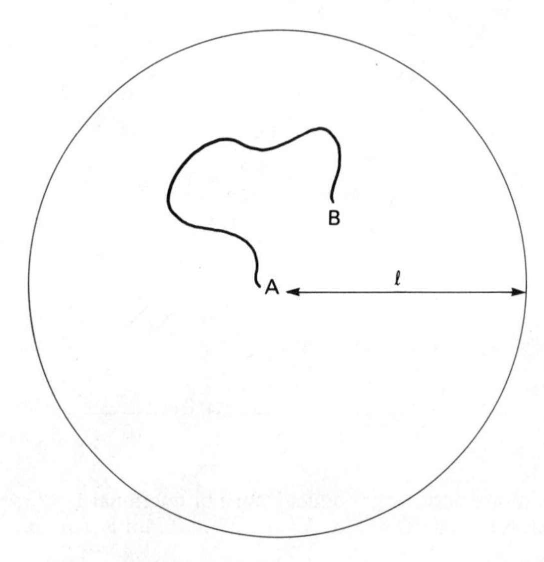

FIG. 2 Schematic representation of an intramolecular reaction. Because B cannot escape from the sphere of radius l equal to the length of the fully extended chain, its average "concentration" in the sphere is $3000\,(4\pi l^3 N)^{-1}$ mol l^{-1}

It is remarkable that in the same year, 1934, two independent approaches, those of Stoll *et al.* and of Kuhn, led to the definition of two quantities which are conceptually quite similar and can be practically identical in many actual cases. In either case the intramolecular reaction is compared to a corresponding intermolecular process. This is the dimerisation reaction of the bifunctional reactant in the definition of the cyclisation constant C; in the case of the effective concentration C_{eff}, k_{inter} must be determined with the aid of an intermolecular model reaction, the choice of which is not always obvious and can possibly lead to conceptual as well as experimental difficulties. It is also worth noting that although these early workers established a firm basis for interpretation of physical as well as of preparative aspects of intramolecular reactions, no extensive use of quantities C and C_{eff} appears to have been made in the chemical literature over more than three decades after their definition. This is in spite of the enormous development of studies in the field of neighbouring group participation and in the closely-related field of intramolecular acid-base and nucleophilic catalysis (Capon, 1964). It was only in the late 1960s and in the early 1970s that it became customary to take as a measure for the efficiency of an intramolecular catalytic group the ratio between the specific rate k_{intra} of the intramolecularly catalysed reaction and the specific rate k_{inter} of an analogous non-cyclisation reaction proceeding through the same mechanism (Jencks, 1969; Capon, 1972). The ratio k_{intra}/k_{inter} is a quantity whose formal meaning is the concentration in mol l^{-1} of the catalytic group to make the rate of the intermolecular model reaction equal to that of the intramolecular reaction. The terms effective concentration, EC and effective molarity, EM, have been given to the ratio k_{intra}/k_{inter}, but the latter is nowadays more widely used (Kirby, 1980; Illuminati and Mandolini, 1981) and will be adopted throughout this chapter.

It does not seem to be widely recognised that the EM is conceptually identical to C_{eff} defined by Kuhn (1934) many years before. However the chemical meaning attached to the EM in the context of intramolecularly catalysed reactions is no longer the actual average concentration of the neighbouring catalytic group, since quite often EM-values for these reactions amount to physically unattainable concentrations of several powers of 10 (Kirby, 1980). A *mis-à-point* of the meaning of the EM for its general use in physical organic studies of intramolecular reactivity has been presented by Illuminati *et al.* (1977), who suggested that the EM should be viewed as a "reduced" intramolecular reactivity, i.e. a reactivity that is corrected for the inherent reactivity of end-groups. The EM-parameter sets on a common scale reactivity data for different cyclisation reactions, thus providing an absolute measure of the ease of cyclisation of a bifunctional chain molecule. Nowadays the EM is a well-defined physical organic quantity, whose utility in the interpretation of ring-closure reactivity data is well recognised by numerous investigators in the field. In his review where some 400 EM's are listed, Kirby (1980) pointed out: "A more likely source of new insights is the comparison of EM's for different reaction types, because most of the data (for intramolecular reactions) have not previously been available in this form". He wrote further: "It is to be hoped that it will rapidly become normal practice to attempt an estimate of the effective molarity as part of any quantitative work on intramolecular reactions".

EQUILIBRIA IN RING FORMATION. MACROCYCLISATION EQUILIBRIA. THE CHELATE EFFECT

Equilibrium studies in ring formation, being much less numerous than kinetic studies, have received but limited attention. Nevertheless, for any reversible cyclisation reaction (12) of a bifunctional chain molecule for which

$$X\sim\sim Y \underset{}{\overset{K_{intra}}{\rightleftarrows}} \text{(cyclic) } Z \qquad (12)$$

$$\sim X + Y\sim \underset{}{\overset{K_{inter}}{\rightleftarrows}} \sim Z\sim \qquad (13)$$

reference is made to an analogous non-cyclisation reaction (13), it is useful to consider the ratio K_{intra}/K_{inter}. This ratio defines the equilibrium effective molarity EM, which corresponds exactly to the EM already discussed in the rate case.

An important source of experimental and theoretical studies of equilibria in ring formation is represented by the field of so-called macrocyclisation equilibria (Flory, 1969). Interest in this field appears to have been restricted so far to chemists conventionally labelled as polymer chemists. Experimental evidence of cyclic oligomer populations of ring-chain equilibrates such as those obtained in polysiloxanes (Brown and Slusarczuk, 1965) may be related to the statistical conformation of the corresponding open-chain molecules (Jacobson and Stockmayer, 1950; Flory, 1969). In these studies experimental results are expressed in terms of molar cyclisation equilibrium constants K_x (14) related to the x-meric cyclic species M_x in equilibrium with the

$$K_x = \frac{[\text{—}M_{y-x}\text{—}][M_x]}{[\text{—}M_y\text{—}]} \qquad (14)$$

linear molecules $\text{—}M_y\text{—}$ and $\text{—}M_{y-x}\text{—}$ (15). Let us now consider the equili-

$$\text{—}M_y\text{—} \rightleftarrows \text{—}M_{y-x}\text{—} + M_x \qquad (15)$$

bria (16) and (17) and the corresponding equilibrium constants given by (18) and (19).

$$\text{—}M_x\text{—} \overset{K_{intra}}{\rightleftarrows} M_x \qquad (16)$$

$$\text{—}M_{y-x}\text{—} + \text{—}M_x\text{—} \overset{K_{inter}}{\rightleftarrows} \text{—}M_y\text{—}M_y\text{—} \qquad (17)$$

Combining (18) and (19) one obtains (20) which shows that the molar cyclisation equilibrium constant K_x related to the cyclic x-mer coincides with the equilibrium EM for the formation of the same ring from the corresponding open-chain x-meric species.

$$K_{intra} = \frac{[M_x]}{[\text{—}M_x\text{—}]} \qquad (18)$$

$$K_{\text{inter}} = \frac{[-M_y-]}{[-M_{y-x}-][-M_x-]} \tag{19}$$

$$\frac{K_{\text{intra}}}{K_{\text{inter}}} = \frac{[-M_{y-x}-][M_x]}{[-M_y-]} \tag{20}$$

Another important source of quantitative evidence of equilibria in ring formation is available in a field which is of primary interest to chemists conventionally labelled as inorganic chemists. It is well known that multidentate ligands, where the donor atoms are connected by chains, give rise to chelate rings on complex formation. The higher stability of complexes with multidentate ligands with respect to those of monodentate ligands is known as the chelate effect. Let us now consider an n-dentate ligand L_n and its unidentate anologue L. The pertinent equilibria, involving in each case complex formation with a metal ion M via the same number n of donor atoms are given in (21) and (22), and the corresponding equilibrium constants in (23) and (24).

$$M + nL \rightleftharpoons M(L)_n \tag{21}$$

$$M + L_n \rightleftharpoons ML_n \tag{22}$$

$$K_{M(L)_n} = \frac{[M(L)_n]}{[M][L]^n} \tag{23}$$

$$K_{ML_n} = \frac{[ML_n]}{[M][L_n]} \tag{24}$$

The advantage of the multidentate ligand over the monodentate ligand is quantitatively expressed, according to Schwarzenbach (1952), by the quantity *Chel* (chelate effect) defined as (25).

$$Chel = \log \frac{K_{ML_n}}{K_{M(L)_n}} \tag{25}$$

The chemical meaning of this quantity is the difference in $-\log[M]$ between 1 M solutions of L_n and L, respectively, in which the total concentration of the metal ion is constant and much smaller than those of the ligands (Anderegg, 1980).

In the case of a bidentate ligand the ratio $K_{ML_2}/K_{M(L)_2}$ has units of concentration in mol l^{-1}. If one assumes that the equilibrium constants for formation of the monoco-ordinated complexes ML and ML $\sim\sim$ L are equal,[2] combination of the equilibrium constants for (26) and (27) shows that the ratio

[2]Neglect of the statistical factor of 2 for L $\sim\sim$ L is irrelevant to the present discussion.

$$\mathrm{ML{\sim\sim}L} \underset{}{\overset{K_{\mathrm{intra}}}{\rightleftarrows}} \mathrm{M}\langle^{\mathrm{L}}_{\mathrm{L}}\rangle \tag{26}$$

$$\mathrm{ML + L} \underset{}{\overset{K_{\mathrm{inter}}}{\rightleftarrows}} \mathrm{M}\langle^{\mathrm{L}}_{\mathrm{L}} \tag{27}$$

$K_{ML_2}/K_{M(L)_2}$ coincides with the EM for the chelate ring formation from ML $\sim\sim$ L. In the general case, the ratio whose common logarithm defines the chelate effect has units of $M^{(n-1)}$ and is equal to the product of the EM's related to the formation of the $(n-1)$ rings on complexation of a metal ion with an open chain n-dentate ligand.

It is therefore remarkable and somewhat curious to note that, both in the rate and equilibrium cases, quantities which are either conceptually identical or very closely related to the EM have been independently defined over a period of some 30 years to describe the quantitative aspects of intramolecular processes by scientists working in different and apparently unrelated areas of chemistry.

3 Hypothetical cyclisation reactions

Since gas-phase reactions are free from complications arising from solvation effects, a convenient starting point for a meaningful analysis of structural effects on reactivity would be the study of cyclisation reactions in the gas phase. Unfortunately, quantitative evidence of this sort is scanty. A section in Winnik's review (Winnik, 1981a) is devoted to cyclisation and the gas-phase conformation of hydrocarbon chains. From the numerous references therein one obtains a substantial body of evidence pointing to a general resemblance of cyclisation reactions in the gas phase with cyclisation reactions in solution. However, as Winnik has pointed out, gas-phase reactions have not been studied so far with the same kind of detail that is possible for reactions in solution. As a result, any attempt at understanding the relations between structure and reactivity in the area of cyclisation reactions must still rely heavily upon solution chemistry data.

It is instructive, however, and even illuminating, to consider the EM's for a few hypothetical gas-phase reactions for which such EM's can be calculated from the available thermodynamic data. Before doing this, however, it is useful to examine briefly the thermodynamic properties of ring *vs* open-chain compounds.

THERMODYNAMIC PROPERTIES OF ALKANE CHAINS

It is well known that such quantities as the standard free energy, enthalpy and entropy display a remarkable tendency to be additive functions of independent contributions of part-structures of the molecule. This property, on which the mathematical simplicity of many extrathermodynamic relationships is largely based, is well illustrated, for example, by the enthalpies of formation at 298°K of several homologous series of gaseous hydrocarbons $Y(CH_2)_mH$, which are expressed by the relation (28) (Stull *et al.*, 1969). In

$$\Delta H_f^\circ = (A + Bm + \delta) \text{ kcal mol}^{-1} \quad (28)$$

this expression A is a constant peculiar to the end-group Y, B is the increment per CH_2 group, and δ is a term which has a small finite value only for the lower members, but rapidly becomes negligibly small for the higher members, as shown graphically in Fig. 3. The coefficient B has the constant value

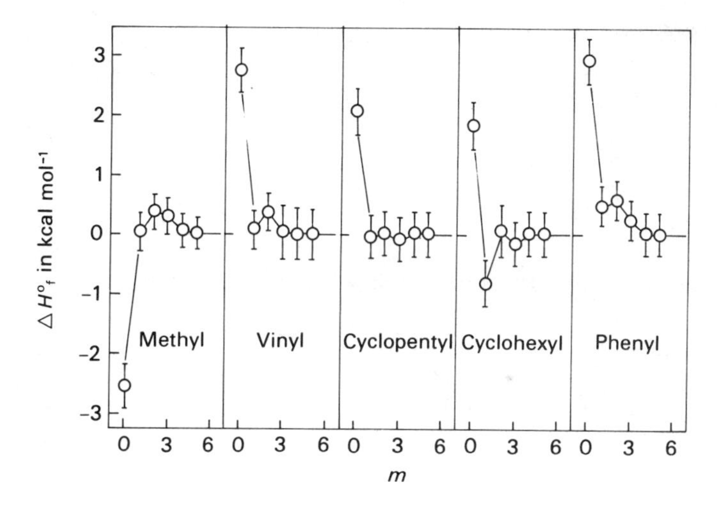

FIG. 3 Deviations from linearity of the enthalpies of formation of several homologous series of hydrocarbons, expressed as the value of δ in (28), and plotted against the number of carbon atoms m in the normal alkyl radical. (Reproduced with permission from Stull *et al.*, 1969)

of -4.926 kcal mol^{-1} (-20.61 kJ mol^{-1}), independent of the nature of the end-group Y. The question of a universal CH_2 increment has been reviewed (Sellers *et al.*, 1978), and its statistical mechanical basis has been discussed by Pitzer (1940). A similar regularity is displayed by the standard entropy, with a CH_2 increment of 9.31 e.u. for S^o_{298}. Although the exact validity of (28) as a fitting function for very precise data has been questioned (Sunner and Wulff, 1980), the essential regularity of properties leads to the important notion that a CH_2 group in an alkane chain retains most of its inherent properties, independent of chain length and of the nature of end-groups. A beautiful example which illustrates the concept of the inherent reactivity of the CH_2 group in an alkane molecule has been reported by Winnik and Maharaj (1979) who measured rate constants and activation parameters for the second-order reaction between photoexcited benzophenone and a series of n-alkanes in CCl_4 solution. A plot of k_2 *vs* alkane chain length is presented in Fig. 4, showing that the increment per CH_2 group is constant within experimental error. Furthermore, the activation energies for each of the n-alkanes are identical, namely, 3.89 kcal mol^{-1}, with a standard deviation of $\pm$ 0.12.

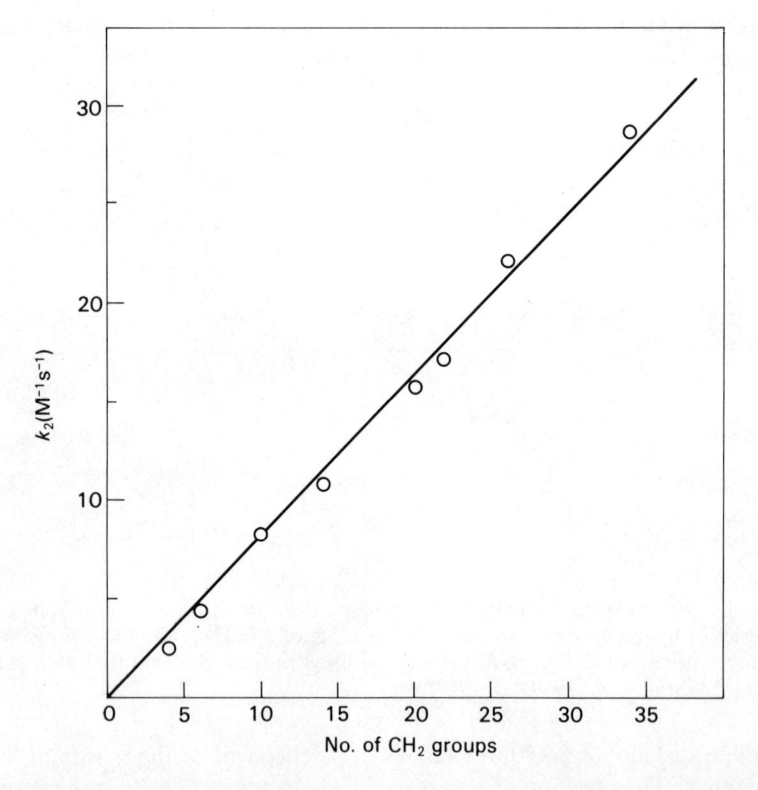

FIG. 4 Photoreaction of benzophenone with the n-alkanes. Plot of the second-order rate constants in CCl_4 at 25°C against the number of CH_2 groups. (Data from Winnik and Maharaj, 1979)

An important property of chain molecules is that a major contribution to the standard entropy is conformational in nature, i.e. is due to hindered internal rotations around single bonds. This property is most relevant to cyclisation phenomena, since a significant change of conformational entropy is expected to take place upon cyclisation. Pitzer (1940) has estimated that the entropy contribution on one C—C internal rotor amounts to 4.43 e.u. A slightly different estimate, namely, 4.52 e.u. has been reported by Person and Pimentel (1953). Thus, it appears that nearly one-half of the constant CH_2 increment of 9.3 e.u. arises from the conformational contribution of the additional C—C internal rotor.

THERMODYNAMIC PROPERTIES OF RING COMPOUNDS

In marked contrast to the n-alkanes, the cycloalkanes exhibit thermodynamic properties where such regularities are no longer present. Heats of formation (ΔH_f°) for a substantial number of cycloalkanes are available from heats of combustion. With the exception of cyclohexane, ΔH_f° is always more positive than the quantity $-4.926n$. The difference between the two quantities leads to a quantitative assessment of the important notion of ring strain. The ΔH_f°-values and strain energy data listed in Table 1 were taken from Skinner and Pilcher (1963). Other references give different but usually comparable

TABLE 1

Standard enthalpies of formation and strain-energies of cycloalkanes[a]

	ΔH_f°	Strain-energy
C_3	12.72	27.5
C_4	6.35	26.1
C_5	−18.46	6.2
C_6	−29.43	0.1
C_7	−28.52	6.0
C_8	−30.03	9.4
C_9	−32.14	12.2
C_{10}	−37.13	12.2
C_{11}	−43.11	11.1
C_{12}	−55.83	3.3
C_{13}	−59.29	4.8
C_{14}	−68.31	0.7
C_{15}	−72.25	1.7
C_{16}	−77.08	1.8

[a] Gas-phase data, 298°K. Energy units kcal mol^{-1}. (Data from Skinner and Pilcher, 1963)

numbers.[3] The discrepancies arise mainly because of the uncertainties in the heats of vaporisation of the higher cycloalkanes (Cox and Pilcher, 1970). The large strain energies for cyclopropane and cyclobutane mainly arise from bond-angle distortions, which are virtually absent in the common rings. The medium rings, which often display abnormal physical and chemical properties when compared with those of the other homologues (Sicher, 1962), have large strain-energies resulting from bond angle deformations, forced adoption of eclipsed conformations, and transannular interactions. These destabilising interactions are substantially relieved in the large rings, on account of their greater flexibility. Although too much emphasis cannot be placed on exact numbers, the large rings (Table 1) show residual strain energies which are consistent with the idea that even large cycloalkanes would have a larger proportion of *gauche* conformations than the corresponding open-chain compounds (Morawetz and Goodman, 1970). An analysis of the various contributions among which the total strain-energy of the C_4- to C_{12}-cycloalkanes is partitioned has been carried out by Allinger *et al.* (1971) by means of molecular mechanics calculations. The results reflect clearly the various types of strain postulated on the basis of the chemical properties of ring compounds (Brown and Ichikawa, 1957). An excellent review on the conformational behaviour of many membered rings is available (Dale, 1976).

Thermochemical data are available for some cyclic *cis*-alkenes, ethers, thioethers, and amines. In spite of their scarcity, they provide an insight into the effect of replacing one or more CH_2 groups with such basic structural units as trigonal carbon atoms and heteroatoms. To be meaningful, the change in ΔH_f° on going from a cycloalkane to a given ring compound with the same number of ring members is to be compared with the corresponding change related to suitably chosen open chain compounds. For example, the effect due to the change cycloalkane → cycloalkanone was calculated for the various ring sizes using (29).

$$[\Delta H_f^\circ(\text{cycloalkanone}) - \Delta H_f^\circ(\text{cycloalkane})] - [\Delta H_f^\circ(\text{5-nonanone}) - \Delta H_f^\circ(\text{nonane})] \qquad (29)$$

Similar calculations were carried out for the remaining ring compounds. The results listed in Table 2 depend somewhat on the particular pair of open-chain reference compounds chosen, but the trends are nevertheless significant.

[3]A notable exception is that of cyclotetradecane for which widely scattered values are available. It was quoted by Liebman and Greenberg (1976) as a virtually strainless molecule, but the exceedingly high value of 15.0 kcal mol^{-1} strain-energy was reported two years later by the same authors (Greenberg and Liebman, 1978).

TABLE 2

Strain energies (kcal mol^{-1}) of ring compounds relative to cycloalkanes from thermochemical data[a]

Ring-size n	3	4	5	6	7	8
Ketone[b]	—	—	0.3	3.4	−3.1	−7.3
cis-Alkene[c]	24.8	−3.9[d]	−2.1	−0.3	−2.2	−5.2
Ether[e]	−0.3	−0.6	−0.5	0.1	—	—
Thioether[f]	−7.8	−6.5	−4.4	−0.4	−0.9	—
Amine[g]	−0.9	—	−0.1	0.0	—	—

[a] Calculated from enthalpies of formation of the gaseous compounds at 298°K. (These were taken from the compilation of Stull *et al.*, 1969, unless otherwise stated)
[b] As based on the pair of open-chain compounds dibutyl ketone/nonane. (Thermochemical data from Pedley and Rylance, 1977)
[c] As based on the pair of open-chain compounds *cis*-3-hexene/hexane
[d] From the ΔH_f°-value of 31.00 kcal mol^{-1} given by Stull *et al.*, 1969. Pedley and Rylance (1977) report a ΔH_f° value of 37.45 kcal mol^{-1}, from which the relative strain of cyclobutene is calculated as +2.6 kcal mol^{-1}
[e] As based on the pair of open-chain compounds dibutyl ether/nonane
[f] As based on the pair of open-chain compounds dibutyl thioether/nonane
[g] As based on the pair of open-chain compounds diethylamine/pentane

The numbers given for the cyclic ethers and cyclic amines require little comment because in all cases they are well below the experimental uncertainties, but in the case of the cyclic thioethers there appears to be a definite tendency for the sulphur heteroatom to stabilise the ring structures, the effect being larger the smaller the ring size.

Noteworthy is the dramatic strain increase upon introduction of a double bond into a 3-membered ring but, strangely enough, no effect of this sort is observed for the next homologue, for which an unexpected strain relief of 3.9 kcal mol^{-1} is found. This might well be due to a wrong ΔH_f°-value, as shown by the fact that a ΔH_f°-value from a different source (see footnote *d* to Table 2) leads to a relative strain of +2.6 kcal mol^{-1}, which is a less unrealistic but in any case modest, figure. It can be concluded, therefore, that the enthalpy changes upon incorporation of a *cis*-double bond into a 4-membered ring and into an alkane chain are not very different from each other.

The relative strain-energies of the 7- and 8-membered ketones and alkenes are worthy of comment; they suggest (*i*) that introduction of trigonal groups into rings where a major source of strain is torsional in nature results in strain relief and (*ii*) that the resulting effect is larger, the larger the ring strain. Although the pertinent figures given in Table 2 seem somewhat exaggerated, the trend is confirmed by the strain-energies calculated by Allinger *et al.*

(1972) for a series of cyclic ketones and by Allinger and Sprague (1972) for cycloalkenes (Table 3). The significant reduction in strain by 3.99 kcal mol^{-1} upon substitution of one CH_2 of cyclodecane with a carbonyl group is due to removal of one of two serious transannular H···H repulsions. The second repulsion is relieved by the presence of the additional carbonyl group of 1,6-cyclodecanedione, for which the strain is reduced by another 3.47 kcal mol^{-1}. Similar considerations hold for *cis*-cyclodecene and *cis*,*cis*-1,6-cyclodecadiene. All these large energy changes are in good agreement with the chemistry of medium rings.

TABLE 3

Strain energies (kcal mol^{-1}) of cyclic ketones and cycloalkenes relative to cycloalkanes from force-field calculations

Ketones[a]	Rel. strain-energy	Alkenes[b]	Rel. strain-energy
Cyclobutanone	1.86	—	—
Cyclopentanone	−1.13	Cyclopentene	−0.60
Cyclohexanone	1.04	Cyclohexene	0.86
Cycloheptanone	−1.53	*cis*-Cycloheptene	−0.69
Cyclo-octanone	−2.55	*cis*-Cyclo-octene	−2.71
Cyclononanone	−3.27	—	—
Cyclodecanone	−3.99	*cis*-Cyclodecene	−5.41
Cyclodecane-1,6-dione	−7.46	*cis*,*cis*-1,6-Cyclodecadiene	−9.66

[a] Data from Allinger *et al.*, 1972
[b] Data from Allinger and Sprague, 1972

Entropy data for the cycloalkanes are available in the limited ring-size range of 3 to 8. These are listed in Table 4, together with the corresponding data for the n-alkanes with the same number of carbon atoms. There are remarkable differences between the two sets of numbers. The entropies of the ring compounds are considerably lower than those of the open-chain compounds, and so is their average rate of increase on increasing the number of carbon atoms. Furthermore, the remarkable regularity displayed by the open-chain alkanes is no longer displayed by the cycloalkanes which show variable CH_2 increments. A rationale for such a behaviour has been presented by O'Neal and Benson (1970) in a thorough discussion of entropies of cyclic compounds. Upon cyclisation, $(n-1)$ internal rotations of a n-alkane are transformed into ring vibrations. This is equivalent to a partial freezing of internal motions, since entropy contributions from the latter are smaller than those from the former. The out-of-plane ring vibrations of the various

TABLE 4

Standard entropies (e.u.) of gaseous cycloalkanes and n-alkanes[a]

	Cycloalkane	n-Alkane
C_3	56.75	64.51
C_4	63.43	74.12
C_5	70.00	83.40
C_6	71.28	92.83
C_7	81.82	102.27
C_8	87.66	111.55

[a] 298°K, 1 Atm standard state. (Data from Stull *et al.*, 1969)

cycloalkanes differ in their degree of "looseness". Cyclopropane, with no out-of-plane vibrations, is a rigid molecule, but cyclobutane, whose single out-of-plane low frequency motion contributes 3.8 e.u. to S°_{298}, and cyclopentane, whose pseudorotation contributes an exceptional 5.8 e.u. to S°_{298}, are flexible molecules. Cyclohexane is again a relatively rigid molecule, with little extra entropy due to out-of-plane vibrations, but, for cycloheptane, cyclooctane, and the larger rings, molecular models suggest extensive freedom of motion resembling concerted partial rotations around the skeletal bonds.

Discussions of entropy changes upon cyclisation of n-alkanes to cycloalkanes, like those for any reaction involving symmetrical molecules, require symmetry considerations. This is because additivity rules for the estimation of thermochemical quantities, as well as current explanations of structure effects on rates and equilibria in terms of additional contributions of part-structures of the reacting molecules to chemical potential changes, have nothing to do with symmetry. The rotational partition function varies inversely with the symmetry number σ, which may be defined as "the number of different values of the rotational co-ordinates which all correspond to one orientation of the molecule, remembering that the identical atoms are indistinguishable" (Mayer and Mayer, 1940). Thus, it is useful to define a symmetry-corrected standard entropy $(S^{\circ})^*$ as (30) and the corresponding

$$(S^{\circ})^* = S^{\circ} + \boldsymbol{R} \ln \sigma \tag{30}$$

symmetry-corrected equilibrium constant K^* as (31). It can be easily shown

$$K^* = \exp\left(-\frac{\Delta H^{\circ}}{\boldsymbol{R}T}\right) \exp \frac{T(\Delta S^{\circ})^*}{\boldsymbol{R}} \tag{31}$$

that K^* is related to the true equilibrium constant K by means of relation (32), where K_{σ} is defined in such a way that for reaction (33) it takes the form

$$K^* = K \cdot K_{\sigma} \tag{32}$$

$$A+B \rightleftharpoons C+D \tag{33}$$

$$K_\sigma = \frac{\sigma_C\, \sigma_D}{\sigma_A \cdot \sigma_B} \tag{34}$$

(34). Symmetry corrections for rate or equilibrium constant are usually expressed in a formally different but perfectly equivalent way by means of statistical factors (Leffler and Grunwald, 1963). If the two reagents of (33) have p_A and p_B equivalent reactive sites, then the rate constant for the forward reaction must be divided by a statistical factor of $p_A \cdot p_B$. Since similar considerations apply to the back reaction as well, the equilibrium constant must be divided by a statistical factor of $p_A \cdot p_B / p_C \cdot p_D$ as in (35). The assign-

$$K^* = \frac{K}{p_A \cdot p_B / p_C \cdot p_D} \tag{35}$$

ment of symmetry numbers to rigid molecules is straightforward (Benson, 1976), but when flexible molecules are involved the question is complex. For the n-alkanes, the currently accepted symmetry number is 18, that is the product of the two-fold external symmetry and the three-fold internal symmetry of each methyl group, i.e. 2 × 3 × 3. According to Benson (1976), σ for cyclopropane is 6, it is 8 for cyclobutane, and 10 for cyclopentane.[4] But it is 6 for cyclohexane, 1 for cycloheptane, and 8 for cyclo-octane. The reason for the choice of $\sigma = 6$ for cyclopropane is obvious. As to the remaining rings, barriers separating conformations are believed to be sufficiently low to cause a time-averaged equivalence of the hydrogens for the C_4 and C_5 rings, but not for the higher homologues. The matter has been discussed in some detail by De Tar and Luthra (1980), who assigned symmetry numbers of $2n$ to all cycloalkanes except cyclohexane for which a value of 6 was preferred.

The question of the choice of the appropriate symmetry number for conformationally labile molecules is a subtle one and is at best an approximation. It is the opinion of this author that, if $\sigma = 18$ for the n-alkanes is accepted as correct, then for the sake of consistency the σ-values for all of the cycloalkanes have to be set at $2n$, as shown by the following example. Consider the hypothetical reaction (36) for which a statistical correction is clearly

$$CH_3(CH_2)_{n-2}CH_3 \overset{K}{\rightleftharpoons} \text{(}(CH_2)_n\text{)} + H_2 \tag{36}$$

required. There are 9 different ways in which H_2 can be formed from the two methyl groups of the alkane molecule. Conversely, there are n equivalent single bonds in the ring molecule across which H_2 can add. Noting that a

[4]For a different view on the symmetry number of cyclopentane see Kabo and Andreevskii (1973) who prefer $\sigma = 1$ for this ring.

statistical factor of 2 must be introduced for the symmetrical hydrogen molecule, the statistically corrected equilibrium constant is $K(9/2n)^{-1}$. Remembering that σ(n-alkane) = 18 and $\sigma(H_2) = 2$, the symmetry number formalism leads to a result consistent with that of the statistical correction approach, namely, $K(2 \cdot 2n/18)$, only if σ(cycloalkane) = $2n$ for all members of the series.

ENTHALPY AND ENTROPY EFFECTS ON RING CLOSURE

Table 5 lists equilibrium data for a new hypothetical gas-phase cyclisation series, for which the required thermodynamic quantities are available from either direct calorimetric measurements or statistical mechanical calculations. Compounds whose tabulated data were obtained by means of methods involving group contributions were not considered. Calculations were carried out by using S°_{298} values based on a 1 M standard state. These were obtained by subtracting 6.35 e.u. from tabulated S°_{298}-values, which are based on a 1 Atm standard state. Equilibrium constants and thermodynamic parameters for these hypothetical reactions are not meaningful as such. More significant are the EM-values, and the corresponding contributions from the enthalpy and entropy terms.

From standard thermodynamic equations one obtains (37), which may be

$$\log EM = -\frac{\Delta H^\circ_{\text{intra}} - \Delta H^\circ_{\text{inter}}}{2.303\,\boldsymbol{R}T} + \frac{\Delta S^\circ_{\text{intra}} - \Delta S^\circ_{\text{inter}}}{2.303\,\boldsymbol{R}} \qquad (37)$$

written in the more compact form (38) upon introduction of the operator θ

$$\log EM = -\frac{\theta\Delta H^\circ}{2.303\,\boldsymbol{R}T} + \frac{\theta\Delta S^\circ}{2.303\,\boldsymbol{R}} \qquad (38)$$

defined in such a way that for any quantity Q related to an intramolecular reaction and to the corresponding intermolecular reaction[5] relation (39)

$$\theta Q = Q_{\text{intra}} - Q_{\text{inter}} \qquad (39)$$

holds. The quantities $\theta\Delta H^\circ$ and $\theta\Delta S^\circ$ may reasonably be called the thermodynamic parameters of EM, and the terms $-\theta\Delta H^\circ/2.303\,\boldsymbol{R}T$ and $\theta\Delta S^\circ/2.303\,\boldsymbol{R}$ respectively the enthalpy and entropy components of $\log EM$. Note that $\theta\Delta H^\circ$ has the meaning of ring strain energy within the precision of the constant CH_2 increment for ΔH°_f of open-chain polymethylene compounds.

[5]Note that, according to the definition of θ, $\log EM = \theta \log k$ or $\theta \log K$.

TABLE 5

Hypothetical gas-phase cyclisation reactions. Effective molarities and thermodynamic parameters[a]

Reaction	Ring-size n	$\theta\Delta H^{\circ}$ kcal mol^{-1}	$\theta(\Delta S^{\circ})^*$ e.u.	$\frac{-\theta\Delta H^{\circ}}{2.303\boldsymbol{RT}}$	$\frac{\theta(\Delta S^{\circ})^*}{2.303\boldsymbol{R}}$	Log EM^*
$CH_3(CH_2)_{n-2}CH_3 \rightleftharpoons (CH_2)_n$ (ring) $+ H_2$	3	27.08	26.13	−19.85	5.71	−14.14
	4	26.04	23.78	−19.09	5.20	−13.89
$2CH_3(CH_2)_2CH_3 \rightleftharpoons CH_3(CH_2)_6CH_3 + H_2$	5	6.06	21.56	−4.44	4.71	+0.27
	6	0.05	13.72	−0.04	3.00	+2.96
	7	5.88	15.13	−4.31	3.31	−1.00
	8	9.28	11.96	−6.80	2.61	−4.19
$CH_3S(CH_2)_{n-3}CH_3 \rightleftharpoons$ S$(CH_2)_{n-1}$ (ring) $+ H_2$	3	19.37	24.12	−14.20	5.27	−8.93
$CH_3(CH_2)_3SCH_3 + CH_3CH_2CH_3 \rightleftharpoons$	4	19.61	20.68	−14.37	4.52	−9.85
	5	2.21	17.23	−1.62	3.77	+2.15
$\rightleftharpoons CH_3(CH_2)_3S(CH_2)_3CH_3 + H_2$	6	0.05	10.96	−0.04	2.40	+2.36
$CH_3(CH_2)_{n-2}SH \rightleftharpoons$ S$(CH_2)_{n-1}$ (ring) $+ H_2$	3	19.96	23.03	−14.26	5.39	−8.87
$CH_3(CH_2)_3SH + CH_3(CH_2)_2CH_3 \rightleftharpoons$	4	19.62	20.56	−14.37	4.81	−9.56
	5	1.76	17.05	−1.29	3.99	+2.70
$\rightleftharpoons CH_3(CH_2)_3S(CH_2)_3CH_3 + H_2$	6	−0.42	10.77	+0.31	2.52	+2.83
$CH_3(CH_2)_{n-2}CH_3 \rightleftharpoons$ HC=CH$(CH_2)_{n-2}$ (ring) $+ 2H_2$	3	53.16	27.19	−38.96	5.94	−33.02
	4	22.89	22.18	−16.78	4.85	−11.93
$2CH_3CH_2CH_3 \rightleftharpoons$ CH_3H_2C(H)C=C(H)CH_2CH_3 $+ 2H_2$	5	4.61	19.15	− 3.38	4.18	+ 0.80
	6	0.42	14.76	− 0.31	3.23	+ 2.92
$CH_3(CH_2)_{n-2}OH \rightleftharpoons$ O$(CH_2)_{n-1}$ (ring) $+ H_2$	3	27.46	25.92	−20.13	5.66	−14.47
$CH_3CH_2OH + CH_3CH_3 \rightleftharpoons CH_3CH_2OCH_2CH_3 + H_2$	4	26.17	21.29	−19.18	4.64	−14.54
$CH_3(CH_2)_{n-2}NH_2 \rightleftharpoons$ HN$(CH_2)_{n-1}$ (ring) $+ H_2$	3	26.56	24.99	−19.47	5.46	−14.01
$CH_3CH_2NH_2 + CH_3CH_3 \rightleftharpoons CH_3CH_2NHCH_2CH_3 + H_2$	5	7.20	20.39	−5.28	4.45	−0.83

[a] Thermochemical data from the compilation of Stull *et al.*, 1969. Entropy values are based on a 1 M standard state. The asterisk denotes symmetry-corrected quantities. Symmetry numbers were chosen as follows: 18 for the n-alkanes, *cis*-3-hexene, dibuthyl sulphide, diethyl ether, and diethyl amine; $2n$ for the cycloalkanes and 2 for all of the remaining ring compounds; 3 for the alkanols, alkanethiols and alkyl amines; 9 for the methyl alkyl

Two sections of Table 5 refer to the formation of thiacycloalkanes by cyclisation of two different series of open chain precursors, namely, methyl alkyl sulphides and alkanethiols. The two sets of log *EM*-values are remarkably similar, showing that to a reasonable approximation the equilibrium EM's for the formation of a ring is a quantity characteristic of that ring. This is again a consequence of the fact that enthalpies of formation and entropies for the simple compounds involved in the reactions considered are to a good approximation the sum of independent contributions from part-structures of the molecules. Were such an additivity rule perfectly obeyed, the effective molarities for the two series given and the corresponding $\theta\Delta H^{\circ}$- and $\theta\Delta S^{\circ}$-values, would be exactly the same.

Inspection of the data reported in Table 5 reveals dramatic structure effects on the EM's, which range from the extremely low value of 10^{-33}M for the cyclisation of propane to cyclopropene, to the high values in the neighbourhood of 10^{2}–10^{3}M for the formation of cyclohexane, cyclohexene, thiacyclopentane, and thiacyclohexane. It seems difficult to attach any physical meaning to a quantity, 10^{-33}M, which corresponds to less than 10^{-10} molecules per litre. One could reasonably say that the dilution to which a sample of propane has to be brought to give "cyclisation" to cyclopropene a chance to compete appreciably with the intermolecular "condensation" to *cis*-3-hexene is one corresponding to a pair of propane molecules in a volume of at least 10 billion litres! On the other hand, cyclisation reactions for which the EM is 10^{3}M are favoured by two orders of magnitude over their intermolecular counterparts even when run at the highest physically attainable concentrations of about 10^{1}M that correspond to the pure liquid samples.

The ease of formation of the various rings appears to be dominated by the strain energies of the ring to be formed. The enthalpy terms are much more sensitive to structural effects than the entropy terms. Indeed, the $\theta(\Delta S^{\circ})^{*}$-values display a remarkable tendency to decrease on increasing n within each cyclisation series, but show relatively insignificant variations when different rings with the same number of ring atoms are compared (Mandolini 1978a). This is graphically shown in Fig. 5, which is a plot of $\theta(\Delta S^{\circ})^{*}$ against the number of skeletal bonds in the bifunctional precursors of the various rings. The straight line was drawn with slope -4.0 e.u. per rotor and intercept 33 e.u. Although considerable scatter is present, there is a remarkable regularity. Cyclobutane, cyclopentane, cycloheptane, and cyclo-octane show significant positive deviations, on account of their extra looseness due to low frequency out-of-plane bending motions (O'Neal and Benson, 1970). If their representative points are omitted, the straight line fits the data with an average deviation of 1.3 e.u. which corresponds to an uncertainty of 0.3 in log *EM*. The drop of 4.0 e.u. per added methylene group is quite close to the average value of about 4.8 e.u. per rotor estimated by O'Neal and Benson

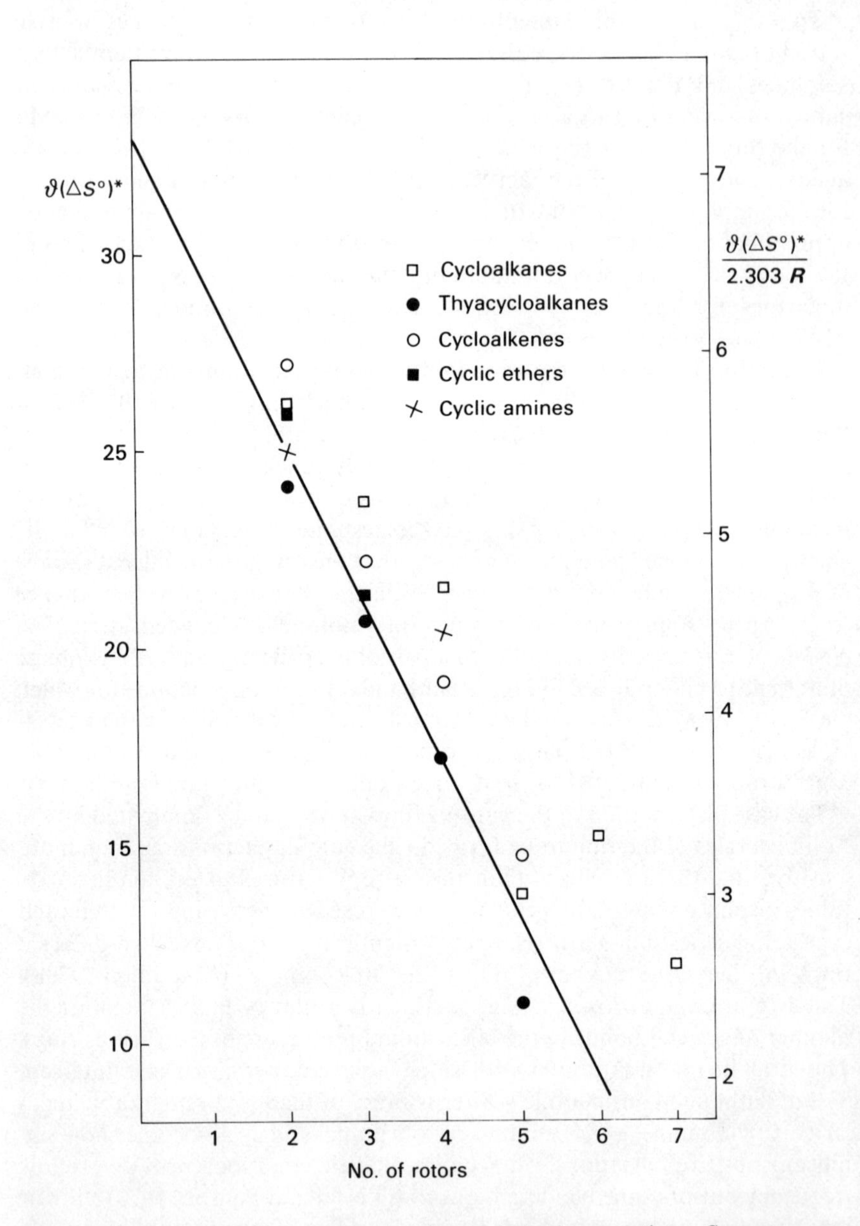

FIG. 5 Entropy effect on simple gas-phase cyclisation reactions. Symmetry corrected θ quantities plotted against the number of single bonds in the open-chain reactants. Data from Table 5

(1970) for the entropy losses on cyclisation of alkane chains to what in these authors' terminology are termed standard rings, i.e. rings whose looseness is comparable to that of the rigid cyclohexane molecule. It is also quite close to the average value of some 4.5 e.u. taken by Page and Jencks (1971) as a representative value for the entropy that may be lost upon freezing an internal rotation. A lower estimate, −2.868 e.u. per rotor, was reported by De Tar and Luthra (1980). This value corresponds to the slope of a straight line fitted to the points for the C_3 to C_8 cycloalkanes neglecting cyclohexane. Because of the well-recognised abnormal looseness of the 4-, 5-, 7-, and 8-membered cycloalkanes, the applicability of the above estimate appears to be of limited general value. Remembering that the methylene increment to S° due to internal rotation is about 4.5 e.u. Fig. 5 shows that cyclisation of short chains, apart from cases where very loose rings are formed, involves a virtual loss of most, if not perhaps all, of the conformational entropy.

The intercept of the straight line of Fig. 5 is most significant to the question of the maximum advantage due to proximity of reactive groups in intramolecular or intracomplex reactions relative to intermolecular reactions. Its meaning will be considered in some detail. The present discussion is derived to a large extent from that of Page and Jencks (1971) and Page (1973) and closely follows that of Mandolini (1978a). The entropy changes in intramolecular reactions can be dissected into chemical and conformational changes (40). Because the decrease of the moment of inertia associated with the cycli-

$$\Delta S^\circ_{\text{intra}} = (\Delta S^\circ_{\text{chem}})_{\text{intra}} + \Delta S^\circ_{\text{conf}} \tag{40}$$

sation of an extended chain is generally small, the corresponding entropy contribution due to change of the rotational partition function can be neglected to a good approximation (Page, 1973). In addition to the contribution due to chemical change, major contributions to $\Delta S^\circ_{\text{inter}}$ arise from changes of translational and (overall) rotational entropy (41). Combining (40) and (41)

$$\Delta S^\circ_{\text{inter}} = (\Delta S^\circ_{\text{chem}})_{\text{inter}} + \Delta S^\circ_{\text{transl}} + \Delta S^\circ_{\text{rot}} \tag{41}$$

and assuming virtually exact cancellation in the $\theta\Delta S^\circ_{\text{chem}}$ term, one obtains (42). The intercept of the straight line of Fig. 5 represents the symmetry cor-

$$\theta\Delta S^\circ = \Delta S^\circ_{\text{conf}} - (\Delta S^\circ_{\text{transl}} + \Delta S^\circ_{\text{rot}}) \tag{42}$$

rected $\theta\Delta S^\circ$ quantity of an ideal gas-phase cyclisation where no conformational entropy is lost. Therefore, to a reasonable degree of approximation, 33 e.u. may be taken as a representative value for the amount of translational plus rotational entropy lost in gas-phase reactions of the A + B → C type in the neighbourhood of ambient temperature.

The values of $\theta(\Delta S^{\circ})^{*}/2.303\ \boldsymbol{R}$ listed in Table 5 are the entropic components of log *EM*. These are the log *EM*-values for ideal strainless cyclisation reactions, i.e. reactions where $\theta\Delta H^{\circ} = 0$. It is of interest to note that, as far as the entropic component is concerned, symmetry corrected effective molarities on the order of 10^{2}–10^{6}M are found. This observation leads to the important conclusion that cyclisation reactions of chains up to about 7 skeletal bonds are entropically favoured over reactions between non-connected 1 M end-groups. The intercept of 33 e.u. corresponds to an effective molarity of $\exp(33/\boldsymbol{R})$ or $10^{7.2}$M, which may be taken as a representative value for the maximum advantage due to proximity of end-groups in intramolecular equilibrium reactions. It compares well with the maximum EM of about 10^{8}M estimated by Page and Jencks (1971).

The present discussion is limited in scope to equilibria as far as the above entropy considerations are concerned. Unfortunately, no kinetic data in the gas phase are available for comparison. According to transition-state theory, the rate of cyclisation is governed by the equilibrium between a ring-shaped transition state and an open-chain reactant. It seems most likely that the shape of the transition state is very similar to that of the cyclic product and consequently, as far as the change of conformational entropy is concerned, going from the open-chain reactant to the transition state is essentially the same as going from the open-chain reactant to the ring product. Similar considerations apply to changes of translational plus rotational entropy involved in the intermolecular model reaction. It seems therefore likely that the arguments presented here for $\theta\Delta S^{\circ}$ will carry over directly to $\theta\Delta S^{\ddagger}$ and consequently to the entropy effects on the rate of ring closure.

Another limitation of the present discussion is that it is concerned with reactions in the gas phase, whereas most of the quantitative information available refers to intramolecular reactions in solution. While these will be dealt with in the following section, we note here that entropies of internal rotation are hardly affected on going from the gas to a condensed phase. Furthermore, Page and Jencks (1971) have noted that equilibrium and activation entropies for typical bimolecular reactions are very similar in the gas and liquid phases. This observation is substantiated by a thorough comparison of equilibrium constants in gas and liquid phases of associative equilibria of hydrocarbons carried out by Stein (1981). The general finding was that the equilibrium constants are very similar in the two phases when solvation effects are not significant and when reactants and products do not have major structural differences. This implies that, in the absence of specific solvation effects, bimolecular rate constants will not in general be significantly different in the gas and liquid phase.

It appears therefore that there are certain aspects of cyclisation phenomena which have a straightforward thermodynamic basis and are likely to be quite general. Arguments based on entropies of structurally simple open-chain and ring compounds can provide a rationale for, as well as a quantitative estimate of, rate or equilibrium enhancements in intramolecular reactions. To the extent that standard entropies of simple substances can be evaluated with high accuracy by means of statistical mechanical calculations, and this has actually been done for many of the compounds involved in the gas-phase reactions considered (Stull *et al.*, 1969), the physical basis of these rate or equilibrium enhancements can be considered as completely understood.

Interpretation of the EM as a "local concentration" effect led in the past to seriously underestimated rate or equilibrium enhancements in the range 10^1M–10^2M. This is the case with Schwarzenbach's interpretation of the chelate effect (Schwarzenbach, 1952) and with Koshland's theory of enzymic catalysis in terms of proximity of reacting groups and orbital steering (Storm and Koshland, 1970, 1972a,b). To explain the rate enhancements in enzymic and some intramolecular reactions which are many powers of 10 larger than the rate factor of about 55.5 due to a "local concentration" effect, a proper spatial orientation of the reacting orbitals was suggested (orbital steering). This theory stimulated a great deal of interest and provoked an enormous controversy in the literature for several years. It has been regarded by many authors as an unnecessary theory for a satisfactory interpretation of available data (for a concise review see Gandour, 1978). Yet many years before Rasmussen (1956) had published an important paper where the entropic basis of the proximity effect was correctly established. He showed that the correct magnitude of the chelate effect could be predicted in a simple manner by the use of empirical equations for the entropies of neutral solutes in water, which were available through the work of Powell and Latimer (1951). Applied to the competition of ethylene diamine and methylamine, Rasmussen showed that the difference in entropy between the two monodentate ligands and the one bidentate ligand amounts to some 18 e.u. which gives rise to a *Chel*-value (25) in the order of 4. Unfortunately, Rasmussen's paper did not attract general attention. It took 15 more years before Page and Jencks (1971) authoritatively set forth their theory of entropic contributions to rate accelerations in enzymic and intramolecular reactions and the chelate effect, a theory where the precise origin of the differences of entropy changes in intra- and intermolecular processes was made clear.

THE *gem*-DIMETHYL EFFECT

It has been recognized for a long time (Ingold, 1921) that alkyl substitution frequently increases the rate of formation of a ring compound from its non-cyclic precursor as well as the concentration of the ring compound at equilibrium. This effect, which for historical reasons may be called the *gem*-dimethyl effect, is by no means confined to methyl groups or to geminal substitution. There are numerous examples of this effect in solution, and excellent discussions are available on the subject (see, for example, Capon and McManus, 1976; Kirby, 1980).

Structural effects have been discussed in terms of (*i*) variations in strain-energy and (*ii*) conformational changes relative to the unsubstituted chains. The intuitive notion has been applied that *gem*-dimethyl groups facilitate ring closure "by constraining the molecule to adopt a configuration in which the potential reaction centres are fairly close together; in the absence of these groups or one of them it seems likely that the molecule would exist preferentially in an extended form" (Silbermann and Henshall, 1957). Essentially the same idea was put forward by Dale (1963) and by Bruice and Pandit (1960). The latter authors suggested the term "profitable rotamer distribution" to indicate the more bent conformers whose population is supposedly increased by alkyl groups. A similar term, "stereopopulation control" was invented by Milstien and Cohen (1970, 1972) in connection with their well-known finding that a "trimethyl lock" can give rise to extraordinary rate enhancements in various cyclisation reactions where interlocking of methyl groups was suggested to produce a conformational freezing into the most productive rotamer.

That rate constants are treated most conveniently by the transition-state theory, which is an extension of theories dealing with equilibrium, is widely accepted by chemists interested in reaction kinetics and mechanisms in solution. It is therefore a little surprising that chemists who do not appear to question the validity of the transition-state theory have based their mechanistic arguments on a putative distinction between productive and non-productive conformations, thus adopting what De Tar and Luthra (1980) have called the reaction-path model as opposed to the Eyring transition-state model. The notion of reactive conformer as applied to ring-closure reactions, although mechanically descriptive and apparently acceptable by chemical intuition, has been criticised (Danforth *et al.*, 1976). It is a part of the general question related to the detailed description of a reaction mechanism where a conformationally labile reactant is involved (Leffler and Grunwald, 1963; Hammett, 1970) which gives rise to what Hammett calls the meaningless question. For quantitative treatment, and even for qualitative discussion, it is clearly sufficient to refer to the difference between the effect of structure on the standard potential of the transition state and on that of the reactant state.

As an explanation of the *gem*-dimethyl effect, Hammond (1965) has suggested that the presence of alkyl groups would hinder the rotation in the open-chain reactant thereby lowering its entropy. This steric hindrance to rotation would cause the entropy loss upon cyclisation to be less in the substituted than in the unsubstituted chain. The correctness of Hammond's suggestion was shown a few years later by Allinger and Zalkow (1960) in a thorough analysis of enthalpy and entropy effects on the hypothetical gas-phase cyclisation reactions of alkyl-substituted hexanes to cyclohexanes, for which the required thermodynamic quantities were available from tables. Considering the reaction of hexane to give cyclohexane at 25°C as the reference reaction for which $\Delta G^\circ = \Delta H^\circ = \Delta S^\circ = 0$, introduction of various alkyl groups on the hexane chain causes in all cases negative ΔG°-values, which is a result of favourable changes of ΔH° and ΔS° of comparable magnitude. This is shown in Table 6, where a selection of Allinger and Zalkow's data is listed. The observed ΔG°-values range from -1.6 to -3.6 kcal mol^{-1}, which correspond to an increase of equilibrium constants and effective molarities for cyclisation by a factor of $10^{1.2}$ to $10^{2.6}$. The ΔS°-values range from 2.2 to 5.7 e.u. showing that steric hindrance to rotation produces in the given compounds entropy changes which are comparable in magnitude to the maximum value upon complete freezing of one internal rotation.

TABLE 6

Thermodynamic quantities for the cyclisation reactions of substituted hexanes to cyclohexanes[a]

Hexane	$-\Delta H^\circ$/kcal mol^{-1}		ΔS°/e.u.		$-\Delta G^\circ$/kcal mol^{-1}	
	Calcd.	Found	Calcd.	Found	Calcd.	Found
n	0.0	0.0	0.0	0.0	0.0	0.0
2-CH_3	0.8	0.9	3.4	3.3	1.8	1.9
3-CH_3	1.6	1.6	2.0	2.2	2.2	2.2
2,2-$(CH_3)_2$	0.0	0.1	6.8	5.7	2.0	1.8
2,3-$(CH_3)_2$[b]	1.6	2.4	3.2	4.1	2.6	3.6
2,3-$(CH_3)_2$[c]	0.0	0.6	4.6	5.0	1.4	2.0
2,4-$(CH_3)_2$[c]	2.4	2.3	3.2	3.6	3.4	3.3
2,5-$(CH_3)_2$[b]	1.6	1.4	4.6	3.8	3.0	2.6
3,3-$(CH_3)_2$	1.6	1.2	4.6	4.1	3.0	2.4
2-C_2H_5	0.8	0.8	3.4	2.7	1.8	1.6

[a] Gas-phase data, 25°C. Entropy changes are corrected for symmetry and optical isomers. (Data from Allinger and Zalkow, 1960)
[b] For closure to *trans* form
[c] For closure to *cis* form

The ΔH°-terms were rationalised and accounted for in terms of increase of *gauche* interactions upon cyclisation in the substituted case relative to the unsubstituted case. The ΔS°-terms were calculated by means of a simplified approximate expression where the number of chain branchings, the symmetry number, and the number of optical isomers were taken into account. The agreement between calculated and observed values is satisfactory. Available data for 5-membered rings show effects in the same direction and of similar magnitude. To the extent that the above ideas apply to rings of other sizes as well as to transition states leading to ring compounds, Allinger and Zalkow's explanation for the effects of alkyl substituents on the formation of 6-membered rings provides a straightforward thermodynamic basis for the interpretation of the *gem*-dimethyl effect in general, thus avoiding the recourse to unnecessary concepts and terms which lie outside the framework of the transition-state theory.

4 Cyclisation reactions in solution

THE EXPERIMENTAL APPROACH TO THE KINETICS OF CYCLISATION

As in many experimental studies of a group of related reactions, the initial difficulty to be overcome in quantitative investigations of ring-closure reactions is to find a series of compounds suitable for accurate quantitative work in a ring-size range as wide as possible. As Bennett (1941) pointed out some 40 years ago, a structurally simple series of open-chain compounds $X(CH_2)_mY$ is desirable, yielding the corresponding rings $(CH_2)_mZ$ irreversibly at convenient rates for kinetic studies and without the complicating interference of side reactions. It is unfortunate that reaction series fulfilling the above requirements for all of the members of the series are not easily found. A major reason for this is that the reactivity of one member of the series may differ by several powers of 10 from that of another, because the ease of ring closure frequently depends in a marked or even dramatic way on the length of the chain. For the same reason, a mechanistically simple reaction is also desirable, to avoid any complication arising from changes of rate-determining step along the series. Even under ideal conditions, complete lack of side reactions for any value of m is rarely met, since the concurrent head-to-tail formation of dimers and higher-order polymers may intervene for certain ring sizes. Whenever possible, polymerisation is suppressed by carrying out a given cyclisation reaction at a concentration which is at least one order of magnitude smaller than the cyclisation constant C. Of course, one is then faced with the serious problem that such a concentration might be prohibitively low for analytical purposes. When this is the case, one is forced to have recourse to indirect methods based on the determination of yields of ring compounds under given sets of experimental conditions.

EARLY STUDIES

The kind and number of difficulties met in kinetic investigations of cyclisation reaction series including the formation of many-membered rings is well-illustrated by a kinetic study of the cyclisation of $Br(CH_2)_{n-1}NH_2$ with $n = 6, 7, 12, 14, 15$, and 17 in 30% aqueous isopropyl alcohol carried out by Salomon (1936a,b) as an extension of a series of pioneering investigations published by Freundlich and his collaborators from 1911 onwards (quoted by Salomon, 1936a). Here the main first-order process is accompanied by second-order as well as by first-order competing reactions. In addition to polymerisation, there is the alkylation of cyclic amine products by the CH_2Br-end of monomers and linear polymers. There is also the reaction with the solvent components, yielding $HO(CH_2)_{n-1}NH_2$, $(CH_3)_2CHO$-$(CH_2)_{n-1}NH_2$, and $CH_2{=}CH(CH_2)_{n-2}NH_2$. The problems due to second-order processes were minimised by running the reactions at increasingly high dilution until a plot of extent of reaction against time was independent of further dilution. For example, the concentration at which reactions of higher order were absent was in the neighbourhood of 5×10^{-3}M for the 14-membered ring, but the extremely low value of 5×10^{-4}M was required by the 12-membered ring, for which only a rough estimate of the specific rate for cyclisation was obtained. The results are shown in Table 7, together with a set of approximate C-values calculated

TABLE 7

Relative rates and approximate cyclisation constants C for the cyclisation of $Br(CH_2)_{n-1}NH_2$[a]

Ring-size n	6	7	12	14	15	17
Rel. rate	*ca* 3000	47	*ca* 0.05	1.0	1.1	2.1
C/M	*ca* 10^2	1.7	*ca* 0.002	0.035	0.04	0.075

[a] Solvent, 30% isopropyl alcohol; temperature, 73.35°C. Data from Salomon, 1936a,b

from the second-order dimerisation rate constants obtained for the higher members of the series. Although no direct measurement could be obtained for the formation of the 10-membered ring, Salomon estimated that the 17-membered ring was formed 10^4 times as fast as the 10-membered one. Thus, a rate minimum was shown to occur in the medium-ring region, a region where the presence of a pronounced yield minimum had been clearly demonstrated by Ziegler cyclisation of dinitriles (Fig. 6). Approximately at the same time as Salomon's investigations, Stoll *et al.* (1934) and Stoll and

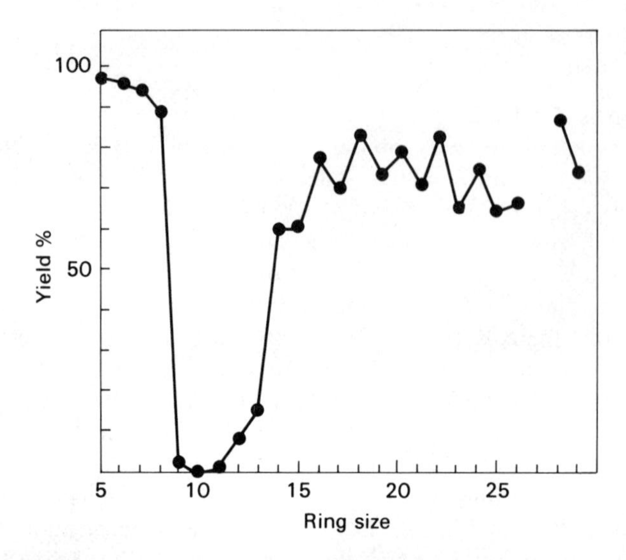

FIG. 6 Yield data as a function of ring size for Ziegler cyclisation of dinitriles. (Data from Ziegler and Hechelhammer, 1937)

Rouvé (1935) published the results of their investigation of the lactonisation of a series of ω-hydroxyacids $HO(CH_2)_{n-2}CO_2H$ in refluxing benzene in the presence of benzenesulphonic acid as catalyst. Approximate reaction rates were obtained in some cases, but the results were expressed as cyclisation constants C calculated by means of (2) from the isolated yields of monomeric lactones (Table 8). Here the reactivity minimum is much more clearly defined than in the bromoamine case. The data reveal a marked dependence of the ease of ring closure on chain length. As in the bromoamine case, there is a remarkable tendency for the data to reach a plateau in the large-ring region, but the C-values are in the neighbourhood of 5×10^{-3}M, whereas those for the bromoamines are one order of magnitude larger. It seems likely that the difference, however, is more apparent than real. The equation used to calculate the C-values is not only incorrect (see discussion on p. 4) but its application relies upon isolated yields of monomeric lactones, for which losses during the isolation step seem inevitable. It appears therefore that the C-values reported in Table 8 are probably underestimated, and are especially so in the medium-ring region, where isolated yields are very low. In spite of these difficulties, the reactivity pattern is most significant, since it reproduces quite closely the basic features of the familiar yield profiles obtained in many ring-closure reactions, which show a deep minimum in the medium-ring region and a wide plateau in the large-ring region (Fig. 6).

TABLE 8

Approximate cyclisation constants C for the lactonisation of $HO(CH_2)_{n-2}\,CO_2H$[a]

Ring-size n	5	7	9	10	11	12	13
C/M	> 0.5	2.8×10^{-3}	*ca* 8×10^{-6}	1×10^{-5}	2.8×10^{-5}	1.4×10^{-4}	2.6×10^{-4}
Ring-size n	14	15	16	17	18	19	24
C/M	1.3×10^{-3}	2.3×10^{-3}	3.8×10^{-3}	4.4×10^{-3}	6.0×10^{-3}	4.9×10^{-3}	3.4×10^{-3}

[a] Solvent, benzene; temperature 80°C. (Data from Stoll and Rouvé, 1935)

Another important investigation of a quantitative nature of the pre-war era is that of Ziegler *et al.* (1937), who studied the kinetics of the base-promoted cyclisation of *o*-(ω-bromoalkoxy)phenols (43) in ethanol solution

$$o\text{-}O^{-}C_6H_4O(CH_2)_{n-4}Br \; [1] \longrightarrow C_6H_4(O)_2(CH_2)_{n-4} \; [2] \qquad (43)$$

in the ring-size range of 6 to 14. Apart from many kinetic complications, which were clearly detected and discussed by the authors, but not completely overcome, a set of approximate half-life times was reported, from which the relative cyclisation rates shown in Table 9 have been calculated. Quite interestingly, the figures indicate a monotonic decrease in the cyclisation rates on

TABLE 9

Approximate relative rates of cyclisation of *o*-bromoalkoxyphenoxides [1] to catechol polymethylene ethers [2] in ethanol[a]

Ring-size *n*	6	7	8	9	10	11	12	13	14
Rel. rate (at 20°C)	*ca* 2300	19	1.0						
Rel. rate (at 78°C)			1.0	0.20	0.078	0.041	0.038	0.034	0.024

[a] Data from Ziegler *et al.*, 1937

increasing the chain length, with no appearance of a reactivity minimumin the medium-ring region. This was attributed by the authors to the presence of two trigonal carbons and two "bare" oxygen atoms, which was suggested to relieve the steric compression of hydrogen atoms crowded against one another in rings of this size. What clearly stems from these data is that the medium-ring region is one where structural effects on the ease of ring closure are most considerable, which is in keeping with the chemistry of medium-sized rings in general (Sicher, 1963). In spite of their approximate nature, these pioneering papers by Salomon, by Stoll and Rouvé , and by Ziegler *et al.* established on a firm basis the technique of kinetic investigation of ring-closure reactions over a wide range of ring sizes while revealing many of the phenomenological features of cyclisation processes. In the following years chemists continued to develop general procedures for the synthesis of rings of all sizes (Belen'kii, 1964), and to elucidate the chemistry and stereochemistry of many-membered rings (Sicher, 1963) made readily available by the application of the acyloin synthesis (Stoll and Rouvé, 1947; Prelog *et al.*, 1947). In spite of the enormous growth of physical organic chemistry in the 1950s and 1960s, and of the great interest devoted to the field of neighbouring group participation (Capon, 1964), little or no information about chain-length effects on cyclisation was obtained in this period, apart from scattered studies of limited scope and reliability (see, for example, Hurd and Saunders, 1952; Davies *et al.*, 1954). No doubt this situation helped to spread the sense of distrust which led Morawetz (1975) to state that "Experimental studies of rates of cyclisation involving the interaction of groups attached to the ends of long-chain molecules are, in general, not possible" because the required "concentrations are so low that analytical methods are unavailable".

Yet, during the last decade, considerable advances have been made towards a quantitative understanding of the structural and energetic factors controlling chain cyclisation. Thanks to the application of modern technology there has been a substantial accumulation of reliable data in the form of accurate kinetic or equilibrium measurements of cyclisation reactions of bifunctional chains, as well as of careful analyses of ring-chain polymerisation equilibria. These will be dealt with in the remaining part of this section.

CYCLISATION OF POLYMETHYLENE AND POLYOXYETHYLENE CHAINS

Lactonisation of ω-bromoalkanoate anions

In this author's opinion, the influence of chain length on the ease of cyclisation of bifunctional chains is the most relevant and fascinating aspect in intramolecular reactions. An illustrative example is provided by the reactivity profile (Fig. 7) related to the lactonisation of ω-bromoalkanoate anions (44), which is the first ring-closure reaction series for which a complete set of

$$\underset{[3]}{Br(CH_2)_{n-2}CO_2^-} \xrightarrow{k_{intra}} \underset{[4]}{\overline{(CH_2)_{n-2}-C(=O)-O}} \qquad (44)$$

rate data has been obtained for all the small, common, and medium rings, and for many of the representative large rings up to ring-size 23 (Galli *et al.*, 1973, 1977; Mandolini, 1978b). The ease of ring formation is extremely sensitive to chain length. There is an increase of 5 powers of 10 on going from the 3- to the 5-membered ring, the latter being located at a sharp maximum. Then the rate decreases rapidly by more than 6 orders of magnitude down to the 8-membered ring, which is located with the higher homologue at a reactivity minimum. The rate then increases again, until a substantial levelling off is observed in the large-ring region. Rate data in the ring-size range 13–23 lie within a factor of 2. Extension of rate measurements to larger rings was prevented by solubility problems, which are an unpleasant consequence of the extremely low solubility of long alkane chains in general. Reaction (44)

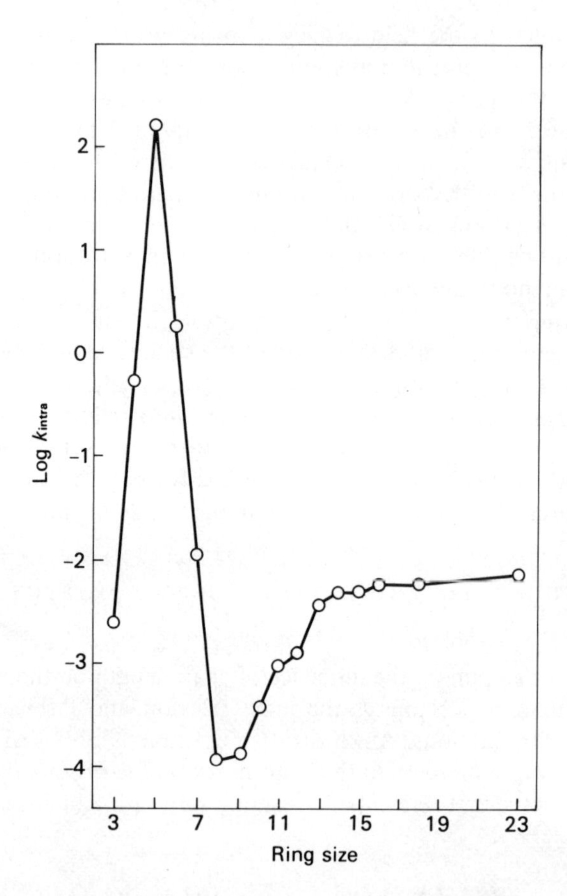

FIG. 7 Reactivity profile for lactone formation (44) in 99% Me_2SO at 50°C

represents a practically ideal case for accurate rate measurements over a wide range of ring sizes. When run in Me_2SO—H_2O (99 : 1, v/v, hereafter referred to as 99% Me_2SO) it occurs smoothly at moderately low temperatures and is virtually free from side-reactions other than polymerisation. The fixed water content (0.56 M) in the solvent, while not altering the essentially aprotic character of the medium, allows the composition of the solvent to remain constant and the results to be reproducible despite the hygroscopic character of Me_2SO. Anions [3] were quantitatively generated *in situ* from solutions of the parent acids by addition of the stoichiometric amount of alkali or tetramethylammonium hydroxide. The kinetics were followed by a sensitive potentiometric technique allowing precise measurement of the released bromide ions at reactant concentrations in the 10^{-3}–10^{-4}M range, where ion pairing effects are unimportant and the observed kinetics proved to be first-order for most of the substrates. The reactions of [3: $n = 8$ and 9], which are the least reactive in the group, were first-order only when performed at the lowest possible concentrations (*ca* 1×10^{-4}M). Consistent with the kinetics, glc product analysis for the reactions of [3: $n = 9$ and 12] showed that the corresponding 9- and 12-membered lactones were formed in virtually quantitative yield under conditions where first-order behaviour had been observed. Deviations from a first-order behaviour occurring at higher concentrations in the reactions of [3: $n = 8$ and 9] were attributed to a competing dimerisation reaction and analysed in terms of the mixed first- and second-order rate eqn. (8). The results obtained with [3: $n = 9$] from two independent determinations illustrate clearly the difficulties arising from the adaptability of multiple-parameter functions. Galli *et al.* (1973) obtained $k_{intra} = 1.06 \times 10^{-4}\,s^{-1}$ and $k_{dim} = 0.038\,l\,mol^{-1}\,s^{-1}$, but the values $k_{intra} = 1.24 \times 10^{-4}\,s^{-1}$ and $k_{dim} = 0.14\,l\,mol^{-1}s^{-1}$ were the results of a subsequent, and probably more reliable determination (Galli and Mandolini, 1977). Clearly the two k_{intra}-values are in fairly good agreement, but the k_{inter}-values are not. This is at least in part a consequence of the fact that in the concentration range investigated the reaction is essentially first-order, with a relatively less important disturbance due to the second-order process. The rates of formation of the 4-, 5-, and 6-membered rings, which were too high to be followed by a conventional kinetic technique, were at first roughly estimated (Galli *et al.*, 1977), but were accurately measured later by a stopped-flow spectrophotometric technique by the device of introducing a visual acid-base indicator into the reaction system (Mandolini, 1978b).

The intermolecular model reaction

In order to translate the specific rates k_{intra} for cyclisation into EM-values, the specific rate k_{inter} for an analogous non-cyclisation is needed. This is a general problem whose experimental and conceptual difficulties have been discussed in detail by Kirby (1980). The general criterion followed is to choose simple, low molecular weight monofunctional reactants, where the immediate environment of the functional groups is as close as possible to that found in the bifunctional reactant. By way of example, the alkylation of butanoate anion with butyl bromide was chosen by Galli *et al.* (1977) as a reasonable intermolecular analogue of the lactonisation of ω-bromoalkanoate ions (44). In order to investigate the limits of application of the model, on account of the possible existence of special effects associated with the chain length and the mutual electronic interactions of end-groups in the bifunctional reactants, Galli and Mandolini (1977) systematically investigated the effect of chain length and of some ω-substituents on the alkylation reaction of straight chain alkanoate ions with primary alkyl bromides (45).

$$RCO_2^- + R'CH_2Br \xrightarrow{k_{inter}} RCO_2CH_2R' + Br^- \qquad (45)$$

The results are listed in Table 10. No alkyl bromide with less than 4 carbon atoms was considered, since the reactions of such nucleophiles as $C_2H_5O^-$, $C_6H_5O^-$, and $(C_2H_5)_3N$ with homologous series of alkyl halides (quoted by Ingold, 1969) are known to follow the reactivity order Me $\gg$ Et $>$ Pr $\geqslant$ Bu $\geqslant$ higher primary. The appearance of a slight rate-enhancing effect on increasing the chain length up to C_{18} in both reactants (entries 1–5) was tentatively attributed, at least in part, to lyophobic interactions of alkyl chains. Since the kinetic effects associated with variations in chain length (entries 1–10) do not exceed a factor of 2, it can be concluded that the choice of the reactant pair in entry 1 as the intermolecular model for reaction (44) is a reasonable one, as far as chain-length effects are concerned.

TABLE 10

Rate data for the ester-forming reactions of alkanoate ions with alkyl bromides in 99% Me_2SO at 50.0°C[a]

Entry no.	Alkanoate ion	Alkyl bromide	k_{rel}
1	$CH_3(CH_2)_2CO_2^-$	$CH_3(CH_2)_3Br$	1.00
2	$CH_3(CH_2)_6CO_2^-$	$CH_3(CH_2)_7Br$	1.42
3	$CH_3(CH_2)_{10}CO_2^-$	$CH_3(CH_2)_{11}Br$	1.48
4	$CH_3(CH_2)_{14}CO_2^-$	$CH_3(CH_2)_{15}Br$	1.77
5	$CH_3(CH_2)_{16}CO_2^-$	$CH_3(CH_2)_{17}Br$	1.88
6	$CH_3(CH_2)_2CO_2^-$	$CH_3(CH_2)_{11}Br$	1.26
7	$CH_3(CH_2)_2CO_2^-$	$CH_3(CH_2)_{15}Br$	1.45
8	$CH_3(CH_2)_2CO_2^-$	$CH_3(CH_2)_{17}Br$	1.39
9	$CH_3CO_2^-$	$CH_3(CH_2)_3Br$	0.96
10	$CH_3(CH_2)_{10}CO_2^-$	$CH_3(CH_2)_3Br$	1.11
11	$Br(CH_2)_7CO_2^-$	$^-O_2C(CH_2)_7Br$	0.46
12	$Br(CH_2)_7CO_2^-$	$CH_3(CH_2)_7Br$	1.11
13	$CH_3(CH_2)_6CO_2^-$	$^-O_2C(CH_2)_7Br$	0.41

[a] Data from Galli and Mandolini, 1977

A more complex situation results when possible electronic effects of end-groups on one another are taken into account. Capon (1964) and Capon and McManus (1976), noting that in intramolecular nucleophilic substitution reactions most of the leaving and neighbouring groups are electron withdrawing, concluded that "these effects act to decrease the nucleophilicity of the neighbouring group and to decrease the tendency of the leaving group to depart". Accordingly, the resulting decrease of the rate of cyclisation was predicted to be greater the shorter the interconnecting chain. In order to provide an estimate for such effects in the case of the ω-bromoalkanoate anions, one might refer to model systems such as (46), (47), and (48). However,

$$Br(CH_2)_{n-2}CO_2^- + Br(CH_2)_{n-2}CO_2^- \rightleftharpoons \left[\begin{array}{c} Br(CH_2)_{n-2}\overbrace{CO_2\cdots CH_2\cdots Br}^{-} \\ | \\ (CH_2)_{n-2} \\ | \\ CO_2^- \end{array} \right]^{\ddagger} \quad (46)$$

$$RCO_2^- + Br(CH_2)_{n-2}CO_2^- \rightleftharpoons \left[\begin{array}{c} \overbrace{RCO_2\cdots CH_2\cdots Br}^{-} \\ | \\ (CH_2)_{n-2} \\ | \\ CO_2^- \end{array} \right]^{\ddagger} \quad (47)$$

$$Br(CH_2)_{n-2}CO_2^- + RCH_2Br \rightleftharpoons \left[\begin{array}{c} Br(CH_2)_{n-2}\overbrace{CO_2\cdots CH_2\cdots Br}^{-} \\ | \\ R \end{array} \right]^{\ddagger} \quad (48)$$

their applicability to the question at hand is ruled out by the following considerations. The fact that the reactant pairs in entries 11 and 13 exhibit the lowest reactivity in the group has been attributed by Galli and Mandolini (1977) to electrostatic repulsion between the CO_2^- negative pole and the negatively charged reaction zones of transition states for (46) and (47). Consequently, it seems unlikely that (46) and (47) are proper intermolecular models for (44), since such electrostatic repulsion is clearly absent in the transition

$$\left[\begin{array}{c} O{=}CO\cdots CH_2\cdots Br \\ \diagdown \quad \diagup \\ (CH_2)_{n-2} \end{array} \right]^{-} \quad [5]$$

state [5] of the latter reaction. Reaction (48) might be believed to model properly the effect of the ω-bromo substituent on the nucleophilicity of the CO_2^- group. Unfortunately, even in this case the model does not work properly. In fact, the effect on the nucleophilicity is not only related to the effect of the bromine (relative to hydrogen) on the chemical potential of the initial state, but also on that of the transition state. Now it is clear that the situation met when the CO_2^- nucleophile reacts with an external electrophile bears no relation to that met in the intramolecular reaction, where the bromine acts as a substituent in the initial state, but is the leaving group in the transition state. In other words, the useful distinction between a reacting and a non-reacting part of the molecule, on which many discussions of structural effects on reactivity are based, does not apply to short bifunctional chains undergoing cyclisation. Consequently, the widely applicable assumption that in a heterolytic bond-making process any change in structure decreasing the electron density on the nucleophile or increasing the electron density on the electrophile will decrease the rate (and vice versa) cannot be extended in a straightforward manner to the effect of end-groups on one another in intramolecular reactions.

It appears therefore that the problem of assessing the mutual electronic influence of end-groups on short chains poses an essentially insoluble question. What remains is to choose a single model reaction involving simple low molecular weight monofunctional reactants, such as the reaction between butanoate anion and butyl bromide, with the obvious advantage that only one k_{inter}-value has to be used in the calculation of the EM for all terms of a given cyclisation series. In this way any factor possibly associated with the electronic influence of end-groups on one another will be included in the calculated EM's.

Activation parameters for the lactone-forming reaction and the Ruzicka hypothesis

A significant step toward the understanding of ring-closure processes was provided by Ruzicka (1935), who correctly recognised that both strain and probability factors must be considered. The strain-energy of the ring to be formed is thought to be reflected by the activation energy for ring closure, and the probability of the chain ends meeting for the reaction to occur is predicted to decrease as the chain gets longer. The Ruzicka hypothesis may be restated using the language of the transition-state theory. The enthalpy of activation $\Delta H^{\ddagger}$ is expected to reflect the enthalpy change ΔH° for ring closure, perhaps in the form of a linear strain-energy relationship, whereas the activation entropy $\Delta S^{\ddagger}$ should include negative contributions arising from reduction of freedom of internal rotations around the single bonds of the chain backbone when the disordered open chain reactant is converted into the ring-shaped transition state.

A confirmation of the essential correctness of the Ruzicka hypothesis is provided by the available parameters of activation for the lactone-forming reaction (44) (Galli *et al.*, 1977; Mandolini, 1978b), shown in Figs 8 and 9 as $\Delta H^{\ddagger}_{intra}$ and $\Delta S^{\ddagger}_{intra}$ *vs* ring-size profiles respectively. In the large-ring region there is a definite tendency for $\Delta H^{\ddagger}_{intra}$ to decrease as the chain gets longer. In terms of the operator θ (p. 21), the quantity $\theta\Delta H^{\ddagger}$ is quite close to zero for the 23-membered ring, thus revealing a virtually strainless situation for this ring size. The highly structured profile shows maxima at ring-sizes 3 and 8, with $\theta\Delta H^{\ddagger}$ values of about 8 kcal mol^{-1} in both cases. The *cis*-conformation of the ester function in the lactone to be formed still predominates for ring-size 8 (Huisgen and Ott, 1959). According to Galli *et al.* (1973), this

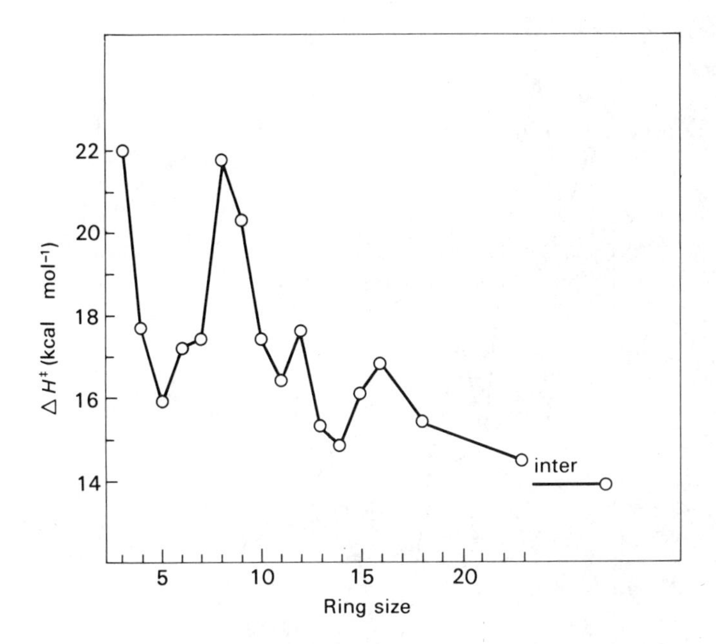

FIG. 8 $\Delta H^{\ddagger}$-profile for the formation of lactones and for the related intermolecular reaction

imposes an extra enthalpic penalty on the lactones with $n \leqslant 8$, which causes the maximum to shift away from ring-sizes 9 and 10 which experience the maximum strain in cycloalkanes (Table 1).

In spite of its irregular shape, the $\Delta S^{\ddagger}{}_{\text{intra}}$-profile clearly demonstrates that the "probability of end-to-end joining" has a definite tendency to decrease as the chain length increases. Contrary to expectations, the $\Delta S^{\ddagger}{}_{\text{intra}}$-values for forming the 3- and 4-membered rings are more negative than that for forming the 5-membered ring. This phenomenon, which seems rather general, has been explained by Mandolini (1978b) as due to a reduced solvation of the initial state in the shortest chains, and will be discussed in more detail later. The irregularities displayed by the $\Delta S^{\ddagger}{}_{\text{intra}}$-profile in the medium- and large-ring region vary in a compensatory manner with those displayed by the related $\Delta H^{\ddagger}{}_{\text{intra}}$-profile, so that the reactivity profile exhibits a more regular pattern. Winnik (1981a) has suggested that monotonic changes in entropy are not to be expected in a homologous series of strained rings, probably on account of changes in the vibrational partition function along the series.

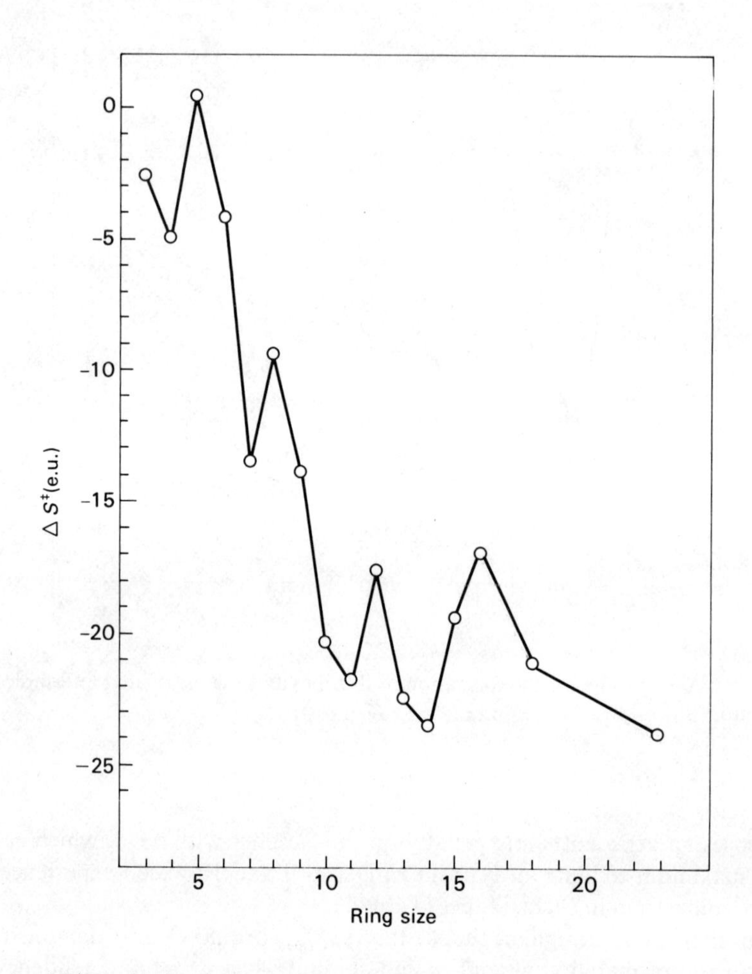

FIG. 9 $\Delta S^{\ddagger}$-profile for the formation of lactones

EM profiles for typical ring-closure reaction series

In addition to the lactone-forming reaction (44), there has been in the last decade a substantial accumulation of accurate kinetic measurements for ring-closure reactions, parts of which have been discussed by Illuminati and Mandolini (1981) and by Winnik (1981a).

A recent example is provided by the kinetics of closure of 1,1-bis(ethoxycarbonyl) cycloalkanes [7] from the anions derived from diethyl ω-bromoalkylmalonates (49a), which have been investigated by Casadei *et al.* (1984) in

$$Br(CH_2)_{n-1}C^-(CO_2Et)_2 \xrightarrow[-Br^-]{k_{intra}} \overbrace{(CH_2)_{n-1}\,C}(CO_2Et)_2 \qquad (49a)$$

[6] [7]

$$BuC^-(CO_2Et)_2 + BuBr \xrightarrow[-Br^-]{k_{inter}} (Bu)_2C(CO_2Et)_2 \qquad (49b)$$

Me_2SO solution for the ring-sizes 4–13, 17, and 21, together with the analogous non-cyclisation reaction (49b). In addition to polymerisation, the kinetics were complicated by a "spontaneous" decay of the carbanionic species [6], which was fast enough to compete with, or even obscure, the slowest reactions in the series. A special competition technique was developed for the 9-, 10-, and 11-membered rings, on account of their EM's being much lower than the concentration of about 1×10^{-4}M at which the kinetics could be conveniently followed. In fact, the EM-values in the order of 10^{-5}–10^{-6}M measured for the above ring sizes are far the lowest values ever recorded.

The reactivity profile for the malonate cyclisation can be conveniently compared with that for the lactone-forming reaction in a plot of log *EM vs* ring size (Fig. 10) since, by definition, the EM-profile for each series has the same shape as the log k_{intra} profile. Furthermore, within the limits of the concept of inherent reactivity of functional groups, the EM's are "corrected" for the reactivity between end-groups as separate entities, thus providing a common scale for comparing cyclisation tendencies in diverse reaction series. This is clearly shown in Fig. 10, where EM-profiles for two additional reaction series are also plotted. These are: (*i*) formation of catechol polymethylene ethers (43) in 99% Me_2SO (Dalla Cort *et al.*, 1980a), and (*ii*) cyclisation of the anions derived from N-tosyl-ω-bromoalkylamines [8] to N-tosylazacycloalkanes (50a) in 99% Me_2SO (Di Martino *et al.*, 1985). The intermolecular model reaction for (43) is the alkylation with BuBr of the guaiacolate

$$Ts\bar{N}(CH_2)_{n-1}Br \xrightarrow[-Br^-]{k_{intra}} \overbrace{TsN\ \ (CH_2)_{n-1}} \qquad (50a)$$

[8] [9]

$$Ts\bar{N}Bu + BuBr \xrightarrow[-Br^-]{k_{inter}} TsN(Bu)_2 \qquad (50b)$$

anion, and that for (50a) is given in (50b). As in the case of the lactone-forming reaction, ring closures of [1] and [8] could be followed at concentrations low enough to render polymerisation negligible relative to cyclisation, with

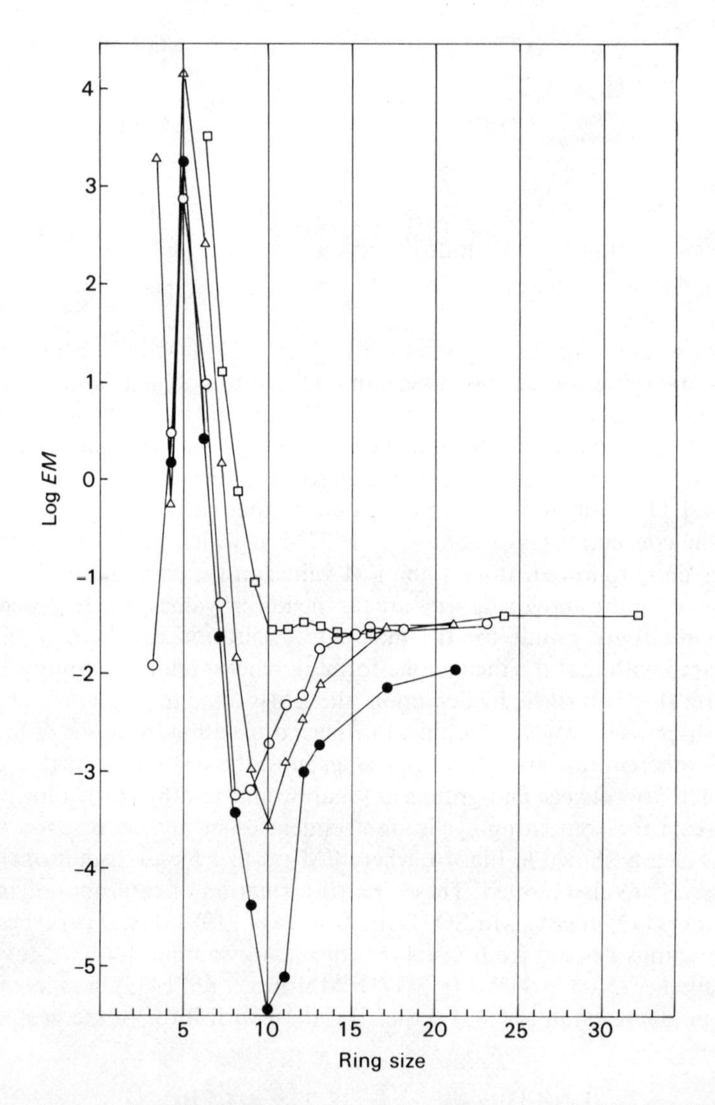

FIG. 10 EM-profiles for the formation of (○) lactones [4] in 99% Me_2SO; (●) 1,1-bis(ethoxycarbonyl) cycloalkanes [7] in Me_2SO; (△) N-tosylazacycloalkanes [9] in 99% Me_2SO; (□) catechol polymethylene ethers [2] in 99% Me_2SO

the sole exception of the reaction of [8: $n = 10$], where even at the lowest accessible concentration of *ca* 7×10^{-5}M, polymerisation could not be suppressed, but only reduced to a relatively minor disturbance, and kinetically treated by means of (8).

Inspection of the 4 profiles plotted in Fig. 10 clearly shows that the structure of the ring to be formed has an important influence on the ease of cyclisation, and that the magnitude of the EM-differences among different ring-closure reactions exhibits a marked dependence on ring size. This is the case for closures of the smallest rings, where the linking chain may prevent the terminal groups X and Y from attaining the required transition-state geometry dictated by the steric requirements of the terminal groups themselves and by the stereoelectronic requirements of the reaction. A qualitative answer to the problem of the relative facility of ring-forming reactions in the ring-size range 3–7 by intramolecular nucleophilic attack at tetrahedral, trigonal, and digonal carbon atoms is provided by Baldwin's rules for ring closure (Baldwin, 1976). A more quantitative approach will be presented in the following sections.

As to the medium rings, the first thing that is apparent is that the reactivity minimum is variable in shape, position, and depth. This can be easily understood on the basis of the simple rule that replacement of one or more methylene groups by trigonal carbons and/or by less-than-tetracovalent atoms can significantly reduce transannular and torsional interactions. Consistently, the deepest minimum is observed for the reactions leading to diethoxycarbonylcycloalkanes [7], which involve closure of all sp^3-carbon chains. Indeed, the reactivity minimum closely corresponds to the maximum in the cycloalkane strain-energies (Table 1), strongly suggesting that a major contribution to the free energy of activation for ring closures arises from the same kind of interactions which are responsible for the large strain-energies of medium-ring cycloalkanes. This provides a further confirmation of the validity of the Ruzicka hypothesis, which was advanced 50 years ago on the basis of scanty experimental evidence and little or no thermochemical data.

The next profile met in Fig. 10 on moving upwards refers to the formation of N-tosylazacycloalkanes [9]. The shape of this profile, with a sharp minimum centred at ring-size 10, closely resembles that of the malonates, but is significantly less deep, probably by virtue of a reduced strain-energy in the N-tosylazacycloalkanes incorporating a tricovalent nitrogen in the ring.

The reactivity profile for the lactone-forming reaction and the effect on its shape of the rate-depressing influence exerted by the *cis*-conformation of the ester function in the 8-membered and smaller rings have already been discussed. In the absence of such an influence the rate minimum would be presumably centred at ring-size 9 (Galli *et al.*, 1973). This is probably a consequence of the presence of the carbonyl group, as suggested by the fact that the strain-energy of cyclononanone is somewhat larger than that of cyclodecanone (Tables 1 and 3).

The last profile refers to the formation of diethers [2] by intramolecular Williamson synthesis. The data are in substantial agreement with earlier semi-quantitative measurements by Ziegler *et al.* (1937). Here the rings being formed have two oxygens and two sp^2-carbons. The strain relief resulting therefrom shows up in the EM-profile as a substantial lack of the familiar reactivity minimum in the medium rings.

In the large-ring region the vertical differences between the different profiles tend to vanish, the related EM's reaching values lying in the fairly limited range of about 0.01 M–0.1 M. First noted by Illuminati *et al.* (1977), and substantiated by additional evidence in subsequent years (Illuminati and Mandolini, 1981), the above regularity is further confirmed by the most recent data. It can be viewed as a general rule underlying the quantitative behaviour of long chains up to 20–30 atoms in cyclisation phenomena. Illuminati and Mandolini (1981) have pointed out that the common EM-level reflects similar losses in conformational entropy in the formation of large rings, irrespective of the chemistry involved at the end of the chains and of the chain length in the range investigated, which is in keeping with the common assumption of polymer chemists that the conformational contribution to the entropy of cyclisation is dominant for large-ring formation (Winnik 1981a).

The rigid group effect

Baker *et al.* (1951) suggested that the ease of formation of many-membered rings should be increased by the presence of a number of atoms in the chain backbone held in the form of a rigid group suitable for ring closure. A *cis*-double bond and an *o*-phenylene unit, where at least 4 atoms are coplanar with angles of about 120°, were considered as typical examples of structural moieties fitting these requirements. The rigid group principle, which has been more or less explicitly interpreted as an entropy effect, has been invoked from time to time to explain the supposedly facile formation of certain many-membered rings. For example, Drewes and Coleman (1972) claimed the formation of a series of macrocyclic monomeric and dimeric phthalate esters to be facile thanks to the presence of the rigid group OOC—C=C—COO, a view that has been even quoted in a textbook on stereochemistry (Potapov, 1978). An answer to the question of the operation of the rigid group effect on large-ring formation is implicit in the above considerations of the essential structural independence of EM's for closure of long chains. A clear-cut demonstration of the lack of the rigid group effect on large-ring formation is provided by an investigation of the kinetics of cyclisation of a number of 12-bromododecyloxyaryloxides (51) incorporating phenylene and

$$\mathrm{Br(CH_2)_{12}OZO^-} \xrightarrow{k_{\mathrm{intra}}} \overline{\mathrm{(CH_2)_{\mathit{m}}\,OZO}} \qquad (51a)$$

$$\mathrm{MeOZO^- + BuBr} \xrightarrow{k_{\mathrm{inter}}} \mathrm{MeOZOBu} \qquad (51b)$$

TABLE 11

EM-values for the formation of cyclophane and naphthalenophane diethers by intramolecular Williamson reaction (51a) in 99% Me_2SO at 25°C[a]

o-phenylene diether, $(CH_2)_{12}$ bridge	3.1×10^{-2}M
m-phenylene diether, $(CH_2)_{12}$ bridge	2.3×10^{-2}M
p-phenylene diether, $(CH_2)_{12}$ bridge	1.5×10^{-2}M
2,7-naphthylene diether, $(CH_2)_{12}$ bridge	1.5×10^{-2}M
1,5-naphthylene diether, $(CH_2)_{12}$ bridge	1.9×10^{-2}M

[a] Based on the non-cyclisation reactions (51b). (Data from Mandolini *et al.*, 1977)

naphthylene moieties Z of varying geometry and size (Mandolini *et al.*, 1977). The results given in Table 11 show EM-values lying within a factor of 2, which clearly indicates that the shape and size of the rigid group spanning the dodecamethylene chain has little effect on the facility of ring closure. It can be concluded that the operation of the rigid group effect on large ring formation can be definitely ruled out, and that no "magic" properties must be attributed to the *o*-phenylene unit. Additional evidence on the ease of formation of meta- and paracyclophane diethers has been reported by Dalla Cort *et al.* (1980b). A comparison with the ortho diethers (Fig. 11) shows that when the number of methylene groups in the bridge is less than 12 the ease of ring closure decreases in a way that is in qualitative agreement with the expectation of an increasing strain-energy upon decreasing the length of the bridge in the meta- and paracyclophane systems.

Another example of ring-closure reaction where an increasing strain factor results from reduction of the length of the connecting chain is the formation

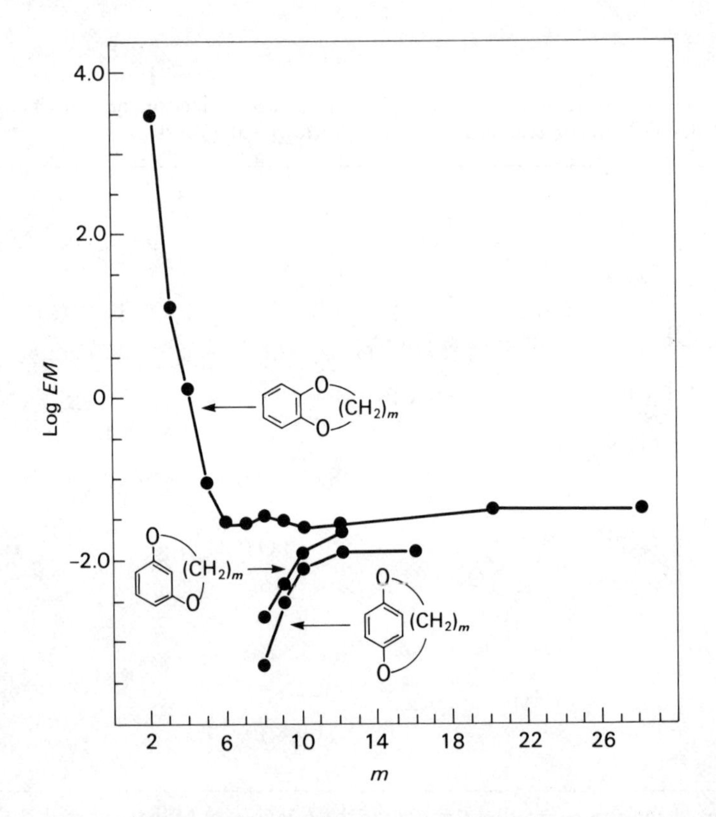

FIG. 11 Plots of log *EM* *vs* number of bridge methylenes for the formation of catechol, resorcinol, and hydroquinone polymethylene ethers by intramolecular nucleophilic substitution in 99% Me2SO. (Reproduced with permission from Dalla Cort *et al.*, 1980)

of thiophenophan-1-ones [11] of varying ring-sizes ranging from 12 to 21 by cyclisation of ω-(2-thienyl)alkanoic acids [10] under homogeneous conditions in CH_3CN solution in the presence of $(CF_3CO)_2O$ and a catalytic

$$\text{2-thienyl}-(CH_2)_mCO_2H \; [10] \longrightarrow [11] \quad (52a)$$

$$CH_3(CH_2)_4CO_2H + \text{2-methylthiophene} \longrightarrow CH_3(CH_2)_4C(=O)-\text{thienyl}-CH_3 \quad (52b)$$

amount of H_3PO_4 (Galli *et al.*, 1980). The effect of chain length on the ease of ring closure is quite close to that found for the meta- and paracyclophane diethers (Fig. 11), showing that the major factor in determining the shape of the reactivity profile is a geometrical one, in spite of the difference in reaction type. It was found, however, that the EM-values for the formation of the large-ring thiophenophan-1-ones [11] lie at least one order of magnitude below the range of EM-values typical for large-ring formation as shown for [12a] and [12b]. The possibility that the particular mechanistic type of the reaction at hand is responsible for this unusual behaviour is ruled out by the fact that the EM-value for the benzothiophen derivative [12c] turns out to be

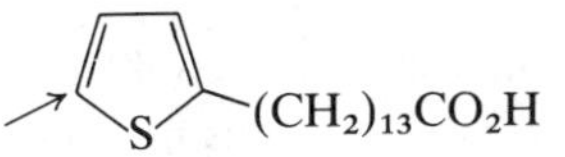

$(CH_2)_{13}CO_2H$

[12a] *EM* 1.5×10^{-3}M

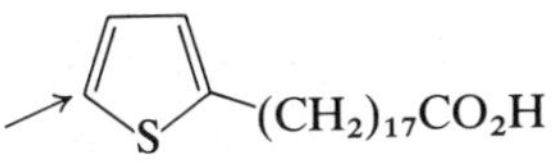

$(CH_2)_{17}CO_2H$

[12b] *EM* 1.7×10^{-3}M

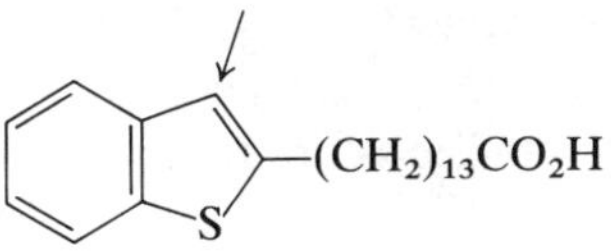

$(CH_2)_{13}CO_2H$

[12c] *EM* 1.1×10^{-2}M

"normal". Galli *et al.* (1980) have explained the low EM-values in the thiophene series as due to inhibition of solvent stabilisation of the developing positive charge, a major share of which is probably concentrated on sulphur, caused by the chain crossing over the thiophene ring plane in the transition state. Since no steric inhibition to solvation is expected to affect the related non-cyclisation reaction (52b), the low EM's may be regarded as resulting from the use of an inappropriate model reaction, which fails to mimic the role of solvation in the activation process for ring closure. This interpretation is supported by the normal EM-value for [12c], whose sulphur atom, being "outside" the ring, cannot suffer from desolvation effects.

The oxygen atom effect

It has long been recognised that replacement of methylene groups in the backbone by oxygen atoms increases the ease of cyclisation of a chain molecule (Hill and Carothers, 1933). Ziegler and Holl (1937) found that the yield of the 10-membered ring from the dinitrile [13] is less than 0.5%, but increases to 5% in the case of [14]. Similarly, the yield of the 13-membered ring from [15] was 12%, but was 70% from [16]. The authors offered an explanation for the effect in terms of strain relief due to reduction of the number of crowded atoms in the interior of the rings being formed. An alternative explanation was presented by Dale *et al.* (1963) in terms of preferred conformations and cyclisation probability, according to which the oxygen atom effect may be predicted to be either favourable or unfavourable to cyclisation depending on the position of the ether linkage. As far as this author is aware, however, no genuine example of cyclisation of long-chain molecules accompanied by an adverse oxygen atom effect has ever been reported. In fact, all of the quantitative evidence available at present points to an interpretation of the phenomenon due to strain relief.

$NC(CH_2)_9CN$ [13] $NC(CH_2)_4O(CH_2)_4CN$ [14] $NC(CH_2)_{12}CN$ [15] $NC(CH_2)_4OCH_2$-$CH_2O(CH_2)_4CN$ [16]

It was Dale (1974) who, in a thorough discussion of the conformational consequence of replacing methylene groups by oxygen atoms in acyclic and cyclic compounds, convincingly argued from a substantial body of experimental evidence that 1,4- and 1,5-CH $\cdots$ O interactions are less unfavourable than the corresponding CH $\cdots$ HC interactions, and that 1,4-O $\cdots$ O interactions are unquestionably attractive. Consequently, one expects transannular CH $\cdots$ HC repulsions in ring compounds to be relieved when replaced by CH $\cdots$ O or O $\cdots$ O interactions, and the magnitude of this effect to be a maximum for the medium rings, where transannular interactions are a major source of strain.

Actually, Paquette and Begland (1966, 1967) obtained strong indications that strain minimisation accompanies the replacement of a methylene group by oxygen in medium-sized rings due to the reduced steric demands of an oxygen atom relative to a methylene group. This view is strongly supported by the recently determined enthalpy of formation of gaseous 1,4,7,10-tetraoxacyclododecane (12-crown-4), which was found by Byström and Månsson (1982) to be twice as great as that of 1,4-dioxan within the experimental error; this indicates that the strain-energy per (CH_2CH_2O) unit of the 12-membered polyether ring is very nearly the same as that of the only slightly strained 1,4-dioxan ring.

To the extent that cyclic transition-state strains parallel ring-product strains, the above considerations provide a proper guideline for prediction and interpretation of the oxygen-atom effect on rates of ring closure. This is clearly shown by a comparison of EM data for the two closely related ring-closure reactions (43) and (53a) in 75% aqueous ethanol (Illuminati *et*

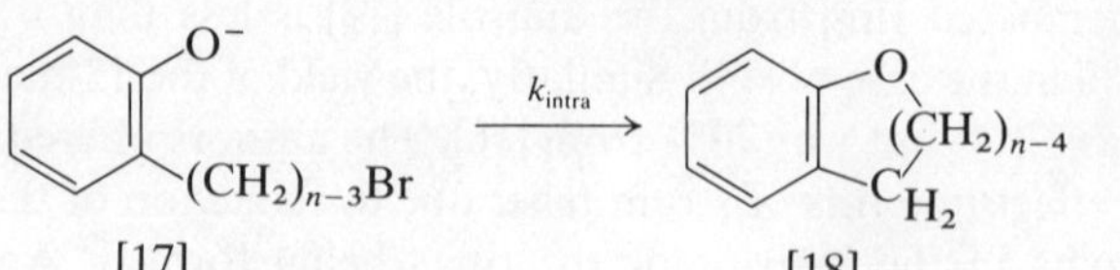

$$o\text{-}O^-C_6H_4(CH_2)_{n-3}Br \; [17] \xrightarrow{k_{intra}} C_6H_4\text{-}O\text{-}(CH_2)_{n-4}\text{-}CH_2 \; [18] \qquad (53a)$$

$$o\text{-}(O^-)C_6H_4CH_2CH_2 + BuBr \xrightarrow{k_{inter}} o\text{-}(OBu)C_6H_4CH_2CH_3 \qquad (53b)$$

al., 1974a, 1975, 1977). Here the oxygen atom effect arises from the replacement of the benzyl methylene of the monoethers [18] by oxygen to afford the diethers [2]. In Table 12 the *EM*(diether)/*EM*(monoether) ratios display a

TABLE 12

The oxygen atom effect as a function of ring size[a]

Ring-size n	6	7	8	9	10	11	14	16
$\frac{EM(\text{diether})}{EM(\text{monoether})}$	0.60	0.31	4.72	5.55	3.06	3.43	1.95	1.40

[a] Data from Illuminati *et al.*, 1974a, 1975, 1977.

definite medium-ring effect which closely parallels the behaviour of the enthalpies of activation (Fig. 12), thus fully confirming the expectation that the oxygen atom effect should be greater the greater the strain in the cyclic transition states. The magnitude of the effect tends to vanish upon increasing the ring size, but it is still appreciable for ring-sizes 14 and 16. This can be taken as an indication that the large rings investigated are still affected by appreciable strain.

A deeper insight into the operation of the oxygen-atom effect as well as a further assessment of its dependence on chain length is provided by an investigation of Illuminati *et al.*, (1981) on the kinetics of formation of benzo-crown ethers (54a) and of the analogous non-cyclisation reaction (54b)

$$o\text{-}(O^-)C_6H_4OCH_2(CH_2OCH_2)_xCH_2Br \; [19] \xrightarrow{k_{intra}} \text{benzo-crown ether} \; [20] \qquad (54a)$$

$$o\text{-}(O^-)C_6H_4OCH_3 + BrCH_2CH_2OEt \xrightarrow{k_{inter}} o\text{-}(OCH_2CH_2OEt)C_6H_4OCH_3 \qquad (54b)$$

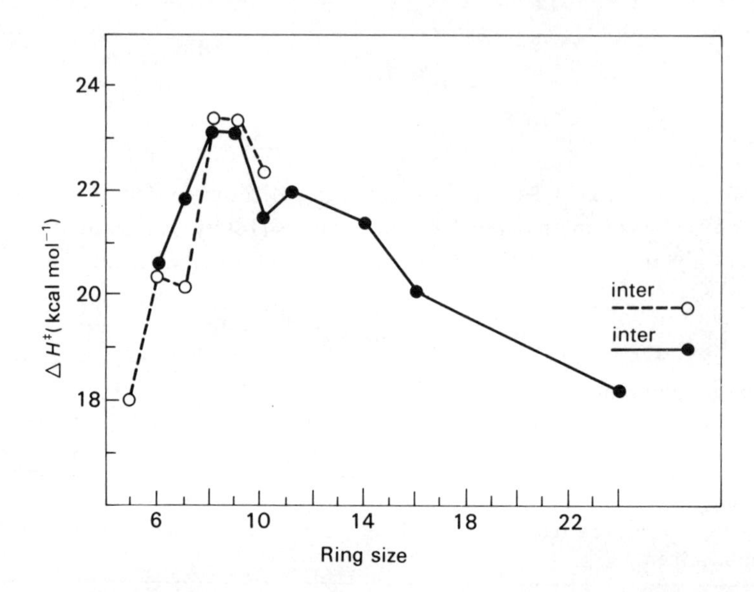

FIG. 12 $\Delta H^{\ddagger}$-profile for the formation of (○) monoethers [18] and (●) diethers [2] and for the corresponding non-cyclisation reactions in 75% aqueous ethanol. Reproduced with permission. (Data from Illuminati *et al.*, 1974a, 1975, 1977)

in 99% Me_2SO at 25°C. Here the phenoxide and the CH_2Br end-groups are linked by poly(oxyethylene) chains for lengths corresponding to $x = 1, 2, 3, 4, 5, 8$, and 14, thus leading to benzo-3$(x + 2)$-crown-$(x + 2)$ ethers [20] of varying ring-sizes in the range of 9 to 48 atoms. It is noteworthy that 48 is the largest ring-size included so far in kinetic studies of cyclisation reactions starting from an open-chain precursor of well-defined molecular weight. The rates of ring-closure of [19] (Fig. 13) exhibit a monotonic drop upon increasing the length of the polyether chain, apart from a small but real inversion on going from ring-size 12 to 15. The EM's vary within a factor of 10, and are within the range 0.1 M–0.01 M, where most of the available data related to the formation of large rings have been found to cluster (Illuminati and Mandolini, 1981). Figure 13 provides a comparison between EM-values for the reactions forming benzo-crown ethers and those for the formation of the corresponding polymethylene diethers [2] under a homogeneous set of experimental conditions. At ring-size 9 the EM-values are quite similar in both series, but in the next higher homologues the ease of ring closure of the poly(oxyethylene) chains is greater than that of the all-carbon chains of similar length by a factor of about 3. When the ring size becomes greater the difference tends to vanish and is in fact negligibly small in the neighbourhood of ring-size 30. This indicates that, at variance with earlier suggestions (Sisido *et al.*, 1978), the

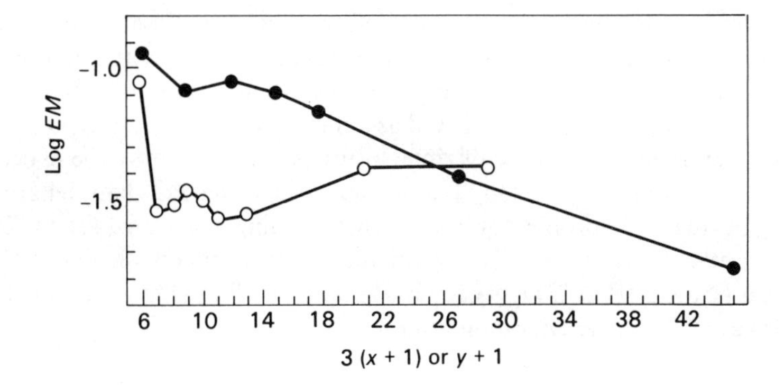

FIG. 13 Log *EM*-values for the formation of (○) catechol polymethylene ethers [2] and (●) benzo-crown ethers [20] in 99% Me_2SO at 25°C plotted against the number of single bonds, *r*, in the chain connecting the end-groups. (Data from Illuminati *et al.*, 1981)

cyclisation behaviour of a long poly(oxyethylene) chain is very much the same as that of a polymethylene one.

The effects observed for the shorter chains are easily understood as a strain relief phenomenon. Indeed, as Illuminati *et al.* (1981) have noted, replacement of suitably placed methylenes of diethers [2] with oxygens to give the corresponding benzo-crown ethers [20] transforms two 1,4-CH · · · O and $(x-1)$ 1,4-CH · · · HC into 1,4-O · · · O interactions, and $2(x-1)$ 1,5-CH · · · HC into 1,5-CH · · · O interactions. As a result, relief of unfavourable CH · · · HC repulsions occurs when $x \geqslant 2$, i.e. when the ring size is $\geqslant 12$. However, at ring-size 9 ($x = 1$) only two presumably weak CH · · · O interactions are relieved as shown in [21] and [22]. Consequently, it seems likely that

[21] [22] [23]

formation of rings [21] and [22] from the corresponding open-chain precursors is accompanied by similar strain-energies, consistent with the close similarity of their EM's; this provides an apparent violation of the rule that the oxygen atom effect should be a maximum in the medium-ring region.

The situation is different when the benzyl methylene of [23] is replaced by oxygen. Relief of one 1,5- and one 1,4-CH · · · HC repulsion is clearly responsible for the significant oxygen-atom effect observed in this case (Table 12).

With respect to the reactions leading to cyclic ethers [2] and [18], where much of the strain factor causing the appearance of a reactivity minimum in the medium-ring region of other series has disappeared, reaction (54a) forming benzo-crown ethers can be viewed as a further step toward a hypothetical reaction series where the ease of ring closure is solely ruled by the expected monotonic drop of the entropy factor upon increasing the chain length, as shown by the ΔS^{+}-profile for the diether-forming reaction (43) in 75% aqueous ethanol (Fig. 14). The slight reactivity inversion on going from ring-size 12 to 15 (Fig. 13) has been interpreted by Illuminati *et al.* (1981) as a mere vestige of a reactivity minimum.

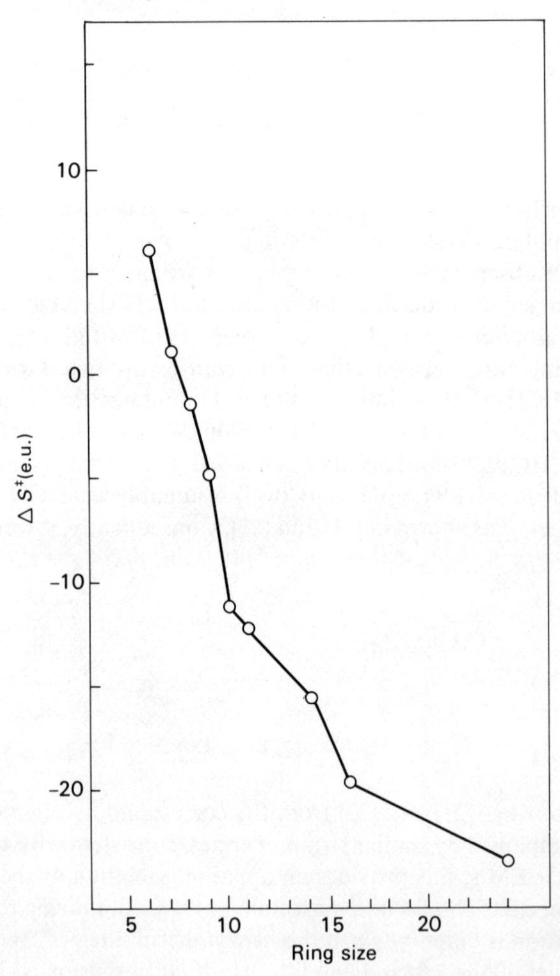

FIG. 14 ΔS^{+}-profile for the formation of catechol polymethylene ethers [2] in 75% aqueous ethanol

The gem-*dimethyl effect*

There is a wealth of quantitative evidence of the *gem*-dimethyl effect in the formation of 3- to 6-membered rings, and excellent discussions are available (see, for example, a very recent paper by Jager *et al.*, 1984, and references cited therein). Much less is known, however, in the medium- and large-ring regions. A clear-cut example of a favourable effect due to methyl substituents on closure of an 8-membered ring was reported by Allinger and Hu (1961), who found a marked increase in the yield of cyclic product by the Ziegler dinitrile condensation. Similar results concerning 9-membered ring formation by the acyloin condensation have been quoted by Dale (1963). As to the large rings, some evidence has been reported by Borgen and Dale (1972), Borgen (1973), Borgen and Gaupset (1974), and Björnstad and Borgen (1975).

Kinetic results for lactone formation from *gem*-dimethyl derivatives of ω-bromoalkanoate ions (Galli *et al.*, 1979) provide the only available set of quantitative data covering a broad spectrum of ring sizes (Table 13). The magnitude of the *gem*-dimethyl effect is significant at ring-size 6, it is still appreciable for $n = 9$, but becomes negligible for $n = 10$ and slightly inverse for $n = 11$. It is again negligibly small, or nearly so, for $n = 16$. No activation parameters were measured for the reactions at hand and, because of the lack of thermochemical data, no information is available as to the effect of alkyl groups on enthalpies and entropies of many-membered rings. Nevertheless, Galli *et al.* (1979) have tentatively discussed the experimental results by extension of the Allinger and Zalkow approach to the *gem*-dimethyl effect on common-ring formation (see p. 29). The entropy advantage due to alkyl substitution is predicted to become less important as the chain gets longer on account of the presumable increase in the looseness of ring structures on increasing ring size (O'Neal and Benson, 1970). On the other hand, a less favourable or unfavourable enthalpy effect is believed to originate from increased strain in the transition state (or ring product) in the medium-ring region, where steric crowding is much more important than in 6-membered rings, but little or no effect of this kind is expected to operate in the formation of the strainless large rings. Accordingly, the inverse effect observed for $n = 11$ is taken as an indication that an adverse strain factor more than offsets a presumably favourable entropy factor, whereas a virtually exact balance occurs for $n = 10$.

The simple pictorial interpretation of the *gem*-dimethyl effect presented by Dale (1963, 1966) was based on the idea that the ease of cyclisation depends more on the conformation of the open-chain reactant than on the tension in a resulting strained ring. According to Dale (1963), "a polymethylene chain tends to be straight zig-zag but, when substituted by a methyl group, it does not matter for the neighbouring C—C bond whether the chain itself or the methyl group continues in the *trans*-position; hence the probability of a bent

TABLE 13

The *gem*-dimethyl effect on the rate of lactone formation[a]

gem-Substituted reactant	Ring-size	k_{rel}[b]
$Br(CH_2)_2—C(CH_3)_2—CH_2CO_2^-$	6	38.5
$Br(CH_2)_5—C(CH_3)_2—CH_2CO_2^-$	9	6.62
$Br(CH_2)_6—C(CH_3)_2—CH_2CO_2^-$	10	1.13
$Br(CH_2)_7—C(CH_3)_2—CH_2CO_2^-$	11	0.61
$Br(CH_2)_6—C(CH_3)_2—(CH_2)_7CO_2^-$	16	1.22

[a] In 99% MeSO at 50°C. (Data from Galli *et al.*, 1979)
[b] Rate of closure of the *gem*-substituted reactant over the unsubstituted one

chain increases". Consequently, one predicts the *gem*-dimethyl effect to be definitely favourable to ring closure in the large-ring region, where complications arising from strain factors are largely absent. It is clear that such a prediction is disproved by the behaviour of the 16-membered lactone, where the *gem*-dimethyl effect is as little as 1.2; this observation is consistent with the general idea that, in the limit of the large rings, the polymethylene chain in ring compounds closely approaches the behaviour of the inner portion of an infinite alkane chain, in that replacement of one or more methylene groups by diverse structural units affects the standard free energy of the chain molecule much in the same way as that of the cyclic transition state or product. The net result is that the free energy change for ring closure is hardly noticeable.

Additional evidence from intramolecular phenomena not involving ring closure

The intramolecular processes considered so far provide examples of reactions where a stable, covalently bonded cyclic product is formed. They can be used, and have actually been used in many instances (Ziegler *et al.*, 1937; Hunsdiecker and Erlbach, 1947; Illuminati *et al.*, 1974b; Galli and Mandolini, 1975, 1978; Mandolini and Masci, 1977; Catoni *et al.*, 1980; Casadei *et al.*, 1981; Kruizinga and Kellogg, 1981; Gargano and Mandolini, 1982; Kimura and Regen, 1983) to prepare synthetically useful amounts of ring products over a wide range of ring sizes. In addition to these "conventional" ring-closure reactions, there is now a remarkable accumulation of quantitative data on intramolecular phenomena published by scientists such as spectroscopists, polymer chemists, photochemists, and photophysicists. For these processes the term ring closure is inappropriate, and the more general term cyclisation is preferred. The main goal of these studies is to relate experimentally measurable quantities to the dynamic and/or conformational behaviour of alkane and other flexible chains. Investigations where experimental data are expressed in the form of EM's, or where the required data for the calculation of EM's are available, are discussed in what follows.

An esr technique permitting the determination of the frequency of electron exchange between groups attached at the ends of chains of varying length has been described by Connor *et al.* (1972). Compounds having structure $N(CH_2)_mN$, where N denotes an α-naphthyl moiety were partially reduced with potassium in hexamethylphosphoric triamide (HMPT) to afford solutions of radical anions $N(CH_2)_mN^{\overset{-}{\cdot}}$ possessing not more than one electron per chain. The observed changes in shape of the esr spectra with the length of the chain were analysed in terms of a degenerate, intramolecular electron-transfer process, $N(CH_2)_mN^{\overset{-}{\cdot}} \rightleftarrows {}^{\overset{-}{\cdot}}N(CH_2)_mN$, and compared with the analogous intermolecular electron exchange between α-butylnaphthalene and its radical anion (Shimada and Szwarc, 1975a). Selected data from these experiments are listed in Table 14. An interesting feature of these data is the monotonic drop of the rate of intramolecular exchange upon increasing the length of the connecting chain, suggesting that the stereochemical requirements for electron transfer between the naphthalene nuclei are not stringent. Indeed, Shimada and Szwarc (1974) obtained evidence that electron exchange between two α-naphthyl groups can efficiently occur on their approach to about 7–9 Å, so that no closed ring need be formed in the intramolecular electron exchange. This is in agreement with the finding that the activation energy reaches a constant value of 5.0 ± 0.4 kcal mol^{-1} for $m \geqslant 5$, which coincides within experimental errors with the value of 4.3 ± 1.0 kcal mol^{-1} for the bimolecular exchange. For the shortest chains the activation energy is substantially higher, namely, 10.5 kcal mol^{-1} for $m = 3$ and 8.9 kcal mol^{-1} for $m = 4$.

TABLE 14

Rate constants and EM's for intramolecular electron exchange in $N(CH_2)_mN^{\overline{\cdot}}$ in HMPT at 30°C[a]

m	k_{intra}/s^{-1}	EM/M^{b}
3	1.3×10^{10}	18
4	2.5×10^{9}	3.5
5	1.4×10^{8}	0.20
6	6.3×10^{7}	0.089
8	4.3×10^{7}	0.061
10	3.9×10^{7}	0.055
12	3.5×10^{7}	0.050
16	1.6×10^{7}	0.022
20	1.0×10^{7}	0.014

[a] Data from Shimada and Szwarc, 1975a
[b] Based on the intermolecular electron transfer between α-butylnaphthalene and its radical anion, $k_{inter} = 7.1 \times 10^{8} M^{-1} s^{-1}$

An extension of these studies utilising N-phthalimido (PI) as an alternative end-group has been carried out by Shimada *et al.* (1975), who determined in five different solvents the frequency of intramolecular electron exchange in a series of radical anions having the formulae $PI(CH_2)_mPI^{\overline{\cdot}}$ and $PICH_2(CH_2OCH_2)_xCH_2PI^{\overline{\cdot}}$. Combination of rate data for intramolecular electron exchange with those for the bimolecular exchange between N-butylphthalimide radical anion and its parent molecule (Shimozato *et al.*, 1975) permits the calculation of the EM's. Values obtained in dimethylformamide (DMF) are shown in Fig. 15. Shimada and Szwarc (1975b) obtained evidence that electron transfer among N-phthalimide units requires a closer approach and more stringent orientational requirements than that among α-naphthyl units. Consequently, the EM for electron exchange in $N(CH_2)_mN^{\overline{\cdot}}$ is about 100 times larger than in $PI(CH_2)_mPI^{\overline{\cdot}}$ when $m = 3$, but the difference between the two series decreases on increasing m, and eventually vanishes for chains in the neighbourhood of 10–16 carbon atoms, where the EM's for both series are in the range of 0.02–0.05 M.

The authors believe that electron exchange occurs at every "intramolecular collision". Thus, their work should provide the first quantitative measurements of the frequency of intramolecular collision between end-groups attached to flexible chains. According to these authors such frequency reflects an intrinsic property of chain molecules, referred to as the dynamic flexibility, which is a measure of the rate of conformational change. It should be distinguished from the static flexibility, which depends on the multitude of the conformational states but not on time. Therefore, either a dynamic or static model applies to intramolecular reactions, depending on whether they operate respectively under diffusion or activation control. Sisido and Shimada (1977) published a theoretical paper which simulated the cyclisation dynamics in $N(CH_2)_mN^{\overline{\cdot}}$, thus supporting the idea that electron exchange occurs under dynamic control. However, Winnik (1981a,b) has convincingly argued that all evidence points to these reactions operating under conformational control.

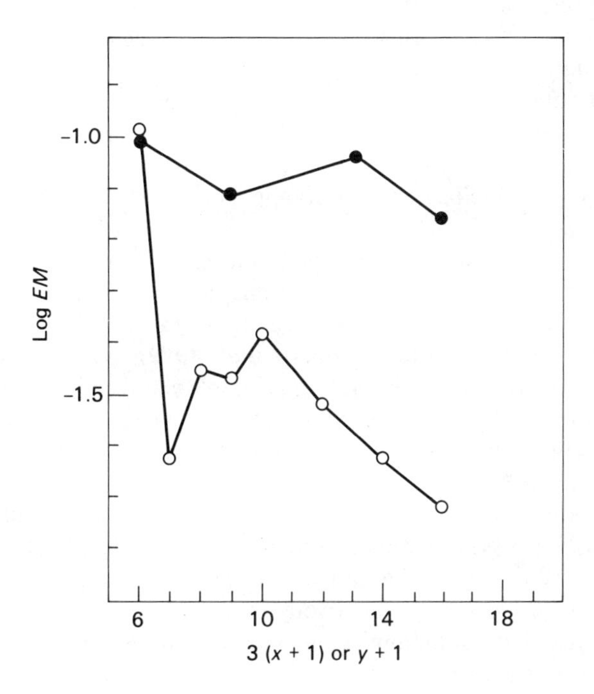

FIG. 15 Log *EM*-values for intramolecular electron-transfer reactions in $PICH_2(CH_2OCH_2)_xCH_2PI^{\overline{\cdot}}$ (●) and $PI(CH_2)_yPI^{\overline{\cdot}}$ (○) in DMF at 20°C as a function of the number of single bonds, *r*, in the chain connecting the end-groups. Reproduced with permission. (Data from Shimada *et al.*, 1975)

The close similarity between the EM-profiles reported in Figs 13 and 15 has been emphasised by Illuminati *et al.* (1981). In spite of the markedly different nature of the reactions involved, which proceed at rates differing by some 10 powers of 10, the EM-values for chains of equal nature and length are not only remarkably similar, but follow profiles which appear to be structured in much the same way. Thus, an undeniable minimum appears in both series in the case of the 6-methylene chain compounds, showing the presence of a scarcely pronounced, but nevertheless clearly detectable, medium-ring effect also in the electron exchange reaction. Even more significant is the fact that when $x > 1$ the frequency of electron exchange for polyoxyethylene chains is some two to three times as fast as that for alkane chains of similar length. But when $x = 1$ the oxygen-atom effect is lacking. As in the case of the reaction leading to [22], the oxygen atom in $PI(CH_2)_2O(CH_2)_2PI^{\overline{\cdot}}$ replaces the only methylene of $PI(CH_2)_5PI^{\overline{\cdot}}$ not involved in interactions with 1,5- and/or 1,4-CH groups. It seems natural therefore to explain the oxygen-atom effect also in the case of intramolecular electron exchange between PI units as a phenomenon due to relief of strain due to 1,5- and/or 1,4-CH $\cdots$ HC repulsions. Although the basic question to be answered for a correct interpretation of the above data is whether the reactions are dynamically or conformationally controlled, it should be emphasised that from a purely phenomenological standpoint the "fast" intramolecular electron exchange and the "slow" Williamson-type ring-closure reactions appear to be affected in the same way by structure.

Excited states of simple tertiary alkyl amines in alkane solvents undergo self-quenching of fluorescence when they encounter an amine molecule in its ground state. The rate of this process is extremely fast, about one encounter in five resulting in deactivation (Halpern *et al.*, 1979). When two dimethylamino groups are attached to the ends of a flexible chain and one of the amino groups becomes electronically excited upon absorption of light, intramolecular self-quenching shortens the decay time of the excited state. Thus, the dimethylamino group was used as a fluorometric probe to test the end-to-end flexibility of alkane chains by Halpern *et al.* (1979). Their results obtained with a series of α,ω-bis(dimethylamino) alkanes $Me_2N(CH_2)_mNMe_2$ in hexane solution at 23° over a wide range of values of m are plotted in Fig. 16. Intramolecular fluorescence quenching in α,ω-diaminoalkanes exhibits a well-defined reactivity minimum in the medium-ring region, which is absent in electron-exchange reactions in $N(CH_2)_mN^{\overline{\cdot}}$ and $PI(CH_2)_mPI^{\overline{\cdot}}$. The different behaviour strongly suggests that more closed ring structures are formed in the transition states for intramolecular fluorescence quenching than in those for intramolecular electron exchange, which is consistent with the view that formation of excimers requires a closer approach of interacting groups than does electron transfer (Shimada and Szwarc, 1975a). Indeed the EM-profile plotted in Fig. 16 showing a sharp minimum centred at ring-size 10, is structured in much the same way as the EM-profiles related to the formation of bis(ethoxycarbonyl)cycloalkanes and N-tosylazacycloalkanes (Fig. 10). The fact that reactions in the former series proceed at rates close to the diffusion-control limit, whereas the latter do not, is clearly irrelevant as far as the EM's are concerned. Presumably, these reactions share in common the fact that the molecular geometry in the transition states may well resemble that of a cycloalkane. This would imply that fluorescence quenching occurs via a transition state whose tightness approaches that required for covalent bond formation in S_N2 processes.

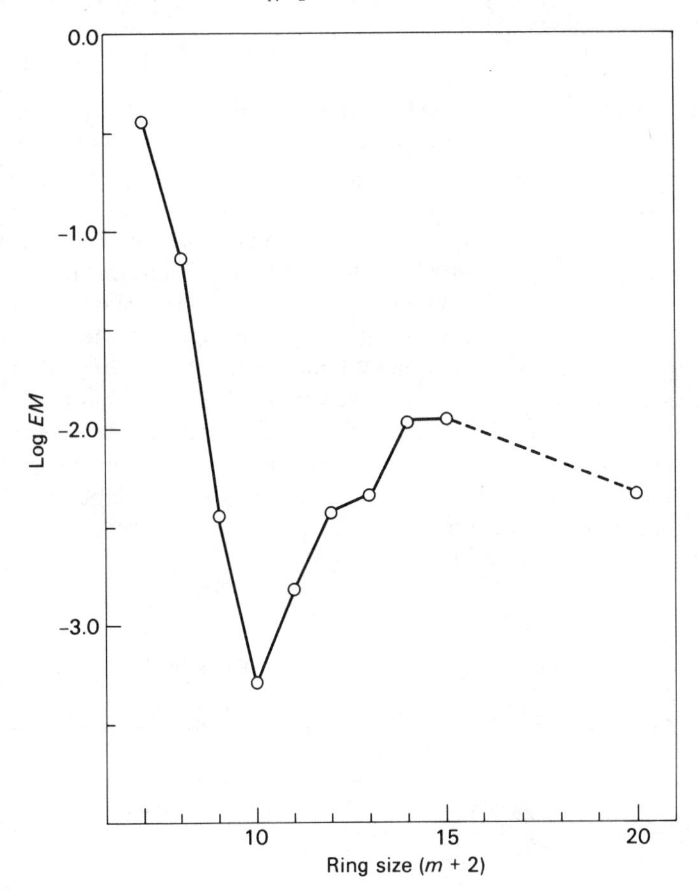

FIG. 16 EM-profile for intramolecular fluorescence quenching in $Me_2N(CH_2)_m NMe_2$ in hexane. (Data from Halpern *et al.*, 1979)

The quenching of benzophenone phosphorescence has been used by Mar and Winnik (1981) as a photochemical probe of hydrocarbon chains in solution. The bimolecular reaction for quenching the triplet state of 4-methoxycarbonylbenzophenone [24] by 1-pentene occurs at rates which are below the diffusion limit by two to three orders of magnitude. Consequently, the intramolecular quenching reactions of ω-alkenyl esters of benzophenone-4-carboxylic acid [25] occurs under conformational control. In [25] the point of

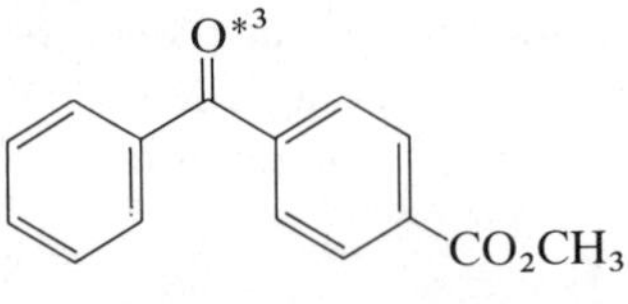

O^{*3} — CO_2CH_3

[24]

O^{*3} — $CO_2(CH_2)_mCH{=}CH_2$

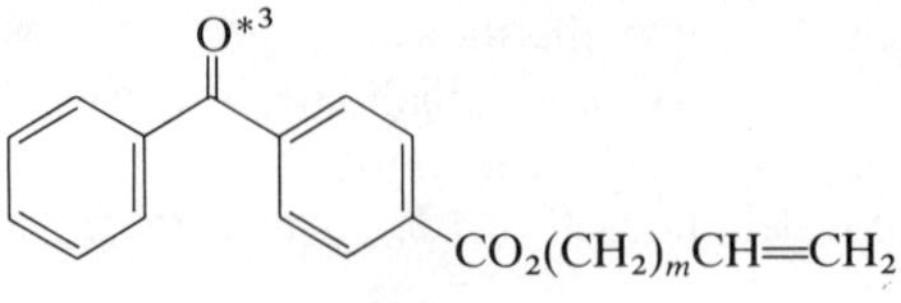

[25]

attachment of the chain is remote from the site of reaction. Molecular models indicate that close contact between the terminal $CH{=}CH_2$ group and the ketone carbonyl requires chains of at least 8 methylene groups. Consistently, there is a sudden increase in reactivity (Table 15) for chains of length m from 6–9. This stereochemical situation closely resembles that found in the formation of hydroquinone polymethylene ethers (Fig. 11). Thus, it is not surprising that the EM's are very similar in the two series for chains of similar length. Measurements of rates of intramolecular phosphorescence quenching in [25] carried out in CCl_4 solution (Winnik, 1981a; Mar *et al.*, 1981) have shown that the activation energy is 3.2 kcal mol^{-1} for $m = 8$, but reaches the constant value of 2.4 kcal mol^{-1} for chains of $m \geqslant 9$, which is somewhat larger than the value of 1.6 kcal mol^{-1} of the corresponding bimolecular reaction. Unfortunately, no activation parameters are available for chains shorter than 8 methylene groups.

TABLE 15

EM-values for intramolecular phosphorescence quenching in [25] as a function of chain-length m[a]

m	4	6	8	9	10	11
EM/M	0.0006	0.0008	0.0147	0.0338	0.0339	0.0432

m	12	13	14	15	18	21
EM/M	0.0493	0.0399	0.0389	0.0335	0.0227	0.0188

[a] Solvent, acetic acid; temperature, 25°C. (Data from Mar and Winnik, 1981)

The photochemistry of [26] was first studied by Breslow and Winnik (1969). The excited triplet state undergoes intramolecular hydrogen abstraction which quenches the excited state and shortens the triplet lifetime. Here cyclisation involves the chain interior, each CH_2 group being inherently

$$O^{*3} \quad CO_2(CH_2)_mCH_3 \longrightarrow OH \quad CO_2 \quad H \qquad (55)$$

[26]

equally reactive (Winnik and Maharaj, 1979) (Fig. 4). An extensive investigation of chain length and solvent effects on (55) has been reported by Winnik *et al.* (1974, 1976). As Fig. 17 clearly shows, no quenching occurs for chains with fewer than 9 CH_2 groups, which indicates that close contact is

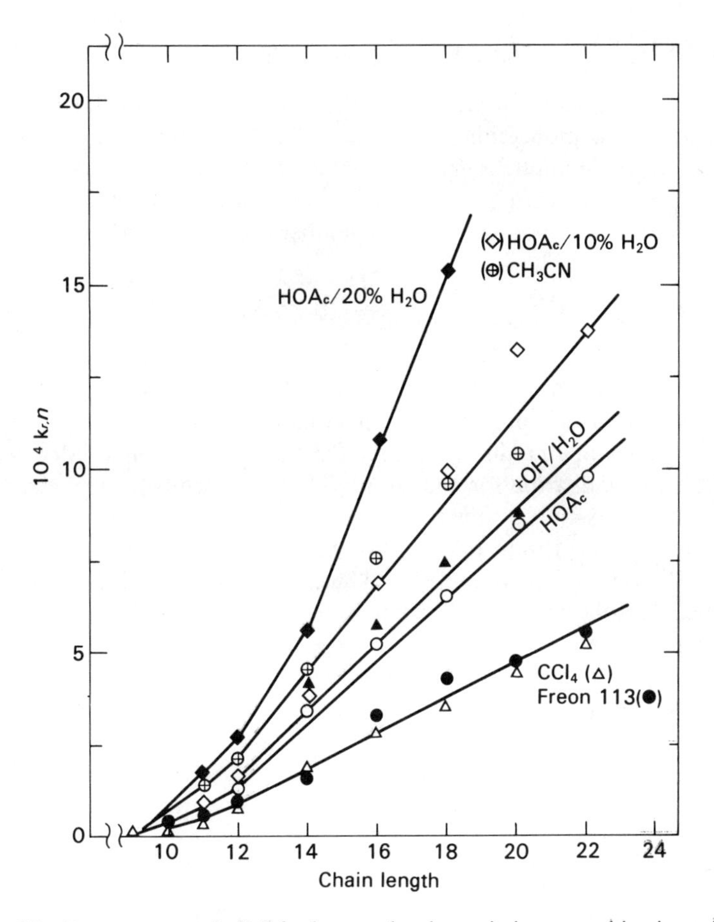

FIG. 17 Rate constants in [26] for intramolecular emission quenching in various solvents. (Reproduced with permission from Winnik, 1977)

required between the alkane chain and the ketone carbonyl group. It also appears that values of k_{intra} increase monotonically with chain length, showing that for sufficiently long chains each CH_2 group affords an approximately constant contribution to the overall reactivity. Polar solvents markedly increase k_{intra} in a way that closely parallels the effect on the rate of the bimolecular reaction between [24] and hexane. Consequently the differences in k_{intra}-values (Fig. 17) largely disappear in the EM-values. It can be calculated that for sufficiently long chains (i.e. $m \geqslant 14$) the average EM value per CH_2 group ranges from 0.04 M in CCl_4 to 0.08 M in 80% CH_3CO_2H. Although they are small, Winnik (1977) believes the above differences to be meaningful. They should reflect solvent effects only on chain shape.

CYCLISATION OF POLYMER CHAINS. THEORY AND EXPERIMENT

Cyclisation of long-chain molecules is a field where theory has far preceded experiment. In his pioneering treatment of flexible chains in terms of the freely-jointed chain model, Kuhn (1934) derived for the "local concentration" C_{eff} of one chain end in the neighbourhood of the other (see p. 7) expression (56) where N_A is Avogadro's number and C_{eff} is given in moles per

$$C_{eff} = \frac{1000}{N_A}\left(\frac{3}{2\pi\langle r^2\rangle}\right)^{3/2} \tag{56}$$

litre when r is given in cm. The physical model suggested by (56) is one corresponding to a particle in a sphere with a radius approximately equal to the mean square chain-end displacement $\langle r^2 \rangle$. In Θ solvents (Morawetz, 1975),[6] where the distribution function $W(r)$ of chain ends for real chains is Gaussian to a reasonable approximation, $\langle r^2 \rangle = C_z Z b^2$, where Z and b are the number and length of real or virtual skeletal bonds, and C_z is the characteristic ratio (Morawetz, 1975). Thus, (56) predicts that in the limit of very long chains the ease of cyclisation decreases with the $-3/2$ power of the chain length. In good solvent media, in which the shape of the chain is expanded over its unperturbed dimensions, no general theory is available. Here C_{eff} is expected to be lower than that calculated from (56).

[6] Following Flory (1969), a Θ solvent is a thermodynamically poor solvent where the effect of the physically occupied volume of the chain is exactly compensated by mutual attractions of the chain segments. Consequently, the excluded volume effect becomes vanishingly small, and the chains should behave as predicted by mathematical models based on chains of zero volume. Chain dimensions under Θ conditions are referred to as unperturbed. The analogy between the temperature Θ and the Boyle temperature of a gas should be appreciated.

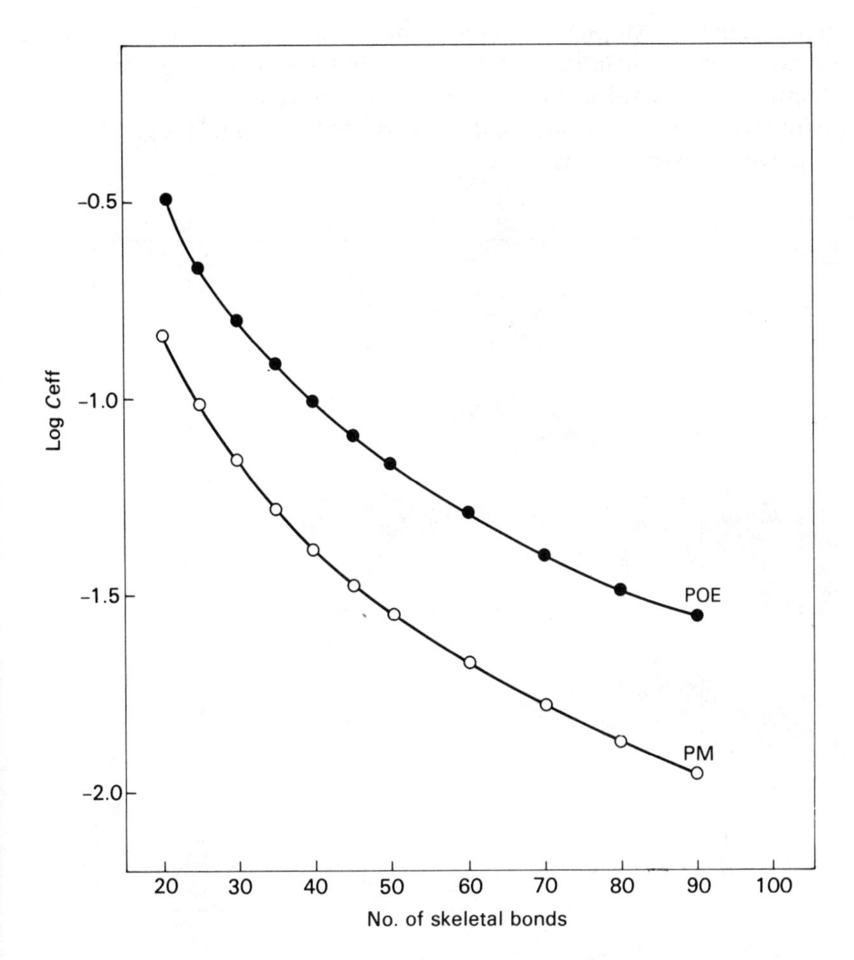

FIG. 18 Plots of log C_{eff} *vs* chain length for polymethylene (PM) and polyoxyethylene (POE) chains under Θ conditions, calculated from (56)

Figure 18 shows plots of log C_{eff} *vs* chain length ranging from 20 to 100 chain segments for polymethylene and polyoxyethylene chains, calculated from (56) and values of C_z given by Flory (1969, p. 170). Comparison with the EM-profiles reported in Figs 10 and 13 shows that there is a general tendency for the polymethylene chains to approach the theoretical predictions. The EM-value for closure of [1: $n = 24$] is 0.041 M in 99% Me_2SO and 0.064 M in 75% EtOH, and that for closure of [1: $n = 32$] is 0.041 M in 99% Me_2SO (Dalla Cort *et al.*, 1980a). The calculated values are 0.10 M and 0.060 M, respectively. The ease of formation of the benzo-crown ethers [20] from [19] in 99% Me_2SO shows the predicted tendency to decrease with the 3/2 power of the chain length for sufficiently long chains (Fig. 19), but the experimental EM-values are smaller than those calculated by a factor of about three. This is not surprising since Me_2SO is no doubt a good solvent for polyoxyethylene chains.

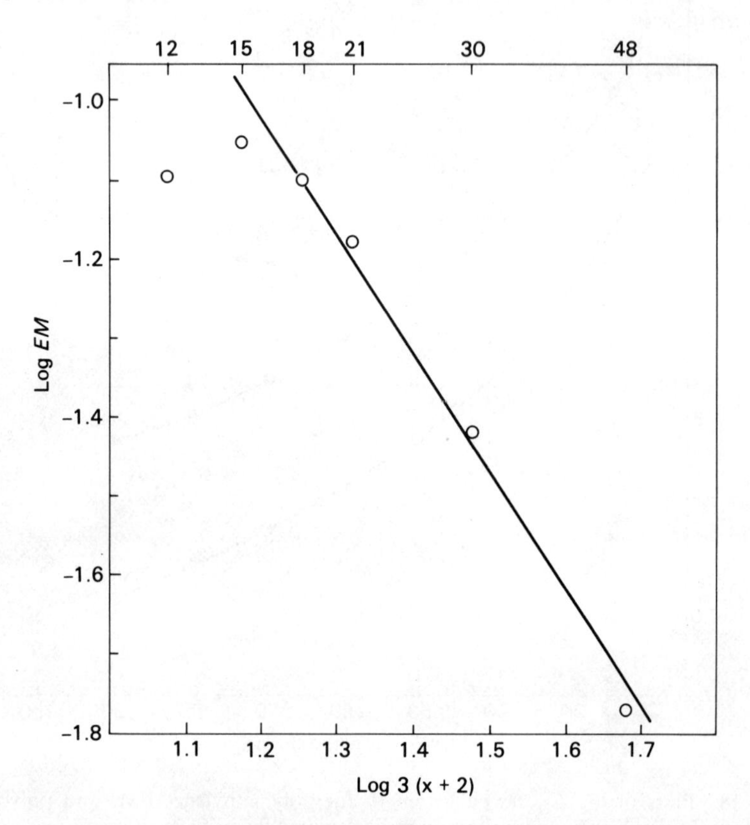

FIG. 19 Closure of benzo-3(x + 2)-crown-(x + 2) ethers [20] in 99% Me_2SO. Log-log plot of EM against number of ring atoms (data from Illuminati *et al.*, 1981). The line was drawn with the theoretical slope of $-3/2$

Quantitative data on the interaction of functional groups attached to the ends of polymer chains with up to 100 chain atoms have been presented by a group of Japanese workers. One of the probe reactions is the pyridine-catalysed hydrolysis of a *p*-nitrophenyl ester group in water solution at pH 6.1. In these experiments, the two reactive groups were attached to the ends of polysarcosine chains [27] or polyoxyethylene chains [28] with varying degrees of polymerisation $\bar{x}$ in the range of 5 to 33 (Sisido *et al.*, 1976, 1978). Suitable

$$\text{NC}_5\text{H}_4\text{—CH}_2\text{NH}\!\left(\text{COCH}_2\underset{\substack{|\\ \text{CH}_3}}{\text{N}}\right)_{\bar{x}}\text{COCH}_2\text{CH}_2\text{CO}_2\text{—C}_6\text{H}_4\text{—NO}_2$$

[27]

$$\text{NC}_5\text{H}_4\!\left(\text{CH}_2\right)_3\text{O}\!\left(\text{CH}_2\text{CH}_2\text{O}\right)_{\bar{x}}\text{COCH}_2\text{CH}_2\text{CO}_2\text{—C}_6\text{H}_4\text{—NO}_2$$

[28]

monofunctional compounds were used to determine the k_{inter}-values for the analogous non-cyclisation reactions, from which the EM-values plotted in Fig. 20 were calculated. From the temperature coefficients of the rate constants determined over the range of 15°C to 35°C, the activation parameters were also obtained. The molecular weight of the synthesised polymer chains [27] and [28] were controlled by choosing appropriate molar ratios of monomer to initiator which produced a Poisson-type distribution of molecular weights. This caused difficulty in the kinetic measurements, as indicated by the marked convexity of first-order plots (Sisido *et al.*, 1976), since initially the reaction proceeds mostly on shorter chains which are more reactive than the larger ones. Though not very precise, the rate constants give a useful indication of the effect of chain length on the ease of cyclisation. The activation parameters, however, for which the precision problem is crucial, should be regarded with suspicion. Indeed, contrary to expectations, both ΔH^{+} and ΔS^{+} values for the reactions of [27] increase on increasing $\bar{x}$, a phenomenon for which the authors themselves could not offer any explanation. A very precise isokinetic relationship with an isokinetic temperature of 93°C was observed, which led to the unrealistic prediction that the intramolecular reaction will proceed faster for longer chains above 93°C.

Another system investigated by the same group is the intrachain charge-transfer complex of polysarcosine chains having a terminal *p*-dimethylaminoanilide group and a terminal 3,5-dinitrobenzoyl group [29]. Equi-

$$(\text{CH}_3)_2\text{N—C}_6\text{H}_4\text{—NH}\!\left(\text{COCH}_2\underset{\substack{|\\ \text{H}_3\text{C}}}{\text{N}}\right)_{\bar{x}}\underset{\substack{\|\\ \text{O}}}{\text{C}}\text{—C}_6\text{H}_3(\text{NO}_2)_2$$

[29]

librium measurements were carried out in chloroform (Sisido *et al.*, 1977) and in ethanol (Takagi *et al.*, 1977) on chains with $\bar{x}$ in the range of 6–21, as

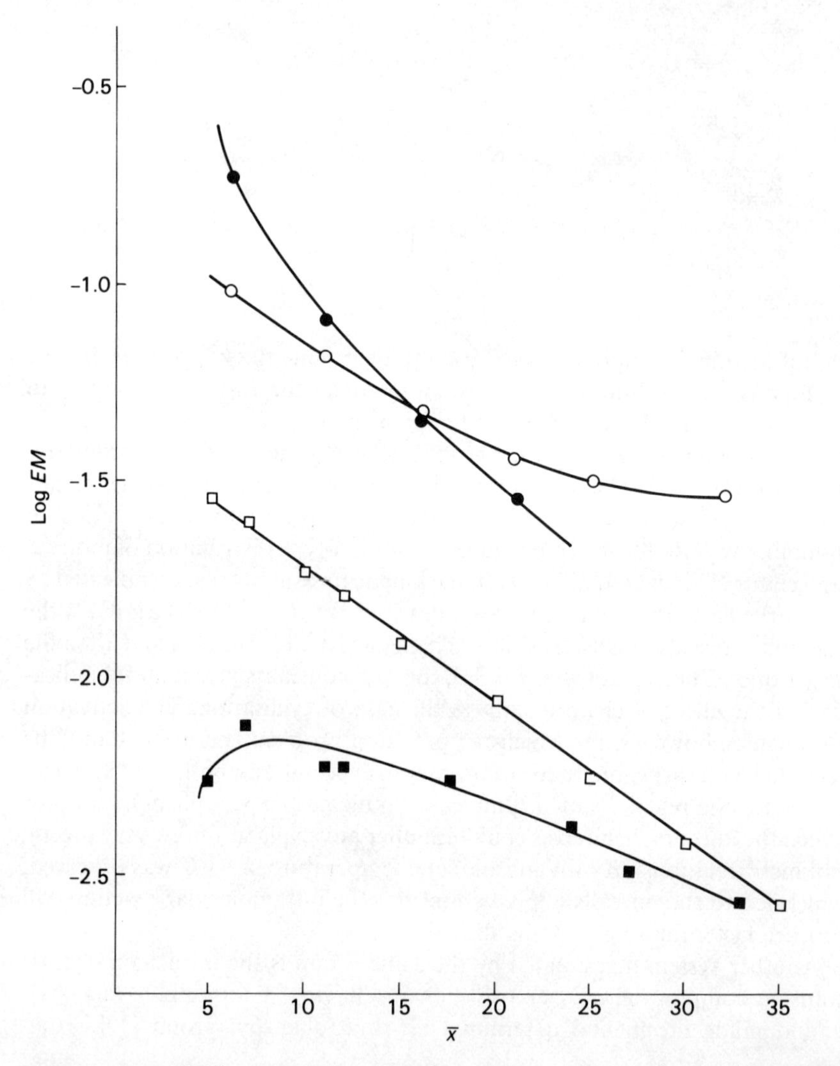

FIG. 20 Log *EM* for interaction of end-groups on polymer chains with average polymerisation degree $\bar{x}$: (*i*) pyridine-catalysed hydrolysis of the *p*-nitrophenyl ester group of [27] (□) and [28] (■) in aqueous solution (data from Sisido *et al.*, 1976, 1978), (*ii*) intramolecular charge-transfer complexes of [29] in chloroform (●) and in ethanol (○). (Data from Sisido *et al.*, 1977; Takagi *et al.*, 1977)

well as on model monofunctional compounds. Equilibrium EM's calculated from the above data, are compared with kinetic EM's in Fig. 20. The reactivity picture is difficult to explain. It is not clear whether the different behaviours of the various polysarcosine compounds reflect solvent effects. Furthermore, the ease of ring closure of the polyoxyethylene chains [28] seem much too low. The authors themselves have reached the admittedly unexpected conclusion that the more flexible polyoxyethylene chains are much less efficient in the intra-chain reaction than the polymethylene ones, which is clearly at variance with theory (Fig. 18) and experimental evidence from other sources (Fig. 13).

As to the chain-length dependence of the EM's plotted in Fig. 20, we note that a log-log plot of EM *vs.* $\bar{x}$ for the reactions of [2] in $CHCl_3$ is linear with a slope of -1.7, i.e. close to the theoretical value of -1.5, but in the other cases the reactivity drop on increasing $\bar{x}$ is much less steep.

A much better agreement between theory and experiment is found in the closely-related field of macrocyclisation equilibria. Investigations of the cyclic populations in ring-chain equilibrates set up in typical polymeric systems such as polyesters, polyethers, polysiloxanes, and polyamides take a major advantage from the relative ease with which the cyclic fraction can be separated from the linear fraction and analysed for the relative abundance of the individual oligomeric rings. This is conveniently done by means of modern analytical techniques such as gas-liquid and gel-permeation chromatography (Semlyen, 1976).

A satisfactory theory of macrocyclisation equilibria for chains obeying Gaussian statistics was presented by Jacobson and Stockmayer (1950) long before the availability of suitable experimental data for the proper testing of the theory. According to this theory, the macrocyclisation equilibrium constant K_x (see p. 10) is related to the density W(0) of the distribution of the end-to-end vector r in the region $r = 0$ through relation (57), where N_A is

$$K_x = \mathrm{W}(0)/N_A\sigma_{Rx} \quad (57)$$

Avogadro's number and σ_{Rx} is the symmetry number of the x-meric cyclic oligomer, which represents a statistical correction for the number of equivalent bonds that can undergo the reverse ring-opening reaction. For Gaussian chains (57) takes the form (58) which is identical to (56), apart from the pres-

$$K_x = \frac{1}{N_A\sigma_{Rx}}\left(\frac{3}{2\pi\langle r^2\rangle}\right)^{3/2} \quad (58)$$

ence of the symmetry number, usually neglected for cyclisations of bifunctional chains which involve little or no change in symmetry number. Since $\langle r^2\rangle$ is proportional to the number of skeletal bonds, and σ_{Rx} is usually x or $2x$, it follows that (58) predicts that K_x varies inversely with the 5/2 power of the chain length. The adverse effect of the symmetry number is striking. For example, a 100-meric ring is disfavoured by a factor of 100 or 200. This is clearly one of the most impressive effects of symmetry on reactivity.

The first extensive investigation of a quantitative nature in the field is that of Brown and Slusarczuk (1965) on polymethylsiloxane. They determined

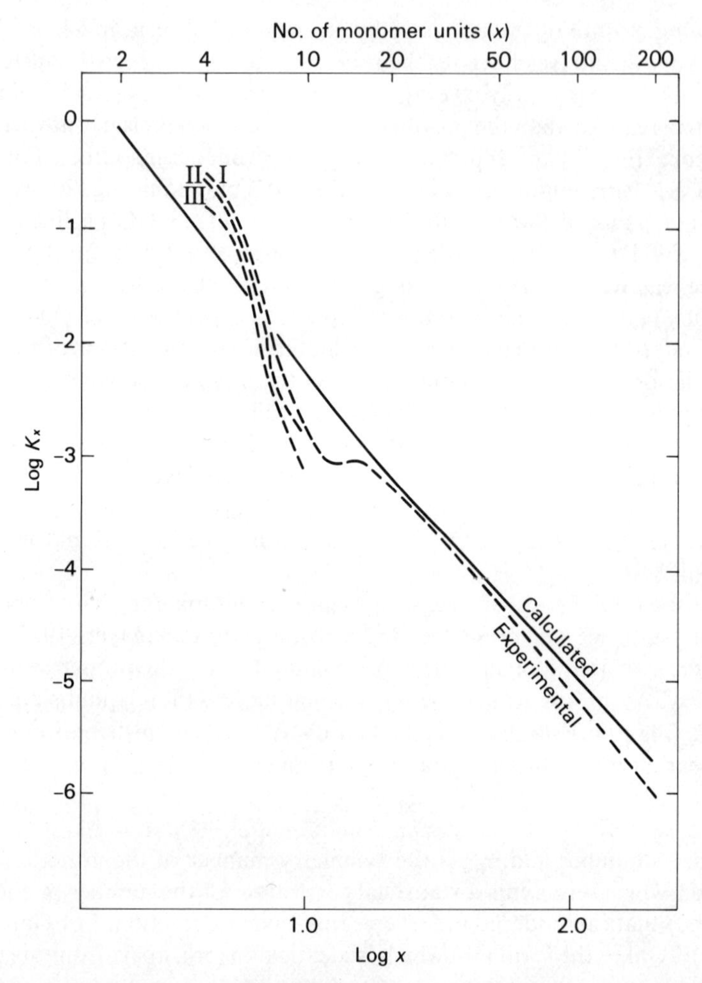

FIG. 21 Macrocyclisation equilibrium constants for polydimethylsiloxane. Experimental data are shown as dashed curves as follows: I, Brown and Slusarczuk; II, Hartung and Camiolo; III, Carmichael and Wigner. The solid curve is calculated from Jacobson–Stockmayer theory. (Reproduced with permission from Flory, 1969)

the concentration of cyclics $\{(CH_3)_2\,SiO\}_x$ with x from 4 to over 200 for polymethylsiloxane equilibrated in toluene at 110°C. K_x-values from their work are plotted in Fig. 21. The reliability of Brown and Slusarczuk's experiments is supported by the results of Hartung and Camiolo [quoted by Flory (1969)] in xylene and of Carmichael and Wigner (1965) for the undiluted polymer, also shown in Fig. 21. The solid line represents K_x-values calculated by Flory and Semlyen (1966) from (58). Clearly, the agreement between theory and experiment leaves little to be desired for chains with $x > 15$ (30-chain atoms).

Noteworthy is that no recourse is made to adjustable parameters. Semlyen (1976) in his review lists a sizeable number of systems where a fair to good agreement is found between observed and calculated values of K_x for sufficiently long-chain compounds. On the other hand, experimental K_x-values for the shorter chains deviate significantly from the calculated ones, the deviations being negative for chains of intermediate size and positive for the shortest ones.

There are several reasons why the behaviour of the shorter chains deviate from the original formulation of the Jacobson and Stockmayer theory (Flory, 1969). First, if the ring size is small enough to induce strain, the enthalpy change for cyclisation (16) will differ from that for the intermolecular process (17). In terms of the θ operator (39), $\theta\Delta H^{\circ}$ will differ from zero and, presumably, be positive. Secondly, (57) is based on the implicit assumption that the relative orientation of the reacting bonds, when they come in close proximity in the cylisation reaction, is random. This independence of orientation and proximity, which leads to the absence of any factor referring to orientation in (57), must fail for short chains. Thirdly, short chains may not follow Gaussian statistics. When this occurs, an appropriate expression for the density of end-to-end vectors is required.

An elaboration of the Jacobson–Stockmayer theory was presented by Flory *et al.* (1976) to take account of restrictions of the relative orientations of terminal bonds when they are forced into proximity in the ring-closure reaction. Unfavourable correlations between the directions of the terminal bonds were disclosed by the calculations for various chains with $10 < Z < 50$, Z being the number of skeletal bonds. Detailed calculations were carried out by Suter *et al.* (1976) for the formation of cyclic siloxanes $\lceil\!\!-[Si(CH_3)_2-O]_x-\!\!\rceil$ in the range $7 \leqslant x \leqslant 30$ by means of exact matrix generation methods and Monte Carlo techniques. It was shown that the discrepancies between theory and experiment are accounted for entirely by departures from Gaussian behaviour for chains with $x > 15$, but only in part for $7 \leqslant x \leqslant 15$. Unfavourable orientational correlations for $n \geqslant 10$ were disclosed by the computations, but their importance is limited. The orientational correlation factor reduces K_x by about half at $x = 12$, but only by an insignificant 10% at $x = 30$, and even less for longer chains. Similar results were obtained by Mutter *et al.* (1976) for homologous sequences of poly(6-aminocaproamide) $\lceil\!\!-[CO(CH_2)_5NH]_x-\!\!\rceil$. The calculations are unreliable for the shortest chains, but a definite indication is obtained that the orientational correlation factor becomes favourable for these chains, in agreement with previous suggestions by Flory and Semlyen (1966) and by Flory (1969). This is an important result providing, at least in qualitative terms, a statistical mechanical basis for comprehending the high ease of closure of short chains. The strict correspondence with the "orbital steering" idea set forth independently by Storm and Koshland (1970, 1972a,b) is evident.

The basic assumption of polymer chemists in their treatments of cyclisation processes of long-chain molecules is that the enthalpy change for cyclisation is equal to that for the intermolecular reaction. In the absence of suitable thermochemical data, the limits imposed by such an approximation are uncertain. Morawetz and Goodman (1970) have pointed out that even large cycloalkanes would have a larger proportion of *gauche* conformations than the corresponding open-chain compounds, and have estimated a limiting value of ring strain-energy as 2 kcal mol^{-1}. Illuminati *et al.* (1977, 1981) have obtained indirect evidence that strain is still significant for ring-sizes 14 and 16, but decreases on increasing the ring size and eventually vanishes at ring-sizes in the neighbourhood of 30 atoms.

An important body of experimental evidence concerning the problem of the strain-energies of large rings has been presented by Yamashita *et al.* (1980), who reported an extensive investigation of ring-chain equilibria of the macrocyclic formals 1,3,6-trioxacyclooctane [30: $m = 1$], 1,3,6,9-tetraoxa-

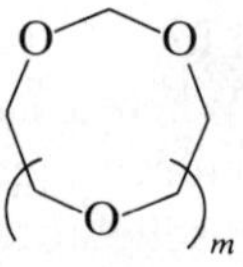

[30]

cycloundecane [30: $m = 2$], 1,3,6,9,12-pentaoxacyclotetradecane [30: $m = 3$] and 1,3,6,9,12,15-hexaoxacycloheptadecane [30: $m = 4$] catalysed by $Et_2O{\cdot}BF_3$ in CH_2Cl_2 in the temperature range −30°C to +30°C. The equilibrium concentrations of cyclic oligomers with polymerisation degrees up to 7–9 were determined for each of the monomeric formals. Log-log plots of K_x

TABLE 16

Thermodynamic parameters in equilibrium polymerisation of [30: $m = 3$] in CH_2Cl_2[a]

x^b	$10^2 K_x$/M	ΔH_p/kcal mol^{-1}	ΔS^o_p/e.u.
1	2.17	−2.7 ± 0.3	−2.0 ± 0.9
2	1.11	−1.8 ± 0.1	+2.4 ± 0.3
3	0.558	0	+10.3 ± 0.2
4	0.255	0	+11.9 ± 0.1
5	0.143	0	+13.0 ± 0.1
6	0.0933	0	+13.9 ± 0.2
7	0.0632	0	+14.6 ± 0.2

[a] Data from Yamashita *et al.*, 1980
[b] Number of monomer units in cyclic oligomers

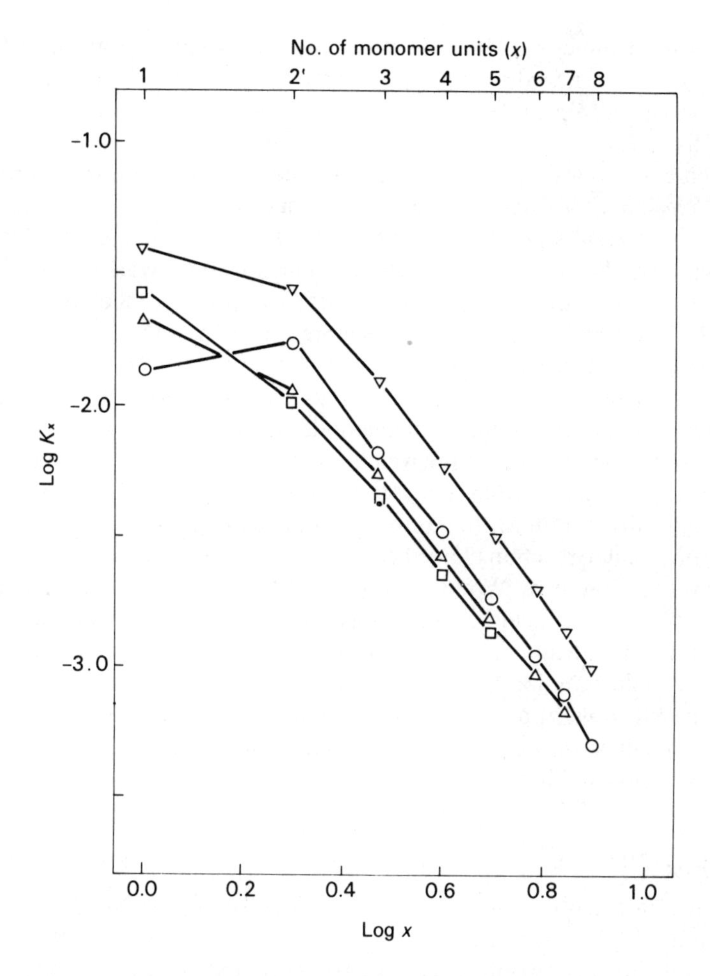

FIG. 22 Macrocyclisation equilibrium constants for macrocyclic formals [30] in CH_2Cl_2 in the presence of $Et_2O{\cdot}BF_3$ as a catalyst: (▽) $m = 1$; (○) $m = 2$; (△) $m = 3$; (□) $m = 4$. (Reproduced with permission from Yamashita *et al.*, 1980)

against the number of monomer units x in cyclic oligomers are shown in Fig. 22. From the temperature coefficients of the K_x-values, heats and entropies of polymerisation for the various cyclic x-mers were obtained. These are shown in Table 16 for equilibrium polymerisation of [30: $m = 3$]. Note that ΔH°_p and ΔS°_p, referring to the transformation of cyclic x-mers into linear polymers, are coincident with $-\theta\Delta H^\circ$ and $\theta\Delta S^\circ$, respectively. The data presented in Table 16 are typical. The enthalpy term is significant for all monomers and still appreciable for dimers. It is noteworthy that even for the

34-membered dimer of [30: $m = 4$] a weak ring-strain energy of 0.7 kcal mol^{-1} with a standard deviation of $\pm$ 0.3 kcal mol^{-1} is revealed. Larger rings are virtually strainless, and the ring-chain equilibrium is solely determined by the entropy term. Consistently, the plots in Fig. 22 are linear with the theoretical slope of -2.5 in the region where $\Delta H_p = 0$, in accordance with Jacobson–Stockmayer theory.

The above results point to a further difficulty in the mathematical treatment of the cyclisation of chains with less than 30 atoms, which are of prevalent interest to organic chemists. Theoretical approaches to the cyclisation of chains of moderate length which are more direct than that of Flory *et al.* (1976) have been developed by Sisido (1971) and by Winnik *et al.* (1974). Individual chains of various conformations are generated by computer, and Monte Carlo or exact enumeration methods are used to calculate the fraction of "cyclised" conformations for which the distance between the end-groups is smaller than a given critical distance r_0. Within limits, the latter can be used as an adjustable parameter to improve the fit to experimental data. A number of intramolecular reactions have been computer-simulated. They have been extensively reviewed by Winnik (1981a) and by Imanishi (1979). There are cases in which agreement between experiment and calculations is very good (see, for instance, Mar *et al.*, 1981). Nevertheless, in spite of their undeniable value for understanding the conformational behaviour of chain molecules, these investigations appear to be of limited utility from the organic chemist's standpoint, since they do not provide simple models that can be used as a basis for chemical thinking.

ENTROPY CHANGES FOR CYCLISATION REACTIONS IN SOLUTION

As shown in the preceding paragraph, theoretical approaches to rates and equilibria of cyclisation reactions of short- to moderate-length chain molecules explicitly taking into account proximity and orientation effects are faced with discouraging conceptual and mathematical difficulties.

An alternative approach, which is usual among physical organic chemists, involves a comparison of changes of thermodynamic quantities for structurally-related reaction series. Such an approach, for which the term extrathermodynamic is widely used (Leffler and Grunwald, 1963), will be adapted here in an appropriate form to the interpretation of chain-length effects on cyclisation rates and equilibria. The experimental basis is provided by cyclisation reactions for which either $\theta\Delta S^{\neq}$- or $\theta\Delta S^{\circ}$-data are available.

It is well known that interpretation of structural effects on reactivity in terms of enthalpy and entropy changes is often complicated, or even overwhelmed, by solvation phenomena. Cyclisation reactions are no exception. This is especially so for systems involving large polarity changes on going from reactants to transition states or products. Therefore, a preliminary discussion of solvent effects on intra- *vs* intermolecular reactions is appropriate.

Solvent effects on EM's

Generally speaking, the influence of solvent on reaction rates (equilibria) is determined by the difference between the effects on the stability of transition states (products) and reactants. According to what Leffler and Grunwald (1963) call the first approximation, the free energy of a solute molecule RX is given by the sum of internal and solvent contributions, as shown in (59). The

$$\overline{G}^{\circ} = G_{int} + (G_R + G_X)_{solv} \tag{59}$$

former is a quantity characteristic of the solute molecule. The latter can be considered as the sum of independent contributions arising from interaction of the non-reacting and reacting parts of the molecule with the surrounding medium. These assumptions are supported by the fact that free energies of transfer between a given solvent and the gas phase tend to be additive functions of the various groups of the solute molecule.

For reactions of bifunctional chains X∼∼Y, two different types of solvent interactions are considered, i.e. (*i*) solvation of the non-reacting part of the molecule affecting the ease of cyclisation through solvent-induced conformational changes, and (*ii*) solvation of the reacting ends affecting the inherent reactivity of the ends themselves.

Early interest in the relation between the shape of chain molecules and solvent in ring-closure reactions can be traced in the work of Salomon (1936a). More recently, the problem has been given much attention by Winnik (1977). Though being small (see p. 64), the corresponding effects on the ease of cyclisation are believed to be of considerable importance in probing the shape of hydrocarbon and other flexible chains in solution.

Here we are mainly concerned with solvation phenomena belonging to the latter category. As a consequence of the approximation underlying (59), the quantities $(G_X)_{solv}$ and $(G_Y)_{solv}$ of a bifunctional compound X∼∼Y are independent of the chain length and equal to the corresponding quantities of monofunctional reactants RX and RY. If the same approximation also holds for transition states or products, two important conclusions are reached. First, chain-length effects on changes in thermodynamic quantities along a given series reflect internal changes, free from solvation contributions. Secondly, EM's and their enthalpic and entropic components (38) are independent of solvent by virtue of exact cancellation of solvent contributions.

The fact that most of the available EM-values for large-ring formation in diverse cyclisation series cluster in the fairly limited range 0.05 M–0.01 M suggests a virtually complete cancellation of contributions arising from the different nature of end groups, including solvation phenomena. Clearly this can be taken as a strong indirect proof of the adequacy of the first approximation.

More direct evidence is derived from investigations of solvent effects on cyclisation reactions (for a short review see Galli *et al.*, 1981). Bruice and Turner (1970) found that transfer from water to 1 M water in Me_2SO causes little changes in the EM's for the formation of the 5-membered cyclic anhydrides [31] and [32]. This was taken as evidence that solvation phenomena contribute very little to rate enhancements of intramolecular reactions. Similar

CO_2Ar, CO_2^- ⟶ O, O, O $+ ArO^-$

EM: 4.8×10^5M iu H_2O
2.6×10^6M iu $Me_2SO(+1\ M\ H_2O)$ [31]

O, CO_2Ar, CO_2^- ⟶ O, O, O, O

EM: 1.3×10^8 M iu H_2O
2.5×10^9 M iu $Me_2SO(+1\ M\ H_2O)$ [32]

views have been expressed by Dafforn and Koshland (1977) who found that some 5-membered lactones and thiolactones are formed at rates following similar trends in the two basically different solvents water and sulpholane.

TABLE 17

Comparison of EM-data in 75% EtOH and 99% Me_2SO at 25°C for the ring-closure reactions of o-$^-OC_6H_4O(CH_2)_{n-4}Br$[a]

Ring-size n	6	7	8	9	10	11	14	16	24
EM(75% EtOH)[b] / EM(99% Me_2SO)	1.0	2.0	1.3	2.2	3.9	1.1	0.62	0.74	1.7

[a] Reaction (43). (Data from Dalla Cort *et al.*, 1980a)
[b] k_{inter} (75% EtOH)/k_{inter} (99% Me_2SO) = 9.7×10^3

An extensive investigation of the influence of solvent over a wide spectrum of ring sizes has been carried out by Dalla Cort *et al.*, (1980a). The EM's for the formation of catechol polymethylene ethers [2] from the anionic precursors [1] are remarkably similar in the protic 75% EtOH and in the essentially aprotic 99% Me_2SO (Table 17), in spite of the fact that in the latter solvent rates are about 10^4 times faster than in the former. These data provide compelling evidence that, to a useful approximation, the free energy change due to solvent reorganisation when the two solvation shells surrounding the end-groups coalesce to a common solvation shell in the transition state is constant along the series and equal to the corresponding change for the analogous non-cyclisation reaction.

Cyclisation of very short chains. Anomalous entropy changes

Entropy changes for the formation of small and common rings are listed in Table 18. Many of the very short chains violate the expected trend toward more negative entropy changes with increasing chain length. Such anomalies, first emphasized by Stirling (1973), find a reasonable explanation in the hypothesis of inadequacy of the first approximation for very short-chain bifunctional molecules (Mandolini, 1978b) because of the close proximity of end-groups. That a series of polar solutes may interact with polar solvents with extraordinarily large changes in enthalpy and entropy, but little change in free energy, is a well recognised phenomenon. Thus, even small variations of solvent interactions with either initial or transition states (products) can result in significant changes in the $\Delta S^{\ddagger}_{solv}$ (ΔS°_{solv})-term. Such effects are expected to be very significant for charged substrates, and to vanish both in reactions where solvation is unimportant and in reactions occurring in the gas phase. Indeed reactions belonging to the latter categories (entries 13–15 in Table 18) exhibit entropy changes regularly decreasing with increasing n.

Intrinsic entropy changes upon cyclisation

Given that the first approximation for solvent interactions seems adequate to a reasonable degree of accuracy, with the possible exception of some of the very short-chain compounds, we can now attempt an extension to available entropy changes for cyclisation reactions in solution of the same treatment as was applied in the preceding section to entropy changes for hypothetical cyclisation reactions in the gas phase.

Figure 23 is a plot of $\theta\Delta S^{\ddagger}$ data, selected for reliability, for the cyclisation of polymethylene compounds occurring in the neighbourhood of room temperature against the number of single bonds in the open-chain reactants. The data refer to reactions differing widely in nature, namely, nucleophilic substitutions at saturated carbon, as carried out either in protic or in dipolar aprotic solvents, electron exchange and photochemical processes. Most of the $\Delta S^{\ddagger}$-data on which Fig. 23 is based are reliable to within ± 1 to ± 2 e.u., which yields uncertainties twice as great in the corresponding $\theta\Delta S^{\ddagger}$-quantities. However, such relatively large experimental uncertainties are not

TABLE 18
Entropy changes for closure of small and common rings

Reaction no.	Substrate	Ring-size n	Solvent	$\Delta S^{\ddagger}$, e.u.
1[a]	$H_2N(CH_2)_{n-1}Cl$	3	50% dioxane	−15
		5		−13
2[a]	$C_6H_5NH(CH_2)_{n-1}Br$	3	60% EtOH	−11
		4		−11
		6		−17
3[a]	$C_6H_5NH(CH_2)_{n-1}Cl$	3	50% dioxane	−17
		5		−15
4[a]	$H_2NC(C_6H_5)H(CH_2)_{n-2}Cl$	3	50% dioxane	−7
		5		−10
5[b]	p-$CH_3C_6H_4S(CH_2)_{n-1}Cl$	3	80% EtOH (w/w)	−24
		5		−20
6[c]	$CH_3O(CH_2)_{n-1}OBs$	5	EtOH	−10.4
		6		−15.1
7[d]	$C_6H_5C(=O)(CH_2)_{n-2}Cl$	4	80% EtOH, Ag^+	−16.0
		5		−15.8
		6		−20.4
		7		−23.1
8[d]	$(CH_3)_2CHC(=O)(CH_2)_{n-2}Cl$	5	80% EtOH, Ag^+	−18.2
		6		−20.8
		7		−22.6
9[e]	o-$^-OC_6H_4(CH_2)_{n-4}Br$	6	75% EtOH	+6.2
		7		+1.2
10[f]	o-$^-OC_6H_4(CH_2)_{n-3}Br$	5	75% EtOH	+4.1
		6		+5.3
		7		−3.2

11[g]	$^{-}O_2C(CH_2)_{n-2}Br$	3	99% Me_2SO	−2.5
		4		−4.9
		5		+0.6
		6		−4.2
		7		−13.6
12[h]	*trans*-$[Pt(NH_3)(NH_2(CH_2)_{n-3}NH_2)Cl_2]$	5	H_2O, $\mu = 0.2$ M	−15.0
		6		−15.0
		7		−17.0
13[i]	$\dot{C}H_2(CH_2)_{n-2}CH{=}CH_2$			
		3	Apolar	−10
		4		−14
		5		−17
		6		−20
		7		−23
14[l,m]	$HO(CH_2)_{n3}OCH_3$	5	CCl_4	·3.0
		6		−6.9
		7		−10.8
15[l,n]	$H_2N(CH_2)_{n-3}NH_3^+$	5	Gas phase	−12.7
		6		−20.6

[a] Data from Bird *et al.*, 1973
[b] Data from Bird and Stirling, 1973
[c] Data from Winstein *et al.*, 1958
[d] Data from Pasto and Serve, 1965. Carbonyl oxygen participation is uncertain for $n = 7$
[e] Data from Illuminati *et al.*, 1974a
[f] Data from Illuminati *et al.*, 1975
[g] Data from Galli *et al.*, 1977 and Mandolini, 1978b including a corrected value of +0.6 e.u. for ring size 5
[h] Ring-closure reactions of monodentate diamine complexes to chelate complexes. (Data from Tobe *et al.*, 1982)
[i] Radical addition to the penultimate carbon (*exo* attack) The solvent is unspecified. (Data from Beckwith, 1981)
[l] ΔS° values for the formation of the hydrogen bonded cyclic forms. The bridging hydrogen atom is included in the calculation of ring size
[m] Data from Kuhn and Wires, 1964
[n] Data from Yamdagni and Kebarle, 1973

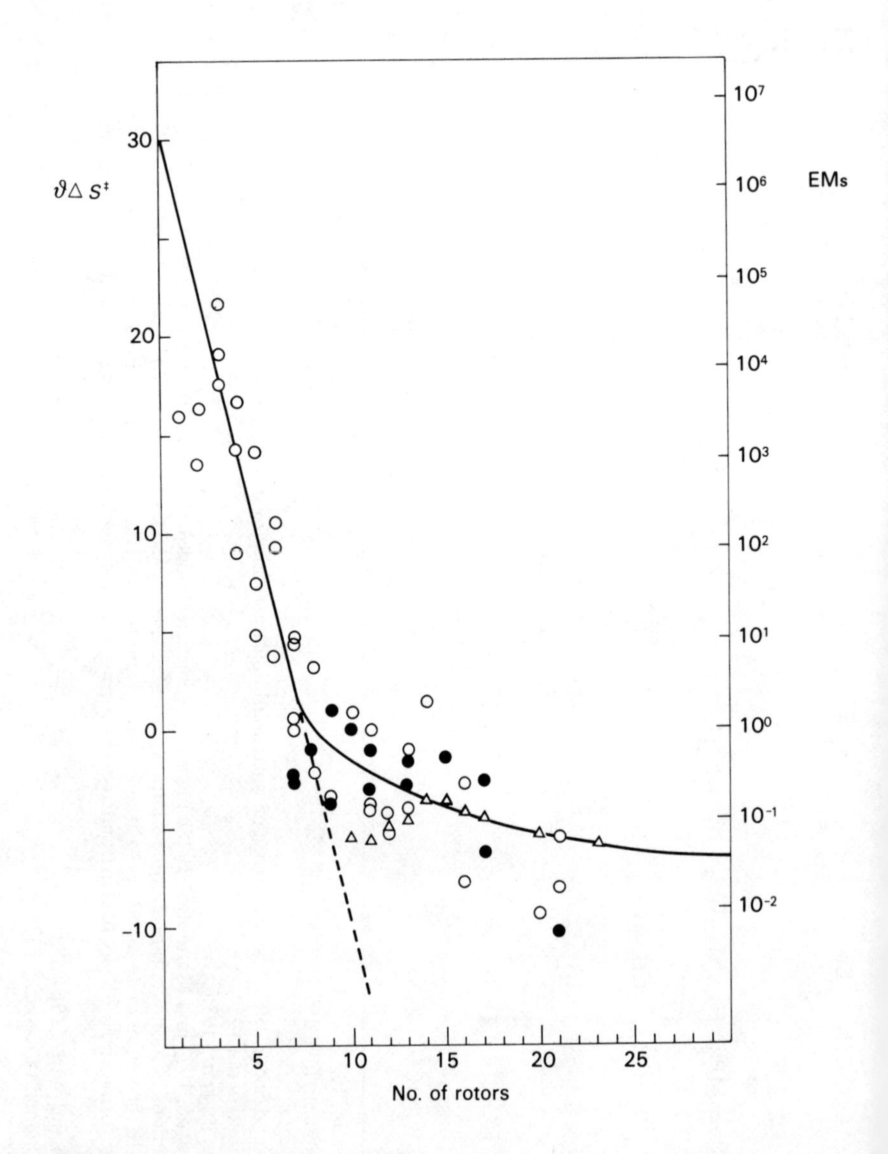

FIG. 23 Entropy effects on intramolecular reactions of polymethylene chains. Plot of $\theta\Delta S^{\ddagger}$ (e.u.) against number of single bonds for: (○) nucleophilic substitutions at saturated carbon; (●) electron-exchange reactions; (△) quenching of benzophenone phosphorescence. The straight line has intercept +30 e.u. and slope −4.0 e.u. per rotor. The right-hand ordinate reports the purely entropic EM's calculated as $\exp(\theta\Delta S^{\ddagger}/\boldsymbol{R})$

believed to be the only source of the considerable scatter present. Contributions from varying degrees of extra looseness due to low frequency out-of-plane bending motions (O'Neal and Benson, 1970) in the ring-shaped transition states are expected to play an important role. In spite of the above limitations, the rather uniform behaviour emerging from Fig. 23 points to an extensive cancellation of contributions due to chemical change and solvation in the entropy component of the EM. By analogy with (42) we can therefore write the approximate relation (60), where transl-rot refers to the motions

$$\theta \Delta S^{\neq} = \Delta S^{\neq}_{\mathrm{conf}} - \Delta S^{\neq}_{\mathrm{transl-rot}} \tag{60}$$

that in the liquid phase correspond to translation plus overall rotation in the gas phase. If the points for the shortest chains are neglected, the behaviour of short chains up to about 7 rotors is reasonably fitted by a straight line with slope -4.0 e.u. per rotor and intercept $+30$ e.u. There is a close similarity with the corresponding plot in Fig. 5, in line with arguments and conjectures presented in Section 3. Thus, cyclisation of short chains corresponds to a virtual freezing of most of the entropy due to internal rotations both in the gaseous and in the condensed phase. The intercept, $+30$ e.u., represents the $\theta \Delta S^{\neq}$-value for an intramolecular reaction taking place with no entropy loss due to freezing of internal rotations. Hence, according to (60), -30 e.u. may be taken as a representative value for the entropy change due to losses of translational-rotational motions in bimolecular reactions occurring in solution in the neighbourhood of room temperature.

The $\theta \Delta S^{\neq}$-value of $+30$ e.u. corresponds to an EM of $\exp(30/\boldsymbol{R})$ or $10^{6.6}$M, which provides an estimate for the maximum advantage of intra- over intermolecular reactions in the absence of strain effect, i.e. when $\theta \Delta H^{\neq} = 0$.

Davis *et al.* (1984) have recently reported a kinetic study of the base-catalysed interconversion of [33] and [34] in 50% aqueous dioxan at 50°C,

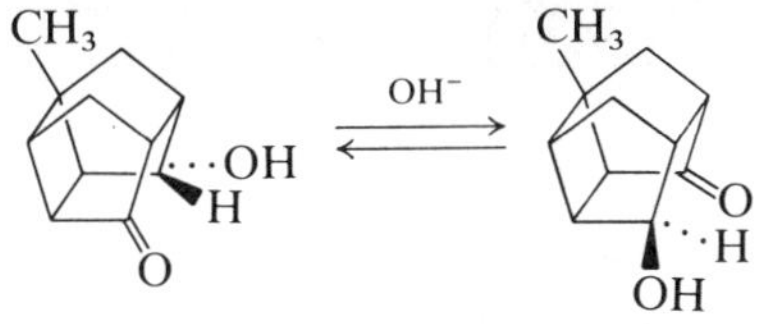

[33] $\overset{OH^-}{\rightleftharpoons}$ [34]

a reaction for which any significant release of strain in the transition state seems unlikely. Since no conformational entropy is obviously lost in the activation process, this system appears to be a practically ideal one for measuring the maximum entropic advantage in intramolecular reactions. The EM for the intramolecular process within the alkoxide ion of [33] was found to be

6.5×10^6 or $10^{6.8}$M), an agreement with the above estimate of $10^{6.6}$M which is impressive.

The entropy data for the longer chains (Fig. 23) exhibit a much smaller dependence on chain length than the shorter ones. Interpolation by means of a single line is difficult on account of the considerable scatter. As one would not expect monotonic entropy changes upon cyclisation of chains yielding strained rings (Winnik, 1981a), it would be of interest to examine entropy data in a region where strain is unimportant (i.e. more than 30 atoms). These are not accessible for polymethylene chains, mainly because of their low solubility, but for the more soluble polyoxyethylene chains the work of Yamashita *et al.* (1980) on polymerisation equilibria of macrocyclic formals [30] is a valuable source of extensive information. Symmetry corrected $\theta\Delta S^o$-data from their work are plotted in Fig. 24. The line was calculated from Jacobson–Stockmayer theory, using the average value of 8 given by Yamashita *et al.* (1980) for the characteristic ratio C_∞. The agreement between theory and experiment is excellent for the very long chains, but significant deviations are present for the lower members of the series.

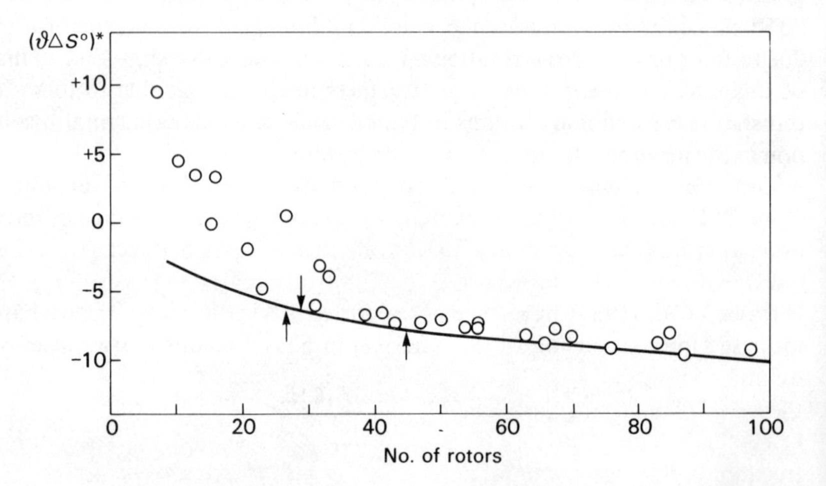

FIG. 24 Plot of symmetry-corrected $\theta\Delta S^o$ (e.u.) against number of single bonds for macrocyclisation equilibria of formals [30], calculated from data reported by Yamashita *et al.* (1980). The curve was calculated from Jacobson–Stockmayer theory. For the meaning of the arrows see text

The close adherence of the calculated curve to the actual behaviour of polyoxyethylene chains is further substantiated by comparison with the kinetic EM's for the formation of the 30- and 48-membered benzo-crown ethers [20] in 99% Me_2SO (Illuminati *et al.*, 1981; see Fig. 13). Assuming that for these large rings the enthalpy term is unimportant, the pertinent $\theta\Delta S^{\ddagger}$-values are readily calculated as $\boldsymbol{R} \ln EM$. These are indicated in Fig. 24 by the two upward arrows.

The ordinate of the downward arrow in Fig. 24 indicates the $\theta\Delta S^{\ddagger}$-value calculated as $\boldsymbol{R} \ln EM$ for the formation of the 32-membered catechol polymethylene ether [2] in 99% Me_2SO (Dalla Cort *et al.*, 1980a), which is the largest polymethylene ring compound for which quantitative cyclisation data are available. It appears therefore that the calculated curve also describes with good precision the behaviour of polymethylene chains. Clearly, one swallow does not make a summer, but we feel that the agreement is not accidental. In spite of the fact that the ease of cyclisation of polyoxyethylene chains is theoretically predicted to be appreciably larger than that of polymethylene chains under Θ conditions[7] (Fig. 18), it seems likely that the generally more soluble polyoxyethylene compounds will more often adopt expanded conformations relative to the unperturbed state. Indeed, the characteristic ratio of 8 on which the curve in Fig. 24 is based is significantly larger than the value of 4.0 (in aqueous K_2SO_4 at 35°C) quoted by Flory (1969), and is close to the value of 6.8 for the polymethylene chain.

The above considerations have provided a useful guideline for drawing the curved part of the full line in Fig. 23. The portion in the range 15–30 rotors was assumed to coincide with the corresponding portion in Fig. 24, and the remaining short gap was filled by means of a smooth curve starting from the point related to the 7-rotor chain, which still lies on the straight line. Combination of the full lines drawn in Fig. 23 and Fig. 24 offers a spectacular panoramic view of the entropy effect on cyclisation reactions of chain compounds with up to nearly 100 skeletal bonds.

For convenience the averaged $\theta\Delta S$-data are tabulated (Table 19) together with the corresponding entropic components of the EM. The latter are the ideal EM-values predicted for a cyclisation reaction for which the strain is unimportant for all of the terms of the series. They can be read directly from the right-hand ordinate in Fig. 23.

The last column in Table 19 lists the entropy losses due to reduction of freedom of internal rotations around the single bonds upon cyclisation. It is of interest to note, for example, that the conformational entropy lost upon cyclisation of an 8-rotor chain amounts to $\frac{3}{4}$ of the corresponding quantity related to cyclisation of a 100-rotor chain.

Of course, too much emphasis cannot be placed on exact figures. It is nevertheless believed that correct orders of magnitude and significant trends are properly represented by the data listed in Table 19.

[7] See footnote 6 (p. 64)

TABLE 19

Entropy changes accompanying cyclisation of bifunctional chains as a function of the number of skeletal single bonds[a]

No. of rotors	$\theta\Delta S$/e.u. [b]	EM[c]	$-\Delta S_{conf}$/e.u. [b,d]
0	+30	3.6×10^{6}	0
1	+26	4.8×10^{5}	4
2	+22	6.4×10^{4}	8
3	+18	8.6×10^{3}	12
4	+14	1.1×10^{3}	16
5	+10	1.5×10^{2}	20
6	+6	2.0×10	24
7	+2	2.7	28
8	0	1.0	30
9	−0.8	6.7×10^{-1}	30.8
10	−1.5	4.7×10^{-1}	31.5
12	−2.7	2.6×10^{-1}	32.7
14	−3.7	1.6×10^{-1}	33.7
16	−4.3	1.1×10^{-1}	34.3
20	−5.3	6.9×10^{-2}	35.3
25	−6.0	4.9×10^{-2}	36.0
30	−6.5	3.8×10^{-2}	36.5
40	−7.3	2.5×10^{-2}	37.3
50	−8.0	1.8×10^{-2}	38.0
75	−9.2	9.8×10^{-3}	39.2
100	−10.2	6.5×10^{-3}	40.0

[a] From the full lines in Figs 23, 24
[b] Either $\theta\Delta S^{\neq}$ or $\theta\Delta S^{\circ}$
[c] Entropic components of the EM from (38), calculated as $\exp(\theta\Delta S/\boldsymbol{R})$
[d] Calculated from (60)

5 EM and transition state structure

One of the major objectives of physical organic chemistry is the detailed description of transition states in terms of nuclear positions, charge distributions, and solvation requirements. A considerable aid to this task is provided for many reaction series by the existence of extrathermodynamic relationships, whose mathematical simplicity largely arises from extensive cancellation of the contribution to the free-energy change from the part of the molecule outside the reaction zone.

For intramolecular reactions the situation is different. Here the free-energy contribution from the non-reacting part of the molecule can be, and quite often is, most significant. Moreover, steric restrictions due to the intervening chain can significantly alter the way in which the end-groups interact in the transition state. Thus, any meaningful description of transition states in intramolecular reactions should consider free-energy changes inside and outside the reaction zone. If a convenient way to separate the two contributions is available, one can gain a considerable degree of understanding of the effect on free-energy changes resulting from incorporation of reaction zones into ring structures of varying geometry and size. This would add considerably to our knowledge of steric and stereoelectronic requirements of reactive groups involved in the intramolecular process, which directly bears on the question of directionality of organic reactions in solution (Menger, 1983).

In the common parlance of physical organic chemists such phrases as product-like or reactant-like transition states are common. The degree of resemblance of transition states to either reactants or products is usually assessed for reaction series obeying the linear-free-energy principle on the basis of suitable reaction constants, such as Brønsted α- and β-values, and Hammett reaction constants ρ. The question is inherently more complex for cyclisation reactions, since they are not expected to follow the linear-free-energy principle.

One expects the conformational entropy lost in the activation process for cyclisation to be very nearly the same as that lost upon formation of the ring product. This idea has been expressed more or less explicitly by several authors in comparisons of kinetic and thermodynamic aspects of cyclisation reactions (Allinger and Zalkow, 1960; Page and Jencks, 1971; Mandolini, 1978a; De Tar and Luthra, 1980). The quantities $\Delta S^{\ddagger}$ and ΔS° for the formation of a given ring may differ widely, due to solvent and chemical contributions to overall changes. However, these are expected to cancel out extensively in the corresponding θ quantities (see previous section). Hence (61) holds.

$$\theta\Delta S^{\ddagger} = \theta\Delta S^{\circ} \tag{61}$$

Another idea, which was originally put forward by Ruzicka (1935) and has ever since provided a common basis for many discussions of reactivity in intramolecular reactions, is that the ring-shaped transition state should be affected by a significant fraction of the strain-energy of the ring being formed. The simplest way to express this idea is by use of (62), where the weighting

$$\theta\Delta H^{\ddagger} = \alpha\theta\Delta H^{\circ} \tag{62}$$

factor α might be taken as a reasonable measure of the degree of resemblance of transition states to ring products. Clearly, one should not expect the same α-value to hold for all of the terms of a cyclisation series, because the relative importance of bond angle deformations, eclipsing interactions and transannular repulsions which determine the total ring-strain energy vary significantly along a series of ring compounds (Allinger *et al.*, 1971). Nevertheless, the extreme simplicity of (62) provides a convenient basis for discussion.

Let EM_K and EM_E be the kinetic and equilibrium EM's, respectively, for the individual members of the same cyclisation reaction, and $\theta\Delta G^{\neq}$ and $\theta\Delta G^{\circ}$ the corresponding free energy quantities. The latter can be expressed as in (63) and (64), combination of which with (61) and (62) gives (65) or (66).

$$\theta\Delta G^{\circ} = \theta\Delta H^{\circ} - T\theta\Delta S^{\circ} \tag{63}$$

$$\theta\Delta G^{\neq} = \theta\Delta H^{\neq} - T\theta\Delta S^{\neq} \tag{64}$$

$$\theta\Delta G^{\neq} = \theta\Delta G^{\circ} - (1-\alpha)\theta\Delta H^{\circ} \tag{65}$$

$$\log EM_K = \log EM_E + \frac{(1-\alpha)\,\theta\Delta H^{\circ}}{2.303\,\boldsymbol{R}T} \tag{66}$$

Equations (65) and (66) show that a linear logarithmic relationship involving the effect of a change in chain length on the kinetic EM and the equilibrium EM of the same cyclisation reaction is impossible, unless $\theta\Delta H^{\circ}$ is a constant throughout the series.

EXTRATHERMODYNAMIC CALCULATIONS OF EM'S

We shall now make use of (66) in the form (67) to calculate extrathermodynamically idealised EM-profiles for reaction series whose cyclisation entropies

$$\log EM = -\frac{\alpha\theta\Delta H^{\circ}}{2.303\,\boldsymbol{R}T} + \frac{\theta\Delta S^{\circ}}{2.303\,\boldsymbol{R}} \tag{67}$$

are given by the values listed in Table 19. The $\theta\Delta H^{\circ}$-values are provided by the cycloalkanes, the only series of ring compounds for which extensive ring-strain energy data are available (Table 1). The dramatic effect of strain on the ease of ring closure is well illustrated in Fig. 25 by the vertical differences between the purely entropic profile (upper plot, $\alpha = 0$) and that related to the fully-formed cycloalkane rings (lower plot, $\alpha = 1$). As long as 6-membered and larger rings are concerned, the plots calculated with $\alpha = 0.60$ and 0.40 (Fig. 26) satisfactorily reproduce the essential features of the experimental EM-profiles related to the formation of 1,1-bis(ethoxycarbonyl)cycloalkanes [7] and N-tosylazacycloalkanes [9], respectively (Fig. 10), whose strain energies are likely to be quite similar to those of the cycloalkanes. The zig-zagging behaviour displayed by the calculated profiles in the region from ring-size 12–15 does not show up in the experimental EM-profiles. This behaviour may be due to the experimental uncertainties of the thermochemical data.

The substantial agreement between calculated and experimental EM-profiles, clearly revealing that significant fractions of the strain-energies of the ring products show up in the transition states, indicates that the former are good models for understanding the strain factors which affect the latter.

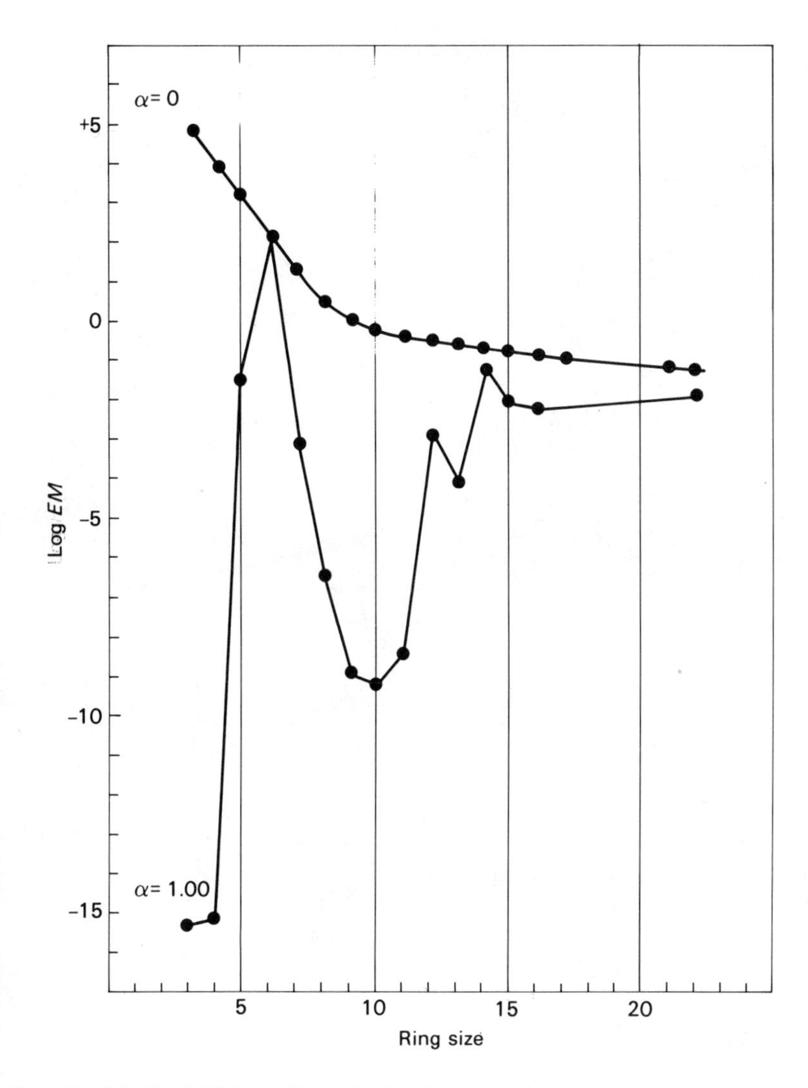

FIG. 25 Idealised EM-profiles calculated extrathermodynamically from (67) for reactions series whose transition states or products have a constant fraction α of the cycloalkane strain energies (see also Fig. 26)

The two reactions upon which the above comparison was made belong to the S_N2 mechanistic type. There are good indications, however, that the same conclusion applies as well to reactions belonging to different mechanistic types. We have already commented (p. 60) on the close similarity between the EM-profile for intramolecuar fluorescence quenching in

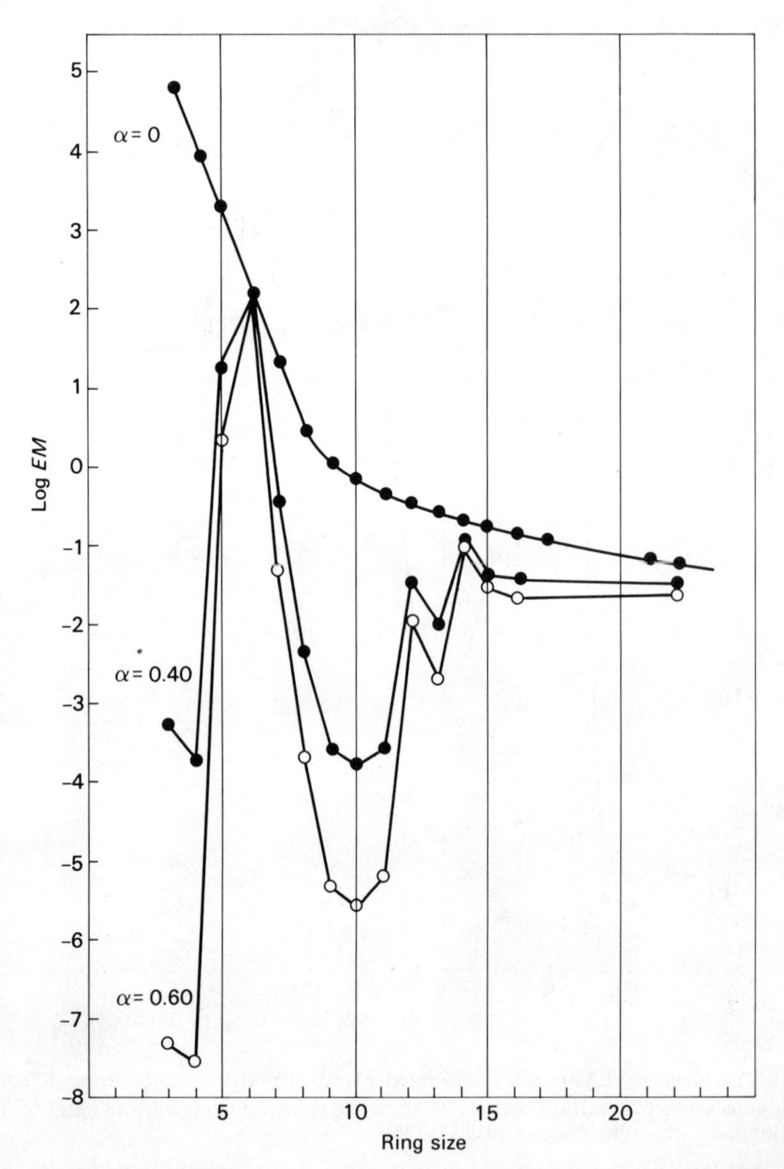

FIG. 26 Idealised EM-profiles calculated extrathermodynamically (see caption to Fig. 25)

$Me_2N(CH_2)_mNMe_2$ (Fig. 16) and that for the formation of N-tosylaza-cycloalkanes [9] (Fig. 10). We have also pointed out (p. 49) that the effect of chain length on the ease of closure of [11] by intramolecular acylation of the thiophene ring is quite close to that found for the meta- and paracyclophane diethers (Fig. 11). Winnik (1981a), noting that the lactonisation of ω-bromoalkanoate ions [3] (S_N2 reaction; Fig. 10) and that of ω-hydroxyalkanoic acids (nucleophilic addition to carbonyl; Table 8) follow remarkably similar reactivity trends, has concluded that the transition states and products are similarly sensitive to medium-ring strain energies, in spite of the different geometries of the reaction zones.

Although quantitative information for reactions belonging to other mechanistic types is lacking, available evidence of a qualitative nature points to common features underlying the cyclisation of relatively long-chain molecules. A notable example is the formation of metal complexes containing large chelate rings which has been concluded to follow rules similar to those followed by organic bifunctional compounds (Ogino and Fujita, 1975; Al-Salem *et al.*, 1979).

Thus, consistent with the basic assumptions of the extrathermodynamic treatment, all of the available evidence suggests that the close resemblance between transition state and ring product is a general property of cyclisation reactions, and that the structure of the reaction zone does not suffer from major changes along a given series.

THE QUESTION OF THE SMALLEST RINGS IN S_N2 REACTIONS

Closures of 3-, 4-, and 5-membered rings exhibit unusual behaviour when compared to higher homologues, as shown, for example, by observing the left-hand ends of the EM-profiles plotted in Fig. 10; these are reproduced very poorly by the calculated values shown in Fig. 26.

That closures of the smallest rings exhibit intriguing peculiarities has been remarked for some time. Stirling (1973), in a provoking and stimulating article, wrote that erroneous "beliefs about the formation of 3-membered rings are promulgated in the literature and in a number of well-known texts ... with a frequency which rivals only the rarity of their substantiation". He laid emphasis on the findings that 3-membered rings are sometimes formed very much more slowly than 5-membered rings, but sometimes are formed very much more rapidly, which was hardly explained in a simple way on the basis of the views currently available at the time on the interplay between strain and probability factors. In a kinetic study of cyclisation and polymerisation of ω-(bromoalkyl) dimethylamines, De Tar and Brooks (1978) remarked that comparison of cyclisation rates with thermodynamic data led to a wrong estimate for 5-membered *vs* 6-membered ring formation by a factor of about 10^4, and for 4-membered *vs* 7-membered ring formation by about 10^{12}. In a subsequent paper, where cyclisation entropies were evaluated extrathermodynamically and activation enthalpies by molecular mechanics, De Tar and Luthra (1980) stressed the view that product rings are not always suitable models for S_N2 ring closures of the smallest rings. More recently, Casadei *et al.* (1984) have analysed the EM-data for cyclisations of [6] to [7] by means of a simplified version of the extrathermodynamic treatment presented in this chapter. It was apparent that only a meagre fraction of the strain-energy of cyclobutane shows up in the estimated $\theta\Delta H^{\ddagger}$-value for the 4-membered ring, and that the 5-membered ring transition state is virtually strainless, in spite of the significant ring-strain energy of cyclopentane.

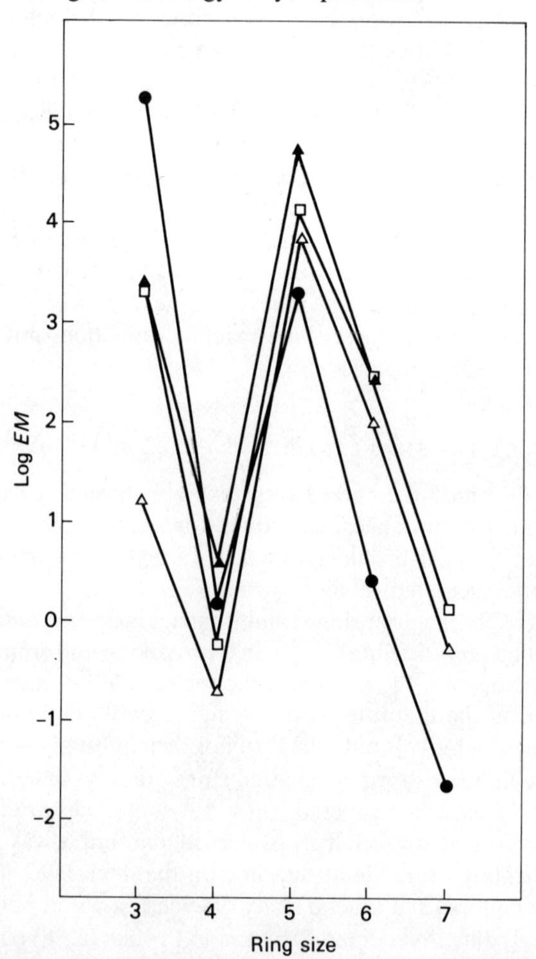

FIG. 27 EM-profiles for S_N2 cyclisation of: (▲) $^{-}O(CH_2)_{n-1}Cl$ and (△) $H_2N(CH_2)_{n-1}Br$ (taken from Kirby, 1980); (□) $Ts\bar{N}(CH_2)_{n-1}Br$ (Di Martino *et al.*, 1985); (○) $(EtO_2C)_2\bar{C}(CH_2)_{n-1}Br$. (Casadei *et al.*, 1984; the point for the 3-membered ring is an estimate for the lower reactivity limit). (Adapted with permission from Di Martino *et al.*, 1985)

Figure 27 shows plots of all the available EM's for closures of small- and common-sized saturated carbocycles and heterocycles by intramolecular nucleophilic displacement. Clearly, α-values as small as 0.1 would be required in order to calculate extrathermodynamically from (67) EM-values comparable to those actually observed for ring-sizes 3 and 4, and an even smaller value would be necessary for ring-size 5. This would lead to the conclusion that the effect of ring strain on cyclisation rates is insignificant. The same conclusion was recently drawn by Benedetti and Stirling (1983), based on rates and activation parameters for the cyclisation of bis-sulphonyl-stabilised carbanions to 3-, 4-, and 5-membered bis-sulphonylcyloalkanes.

On the other hand, Di Martino *et al.* (1985) have argued that when closures of rings of equal size are compared, the strain energies of the ring products do have a profound influence on the ease of ring closure. They based their arguments mainly on the EM's for closure of 3-membered rings, which exhibit the most pronounced sensitivity to structural effects. Combination of the data plotted in Fig. 27 with the EM-value of 1.23×10^{-2}M for the cyclisation of $BrCH_2CO_2^-$ (Galli *et al.*, 1977; see Fig. 10) indeed shows EM's spanning a range of more than 7 orders of magnitude, whereas the EM's associated with the closing of 4- and 5-membered rings range over about one order of magnitude.

The strain-energies of cyclopropane, oxirane, and aziridine are quite similar to each other (Table 2). Whereas the presence of an N-tosyl group would not affect the stability of the aziridine ring to a significant extent, protonation at nitrogen decreases the inherent stability of the 3-membered nitrogen heterocycle by some 4 kcal mol^{-1}, while having practically no effect on its higher homologues (Rozeboom *et al.*, 1982). A significant stabilisation from the ethoxycarbonyl substituents is predicted for the cyclopropane ring, owing to its well-established ability to enter into π-type conjugation with adjacent π-electron systems such as carbonyl, phenyl, etc. but little or no effect of this sort is expected to operate in the 4-ring case (Hoffmann, 1970; Hoffmann and Davidson, 1971). Hence the following stability order among 3-membered rings was inferred (Di Martino *et al.*, 1985):

1,1-bis(ethoxycarbonyl)cyclopropane > N-tosylaziridine ~
~ oxirane > aziridinium ion > 2-ethanolide

where the position of the last ring follows from its possessing a highly-strained α-lactone structure. Thus, there is a close correspondence between stability order and ease of ring closure of these 3-membered rings. The magnitude of the overall effect on the ring-strain energies of the above rings is unknown. It can nevertheless be concluded that the observed effect of 10^7 in the EM's, which is equivalent to a change in $\theta\Delta G^{\neq}$ of nearly 10 kcal mol^{-1}, must at any rate correspond to a sizeable fraction of the unknown variation in ring-strain energies. Consistently, the 4- and 5-membered rings, for which minor effects on the strain energies are expected, exhibit an insensitivity to structural effects.

We are therefore faced with apparently conflicting evidence and contrasting conclusions as to the role of ring strain on the ease of formation of the smallest rings, and to the question of whether the ring product is a good model for the transition state. That the contradiction is more apparent than real can be tentatively shown on the basis of the classical interpretation of steric retardation in bimolecular substitution (Ingold, 1969), according to which a major contribution to the activation energy derives from mutual compression of atoms not directly bonded together. For a primary alkyl substrate such non-bonded compressions arise from interactions of both entering (N) and leaving (L) groups with substituents attached to the α and β carbon atoms [35]. A glance at the naïve picture of the 3-membered transition

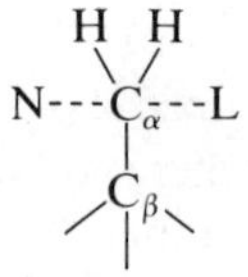

[35] 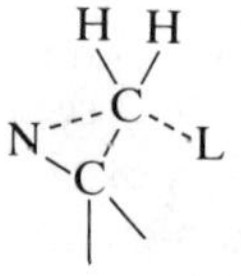 [36] 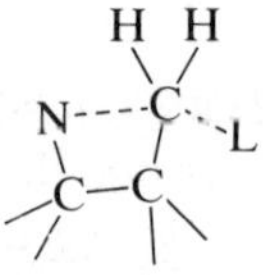 [37]

state [36] clearly suggests that such interactions should be significantly relieved. Since N is bonded to C_β, the only atoms against which it exerts a compression are the α-hydrogen atoms. Moreover, N is obviously forced to follow a bent trajectory, which keeps it further apart from the α-hydrogens than in [35]. It seems likely also that the leaving group L would take advantage of this geometrical situation, in that it can follow a more favourable bent trajectory, as the β-carbon and the atoms attached to it are perforce kept far away. One can expect this reduction of steric compression still to be significant for ring-size 4 [37], and to fall off rapidly with increasing ring size.

If the above hypothesis is correct, the basic assumption of the extrathermodynamic treatment, namely, that there is exact cancellation of chemical contributions in the quantity $\theta\Delta H^{\ddagger}$, no longer holds. Reduction of the contribution to the activation enthalpy from steric compression ($\Delta H^{\ddagger}_{sc}$) should be accounted for by introducing an appropriate $\theta\Delta H^{\ddagger}_{sc}$-term besides $\alpha\theta\Delta H^{\circ}$ into (62), giving (68). Now, if we make the reasonable assumptions that the

$$\theta\Delta H^{\ddagger} = \alpha\theta\Delta H^{\circ} + \theta\Delta H^{\ddagger}_{sc} \tag{68}$$

magnitude of the (negative) $\theta\Delta H^{\ddagger}_{sc}$ quantity, (*i*) is significant with respect to $\alpha\theta\Delta H^{\circ}$, (*ii*) decreases in the order $3 > 4 > 5$, and (*iii*) becomes negligibly small from ring-size 6 onwards, its addition to $\theta\Delta H^{\circ}$ in (68) will yield for the smallest rings $\theta\Delta H^{+}$-values, and hence EM-values that reflect the strain-energies of the ring products quite poorly, in a way which is consistent with experimental findings.

On the other hand, if the ease of formation of rings of equal size from different reactions is compared (e.g. 3-membered rings) the correct sensitivity to variations in ring-strain energy will show up in the measured EM's, provided that the magnitude of $\theta\Delta H^{+}_{sc}$ does not suffer from major variations along the series of given rings.

Additional evidence from S_N2 ring opening reactions

Since a ring-opening reaction involving cleavage of a ring bond by attack from an external nucleophile is just the reverse of an S_N2 ring-closure reaction, one might reasonably expect the two processes to share many common features, since both reactions are concerned with strained, ring-shaped transition states and their relation to the strain energies of the corresponding stable ring compounds.

The reaction of cyclic dimethylammonium ions [38] with sodium methoxide in methanol (69a) is the only S_N2-type ring-opening reaction for which

$$\overbrace{\underbrace{(CH_2)_{n-1}\ N^+}}(CH_3)_2 + CH_3O^- \longrightarrow CH_3O(CH_2)_{n-1}N(CH_3)_2 \qquad (69a)$$

[38]

$$[CH_3(CH_2)_3]_2N^+(CH_3)_2 + CH_3O^- \longrightarrow CH_3O(CH_2)_3CH_3 + CH_3(CH_2)_3N(CH_3)_2 \qquad (69b)$$

kinetic data are available over a wide range of ring sizes, including small, common, medium, and large rings (Cerichelli *et al.*, 1980; Cospito *et al.*, 1981; Di Vona *et al.*, 1985). The pertinent data, which are listed in Table 20 together with the rate constant of the model reaction (69b), show a definite tendency for the highly-strained rings to be very reactive, thus suggesting that relief of ring strain in the activation process is an important factor. In the absence of ring-strain energies for compounds [38] we are again forced to resort to the cycloalkanes as models for the effect of ring size on strain-energy. A closer comparison between cycloalkane strain-energies (Table 1) and rate data (Table 20) reveals, however, that the relation of reactivity to strain is not a simple one. For example, the 5-membered ring is about 50 times more reactive than the 7-membered one, but cyclopentane and cycloheptane are equally strained. Furthermore, the huge rate enhancements observed for the 3- and 4-membered rings, whose strain-energies are about twice as great as those of the medium rings, are not proportional to the rate enhancements observed for the 9- and 11-membered rings. These anomalies are well-illustrated by a plot of free energies of activation relative to ring

TABLE 20

Second-order rate constants for the reaction of N,N-dimethyl cyclic ammonium ions [38] with sodium methoxide in methanol at 130°C[a]

Ring-size n	$k/M^{-1}s^{-1}$	k_{rel}
3	1.0×10^{4} [b]	3.4×10^{9}
4	8.5[b]	2.9×10^{6}
5	3.14×10^{-4}	1.07×10^{2}
6	2.51×10^{-6}	0.85
7	6.04×10^{-6}	2.05
8	1.64×10^{-5}	5.57
9	5.75×10^{-5}	19.6
11	4.08×10^{-5}	13.9
13	6.84×10^{-6}	2.33
16	4.60×10^{-6}	1.56
Open chain	2.94×10^{-6}	1.00

[a] Data from Cerichelli *et al.*, 1980; Cospito *et al.*, 1981; Di Vona *et al.*, 1984
[b] Extrapolated from data obtained at lower temperatures

size 6 ($\Delta G^{\ddagger}_{rel}$) against cycloalkane strain-energies (Fig. 28). The scattered behaviour of the 3-, 4-, and 5-membered rings is in marked contrast with the definitely linear trend displayed by the higher homologues. Two important observations follow from the reactivity picture outlined above.

First, since it seems likely that $\Delta S^{\ddagger}$ should not suffer from major variations along the series, to a reasonable approximation $\Delta G^{\ddagger}_{rel} = \Delta H^{\ddagger}_{rel}$ for any term of the series. The approximately linear relationship with slope -0.2 displayed by rings with $n \geqslant 6$ provides therefore a strong indication that the effect of ring size on reactivity is solely due to relief of initial state strain in the transition state. The magnitude of the slope indicates that a significant fraction, namely, about 80%, of the strain-energy of the parent rings is still present in the corresponding transition states. This is an important result, providing independent evidence of the essential validity of hypothesis (62) as applied to the extrathermodynamic treatment of ring-closure reactions.

Secondly, the strong deviations of the 3-, 4-, and 5-membered rings from the linear trend, in the direction of higher reactivity than predicted, clearly indicates that the corresponding transition states take advantage from a "special discount", whose magnitude grades from ring-size 3 to ring-size 5. This is clearly in close conformity with the anomalous behaviour found in the corresponding ring-closure reactions. Thus, as Di Vona *et al.* (1985) have pointed out, ring-strain relief is far from being the only factor controlling the reactivity of small rings in nucleophilic ring-opening reactions. The widely held belief that these rings are very reactive simply because they are highly strained is erroneous.

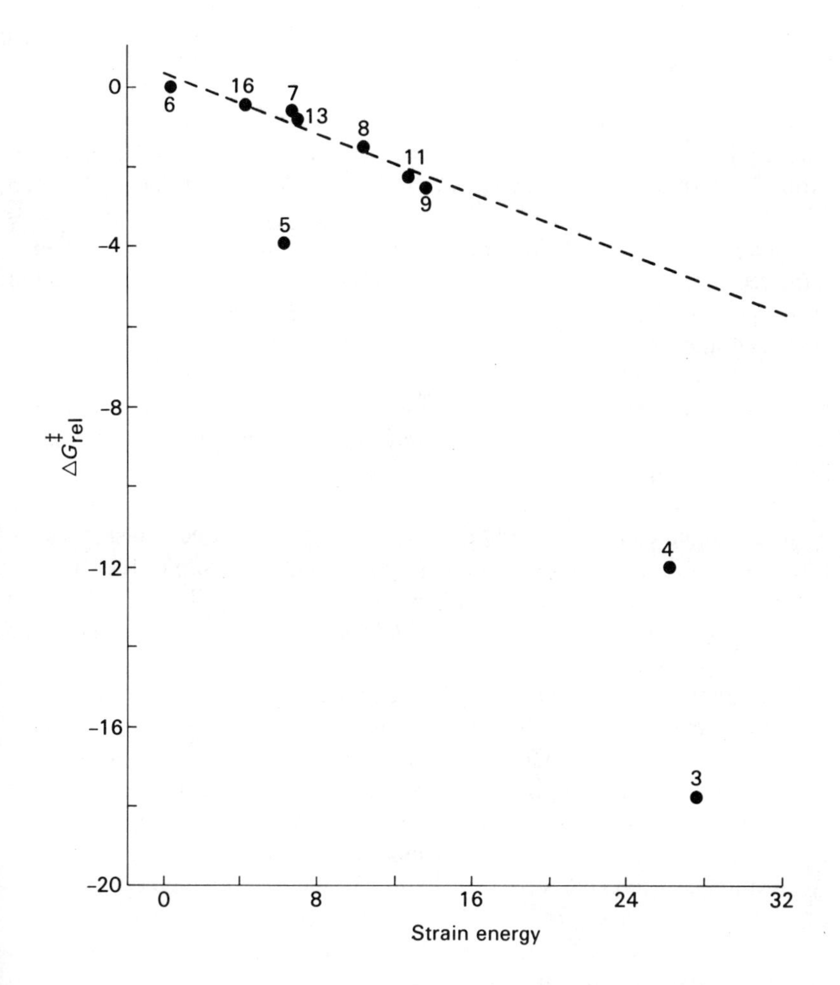

FIG. 28 Nucleophilic ring opening of cyclic dimethylammonium ions [38]. Free energies of activation relative to ring-size 6 against cycloalkane strain-energies. Energy units kcal mol^{-1}. (Reproduced with permission from Di Vona *et al.*, 1985)

MISCELLANEOUS REACTIONS

A point which emerges very clearly from the above discussion of structural effects on S_N2 ring-closure reactions, is that there is a special effect, namely, a substantial reduction of non-bonded interactions in the transition states for closing of the smallest rings. This effect is held responsible for remarkably high kinetic EM's for the 3- and 4-membered rings in spite of their extremely low equilibrium EM's, and unusually high rate ratios of 5- *vs* 6-membered ring formation, the advantage of the latter being in the order of 2 or more powers of 10 in typical cases.

We may now enquire whether and to what extent similar behaviour is displayed by ring-closure reactions other than nucleophilic displacement at saturated carbon. It should be stressed, however, that only a meagre fraction of the amount of information available for S_N2 ring-closure reactions is available for reactions belonging to different mechanistic types. The very imperfect knowledge of the factors which control rates of ring formation, even in synthetically important systems, has amply justified the search for empirical and semi-empirical rules, such as the well-known Baldwin's rules (Baldwin, 1976). These rules, whose physical basis is provided by the stereochemical requirements of the transition states for the various ring-closure processes, have presumably been derived from an admixture of phenomenological information and directionality arguments. A reaction is classified as favoured if the nature and length of the chain connecting the end-groups in X~~Y is such as to allow X and Y to attain the required transition-state geometry. In these rules a numerical prefix indicates the size of the ring; the prefix *exo* is used when the breaking of the bond is exocyclic, and *endo* when it is endocyclic; finally, the suffixes *tet, trig, dig* indicate the geometry of the carbon atom undergoing attack.

In order to define more precisely the significance of the term favoured, we propose to classify as favoured a reaction whose EM is greater than 0.1 M. This simply means that when such a reaction is run at an initial concentration of 0.1 M, it is favoured over the corresponding intermolecular reaction. The admittedly arbitrary limit of 0.1 M derives from the simple consideration that such a concentration is still acceptable in ordinary preparative work. Accordingly, all but one of the S_N2 ring-closure reactions whose pertinent data are plotted in Fig. 27 are classified as favoured processes, since their EM's are greater than 0.1 M. This is consistent with Baldwin's rule that 3- to 7-*exo-tet* are all favoured processes.

Next to tetrahedral systems, trigonal systems are considered as reaction centres in Baldwin's rules. These predict that 3- to 7-*exo-trig* reactions are all favoured processes. Very accurate EM's have been reported by Bruice and Benkovic (1963) for intramolecular nucleophilic attack on carbonyl in [39] and [40]. When Ar is varied from C_6H_5 to p-$NO_2C_6H_4$, the EM of [39]

$Me_2N(CH_2)_3CO_2Ar$ [39] $Me_2N(CH_2)_4CO_2Ar$ [40] $Ar = C_6H_5, p$-ClC_6H_4, m-$NO_2C_6H_4, p$-$NO_2C_6H_4$

ranges from 1.26×10^3 to 5.40×10^3M and that of [40] from 0.49×10^3 to 2.33×10^3M. Clearly, these are highly favoured processes, as predicted by

Baldwin's rules. It is noteworthy that the EM_5/EM_6 ratio is reduced to a factor as small as about 2, which is less than the intrinsic entropic advantage of 5- over 6-membered ring formation. Kirby (1980) in his review lists a large number of EM data for intramolecular nucleophilic additions to carbonyl. Probably because these data derive from laboratories of chemists mainly interested in intramolecular nucleophilic catalysis and its relevance to understanding enzymic catalysis, the great majority of them refer to reactions occurring via 5- and 6-membered transition states. The only example where a 4-membered transition state is involved is (70), whose kinetics were studied

$$Et_2NCH_2SC(=O)CH_3 \rightleftharpoons Et_2\overset{+}{N}(Ac)CH_2S^- \longrightarrow Et_2NAc + (CH_2S)_x \quad (70)$$

in a wide range of solvents including nitromethane, acetonitrile, chloroform, acetone, dimethoxyethane, and hexane (Searles and Nukina, 1965). Strangely enough, the EM ranges from a value of 0.029 M in nitromethane to one of 22 M in acetonitrile. Apart from this undoubtedly complex, and somewhat intriguing example, we are aware of no instance of nucleophilic addition to carbonyl via 3- and 4-membered transition states that has been reported in quantitative studies. We therefore suspect 3- and 4-*exo-trig* reactions to carbonyl to be disfavoured, rather than favoured processes.

TABLE 21

Rate constants at 65°C for ring closure of ω-alkenyl radicals in the *exo*-mode (71)[a]

Radical	k/s^{-1}
3-Butenyl	1.8×10^4
4-Pentenyl[b]	7×10^{-1}
5-Hexenyl	3.6×10^5
6-Heptenyl	1.1×10^4
7-Octenyl	3.0×10^2

[a] Data from Beckwith, 1981
[b] Too slow to measure. The rate constant is estimated from data for the reverse reaction

In addition to nucleophilic reactions, Baldwin's rules also apply to homolytic and cationic processes. Table 21 lists rate constants for ring closure of lower ω-alkenyl radicals (71), in which intramolecular addition to the double bond occurs in the *exo*-mode (Beckwith, 1981). It is unfortunate that EM-

$$\cdot CH_2(CH_2)_{m-2}CH{=}CH_2 \xrightarrow{exo} (CH_2)_{m-1}\,\text{C}{<}^{CH_2^\bullet}_{H} \quad (71)$$

data are not available for this interesting ring-closure series. It is nevertheless clear that closure of the 4-membered ring is a disfavoured process, in contrast to Baldwin's rules. In fact, the reaction is too slow to measure (see footnote *b* to Table 21). On the other hand, the formation of the 3-membered ring, whose rate is comparable to that of the 6-membered homologue appears to be highly favoured. This is probably due to strong stabilisation of the incipient cyclopropane ring from conjugation with the singly occupied localised orbital (Bischof, 1980). In other words, the high reactivity of the 3-butenyl radical is presumably due to its homoallylic nature. We also note that the 5-membered ring is formed some 30 times more rapidly than the 6-membered one, but no simple explanation for this phenomenon seems available.

It is well known (March, 1977) that an important synthetic application of Friedel Crafts alkylation and acylation is to effect ring closure. These reactions are most successful when 6-membered rings are formed, but 5- and 7-membered rings can be sythesised as well, although less readily. Quantitative data are not available, but the indication is strong that formation of 6-membered rings is much easier than formation of 5-membered rings. This provides a useful piece of information concerning ring closure via cationic attack on trigonal carbon. It has been suggested (Sundberg and Laurino, 1983) that the stereoelectronic requirements are such that the electrophilic carbon must attack from a direction very nearly perpendicular to the plane of the ring, which would lead to a more strained transition state in the 5-ring case.

The effect of ring size on the kinetics of the chelate effect has been studied by Romeo *et al.* (1977, 1978). In aqueous solution at 30°C the EM for closing of the chelate complex [42: $n = 5$] from the corresponding complex with the monodentate form of 1,2-diaminoethane [41: $n = 5$] is 1260 M. The EM decreases to 468 M with 1,3-diaminopropane ($n = 6$) and to 6.0 M with 1,4-diaminobutane ($n = 7$).

$$\text{Cl}_2\text{Pt(DMSO)}\text{NH}_2(\text{CH}_2)_{n-3}\text{NH}_2 \underset{+\,\text{Cl}^-}{\overset{-\text{Cl}^-}{\rightleftharpoons}} \left[\text{Cl(DMSO)Pt}(\text{NH}_2(\text{CH}_2)_{n-3}\text{NH}_2)\right]^+$$

[41] [42]

In a closely related and probably more accurate investigation, the same group of authors (Tobe *et al.*, 1982) have studied the kinetics of cyclisation of [43] to [44] in aqueous solution for $n = 5$, 6, and 7. The EM was found to be

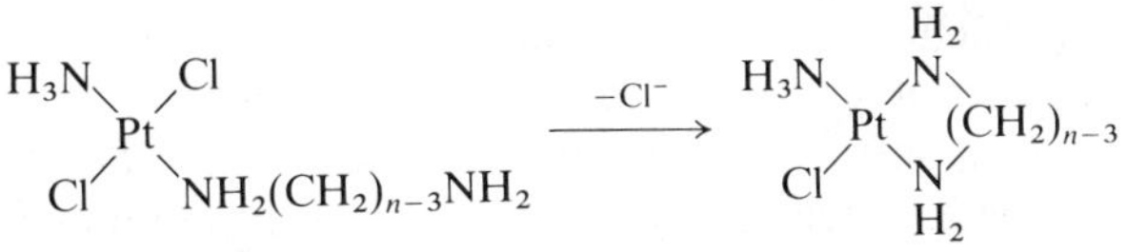

$$\underset{[43]}{H_3N\diagdown\ \diagup Cl \atop Cl\diagup Pt \diagdown NH_2(CH_2)_{n-3}NH_2} \xrightarrow{-Cl^-} \underset{[44]}{H_3N\diagdown\ \diagup N(H_2) \atop Cl\diagup Pt \diagdown N(H_2)}\ (CH_2)_{n-3}$$

1010 for $n = 5$, 77 for $n = 6$, and 0.505 M for $n = 7$. Thus, in either series the ease of ring closure of chelate rings decreases in the order $5 > 6 \gg 7$, although somewhat more spaced values are observed in the latter system. The equilibrium constants for the ring-closure reactions [41] ⇄ [42] + Cl^- have also been measured (Romeo *et al.*, 1978). The pertinent values are 8.8×10^7, 6.6×10^7, and $2.1 \times 10^5 M^{-1}$ for $n = 5$, 6, and 7, respectively. Equilibrium data for [43] ⇄ [44] + Cl^- are not available, but the literature data cited by Romeo *et al.* (1978) clearly indicate that complexes of divalent metal ions with 1,2-diaminoethane are more stable than those with 1,3-diaminopropane. Moreover, in a thorough discussion of the relations between the chelate effect and the ring size, Anderegg (1980) has listed thermodynamic data of complex formation between divalent metal ions and ligand [45], showing that almost invariably the stability of chelate rings decreases with increasing n in the order $5 > 6 \gg 7$.

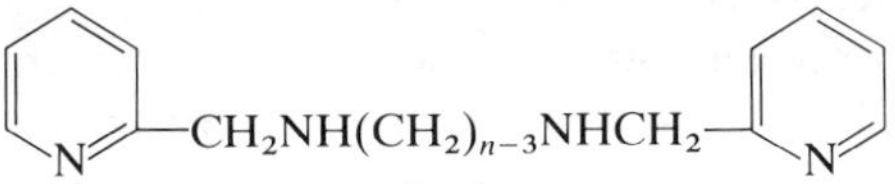

$$C_5H_4N\text{—}CH_2NH(CH_2)_{n-3}NHCH_2\text{—}C_5H_4N$$

[45]

Although the scope of these studies is somewhat limited by the restricted ring-size range investigated, the definite tendency for the kinetic EM's in chelate ring formations to parallel the stabilities of the ring products is noteworthy. It would seem to indicate that, unlike the S_N2 case, specific effects associated with the transition state for closing of 5-membered chelate rings are lacking.

THE TIGHT AND LOOSE TRANSITION STATE HYPOTHESIS. INTRAMOLECULAR PROTON TRANSFER

Page and Jencks (1971; see also Page, 1973), correctly recognised the entropic contributions to rate accelerations in intramolecular reactions as due to the significant losses of translational plus rotational entropy in the corresponding intermolecular counterparts. They reasoned that a partial compensation for these losses arises from low-frequency motions in products and transition states, which can reduce to a large extent the entropic advantage of intra- over intermolecular reactions. This means that, other things being equal, intramolecular reactions proceeding via loose transition states will exhibit EM's lower than when tight transition states are involved.

Experimental support to what might be called the tight and loose transition state hypothesis is believed to be provided by intramolecular reactions involving general acid or base catalysis, whose distinguishing feature is that the proton being transferred is a part of a cyclic transition state. For these reactions EM's are usually small, typically in the order of 10^1M or less, even when 5- to 7-membered transition states are formally involved[8], whereas reactions involving intramolecular nucleophilic catalysis exhibit high EM's in the order of 10^3–10^5M, or even higher in special cases. As pointed out by Kirby (1980), the differences are large enough to provide a useful criterion of mechanism. The relative ineffectiveness of intramolecular reactions involving proton transfer has been attributed to the inherent looseness of transition states for proton-transfer processes (Page and Jencks, 1971; Page, 1973).

The tight and loose transition-state hypothesis is in contrast with the assumption that there is extensive cancellation of contributions due to chemical change in the entropic component of the EM (p. 81). Indeed, the uniform behaviours displayed by $\theta\Delta S$-data for reactions widely differing in nature (Figs 5, 23, and 24) clearly shows that no matter how loose a transition state or product is, the entropy contribution from such looseness will be cancelled out extensively by virtue of the operator θ.

A close comparison between intramolecular proton transfer and intramolecular nucleophilic attack at carbon is provided by the varying amounts of olefins [46] which accompany the ring-closure reactions of o-$^{-}OC_6H_4O(CH_2)_{n-4}Br$ [1] to the corresponding catechol polymethylene ethers [2] (Illuminati *et al.*, 1975).

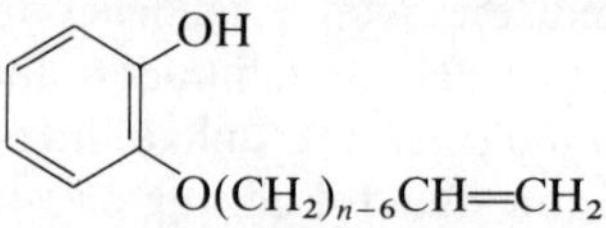

OH

$O(CH_2)_{n-6}CH{=}CH_2$

[46]

Olefins [46] have been shown to derive from unimolecular β-elimination of HBr from the CH_2CH_2Br end of [1] promoted by the distal phenoxide end (Dalla Cort *et al.*, 1983), as depicted in [47]. Rate constants and EM's have been measured in 99% Me_2SO at 25°C for reactions occurring via 7-, 8-, 9-,

[8] The possibility of intervention of a solvent bridge between the proton and the basic site, where the solvent molecule acts simultaneously as a proton donor and proton acceptor, should always be considered. Here as elsewhere, the operational recognition of what Hammett (1970) calls the stoichiometric involvement of solvent is not a simple task.

$$\left[C_6H_4 \begin{matrix} O\cdots H\cdots CH \overset{\cdots}{=} CH_2\cdots Br \\ | \\ O\text{———}(CH_2)_{n-6} \end{matrix} \right]^-$$

[47]

10-, and 14-membered ring transition states (Dalla Cort *et al.*, 1983), the hydrogen undergoing transfer being included in the computation. There seems to be no question of the intervention of a bridging solvent molecule in the virtually aprotic 99% Me_2SO.

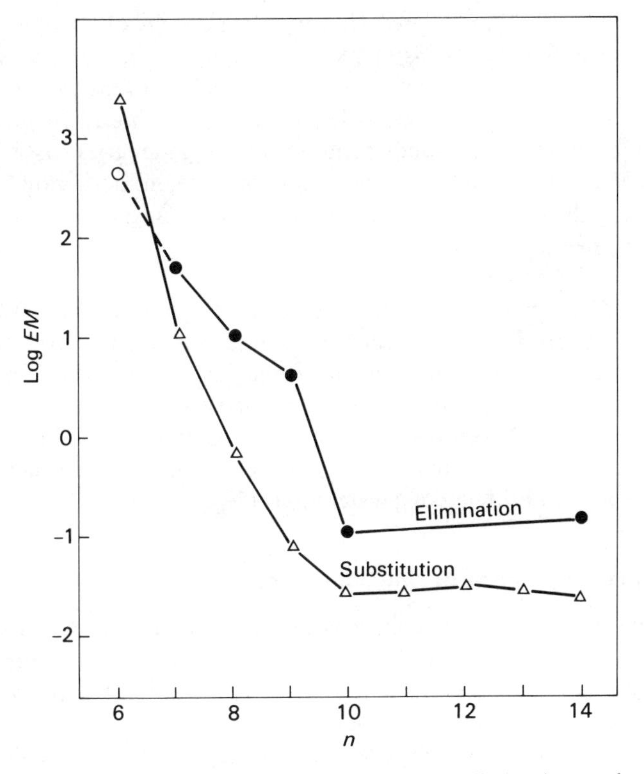

FIG. 29 EM-profiles for competing intramolecular elimination and substitution from $o\text{-}^{-}OC_6H_4O(CH_2)_{n-4}Br$ [1] in 99% Me_2SO as a function of the size n of the cyclic transition states. The point for the elimination reaction where $n = 6$ is an estimate for the upper reactivity limit. (Reproduced with permission from Dalla Cort *et al.*, 1983)

A comparison with the competing displacement reaction leading to ring formation is provided by Fig. 29. It is remarkable that no entropic handicap is revealed for intramolecular proton transfer relative to intramolecular nucleophilic attack, in spite of the presumably greater looseness of the former. In fact, apart from the case where $n = 6$ where no evidence for elimination could be obtained, the EM's for elimination are not only higher than the EM's for substitution, but are also rather close to the values estimated extrathermodynamically (Table 19) on purely entropic grounds. The objection that such a behaviour might be ascribed to a tight transition state for proton transfer from carbon is in contrast with available data (Kirby, 1980) showing that there seems to be no general tendency for intramolecular proton transfer to occur more efficiently from carbon than from oxygen or nitrogen.

The case where a 7-membered transition state is involved is worthy of comment, since its geometry is not large enough to accommodate a linear C· · ·H· · ·O arrangement. Consistent with current views on the directional requirements of proton transfers (Menger, 1983), the high EM provides a strong indication that bent donor-proton-acceptor geometries cost relatively little. On the other hand, the fact that elimination does not compete apprcciably with substitution when $n = 6$ indicates a higher cost for a more bent arrangement, such as that which can be accommodated in a 6-membered transition state.

Noting that much of the supporting evidence for exceedingly low EM's in intramolecular proton transfer reactions is based on systems involving 5- to 7-membered cyclic transition states, we are inclined to interpret the general ineffectiveness of intramolecular general acid-base catalysis as a strain phenomenon rather than an entropy phenomenon. Another possibility is that, at least in some cases, the actual ring size of the transition state is larger than believed because of bridging of a water molecule.

6 EM and the synthesis of ring compounds

In principle, any reaction between separate molecules R′X and R″Y, leading to a stable, covalently bonded product R′ZR″, can be translated into a cyclisation reaction for the preparation of a ring product ⌐Z¬ from a suitable precursor |X∼∼Y.| However, as was clearly apparent from the very early efforts to synthesise many-membered ring compounds through cyclisation of long-chain bifunctional reactants (Ruzicka *et al.*, 1926a,b) and later confirmed by an impressive mass of experimental information accumulated by synthetic organic chemists over nearly 60 years (Rossa and Vögtle, 1983), these reactions are accompanied by a number of disadvantages, even when the chemistry is clean, i.e. when virtually quantitative yields of products R′ZR″ are obtained under a convenient set of experimental conditions.

A major reason for these disadvantages is that polymerisation can, and actually does to a significant extent in many instances, compete with ring closure. That dilution favours cyclisation was first recognised by Ruggli (1912), but the first convenient application of this principle to the synthesis of many-membered rings was carried out some 20 years later by Ziegler *et al.* (1933) in the first example of what is commonly known as the Ziegler high dilution technique. It is the purpose of this section to discuss the importance of the dilution factor in the light of the available knowledge of phenomenological features of ring-closure reactions.

Because yields of ring products depend on competition between cyclisation and polymerisation, the relevant quantity is the cyclisation constant C (Section 2). However, since cyclisation constants are generally unknown, arguments will be mainly based on the more readily available EM-values. The possible difference in magnitude between the two quantities is relatively unimportant for the purposes of the present discussion.

RING CLOSURE UNDER BATCHWISE CONDITIONS

The simplest way to carry out a cyclisation reaction is one where the bifunctional reactant M is charged over a relatively short time into the reaction medium, where proper reaction conditions are set. Here short time means that the reaction proceeds to a negligible extent during the addition time. In the absence of competing reactions other than polymerisation, the outcome of a batchwise cyclisation experiment is almost entirely determined by the ratio between $[M]_0$ and C, as shown by the plots reported in Fig. 1.

Let us now choose an initial concentration in the range 0.1 M–1 M, which is convenient for preparative purposes, and draw a horizontal line at log $EM = 0$ in Fig. 10. We see that the representative points for most of the rings from 3- to 7-membered lie above the line. For these rings polymerisation is no problem, but for those whose representative points are below the line polymerisation predominates over cyclisation. The situation can be somewhat improved by lowering the initial concentration to 10^{-2}M, which is still acceptable for preparations on a millimolar scale, to include in the allowed region most of the large rings. But for rings whose EM's are lower than 10^{-2}M the situation is apparently hopeless, since concentrations of, say, 10^{-3}M or 10^{-4}M are of no practical value in synthetic work.

RING CLOSURE UNDER INFLUXION CONDITIONS

The difficulties encountered in the synthesis of ring systems with low EM's are partly overcome by the well-known Ziegler high dilution technique (Ziegler *et al.*, 1933). Here the bifunctional reactant M is introduced very slowly into the reaction medium in order to prevent it accumulating. Since the term high dilution is often used for experiments carried out under batch-wise conditions, the term influxion has been recommended to avoid confusion (Galli and Mandolini, 1982).

The rate of feed v_f (in mol l^{-1} s^{-1}) of the reactant into the reaction medium is now the critical parameter to adjust for a favourable outcome of a cyclisation experiment. A kinetic treatment of the open system under influxion incurs the same difficulties already discussed for the closed system in Section 2. However, when the higher-order polymerisation terms are relatively unimportant and the overall process is described to a useful approximation by (6), an exact mathematical solution is possible (Galli and Mandolini, 1975). After a relatively short initial time[9], the concentration of M reaches a steady value $[M]_{st}$ given by (72), where β defined by (73) is a dimensionless parameter

$$[M]_{st} = \frac{k_{intra}}{4\,k_{dim}} \, [(1 + 8\beta)^{\frac{1}{2}} - 1] \tag{72}$$

$$\beta = v_f \, k_{dim}/k^2_{intra} \tag{73}$$

which is a measure of the feed rate in units of k^2_{intra}/k_{dim}. Under typical Ziegler high dilution conditions, where the total feeding time is much greater than the time required to reach the steady state, the (normalised) yield of ring product is expressed by relation (74), which is shown graphically in Fig. 30. It

$$\text{Yield} = \frac{2}{1 + (1 + 8\beta)^{\frac{1}{2}}} \tag{74}$$

is clear that the necessary condition to be fulfilled for an influxion cyclisation to be successful is that the feed rate is low enough to render the parameter β much smaller than unity. This condition may be represented by (75). Equa-

$$v_f \ll \frac{k^2_{intra}}{k_{dim}} \approx k_{inter} \, EM^2 \tag{75}$$

tion (74) has been used as a valuable guide for adjusting appropriate synthetic conditions in systems where the required kinetic parameters were known (Galli and Mandolini, 1975; Catoni *et al.*, 1980). The results were in general agreement with predictions.

Synthetic experiments in which the required kinetic information is available are admittedly more the exception than the rule, but an approximate knowledge, or even a rough estimate, of the relevant kinetic parameters can

[9] The integrated equation relating the concentration of M to time (Galli and Mandolini, 1975) shows that the steady state is reached within a time interval which is not greater than that required to fulfil the condition $\exp(-k_{intra}t) \ll 1$, i.e. the time required to obtain a practically complete conversion of M into the ring product under first-order conditions.

prove to be most useful for properly orienting the search for convenient reaction conditions (Mandolini and Masci, 1977; Galli and Mandolini, 1978; Casadei *et al.*, 1981; Gargano and Mandolini, 1982).

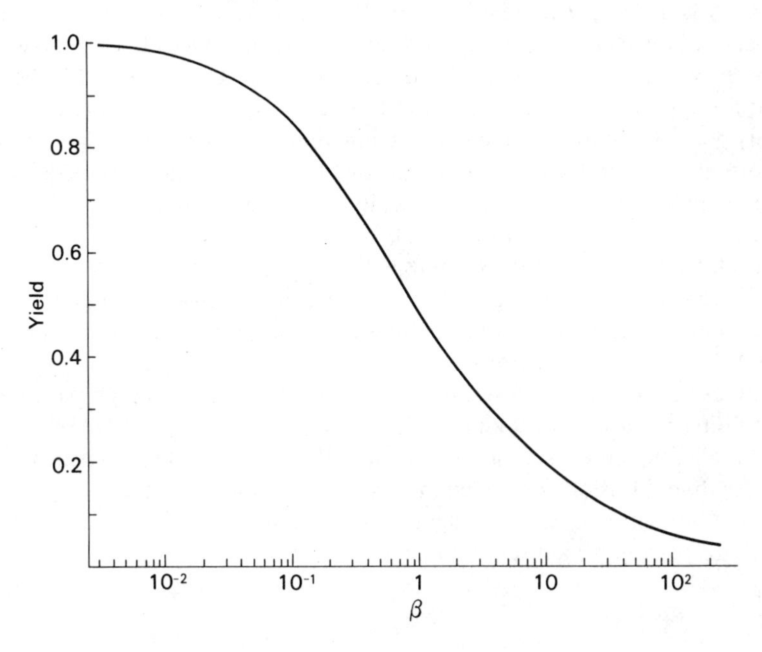

FIG. 30 Cyclisation of a bifunctional reactant under influxion conditions. Yield of monomeric ring product as a function of the feed rate parameter β, calculated from (74). (Reproduced with permission from Galli and Mandolini, 1975)

In any event, (74) and (75) can be used as a basis for a general discussion of the synthesis of many-membered rings under influxion conditions. Let us consider the problem of translating into a general cyclisation procedure an intermolecular reaction which has been found to proceed in high yield under a given set of experimental conditions. Influxion procedures are time-consuming, the duration of the process being controlled by the feed rate v_f and by the total amount of material to be used per unit volume of solvent. The maximum feed rate which fulfils condition (75) may be reasonably set at that given by (76), which leads to a yield of 86% calculated from (74). The total

$$v_f = 0.1\, k_{inter}\, EM^2 \tag{76}$$

addition time in hours required to convert a given number of moles of reactant per unit volume, calculated from (76), is given by (77). If the amount of

$$\text{Total addition time} = \frac{\text{moles of reactant}}{360 k_{inter}\, EM^2} \tag{77}$$

material to be converted is 0.1 mol per litre of solvent, and the specific rate k_{inter} of the intermolecular reaction is $1\ M^{-1}\ s^{-1}$, the total addition time calculated from (77) is 2.8 h for a ring whose *EM* is 10^{-2}M, but increases to 280 h when the *EM* decreases to 10^{-3}M. This unpleasant result is a consequence of the fact that the total addition time varies inversely with the second power of the *EM*. When the *EM* decreases further to, say, 10^{-5}M, the time required is 2.8×10^5h or 3.2 years! For such a highly disfavoured ring, the only possible solution is to choose an inherently faster reaction. Yet an intermolecular reaction with a specific rate of $1\ M^{-1}\ s^{-1}$ is usually considered fast for synthetic purposes, being practically complete within less than 30 s when the concentrations of both reactants are 1 M. Nevertheless, it is practically unsuitable for the synthesis of rings with *EM*'s lower than 10^{-2}M.

To return to the ring with $EM = 10^{-5}$M, the k_{inter}-value required to keep the total addition time within the reasonable limit of 24 h is in the order of $10^5 M^{-1}\ s^{-1}$. Since reactions as fast as these are rarely met among synthetically useful reactions, the well-known difficulties met in the preparation of medium rings are easily understood.

Finally, we note that the obvious fact that chemical reactions cannot be faster than diffusion controlled makes it impossible to synthesise by cyclisation of bifunctional reactants significant amounts of rings with $EM \leqslant 10^{-7}$M within a reasonable amount of time.

Acknowledgements

Many of the ideas reported in this chapter stemmed from endless discussions with Professor G. Illuminati,* Dr C. Galli, and Dr B. Masci, who have co-authored most of the author's work on ring-closure reactions. To these colleagues the author wishes to express his sincere thanks. Thanks are also due to Miss G. Gigli for her skilful contribution in the preparation of the typescript.

***Note Added in Proof**

By the sudden and unexpected death of Professor Gabriello Illuminati on 22nd January 1986 the Italian chemical community has lost one of its most distinguished and authoritative members. I dedicate this chapter to his memory, hoping that this will help to keep alive the remembrance of the man who has contributed more than anyone else to the introduction and development of Physical Organic Chemistry in Italy.

References

Allinger, N. L. and Hu, S. (1961). *J. Am. Chem. Soc.* **83**, 1664
Allinger, N. L. and Sprague, J. T. (1972). *J. Am. Chem. Soc.* **94**, 5734
Allinger, N. L. and Zalkow, V. (1960). *J. Org. Chem.* **25**, 701
Allinger, N. L., Tribble, M. T. and Miller, M. A. (1972). *Tetrahedron,* **28**, 1173
Allinger, N. L., Tribble, M. T., Miller, M. A. and Wertz D. H. (1971). *J. Am. Chem. Soc.* **93**, 1637
Al-Salem, N. A., Empsall, H. D., Markham, R., Shaw, B. L. and Weeks, B. (1979). *J. C. S. Dalton,* 1972
Anderegg, G. (1980). *In* "Advances in Molecular Relaxation and Interaction Processes", Vol. 18, p. 79. Elsevier Scientific Publishing, Amsterdam
Baker, W., McOmie, J. F. W. and Ollis, W. D. (1951). *J. Chem. Soc.* 200
Baldwin, J. E. (1976). *J. C. S. Chem. Commun.* 734
Beckwith, A. L. J. (1981). *Tetrahedron,* **37**, 3073
Belen'kii, L. I. (1964). *Russ. Chem. Rev.* **33**, 551
Benedetti, F. and Stirling, C. J. M. (1983). *J. C. S. Chem. Commun.* 1374
Bennett, G. M. (1941). *Trans. Faraday Soc.* **37**, 794
Benson, S. W. (1976). "Thermochemical Kinetics", 2nd edn. John Wiley and Sons, New York
Billmeyer, F. W. Jr (1971). "Textbook of Polymer Science", 2nd edn. John Wiley and Sons, New York
Bird, R. and Stirling, C. J. M. (1973). *J. C. S. Perkin II,* 1221
Bird, R., Knipe, A. C. and Stirling, C. J. M. (1973). *J. C. S. Perkin II,* 1215
Bischof, P. (1980). *Helv. Chim. Acta* **63,** 1434
Björnstad, S. L. and Borgen, G. (1975). *Acta Chem. Scand.* **B29**, 13
Borgen, G. (1973). *Acta Chem. Scand.* **27**, 1840
Borgen, G. and Dale, J. (1972). *Acta Chem. Scand.* **26**, 952
Borgen, G. and Gaupset, G. (1974). *Acta Chem. Scand.* **B28**, 816
Breslow, R. and Winnik, M. A. (1969). *J. Am. Chem. Soc.* **91**, 3083
Brown, H. C. and Ichikawa, K. (1957). *Tetrahedron* **1**, 221
Brown, J. F. and Slusarczuk, G. M. J. (1965). *J. Am. Chem. Soc.* **87**, 931
Bruice, T. C. and Benkovic, S. J. (1963). *J. Am. Chem. Soc.* **85**, 1
Bruice, T. C. and Pandit, U. K. (1960). *J. Am. Chem. Soc.* **82**, 5858
Bruice, T. C. and Turner, A. (1970). *J. Am. Chem. Soc.* **92**, 3422
Byström, K. and Månsson, M. (1982). *J. C. S. Perkin II,* 565
Capon, B. (1964). *Quart. Rev.* **18**, 45
Capon, B. (1972). *Essays Chem.* **3**, 148
Capon, B. and McManus, S. P. (1976). "Neighboring Group Participation". Plenum Press, New York
Carmichael, J. B. and Winger, R. (1965). *J. Polymer Sci. A* **3**, 971
Casadei, M. A., Galli, C. and Mandolini, L. (1981). *J. Org. Chem.* **46**, 3127
Casadei, M. A., Galli, C. and Mandolini, L. (1984). *J. Am. Chem. Soc.* **106**, 1051
Catoni, G., Galli, C. and Mandolini, L. (1980). *J. Org. Chem.* **45**, 1906
Cerichelli, G., Illuminati, G. and Lillocci, C. (1980). *J. Org. Chem.* **45**, 3952
Connor, H. D., Shimada, K. and Szwarc, M. (1972). *Macromolecules,* **5**, 801
Cospito, G., Illuminati, G., Lillocci, C. and Petride, H. (1981). *J. Org. Chem.* **46**, 2944
Cox, J. D. and Pilcher, G. (1970). "Thermochemistry of Organic and Organometallic Compounds". Academic Press, London
Dafforn, G. A. and Koshland, D. E. (1977). *J. Am. Chem. Soc.* **99**, 7246
Dale, J. (1963). *J. Chem. Soc.* 93

Dale, J. (1966). *Angew. Chem. Internat. Edn.* **5**, 1000
Dale, J. (1974). *Tetrahedron,* **30**, 1683
Dale, J. (1976). *Topics Stereochem.* **9**, 199
Dale, J., Hubert, A. J. and King, G. S. D. (1963). *J. Chem. Soc.* 73
Dalla Cort, A., Illuminati, G., Mandolini, L. and Masci, B. (1980a). *J. C. S. Perkin II,* 1774
Dalla Cort, A., Mandolini, L. and Masci, B. (1980b). *J. Org. Chem.* **45**, 3923
Dalla Cort, A., Mandolini, L. and Masci, B. (1983). *J. Org. Chem.* **48**, 3979
Danforth, C., Nicholson, A. W., James, J. C. and Loudon, G. M. (1976). *J. Am. Chem. Soc.* **98**, 4275
Davies A. G., Davies, M. and Stoll, M. (1954). *Helv. Chim. Acta,* **37**, 1351
Davis, A. M., Page, M. I., Mason, S. C. and Watt, I. (1984). *J. C. S. Chem. Commun.* 1671
De Tar, D. F. and Brooks, W. (1978). *J. Org. Chem.* **43**, 2245
De Tar, D. F. and Luthra, N. P. (1980). *J. Am. Chem. Soc.* **102**, 4505
Di Martino, A., Galli, C., Gargano, P. and Mandolini, L. (1985). *J. C. S. Perkin II,* 1345
Di Vona, M. L., Lillocci, C. and Illuminati, G. (1985). *J. C. S. Chem. Commun.* 380
Drewes, S. E. and Coleman, P. C. (1972). *J. C. S. Perkin I,* 2148
Fersht, A. R. and Kirby, A. J. (1980). *Chem. Brit.* **16**, 136
Fife, T. H. (1975). *Adv. Phys. Org. Chem.* **11**, 1
Flory, P. J. (1969). "Statistical Mechanics of Chain Molecules". Wiley-Interscience, New York
Flory, P. J. and Semlyen, J. A. (1966). *J. Am. Chem. Soc.* **88**, 3209
Flory, P. J., Suter, U. W. and Mutter, M. (1976). *J. Am. Chem. Soc.* **98**, 5733
Galli, C. and Mandolini, L. (1975). *Gazz. Chim. Ital.* **105**, 367
Galli, C. and Mandolini, L. (1977). *J. C. S. Perkin II,* 443
Galli, C. and Mandolini, L. (1978). *Org. Synth.* **58**, 98
Galli, C. and Mandolini, L. (1982). *J. C. S. Chem. Commun.* 251
Galli, C., Illuminati, G. and Mandolini, L. (1973). *J. Am. Chem. Soc.* **95**, 8374
Galli, C., Illuminati, G., Mandolini, L. and Tamborra, P. (1977). *J. Am. Chem. Soc.* **99**, 2591
Galli, C., Giovanelli, G., Illuminati, G. and Mandolini, L. (1979). *J. Org. Chem.* **44**, 1258
Galli, C., Illuminati, G. and Mandolini, L. (1980). *J. Org. Chem.* **45**, 311
Galli, C., Illuminati, G., Mandolini, L. and Masci, B. (1981). *In* "Advances in Solution Chemistry" (I. Bertini, L. Lunazzi and A. Dei, eds), p. 319. Plenum Publishing
Gandour, R. D. (1978). *In* "Transition States of Biochemical Processes" (R. D. Gandour and R. L. Schowen, eds) p. 355. Plenum Press, New York
Gargano, P. and Mandolini, L. (1982). *Gazz. Chim. Ital.* **112**, 31
Greenberg, A. and Liebman, J. F. (1978). "Strained Organic Molecules". Academic Press, New York
Guthrie, J. P. (1976). *In* "Applications of Biochemical Systems in Organic Chemistry" (J. B. Jones, C. J. Sih and D. Perlman, eds) Part II, p. 627. John Wiley and Sons, New York
Halpern, A. M., Legenza, M. W. and Ramachandran, B. R. (1979). *J. Am. Chem. Soc.* **101**, 5736
Hammett, L. P. (1970). "Physical Organic Chemistry", 2nd edn. McGraw-Hill, New York

Hammond, G. S. (1956). *In* "Steric Effects in Organic Chemistry" (M. S. Newman, ed) p. 468. John Wiley and Sons, New York
Hill, J. W. and Carothers, W. H. (1933). *J. Am. Chem. Soc.* **55**, 5031
Hoffmann, R. (1970). *Tetrahedron Lett.* 2907
Hoffmann, R. and Davidson, R. B. (1971). *J. Am. Chem. Soc.* **93**, 5699
Huisgen, R. and Ott, H. (1959). *Tetrahedron,* **6**, 253
Hunsdiecker, H. and Erlbach, H. (1947). *Chem. Ber.* **80**, 129
Hurd, C. D. and Saunders, W. H. (1952). *J. Am. Chem. Soc.* **74**, 5324
Illuminati, G. and Mandolini, L. (1981). *Acc. Chem. Res.* **14**, 95
Illuminati, G., Mandolini, L. and Masci, B. (1974a). *J. Am. Chem. Soc.* **96**, 1422
Illuminati, G., Mandolini, L. and Masci, B. (1974b). *J. Org. Chem.* **39**, 2598
Illuminati, G., Mandolini, L. and Masci, B. (1975). *J. Am. Chem. Soc.* **97**, 4960
Illuminati, G., Mandolini, L. and Masci, B. (1977). *J. Am. Chem. Soc.* **99**, 6308
Illuminati, G., Mandolini, L. and Masci, B. (1981). *J. Am. Chem. Soc.* **103**, 4142
Imanishi, Y. (1979). *J. Polym. Sci. Macromol. Rev.* **14**, 1
Ingold, C. K. (1921). *J. Chem. Soc.* **119**, 305
Ingold, C. K. (1969). "Structure and Mechanism in Organic Chemistry", 2nd edn. Cornell University Press, Ithaca
Jacobson, H. and Stockmayer, W. H. (1950). *J. Chem. Phys.* **18**, 1600
Jager, J., Graafland, T., Schenk, H., Kirby, A. J. and Engberts, J. B. F. N. (1984). *J. Am. Chem. Soc.* **106**, 139
Jencks, W. P. (1969). "Catalysis in Chemistry and Enzymology". McGraw-Hill, New York
Jencks, W. P. (1975). *Adv. Enzymol.* **37**, 219
Kabo, G. Ya. and Andreevskii, D. N. (1973). *Zh. Strukt. Kim.* **14**, 1043
Kimura, Y. and Regen, S. L. (1983). *J. Org. Chem.* **48**, 1533
Kirby, A. J. (1980). *Adv. Phys. Org. Chem.* **17**, 183
Kruizinga, W. H. and Kellogg, R. M. (1981). *J. Am. Chem. Soc.* **103**, 5183
Kuhn, W. (1934). *Kolloid. Z.* **68**, 2
Kuhn, L. P. and Wires, R. A. (1964). *J. Am. Chem. Soc.* **86**, 2161
Leffler, J. E. and Grunwald, E. (1963). "Rates and Equilibria of Organic Reactions". John Wiley and Sons, New York
Liebman, J. F. and Greenberg, A. (1976). *Chem. Rev.* **76**, 311
Mandolini, L. (1978a). *Abstracts of the EUCHEM Conference on Ring-Closure Reactions and Related Topics*, Castel Gandolfo, Italy
Mandolini, L. (1978b). *J. Am. Chem. Soc.* **100**, 550
Mandolini, L. and Masci, B. (1977). *J. Org. Chem.* **42**, 2840
Mandolini, L., Masci, B. and Roelens, S. (1977). *J. Org. Chem.* **42**, 3733
Mar, A. and Winnik, M. A. (1981). *Chem. Phys. Lett.* **77**, 73
Mar, A., Fraser, S. and Winnik, M. A. (1981). *J. Am. Chem. Soc.* **103**, 4941
March, J. (1977). "Advanced Organic Chemistry. Reactions, Mechanisms, and Structure", 2nd edn. McGraw-Hill, New York
Mayer, J. E. and Mayer, M. G. (1940). "Statistical Mechanics". John Wiley and Sons, London
Menger, F. M. (1983). *Tetrahedron,* **39**, 1040
Milstien, S. and Cohen, L. A. (1970). *Proc. Nat. Acad. Sci. USA,* **67**, 1143
Milstien, S. and Cohen, L. A. (1972). *J. Am. Chem. Soc.* **94**, 9158
Morawetz, H. (1975). "Macromolecules in Solution", 2nd edn. John Wiley and Sons, New York
Morawetz, H. and Goodman, N. (1970). *Macromolecules,* **3**, 699

Mutter, M., Suter, U. W. and Flory, P. J. (1976). *J. Am. Chem. Soc.* **98**, 5745
Ogino, H. and Fujita, J. (1975). *Bull. Chem. Soc. Jpn.* **48**, 1836
O'Neal, H. E. and Benson, S. W. (1970). *J. Chem. Eng. Data* **15**, 266
Page, M. I. and Jencks, W. P. (1971). *Proc. Nat. Acad. Sci. USA* **68**, 1678
Page, M. I. (1973). *Chem. Soc. Rev.* **2**, 295
Paquette, L. A. and Begland, R. W. (1966). *J. Am. Chem. Soc.* **88**, 4685
Paquette, L. A. and Begland, R. W. (1967). *J. Org. Chem.* **32**, 2723
Pasto, D. J. and Serve, M. P. (1965). *J. Am. Chem. Soc.* **87**, 1515
Pedley, J. B. and Rylance, J. (1977). "Sussex — N.P.L. Computer Analysed Thermochemical Data: Organic and Organometallic Compounds". University of Sussex
Perkin, W. H. (1929). *J. Chem. Soc.* 1347
Person, W. B. and Pimentel, G. C. (1953). *J. Am. Chem. Soc.* **75**, 532
Pitzer, K. S. (1940). *J. Chem. Phys.* **8**, 711
Potapov, V. M. (1978). "Stereochemistry" (English translation) p. 366. Mir Publishers
Powell, R. E. and Latimer, W. M. (1951). *J. Chem. Phys.* **19**, 1139
Prelog, V., Frenkiel, L., Kobelt, M. and Barman, P. (1947) *Helv. Chim. Acta,* **30**, 1741
Rasmussen, S. E. (1956). *Acta Chem. Scand.* **10**, 1279
Romeo, R., Lanza, S. and Tobe, M. L. (1977). *Inorg. Chem.* **16**, 785
Romeo, R., Lanza, S., Minniti, D. and Tobe, M. L. (1978). *Inorg. Chem.* **17**, 2436
Rossa, L. and Vögtlc, F. (1983). *Top. Curr. Chem.* **113**, 1
Rozeboom, M. D., Houk, K. N., Searles, S. and Seyedrezai, S. E. (1982). *J. Am. Chem. Soc.* **104**, 3448
Ruggli, P. (1912). *Liebigs Ann. Chem.* **392**, 92
Ruzicka, L., Stoll, M. and Schinz, H. (1926a). *Helv. Chim. Acta,* **9**, 249
Ruzicka, L., Brugger, W., Pfeiffer, M., Schinz, M. and Stoll, M. (1926b). *Helv. Chim. Acta* **9**, 499
Ruzicka, L. (1935). *Chem. Ind. (London),* 2
Salomon, G. (1936a). *Trans. Faraday Soc.* **32**, 153
Salomon, G. (1936b). *Helv. Chim. Acta,* **19**, 743
Schwarzenbach, G. (1952). *Helv. Chim. Acta,* **35**, 2344
Searles, S. and Nukina, S. (1965). *J. Am. Chem. Soc.* **87**, 5656
Sellers, P., Stridh, G. and Sunner, S. (1978). *J. Chem. Eng. Data,* **23**, 250
Semlyen, J. A. (1976). *Adv. Polym. Science* **21**, 42
Shimada, K. and Szwarc, M. (1974). *Chem. Phys. Lett.* **28**, 540
Shimada, K. and Szwarc, M. (1975a). *J. Am. Chem Soc.* **97**, 3313
Shimada, K. and Szwarc, M. (1975b). *Chem. Phys. Lett.* **34**, 503
Shimada, K., Shimozato, Y. and Szwarc, M. (1975). *J. Am. Chem. Soc.* **97**, 5834
Shimozato, Y., Shimada, K. and Szwarc, M. (1975). *J. Am. Chem. Soc.* **97**, 5831
Sicher, J. (1962). *Progr. Stereochem.* **3**, 202
Silbermann, W. E. and Henshall, T. (1957). *J. Am. Chem. Soc.* **79**, 4107
Sisido, M. (1971). *Macromolecules,* **4**, 737
Sisido, M. and Shimada, K. (1977). *J. Am. Chem. Soc.* **99**, 7785
Sisido, M., Mitamura, T., Imanishi, Y. and Higashimura, T. (1976). *Macromolecules,* **9**, 316
Sisido, M., Takagi, H., Imanishi, Y. and Higashimura, T. (1977). *Macromolecules,* **10**, 125
Sisido, M., Yoshikawa, E., Imanishi, Y. and Higashimura, T. (1978). *Bull. Chem. Soc. Jpn.* **51**, 1464
Skinner, H. A. and Pilcher, G. (1963). *Quart. Rev.* **17**, 264

Stein, S. E. (1981). *J. Am. Chem. Soc.* **103**, 5685
Stirling, C. J. M. (1973). *J. Chem. Ed.* **50**, 844
Stoll, M. and Rouvé, A. (1935). *Helv. Chim. Acta,* **18**, 1087
Stoll, M. and Rouvé, A. (1947). *Helv. Chim. Acta,* **30**, 1822
Stoll, M., Rouvé, A. and Stoll-Comte, G. (1934). *Helv. Chim. Acta,* **17**, 1289
Storm, D. R. and Koshland, D. E. (1970). *Proc. Nat. Acad. Sci. USA,* **66**, 445
Storm, D. R. and Koshland, D. E. (1972a). *J. Am. Chem. Soc.* **94**, 5805
Storm, D. R. and Koshland, D. E. (1972b). *J. Am. Chem. Soc.* **94**, 5815
Stull, D. R., Westrum, E. F. and Sinke, G. C. (1969) "The Chemical Thermodynamics of Organic Compounds". John Wiley, New York
Sundberg, R. J. and Laurino, J. P. (1983). *J. Org. Chem.* **49**, 249
Sunner, S. and Wulff, C. A. (1980). *J. Chem. Thermodynamics,* **12**, 505
Suter, U. W., Mutter, M. and Flory, P. J. (1976). *J. Am. Chem. Soc.* **98**, 5740
Takagi, H., Sisido, M., Imanishi, Y. and Higashimura, T. (1977). *Bull. Chem. Soc. Jpn.* **50**, 1807
Tobe, M. L., Schwab, A. P. and Romeo, R. (1982). *Inorg. Chem.* **21**, 1185
Winnik, M. A. (1977). *Acc. Chem. Res.* **10**, 173
Winnik, M. A. (1981a). *Chem. Rev.* **81**, 491
Winnik, M. A. (1981b). *J. Am. Chem. Soc.* **103**, 708
Winnik, M. A. and Maharaj, U. (1979). *Macromolecules,* **12**, 902
Winnik, M. A., Trueman, R. E., Jackowski, G., Saunders, D. S. and Whittington, S. G. (1974). *J. Am. Chem. Soc.* **96**, 4843
Winnik, M. A., Basu, S., Lee, C. K. and Saunders, D. S. (1976). *J. Am. Chem. Soc.* **98**, 2928
Winstein, S., Allred, E., Heck, R. and Glick, R. (1958) *Tetrahedron,* **3**, 1
Yamashita, Y., Mayumi, J., Kawakami, Y. and Ito, K. (1980). *Macromolecules,* **13**, 1075
Yamdagni, R. and Kebarle, P. (1973). *J. Am. Chem. Soc.* **95**, 3504
Ziegler, K., Eberle, H. and Ohlinger, H. (1933). *Justus Liebigs Ann. Chem.* **504**, 94
Ziegler, K. and Hechelhammer, W. (1937). *Justus Liebigs Ann. Chem.* **528**, 114
Ziegler, K. and Holl, H. (1937). *Justus Liebigs Ann. Chem.* **528**, 143
Ziegler, K., Lüttringhaus, A. and Wohlgemuth, K. (1937). *Justus Liebigs Ann. Chem.* **528**, 162
Ziegler, K. (1955). *In* "Methoden der Organischen Chemie" (Houben-Weyl), (E. Müller, ed.) Vol. 4/2. Georg Thieme

Mechanisms of Proton Transfer between Oxygen and Nitrogen Acids and Bases in Aqueous Solution

FRANK HIBBERT

Department of Chemistry, King's College London,
Strand, London WC2R 2LS

1 Introduction 113
2 Simple proton transfers of oxygen and nitrogen acids 115
3 Proton transfer along hydrogen bonds 127
  Potential energy functions 127
  Experimental techniques and results 129
4 Proton removal from intramolecular hydrogen bonds 148
  Mechanism 149
  Proton transfer reactions of hindered diaminonaphthalenes 165
  Proton transfer from hydrogen-bonded phenylazoresorcinols 177
5 Hindered proton transfer from molecular cavities 184
6 Multiple proton transfers 190
7 Future work 205
References 205

1 Introduction

Acid and base catalysis of a chemical reaction involves the assistance by acid or base of a particular proton-transfer step in the reaction. Many enzyme catalysed reactions involve proton transfer from an oxygen or nitrogen centre at some stage in the mechanism, and often the role of the enzyme is to facilitate a proton transfer by acid or base catalysis. Proton transfer at one site in the substrate assists formation and/or rupture of chemical bonds at another site in the substrate. To understand these complex processes, it is necessary to understand the individual proton-transfer steps. The fundamental theory of simple proton transfers between oxygen and nitrogen acids and

ADVANCES IN PHYSICAL ORGANIC CHEMISTRY
VOLUME 22 ISBN 0 12 033522 0

bases was established 20–30 years ago by Eigen (Eigen, 1963, 1964, 1967; Eigen and de Maeyer, 1963; Eigen *et al.*, 1964) and will be reviewed briefly in Section 2. Simple proton transfers involving oxygen and nitrogen acids or bases in the thermodynamically favourable direction have rate coefficients with diffusion-limited values. In the reverse direction the magnitude of the rate coefficient is determined by the value of the equilibrium constant of the reaction. The transition state of the reaction resembles an encounter complex of the acidic and basic reactants. These simple ideas are of great usefulness in discussing the relevance and importance of proton transfers as elementary steps in multi-step reactions and in estimating rate coefficients for these steps. However, an increasing number of categories of proton transfer of oxygen and nitrogen acids are being discovered that do not fit this simple picture, and many of these more complex proton transfers have been proposed as components of multi-step mechanisms. In a sense the more complex proton transfers are of greater chemical interest than processes controlled by the rate of encounter between reactant molecules. Eigen recognised several of the categories of proton transfer that would not fit into the simple picture, and his ideas have stimulated much important research in this area. Developments in the 20 years following Eigen's review of the subject (Eigen, 1964) have been quite striking and the purpose of the present review is to discuss examples of those proton transfers between oxygen and nitrogen having rates which differ from the values expected for a diffusion-controlled process. Mostly, coverage will be restricted to reactions in aqueous solution. Many of the reactions require fast reaction methods and these have now become established techniques of the kineticist (Friess *et al.*, 1963; Hammes, 1974; Bernasconi, 1976).

In diffusion-controlled proton transfer between oxygen and nitrogen acids or bases, formation of a hydrogen bond between the acid and base is thought to be a necessary step prior to proton transfer. However, as will be described in Sections 3 and 4, hydrogen-bond formation can sometimes play a very different role. The presence of an intramolecular hydrogen bond in an acid considerably modifies the acid-base behaviour, and rates of proton transfer are different from the diffusion-limited values. For example, the value of the unimolecular rate coefficient for proton transfer between acidic and basic groups involved in an intramolecular hydrogen bond (1) usually exceeds the value of the pseudo first-order rate coefficient that can be achieved for an intermolecular diffusion-controlled process. However, proton transfer out of an intramolecular hydrogen bond to an external base (2) often occurs at a rate which is well below the diffusion limit. Reactions of the type shown in (1) and (2) are discussed in Sections 3 and 4 respectively.

$$\overline{X\text{—}H\cdots{}^{-}Y} \rightleftharpoons \overline{X^{-}\cdots H\text{—}Y} \qquad (1)$$

$$\text{X—H}\cdots{}^{-}\text{Y} + \text{B}^{-} \rightleftharpoons \text{X}^{-} \quad {}^{-}\text{Y} + \text{BH} \qquad (2)$$

Examples of dramatic reductions in rates of proton removal are found if the acidic proton is protected within a molecular cavity and if an unfavourable conformational change is required to expose the proton to attack by base. This type of complex proton transfer is described in Section 5.

The transfer of a proton between an acidic and a basic group within the same molecule is often more complex than the process shown in (1). The proton may be transferred along hydrogen-bonded solvent molecules between the acidic and basic groups if these are too remote to permit formation of an intramolecular hydrogen bond. Alternatively, two intermolecular proton transfers with an external acid or base may be necessary. Tautomerisation of oxygen and nitrogen acids and bases (3) will be described in Section 6. The reactions are usually quite rapid and fast reaction

$$\text{H—X} \quad {}^{-}\text{Y} \rightleftharpoons \text{X}^{-} \quad \text{Y—H} \qquad (3)$$

techniques are necessary to follow them. In contrast, tautomerisations in which one of the groups is a carbon acid or carbon base are usually slow and have been studied using traditional kinetic methods (Toullec, 1982). Tautomerisations at oxygen and nitrogen centres are important in some biological reactions. The simultaneous or stepwise transfer of two protons has also been suggested as a mechanistic step in certain chemical and enzymic reactions. For example bifunctional acid-base catalysis and the charge relay mechanism of serine proteases are thought to involve proton transfers similar to that shown in (4). These topics will be dealt with briefly in Section 6. Throughout, emphasis will be placed on the rate and mechanism of proton-transfer reactions.

$$\begin{matrix} \text{X} & & {}^{-}\text{Y} \\ | & & \\ \text{H} & & \text{H} \\ & & | \\ {}^{-}\text{A} & \text{———} & \text{B} \end{matrix} \rightleftharpoons \begin{matrix} \text{X}^{-} & & \text{Y} \\ & & | \\ & & \text{H} \\ \text{H} & & \\ | & & \\ \text{A} & \text{———} & \text{B}^{-} \end{matrix} \qquad (4)$$

2 Simple proton transfers of oxygen and nitrogen acids

The proton-transfer behaviour of most oxygen and nitrogen acids follows an extremely simple pattern (Eigen, 1964). Our purpose in this section will be to provide a brief summary of this behaviour, to point out features that are not yet fully understood, and to consider some recent developments.

A normal proton transfer was defined by Eigen as one whose rate in the thermodynamically favourable direction was diffusion-controlled (Eigen, 1964). By use of relaxation techniques Eigen was able to show that many proton transfers involving oxygen and nitrogen acids and bases were in this category. If the reactions (5) of an acid (HA) with a series of bases (B^-) shows normal proton-transfer behaviour, the rate coefficients in the forward

$$HA + B^- \underset{k_r}{\overset{k_f}{\rightleftharpoons}} A^- + BH \tag{5}$$

and reverse directions will vary with the equilibrium constant (K) for the reaction according to Fig. 1. In Fig. 1, log k_f and log k_r are plotted against ΔpK the difference in acidity of the acids HA and BH, defined by (6). For

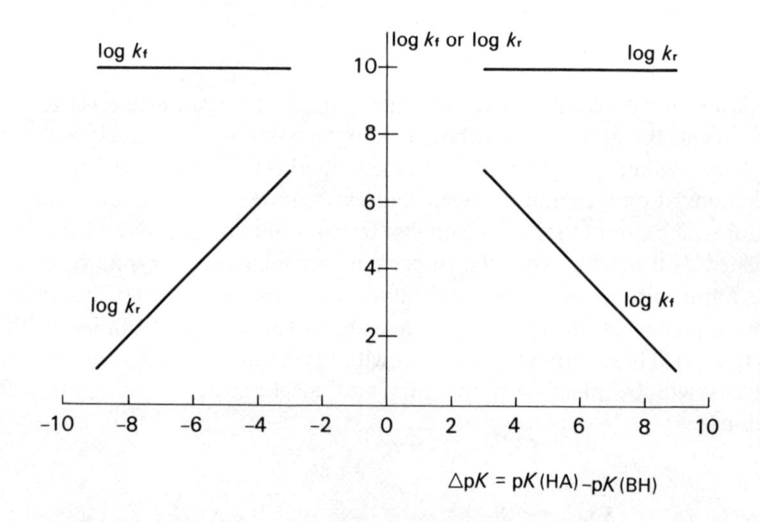

FIG. 1 Variation of forward (k_f) and reverse (k_r) rate coefficients with ΔpK for a normal proton transfer from an acid HA to a base B^-

reactions with $\Delta pK < 0$, reaction in the forward direction is thermodynamically favourable and for normal proton-transfer behaviour the value of k_f will be diffusion limited (*ca* 1×10^9 to $5 \times 10^{10} dm^3 mol^{-1} s^{-1}$) and therefore independent of ΔpK in this region. It follows from (7) that the rate coefficient

$$\Delta pK = pK(HA) - pK(BH) = -\log_{10} K \tag{6}$$

$$K = k_f/k_r; \quad -\log k_f + \log k_r = \Delta pK \tag{7}$$

k_r will be inversely proportional to K, i.e. log k_r will increase linearly with ΔpK. When $\Delta pK > 0$ reaction in the reverse direction is thermodynamically favourable and now it is k_r which will have attained the diffusion limit. These conclusions are summarised in Fig. 1 in which the slopes of plots of log k_f or log k_r against ΔpK are either zero or unity. Many oxygen and nitrogen acids follow this pattern of behaviour and for such reactions the values of the forward and reverse rate coefficients for ionisation in the presence of a base (B^-) can be estimated from the value of the equilibrium constant of the reaction. The value of the rate coefficient for a proton transfer occurring as part of a multi-step reaction sequence can be deduced in this way and the result may be useful in reaching mechanistic conclusions.

When the acids HA and BH are of similar acidity (ΔpK *ca* 0) the interpretation of the kinetic behaviour is less straightforward. It is usual to interpret the reaction by the mechanism in (8). The intermediates I_1 and I_2 are assumed to be hydrogen-bonded complexes, so that the proton-transfer step

$$\mathrm{HA + B^-} \underset{k_{-a}}{\overset{k_a}{\rightleftharpoons}} \underset{I_1}{\mathrm{AH \cdots B^-}} \underset{k_{-p}}{\overset{k_p}{\rightleftharpoons}} \underset{I_2}{\mathrm{A^- \cdots HB}} \underset{k_{-d}}{\overset{k_d}{\rightleftharpoons}} \mathrm{A^- + BH} \tag{8}$$

consists of movement of the proton along an intermolecular hydrogen bond, together with the necessary solvation changes. If the stationary state condition applies to the intermediates I_1 and I_2, the observed rate coefficients for this mechanism are given by (9) and (10). When $k_p \gg k_{-a}$ and $k_d \gg k_{-p}$,

$$k_f = k_a k_p k_d/(k_p k_d + k_{-a}k_d + k_{-p}k_{-a}) \tag{9}$$

$$k_r = k_{-a}k_{-p}k_{-d}/(k_p k_d + k_{-a}k_d + k_{-p}k_{-a}) \tag{10}$$

formation of the hydrogen-bonded complex (I_1) from reactants is rate-limiting and (11) applies. If $k_{-a} \gg k_p$ and $k_d \gg k_{-p}$, the proton-transfer step is rate-limiting and (12) is obtained. When $k_{-a} \gg k_p$ and $k_{-p} \gg k_d$, breakdown of the encounter complex to products is rate-limiting (13). Before deciding whether these ideas can be used to explain results for proton transfer to and

$$k_f = k_a \text{ and } k_r = (k_{-p}/k_p)(k_{-d}/k_d)k_{-a} \tag{11}$$

$$k_f = (k_a/k_{-a})k_p \text{ and } k_r = (k_{-d}/k_d)k_{-p} \tag{12}$$

$$k_f = (k_a/k_{-a})(k_p/k_{-p})k_d \text{ and } k_r = k_{-d} \tag{13}$$

from oxygen and nitrogen acids in aqueous solution it is useful to consider results for a typical proton transfer at ΔpK *ca* 0. Rate coefficients for reaction of acetic acid with a series of bases of varying pK(BH) are given in Table 1 and Fig. 2. The data were obtained (Ahrens and Maass, 1968) by the

TABLE 1

Kinetic results for proton transfer from acetic acid to bases in aqueous solution[a,b]

B^-	ΔpK^c	$10^{-8}k_f/dm^3mol^{-1}s^{-1}$	$10^{-8}k_r/dm^3mol^{-1}s^{-1}$
$Cl_2CHCO_2^-$	3.27	0.058	110
$ClCH_2CO_2^-$	1.90	0.24	21
HCO_2^-	1.00	4.9	10
$CH_3CH_2CO_2^-$	−0.12	3.9	2.9
$HO_2CCH_2CO_2^-$	−0.95	5.6	0.62
$(CH_3)_2AsO_2^-$	−1.44	17	0.62
$H_2PO_4^-$	−2.45	48	0.17

[a] Reaction:

$$CH_3COOH + B^- \underset{k_r}{\overset{k_f}{\rightleftharpoons}} CH_3CO_2^- + BH$$

Temperature 293K, ionic strength 1.0 mol dm^{-3}

[b] Data from Ahrens and Maass, 1968

[c] $\Delta pK = pK(CH_3COOH) - pK(BH)$

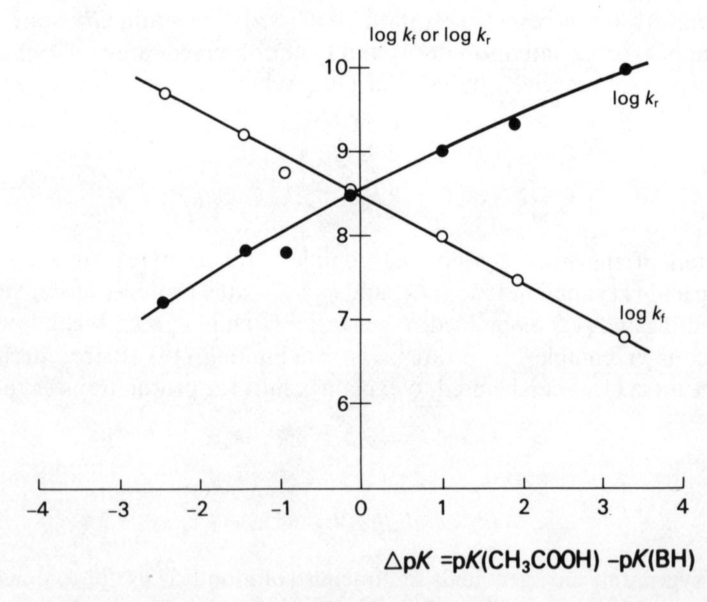

FIG. 2 Variation of forward (k_f)and reverse (k_r) rate coefficients with ΔpK for proton transfer from acetic acid to various bases in aqueous solution at 293K and ionic strength 1.0 mol dm^{-3}

ultrasonic absorption technique (Eggers, 1957; Hammes, 1974; Bernasconi, 1976). If it is assumed that the proton-transfer step in mechanism (8) occurs rapidly compared with the diffusion steps, eqns (14) and (15) are deduced from (9) and (10). In (14) and (15), K is the overall equilibrium constant for

$$k_f = k_a/(1 + k_a/Kk_{-d}) \tag{14}$$

$$k_r = k_{-d}/(1 + Kk_{-d}/k_a) \tag{15}$$

the reaction. For a reaction in which $K \gg 1$, (14) and (15) reduce to the relations in (11), and when $K \ll 1$ (14) and (15) reduce to the relations in (13). If the value $k_a = k_{-d} = 6 \times 10^9\ \mathrm{dm^3\ mol^{-1}\ s^{-1}}$ is taken, the variation of log k_f and log k_r with ΔpK shown in Fig. 3 is obtained. The theoretical plots in Fig. 3 are in rather poor agreement with the experimental plots in Fig. 2 for acetic acid. At $\Delta pK = 0$ the value of k_f and k_r in Fig. 3 is a factor of two

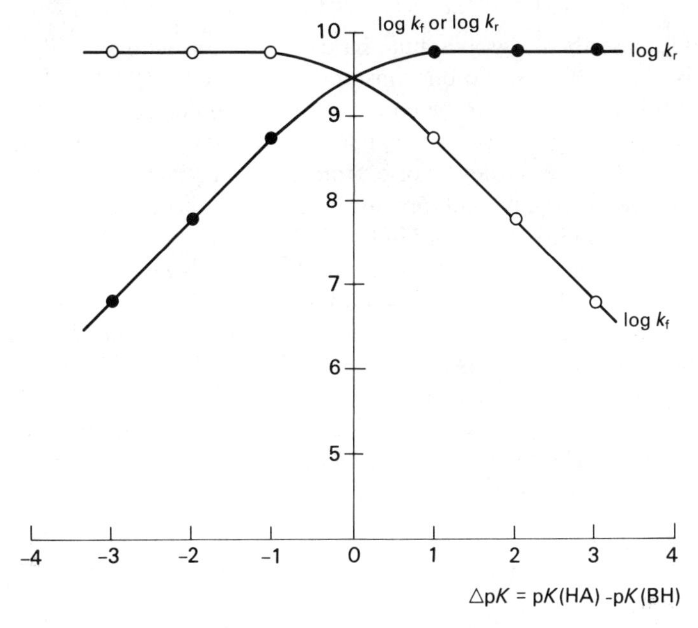

FIG. 3 Theoretical variation of forward (k_f) and reverse (k_r) rate coefficients for reaction between an acid (HA) and base (B^-) for which diffusion steps are rate limiting

below the limiting values at high or low ΔpK whereas in Fig. 2 the value of k_f and k_r at $\Delta pK = 0$ is at least 10 times lower than the limiting rate coefficients at high or low ΔpK. In Fig. 3, the region where the slopes of the plots of log k_f and log k_r against ΔpK differ from zero or unity is in the range

$-1 < \Delta pK < 1$. In Fig. 2 the region of curvature is much broader and extends beyond $-4 < \Delta pK < +4$. One explanation for the poor agreement between the predictions in Fig. 3 and the behaviour observed for ionisation of acetic acid is that in the region around $\Delta pK = 0$, the proton-transfer step in mechanism (8) is kinetically significant. In order to test this hypothesis and attempt to fit (9) and (10) to experimental data, it is necessary to assume values for the rate coefficients for the formation and breakdown of the hydrogen-bonded complexes in mechanism (8) and to propose a suitable relationship between the rate coefficients of the proton-transfer step and the equilibrium constant for the reaction. There are various ways in which the latter can be achieved. Experimental data for proton-transfer reactions are usually fitted quite well by the Brønsted relation (17). In (17), G_B is a

$$HA + B^- \underset{}{\overset{k_f}{\rightleftarrows}} A^- + HB \tag{16}$$

$$k_f = G_B(1/K(BH))^\beta; \log k_f = \log_{10} G_B - \beta \log_{10} K(BH) \tag{17}$$

constant for a series of closely similar bases B^- reacting with an acid HA and $K(BH)$ is the acid dissociation constant of the acid HB. The Brønsted exponent β is often found to be constant over a limited range of values of $K(BH)$. Although (17) is an empirical relation, the view is widely held that the value of β is related to the degree of proton transfer in the transition state for the reaction and usually, but not always, lies in the range $0 \leqslant \beta \leqslant 1.0$ (Kresge, 1973, 1975; Bell, 1973; Hibbert, 1977). For the mechanism shown in Scheme 1, a satisfactory fit to the experimental data (Fischer *et al.*, 1980) was obtained using $\log k_p = 10.3 + \beta \Delta pK$, $\log k_{-p} = 10.3 - \beta \Delta pK$, $\Delta pK = pK(BH) - 2.85$, and $\beta = 0.5$ with the numerical values $k_a = k_{-d} = 1.0 \times 10^{10}\ dm^3\ mol^{-1}\ s^{-1}$ and $k_{-a} = k_d = 1.0 \times 10^{11}\ s^{-1}$. The expressions for $\log k_p$ and $\log k_{-p}$ are forms of the Brønsted relation, and the value of β implies a transition state for the reaction in which the proton is half-transferred. The fit to the experimental data is not unique and other values have been used (Sayer and Jencks, 1973).

$$HO\overset{+}{N}H_2\text{—}CPh_2\text{—}NMe_2 + B^- \underset{k_{-a}}{\overset{k_a}{\rightleftarrows}} HO\overset{+}{N}H(\cdots H\cdots B^-)\text{—}CPh_2\text{—}NMe_2$$

$$HO\overset{+}{N}H(\text{—}H\cdots B^-)\text{—}CPh_2\text{—}NMe_2 \underset{k_{-p}}{\overset{k_p}{\rightleftarrows}} HONH(\cdots HB)\text{—}CPh_2\text{—}NMe_2$$

$$HONH(\cdots HB)\text{—}CPh_2\text{—}NMe_2 \underset{k_{-d}}{\overset{k_d}{\rightleftarrows}} HONH\text{—}CPh_2\text{—}NMe_2 + HB$$

Scheme 1

A rather more elegant procedure with which to predict the variation of k_p and k_{-p} with ΔpK for the purposes of fitting mechanism (8) and eqns (14) and (15) to experimental data is provided by Marcus Theory (Murdoch, 1980; Albery, 1980). The theory was originally developed for electron-transfer reactions (Marcus, 1964) and applied to proton transfers in a modified form (Marcus, 1968, 1969, 1975; Cohen and Marcus, 1968). The equations of Marcus Theory have been derived in various ways (German *et al.*, 1971; Koeppl and Kresge, 1973); for example, one derivation has been obtained (Murdoch, 1972) from the Hammond postulate (Hammond, 1955) and from Leffler's principle (Leffler, 1953; Leffler and Grunwald, 1963). Comparisons have been made (Murdoch, 1983a) with other procedures for interpreting substituent effects on the rates of proton-transfer reactions. The theory is considered to be of widespread importance in the correlation of rate-equilibrium data for proton-transfer reactions (Kreevoy and Konasewich, 1971; Kreevoy and Oh, 1973; Albery *et al.*, 1972) and has been applied to other reactions (Murdoch, 1983a,b; Pellerite and Brauman, 1983), including methyl and other alkyl transfers (Albery and Kreevoy, 1978; Lewis and Hu, 1984). According to Marcus Theory the Gibbs energy of activation of a proton transfer ($\Delta G^{\ddagger}$) is given by (18) in which ΔG° is the standard Gibbs energy for the reaction and $\lambda/4$ is the Gibbs energy of activation of the reaction for which $\Delta G^{\circ} = 0$. The term $\lambda/4$ is usually called the intrinsic

$$\Delta G^{\ddagger} = (\lambda + \Delta G^{\circ})^2/4\lambda \tag{18}$$

barrier, that is the Gibbs energy of activation when the standard Gibbs energy difference between reactants and products makes no contribution to the Gibbs energy of activation in either the forward or reverse direction. The intrinsic barrier ($\lambda/4$) for the reaction shown in (19) is usually obtained by extrapolation or interpolation of experimental data and is thought to correspond to the average of the Gibbs energies of activation, $\lambda_{AA}/4$ and $\lambda_{BB}/4$, for the identity reactions (20) and (21), as in (22). A work term which

$$HA + B^- \rightleftharpoons A^- + BH \tag{19}$$

$$HA + A^- \rightleftharpoons A^- + HA \tag{20}$$

$$BH + B^- \rightleftharpoons B^- + BH \tag{21}$$

$$\lambda = (\lambda_{AA} + \lambda_{BB})/2 \tag{22}$$

represents the Gibbs energy for bringing together the reactants is often included in (18) together with a similar term for the products.

For proton transfer from an acid (HA) to a series of bases (B^-) the Gibbs energy of activation for the reaction is predicted to vary with the standard Gibbs energy for the proton transfer according to (18) The value of the

Brønsted exponent for the reaction is given by (23) and is predicted to vary with ΔG° as in (24).

$$\beta = \mathrm{d}\Delta G^{\ddagger}/\mathrm{d}\Delta G^{\circ} = (1 + \Delta G^{\circ}/\lambda)/2 \tag{23}$$

$$\mathrm{d}\beta/\mathrm{d}\Delta G^{\circ} = 1/2\lambda \tag{24}$$

Equation (18) can be used to examine the effect of changing the values of the rate coefficients for the proton-transfer step in (8) on the fit of equations (9) and (10) to experimental data for proton transfer from acetic acid to a series of bases (Fig. 2). It is necessary to choose suitable values for formation and dissociation of the hydrogen-bonded complexes I_1 and I_2 and we shall use $k_a = k_{-d} = 6 \times 10^9 \mathrm{dm}^3 \mathrm{mol}^{-1} \mathrm{s}^{-1}$ and $k_{-a} = k_d = 6 \times 10^{10} \mathrm{s}^{-1}$. For a given value of the intrinsic barrier $\lambda/4$, (18) can be used to calculate values of k_p and k_{-p} corresponding to different values of ΔG° or ΔpK. These results can be used to calculate values of k_f and k_r from (9) and (10). The results of such calculations for a reaction temperature of 298K using values for $\lambda/4$ of 5, 20, and 40 kJ mol^{-1} are given in Fig. 4 and show how the proton-transfer step modifics thc overall kinetics. With $\lambda/4 = 5$ kJ mol^{-1} the variation of k_f and k_r with ΔpK is identical with the variation of k_f and k_r given in Fig. 3 which was constructed by assuming that the proton-transfer step is kinetically insignificant. The predicted variation of k_f and k_r with ΔpK for a value $\lambda/4 = 20$ kJ mol^{-1} closely resembles the observed variation in rate coefficients for proton transfer from acetic acid. With $\lambda/4 = 40$ kJ mol^{-1} the variation of k_f and k_r with ΔpK resembles more closely the results observed for proton transfer from a carbon acid (Kresge, 1973; Hibbert, 1977). Results similar to those for acetic acid have been fitted (Albery, 1980) using $k_a = k_{-d} = 7 \times 10^9 \mathrm{dm}^3 \mathrm{mol}^{-1} \mathrm{s}^{-1}$, $k_{-a} = k_d = 6 \times 10^{10} \mathrm{s}^{-1}$, and $\lambda/4 =$ 19 kJ mol^{-1}. For the reaction with $\Delta G^{\circ} = 0$ or $\Delta pK = 0$, a Gibbs energy of activation for the proton-transfer step of *ca* 20 kJ mol^{-1} is considered to be a reasonable value for proton transfer along an intermolecular hydrogen bond with the accompanying solvent reorganisation. A Gibbs energy of activation of this magnitude corresponds to a rate coefficient at 298K of $k_p = k_{-p} = 1.9 \times 10^9 \mathrm{s}^{-1}$. Rate coefficients for proton transfer along intramolecular hydrogen bonds are discussed in Section 3 and, as will be seen, appear to have larger values than this. When the Marcus Theory is applied to data for proton transfer to and from carbon, much higher intrinsic energy barriers are obtained. These may be a consequence of the electronic and heavy atom reorganisation which accompanies proton transfer from carbon (Bell, 1973; Kresge, 1973; Hibbert, 1977).

One of the reasons for the quite high intrinsic energy barrier ($\lambda/4$ *ca* 20 kJ mol^{-1}) which appears to be necessary to explain the observed variation of k_f and k_r with ΔpK for the ionisation of acetic acid is thought to be the

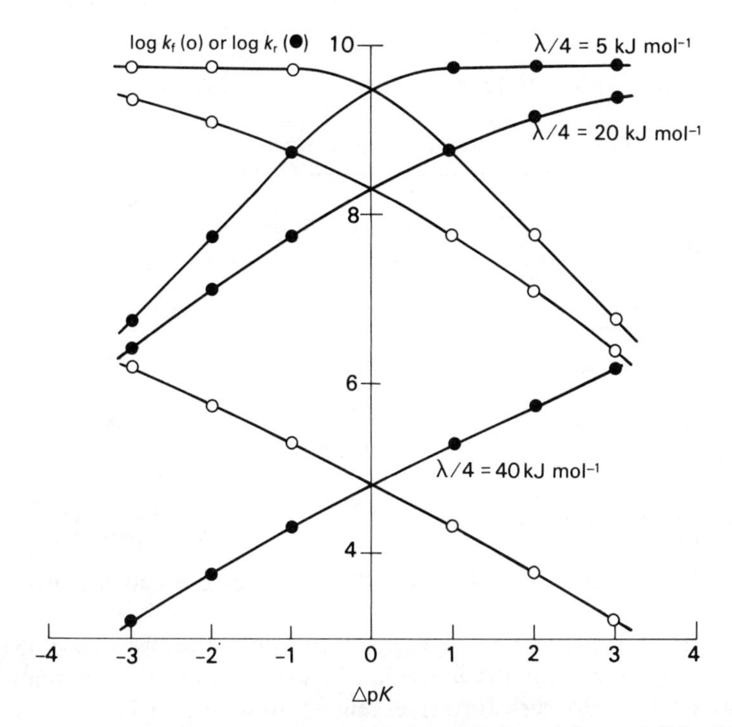

FIG. 4 Predictions of Marcus Theory for the variation of forward (k_f) and reverse (k_r) rate coefficients with ΔpK for reactions with different intrinsic barriers ($\lambda/4$)

solvent reorganisation which accompanies the proton transfer. The question as to whether solvent reorganisation is concerted with the motion of the proton along the hydrogen bond is an interesting one and has been examined theoretically (Kurz and Kurz, 1972). A reasonable alternative to concerted proton transfer and solvent reorganisation is a stepwise process in which solvent reorganisation to a configuration appropriate to the transition state occurs before proton transfer. In some cases the proton transfer will occur through solvent molecules intervening between the acid and base species. The direct proton transfer (25) will compete with the indirect process (26) and in many cases the relative contributions of the two paths have been determined

$$XH^+ + X \underset{}{\overset{k_d}{\rightleftharpoons}} X + HX^+ \tag{25}$$

$$XH^+ + \underset{\substack{|\\H}}{O}{-}H + X \overset{k_i}{\rightleftharpoons} X + H{-}\underset{\substack{|\\H}}{O} + XH^+ \tag{26}$$

from nmr spectroscopic measurements. Proton transfer involving amines has been particularly well studied (Grunwald and Ku, 1968; Grunwald and

TABLE 2

Rate coefficients for direct (k_d) and indirect (k_i) proton transfer for amines[a,b]

X	$10^{-8}k_d/dm^3mol^{-1}s^{-1}$	$10^{-8}k_i/dm^3mol^{-1}s^{-1}$
NH_3	12.7 ± 0.5	0.50 ± 0.04
CH_3NH_2	4.4 ± 0.4	5.8 ± 0.4
$(CH_3)_2NH$	0.5 ± 0.4	9.9 ± 0.3
$(CH_3)_3N$	0.0 ± 0.3	3.7 ± 0.3

[a] Reaction:

$$XH^+ + X \underset{}{\overset{k_d}{\rightleftharpoons}} X + HX^+$$

$$XH^+ + \underset{\displaystyle H}{\overset{|}{O}} - H + X \overset{k_i}{\rightleftharpoons} X + H - \underset{\displaystyle H}{\overset{|}{O}} + HX^+$$

[b] Data from Grunwald and Ku, 1968

Ralph, 1971; Grunwald and Eustace, 1975) and some results are given in Table 2.

The important question as to whether the proton-transfer step in (8) is kinetically significant for the ionisation of oxygen and nitrogen acids in aqueous solution has been further examined by studies of kinetic solvent isotope effects. From the analysis given in Fig. 4 it was tentatively concluded that for the ionisation of acetic acid in the presence of carboxylate bases, proton transfer was partially rate limiting for reactions for which ΔpK is *ca* 0. Rate coefficients for proton transfer between 4-nitrophenol and imidazole (27) have been measured (Chiang *et al.*, 1982) using the laser temperature-jump technique (Bannister *et al.*, 1984). The reaction occurs with half-lives as short as 0.2 μs and is too rapid for study with commercial temperature-jump equipment which uses Joule heating (Caldin, 1975). Proton transfer

$$NO_2-C_6H_4-OH + N\text{(imidazole)}NH \overset{k}{\rightleftharpoons} NO_2-C_6H_4-O^- + HN^+\text{(imidazole)}NH \qquad (27)$$

from 4-nitrophenol (pK 7.16) to imidazole (6.99) is close to the region $\Delta pK \sim 0$ and occurs with rate coefficients $k(H_2O) = (2.71 \pm 0.29) \times 10^8\ dm^3\ mol^{-1}\ s^{-1}$ and $k(D_2O) = (0.98 \pm 0.04) \times 10^8\ dm^3\ mol^{-1}\ s^{-1}$. The kinetic solvent isotope effect $k(H_2O)/k(D_2O)$ of 2.8 ± 0.3 is larger than that expected for a process controlled by diffusion. However, the primary kinetic hydrogen-isotope effect for proton removals from oxygen and nitrogen through transition states in which the proton is half-transferred could be as large as 11 and 9 respectively (Bell, 1973; More O'Ferrall, 1975) and the somewhat lower observed value may suggest that proton transfer is only partially limiting. Away from $\Delta pK \sim 0$ where the contribution from the proton-transfer step is much smaller, a lower isotope effect would be expected. This interesting prediction has not yet been tested for the reaction of 4-nitrophenol but kinetic solvent isotope effects for proton transfer from a different oxygen acid have been measured (Bergman *et al.*, 1978) for reactions over a range $-2 < \Delta pK < 6$. The proton transfer in (28) involving reaction of the adduct of 4-methoxybenzaldehyde and methoxylamine with a series of carboxylic acids and protonated amines (HA) of varying pK

$$CH_3O\overset{+}{N}H_2\text{—}\underset{H}{\overset{Ph}{C}}\text{—}O^- + HA \underset{}{\overset{k}{\rightleftarrows}} CH_3O\overset{+}{N}H_2\text{—}\underset{H}{\overset{Ph}{C}}\text{—}OH + A^- \qquad (28)$$

was studied in H_2O and D_2O solvents and the results are given in Fig. 5. The value of the kinetic solvent isotope effect passes through a maximum of $k(H_2O)/k(D_2O)$ *ca* 2.8 for reactions in the region of $\Delta pK \sim 0$. It has been proposed that the largest contribution to this overall isotope effect is made by a primary kinetic isotope effect, and secondary and medium kinetic isotope

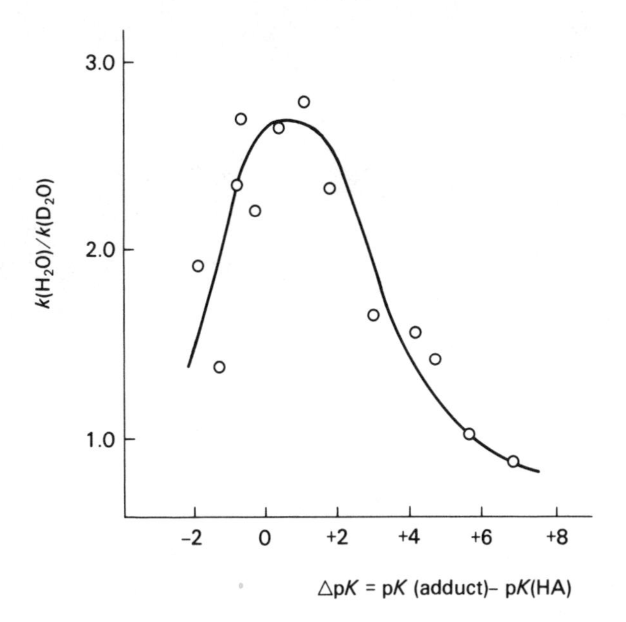

FIG. 5 Variation in kinetic solvent isotope effect, $k(H_2O)/k(D_2O)$, for the normal proton-transfer reaction (28)

effects are thought to be small and roughly independent of Δp*K* (Bergman *et al.*, 1978). A maximum in the value of the kinetic solvent isotope effect has also been observed for the proton transfer given in (29) as the base strength of the carboxylate ion catalyst was varied from weak to strong bases (Fischer *et al.*, 1980). The maximum $k(H_2O)/k(D_2O) \simeq 2.8$ occurred for proton

$$HO\overset{+}{N}H_2\text{—}C(Ph)_2\text{—}NMe_2 + RCO_2^- \underset{}{\overset{k}{\rightleftharpoons}} HONH\text{—}C(Ph)_2\text{—}NMe_2 + RCO_2H \quad (29)$$

transfer to bases such that $\Delta pK \sim 0$. A somewhat larger maximum value of $k(H_2O)/k(D_2O) \simeq 3.8$ at $\Delta pK \sim 0$ has been observed for the reaction shown in (30) with a series of protonated amines (HA) (Cox and Jencks,

$$CH_3O\overset{+}{N}H_2\text{—}C(OPh)(CH_3)\text{—}O^- + HA \overset{k}{\rightleftharpoons} CH_3O\overset{+}{N}H_2\text{—}C(OPh)(CH_3)\text{—}O\text{—}H + A^- \quad (30)$$

1978), and these data are given in Fig. 6. The observed maxima in kinetic solvent isotope effects in proton transfers involving oxygen and nitrogen

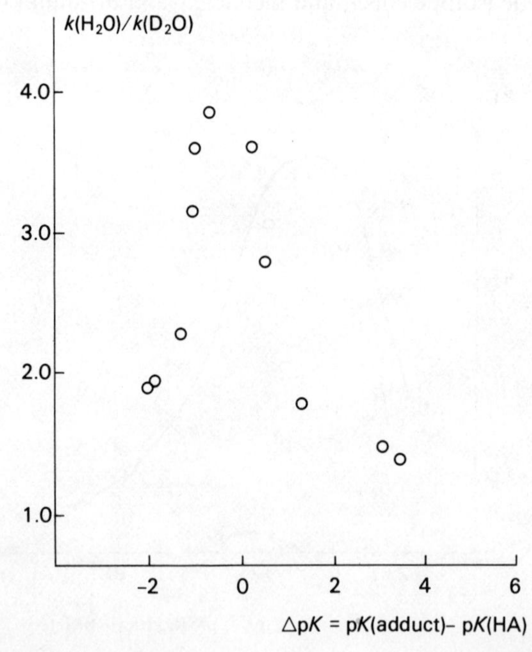

FIG. 6 Variation in kinetic solvent isotope effect, $k(H_2O)/k(D_2O)$, for the normal proton-transfer reaction (30)

acids can be explained (Bergman *et al.*, 1978; Cox and Jencks, 1978; Fischer *et al.*, 1980) by assuming a constant value for the isotope effect on the proton-transfer step in mechanism (8) of magnitude in the range $k_p(H_2O)/k_p(D_2O)$ 3 to 5. The overall isotope effect varies as ΔpK changes from a negative value to zero and to a positive value because the rate-limiting step changes from an encounter step, to partially rate-limiting proton transfer, and then again to an encounter step. There is no evidence to show whether the isotope effect on the proton-transfer step varies with ΔpK. An alternative explanation for the observed maximum in the kinetic solvent isotope effect supposes that solvation changes accompanying the proton transfer, and which may vary with ΔpK, could be important (Fischer *et al.*, 1980). The sharp low maxima that are observed for proton transfer from nitrogen or oxygen are in contrast to the broad and substantial maxima observed in isotope effects for proton transfer from carbon (More O'Ferrall, 1975). However, even in proton transfer from carbon the true explanation for the isotope effect maximum remains uncertain (Bell, 1973, 1981; More O'Ferrall, 1975).

3 Proton transfer along hydrogen bonds

In Section 2, proton transfer between an acid (HA) and base (B^-) in aqueous solution was assumed to take place in three steps with the actual proton transfer occurring along an intermolecular hydrogen bond between the reaction partners (8). A value of the Gibbs energy of activation of *ca* 20 kJ mol^{-1} corresponding to a rate coefficient of *ca* $1.9 \times 10^9\ s^{-1}$ was proposed for the proton transfer when the p*K*-values of HA and HB were similar. In this section we will examine data which provide information about proton transfer along inter- and intramolecular hydrogen bonds. Comprehensive reviews of hydrogen bonding are available (Joesten and Schaad, 1974; Schuster *et al.*, 1976a,b,c; Ratajczak and Orville-Thomas, 1980, 1981, 1982) and attention will be restricted to a consideration of the potential energy functions governing the movement of the bridging proton and to the rate at which a proton is transferred along an inter- or intramolecular hydrogen bond. Although this review is largely concerned with proton transfer in aqueous media, relevant results from studies in gaseous, solid, and non-aqueous solvents are referred to in this section.

POTENTIAL ENERGY FUNCTIONS

Recently, new experimental techniques have been added to the existing methods of obtaining information about the potential energy function of a proton in a hydrogen bond X$\cdots$H$\cdots$Y. Four types of potential function are

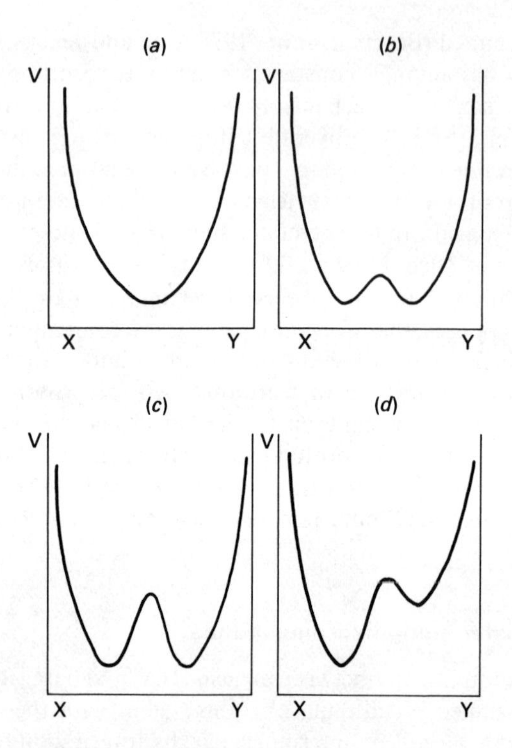

FIG. 7 Potential functions for protons in hydrogen bonds

drawn in Fig. 7. In Fig. 7*a* the function describes a proton which vibrates about an average position at the mid-point between the two atoms X and Y in a single minimum potential well. In Figs 7*b*, 7*c*, and 7*d* the potential well is a double minimum with a central barrier to the motion of the proton. In general the potential energy (V) is given by (31), in which x is the distance from the central maximum and a, b, and c are constants for a particular hydrogen bond (Laane, 1970; Wood, 1973; Kreevoy and Liang, 1980). For homoconjugates (X = Y), b is zero and a symmetrical function is obtained

$$V = ax^2 + bx^3 + cx^4 \tag{31}$$

for which $a < 0$ gives a double minimum and $a \geqslant 0$ gives a single minimum. For heteroconjugates ($X \neq Y$), b is positive and the asymmetric function of Fig. 7*d* is obtained. The positions of the minima correspond to average positions of the proton, one minimum near X and the other near Y. These situations can be represented by the equilibrium (32) which may apply to inter- and intramolecular hydrogen bonds. In a particular case it is of interest

$$\mathrm{X-H\cdots{}^{-}Y \rightleftharpoons X^{-}\cdots H-Y} \tag{32}$$

to establish which is the predominant species and at what rate the interconversion occurs. The former is determined by the relative heights of the minima and the latter by the height of the barrier between them (unless the proton can tunnel through the barrier). If the rate of equilibration is rapid compared with the timescale of the observation, the properties of the system as measured by the observation will be an average of the properties of the separate species in (32) or (33), relating to the corresponding equilibrium for

$$X-H\cdots{}^{-}X \rightleftharpoons X^{-}\cdots H-X \qquad (33)$$

a homoconjugate complex. In this case the system may appear to be governed by a single-minimum potential function. Several experimental techniques are now available to distinguish between these various possibilities and in some cases to locate the proton in the hydrogen bond. These methods do not always provide consistent answers and different results are sometimes obtained for the same hydrogen bond in gaseous, liquid, or solid phases. Some of the techniques provide information about the rate at which the equilibria (32) and (33) are established. In this section the most important techniques will be briefly described and the results which have been obtained for inter- and intramolecularly hydrogen-bonded species will be discussed.

EXPERIMENTAL TECHNIQUES AND RESULTS

X-ray and neutron diffraction

The location of protons in intermolecularly hydrogen-bonded carboxylate-carboxylic acid complexes, for example sodium hydrogen bis(acetate) [1],

$$\left[CH_3-\overset{\overset{O}{\|}}{C}-O\cdots H\cdots O-\overset{\overset{O}{\|}}{C}-CH_3\right]^{-}$$

[1]

has been widely investigated in the solid state by X-ray and neutron-diffraction methods. The oxygen-to-oxygen distance in sodium hydrogen bis(acetate) was found to be 244.5 pm, with the proton placed centrally (Barrow *et al.*, 1975; Stevens *et al.*, 1977; Muir and Speakman, 1979). In potassium hydrogen bis(acetate), however, the hydrogen bond may be unsymmetrical and, if this is correct, the structure of the solids must be partly determined by the nature of the cation (Currie, 1972). In potassium hydrogen bis(trifluoroacetate) the hydrogen bond is thought to be linear and symmetrical (Golic and Speakman, 1965; Macdonald *et al.*, 1972). The $O\cdots H\cdots O$ distance is estimated as 243.7 pm with a root mean square amplitude of vibration of the proton of 17 pm. It is the same in potassium deuterium bis(trifluoroacetate). In potassium hydrogen bis(phenylacetate) the distance is 245.4 pm with a root mean square amplitude of 18 pm for the proton (Bacon *et al.*, 1977). These results mean that the proton may be vibrating in a single minimum potential or in a double minimum potential well with small separation of the minima between which the proton can move. The results of infra-red spectroscopic studies are apparently compatible with a single-minimum potential (Novak, 1974), although a more recent interpretation of infra-red and nmr studies favours a double-minimum potential well for the proton in solutions of carboxylate–carboxylic acid complexes (Kreevoy and Liang, 1980).

In the hydrogen bis(quinuclidone) cation [2] the N···H···N distance is short (263.5 pm) compared with the van der Waals contact distance (*ca* 300 pm) and the proton is placed centrally (Rozière *et al.*, 1982). However, the N···H···N hydrogen bond in hydrogen bis(9-ethylguanine) [3] (bond length 263.7 pm) is thought to be asymmetric (Mandel and Marsh, 1975).

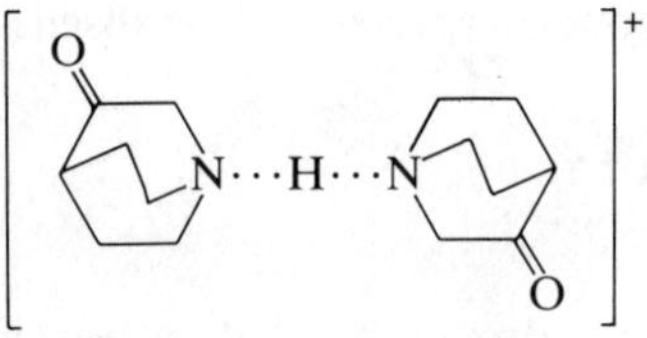

[2]

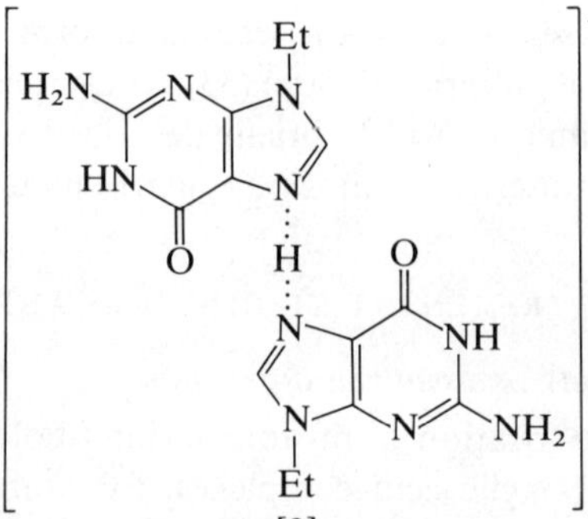

[3]

There have been numerous X-ray and neutron diffraction studies of the structure of the enol form of β-diketones in the solid state. The results are not always helpful in reaching conclusions about the possibility of a single- or a double-minimum potential function for the intramolecularly hydrogen-bonded proton in these molecules. In an early X-ray crystal structure determination for the enol form of 1,3-diphenyl-1,3-propanedione [4][1] it was concluded (Williams, 1966) that the hydrogen bond was asymmetric. However, the difference (2.8 pm) between the two C—C bond lengths and between the two C—O bond lengths (2.5 pm) is similar to the experimental uncertainty on each length. The data obtained by Hollander (Hollander *et al.*, 1973) which are given in structure [5] are even closer to those expected for a symmetrical structure. However, from the results of a neutron diffraction study, the conclusion has been reached that the proton is placed asymmetrically and that the difference between the two C—O bond lengths and

[1] All bond lengths are given in pm.

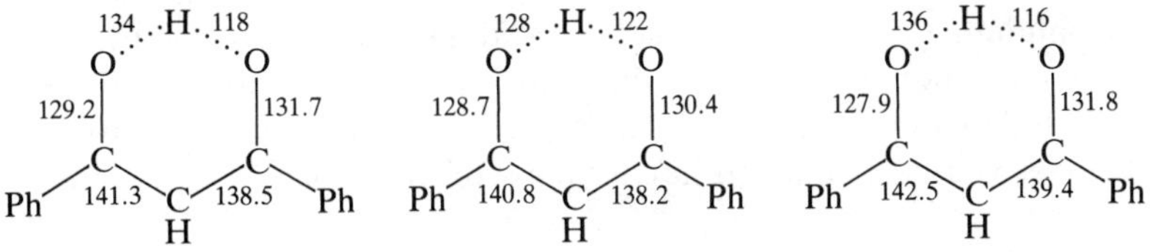

[4] [5] [6]

between the two C—C bond lengths is significant; thus a unique enol tautomer with bond lengths as shown in [6] was proposed (Jones, 1976). It was argued that the data were not compatible with the presence in the crystal of a mixture of two identical enol tautomers as in (34), and an average structure arising from a particular packing of the two tautomers. However,

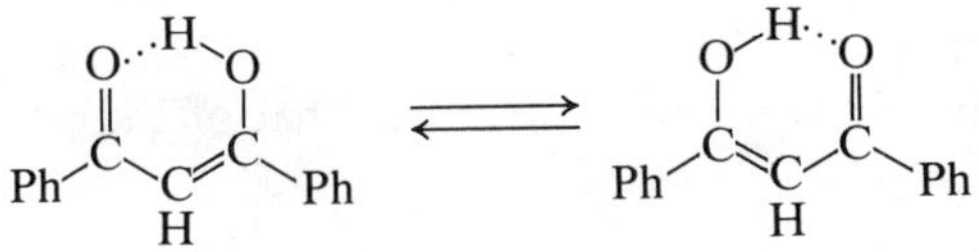

(34)

the asymmetric structure which is found may be a result of molecular packing forces within the solid. Symmetrical structures with equal C—C bond lengths within experimental error were proposed from crystal structures of the enol forms of 1,3-bis(3-bromophenyl)propane-1,3-dione (Williams *et al.*, 1962), 1,3-bis(3-chlorophenyl)propane-1,3-dione (Engebretson and Rundle, 1964) and 3,4-diacetylhexane-2,5-dione (Williams, 1966). Symmetrical structures [7] and [8] were proposed for the enol of pentane-2,4-dione (acetylacetone) in the vapour phase from electron-diffraction studies (Lowrey *et al.*, 1971; Andreassen and Bauer, 1972), whereas the unsymmetrical structure [9] was proposed from an X-ray determination of the enol of pentane-2,4-dione contained in the crystal lattice of a drug complex (Camerman *et al.*, 1983). The evidence from a variety of structural determinations of

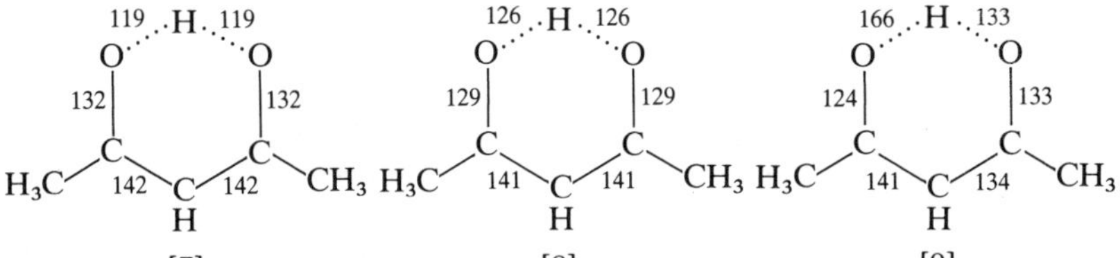

[7] [8] [9]

the enol forms of β-diketones has been summarised by Emsley (1984) who has concluded that the intramolecular hydrogen bond is short, non-centred, and non-linear. The potential function for the proton is accordingly of the double-minimum type, having well-separated minima with an energy barrier between them not much higher than the ground-state vibrational levels. Other techniques for investigating the potential function of the enol forms of β-diketones will be mentioned in later parts of this review.

Infra-red, microwave, and X-ray photoelectron spectroscopy

Infra-red and ultra-violet spectroscopy has been widely used for investigating the structure of intermolecularly hydrogen-bonded complexes in the solid state (Novak, 1974) and in solution (Zundel, 1976, 1978; Clements *et al.*, 1971a,b,c; Pawlak *et al.*, 1984). By analysing the infra-red spectra of equimolar liquid mixtures of amines with formic or acetic acid, the relative importance of structures [10] and [11] was estimated (Lindemann and Zundel, 1977). It was proposed that [10] and [11] make equal contributions to the observed structure of the complex when the pK_a-value of the carboxylic acid is approximately two units lower than that of the protonated amine.

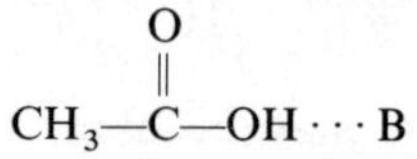

$$CH_3{-}C({=}O){-}OH \cdots B$$

[10]

$$CH_3{-}C({=}O){-}O^- \cdots HB^+$$

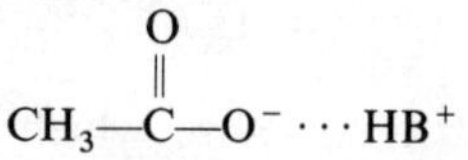

[11]

Hence for this complex the proton is held roughly equally between the carboxylate and amine bases in a double-minimum potential well.

Structures for the hydrogen-bonded species [12], [13], and [14] have been suggested on the basis of their infra-red spectra in solution (Clements *et al.*,

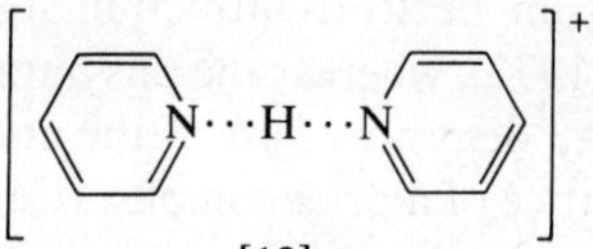

$$[C_5H_5N \cdots H \cdots NC_5H_5]^+$$

[12]

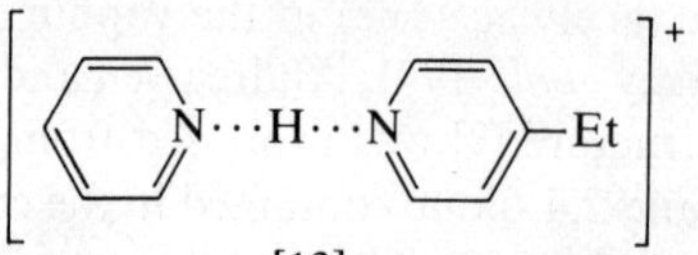

$$[C_5H_5N \cdots H \cdots NC_5H_4{-}Et]^+$$

[13]

$$[C_5H_5N \cdots H \cdots NC_5H_2(Me)_3]^+$$

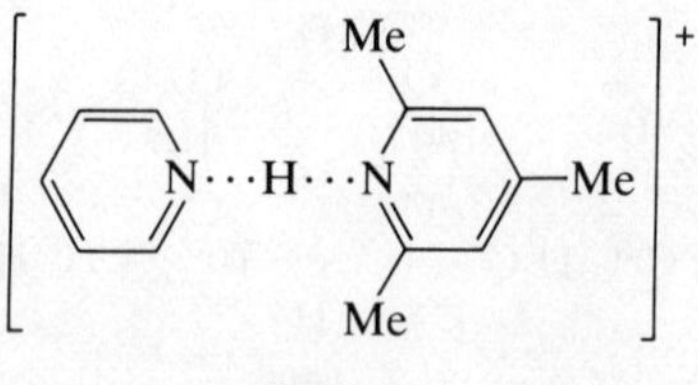

[14]

1971a,b,c). For homoconjugate complexes such as [12] or for heteroconjugate complexes like [13] between partners of similar basicities, it is thought that the proton moves in a double-minimum potential well with a low central barrier. In complexes between bases which differ in p*K* by several units, such as [14] an unsymmetrical single-minimum potential is proposed. As the base is varied and the p*K*-difference changes, this is thought to change smoothly from the symmetrical double minimum to the unsymmetrical single minimum.

Microwave rotational spectra of intramolecularly hydrogen-bonded species in the gas phase have been analysed to provide details of the symmetry of the species and thereby to determine whether the potential energy is best represented by a single- or double-minimum function. The spectrum of 6-hydroxy-2-formylfulvene is compatible with a structure of C_{2v} symmetry in which the proton is centred in the hydrogen bond [15] or with a rapidly

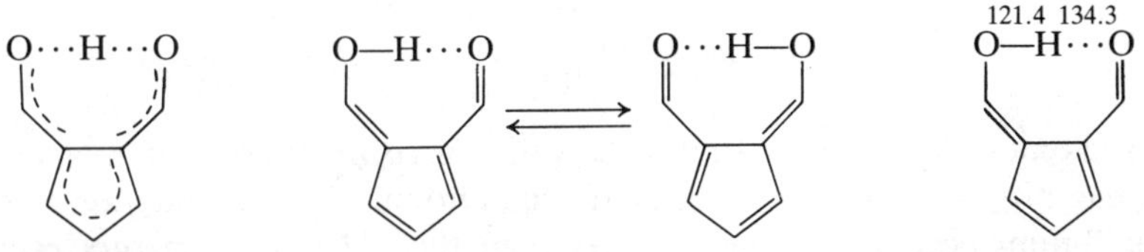

121.4 134.3

[15] [16] [17]

interconverting pair of isomers [16] each with C_s symmetry (Pickett, 1973). If a pair of isomers is present the rate coefficient for equilibration must exceed $2 \times 10^{12} s^{-1}$. The infra-red spectrum was interpreted in terms of two rapidly interconverting tautomers whereas on the much longer timescale of an nmr observation, the molecule appears to possess C_{2v} symmetry (Hafner *et al.*, 1964). X-ray and neutron diffraction studies of the solid give a structure with an unsymmetrical hydrogen bond (Fuess and Lindner, 1975), as in [17].

Detailed analysis of the microwave spectrum of malondialdehyde (Baughcum *et al.*, 1981, 1984) has led to the conclusion that the molecule is best represented by a rapid tautomeric equilibrium (35) of enol forms. The results

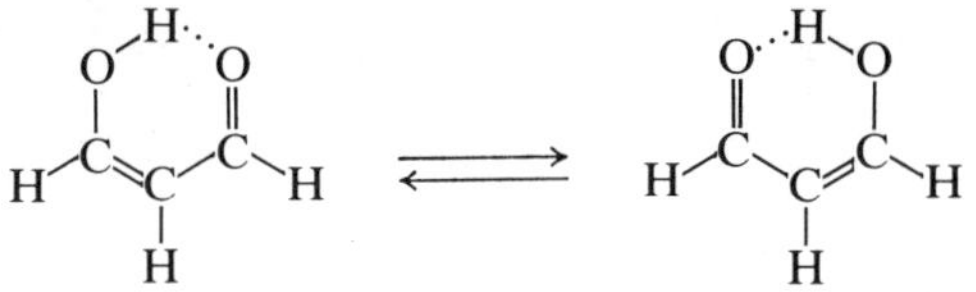

(35)

of quantum mechanical calculations (Karlström *et al.*, 1976; Bouma *et al.*, 1978; Bicerano *et al.*, 1983) support this conclusion. The function describing the proton motion is therefore of the double-minimum type and the proton is thought to tunnel through the central energy barrier with a period of oscillation of 0.8×10^{-12}s. Proton tunnelling through a central barrier is considered to be a feasible process for tautomeric interconversion of other inter- or intramolecularly hydrogen-bonded species (de la Vega, 1982). Nmr data for malondialdehyde in a 50% (v/v) mixture of $CFCl_3$—CD_2Cl_2

indicate that on the nmr timescale, equilibration between the tautomers in (35) occurs rapidly (Brown *et al.*, 1979). This means that if the equilibrium occurs through a transition state which resembles the C_{2v} form [18] the difference in energy between the C_s and C_{2v} tautomers is less than *ca* 25 kJ mol^{-1}. The use of nmr spectroscopy in studies of this type will be discussed more fully in the next part of Section 3.

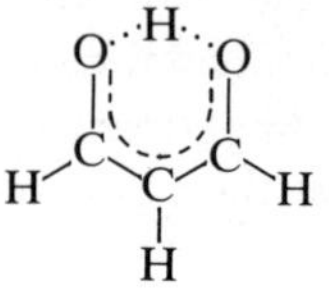

[18]

The timescale of a microwave observation is *ca* 10^{-12}s so that an average of the properties of the species in equilibrium (35) is obtained if the equilibrium occurs in a time shorter than this. The X-ray photoelectron spectra of intramolecularly hydrogen-bonded species in the gas phase have been studied in an attempt to obtain an instantaneous picture of the structure of these molecules. In this technique the ionisation of core electrons which occurs within 10^{-16}s is observed. For malondialdehyde, 6-hydroxy-2-formylfulvene, 2-hydroxy-1,1,1,5,5,5-hexafluoropent-2-ene-4-one, 9-hydroxyphenalenone [19], and tropolone [20], two peaks are observed in the O_{1s} region of the photoelectron spectrum (Brown *et al.*, 1979). If these molecules existed in the C_{2v} form with a symmetrical hydrogen bond and equivalent oxygen

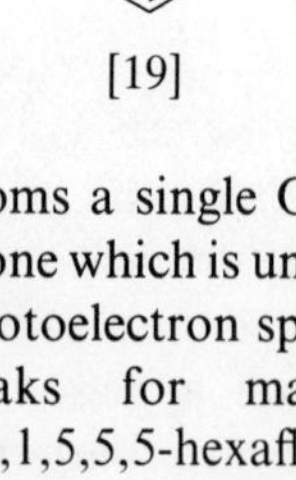

[19] 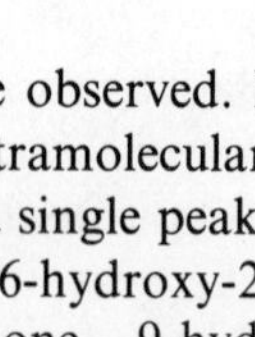 [20] 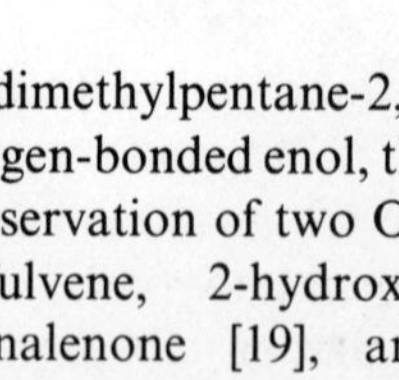 [21]

atoms a single O_{1s} peak would be observed. For 3,3-dimethylpentane-2,4-dione which is unable to form an intramolecularly hydrogen-bonded enol, the photoelectron spectrum contains a single peak. The observation of two O_{1s} peaks for malondialdehyde, 6-hydroxy-2-formylfulvene, 2-hydroxy-1,1,1,5,5,5-hexafluoropent-2-ene-4-one, 9-hydroxyphenalenone [19], and tropolone [20] is compatible with the presence of oxygen atoms made non-equivalent by the unsymmetrical hydrogen bond. The photoelectron spectrum of protonated 1,8-bis-(dimethylamino)naphthalene [21] contains two peaks in the region expected for the ionisation of N_{1s} electrons and this is compatible with the hydrogen-bonded proton being located closer to one nitrogen atom (Haselbach *et al.*, 1972). The nmr spectrum of protonated 1,8-bis(dimethylamino)naphthalene shows that, on the longer timescale of such an observation, the equilibrium between two C_s tautomers is relatively rapid. (Alder *et al.*, 1968).

Nmr spectra of hydrogen-bonded species

If the equilibrium in (36) is established rapidly on the nmr timescale, the observed nmr spectrum will correspond to an average of the spectra of the

$$X{-}H\cdots Y^- \rightleftharpoons X^-\cdots H{-}Y \qquad (36)$$

individual species in the equilibrium. On the basis of the spectrum it is then not possible to determine whether the hydrogen bond is best described by a double-minimum potential function with rapid interconversion of two species or whether a single species with a single-minimum potential function is a more appropriate description. For example, in solution or in a liquid melt, the proton-decoupled ^{13}C spectrum of tropolone consists of four lines (Jackman *et al.*, 1980). In the absence of other evidence this could be explained by a rapid equilibration between the tautomers in equation (37) or by a single species [22] with a symmetrical hydrogen bond. Quite recently two

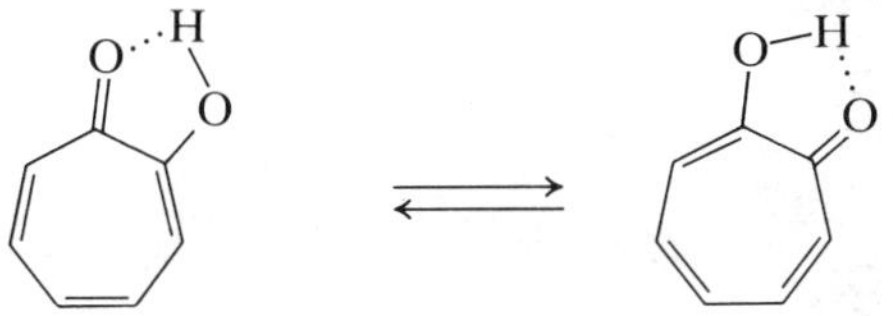

(37)

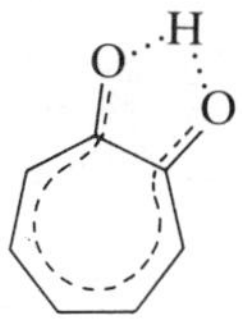

[22]

nmr methods have been developed which will distinguish between these possibilities. The methods have been applied to inter- and intramolecularly hydrogen-bonded species and depend upon a difference in the properties of a hydrogen-bonded proton in a single-minimum potential well and the time-averaged properties of a proton in a double-minimum potential. In one procedure the hydrogen-isotope effect (H, D, T) on the chemical shift of the hydrogen-bonded proton is measured. In the other procedure ^{13}C and ^{2}H

TABLE 3

Chemical shifts of protons in hydrogen-bonded complexes

Complex	Solvent	δ/ppm	$\Delta\delta = \delta^{H} - \delta^{D}$	Reference
$[O_2N-C_6H_4-O\cdots H\cdots O-C_6H_4-NO_2]^-$	CH_3CN	16.9		Kreevoy and Liang, 1980
$[CF_3-C(=O)-O\cdots H\cdots O-C(=O)-CF_3]^-$	CH_3CN	19.0		Kreevoy and Liang, 1980
$[H-C(=O)-O\cdots H\cdots O-C(=O)-H]^-$	$H_2O/Na^+HCO_2^-$	14.1	+0.64	Fenn and Spinner, 1984
$[F\cdots H\cdots F]^-$	CH_3CN H_2O	16.4 16.3	−0.30	Gunnarsson *et al.*, 1976 Fujiwara and Martin, 1974
$[Me_2N\cdots H\cdots NMe_2]^+$ (1,8-naphthalene)	CF_3COOH	19.5	+0.66	Alder *et al.*, 1968 Altman *et al.*, 1978

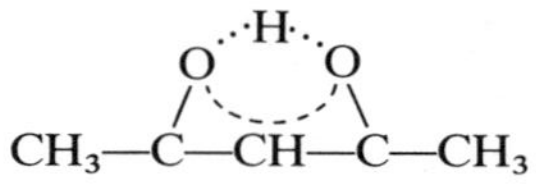 $CH_3—C(O)—CH—C(O)—CH_3$ (O···H···O)	C_6D_6 neat liquid	16.1	+0.61 +0.5	Altman *et al.*, 1978 Egan *et al.*, 1977
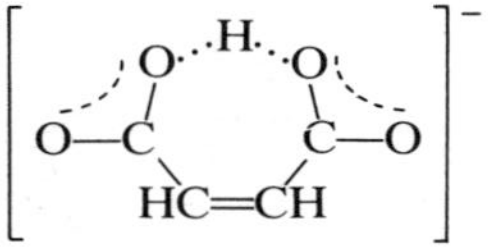 $[O—C(O)—HC{=}CH—C(O)—O]^-$ (O···H···O)	CH_2Cl_2	20.5	−0.03	Altman *et al.*, 1978
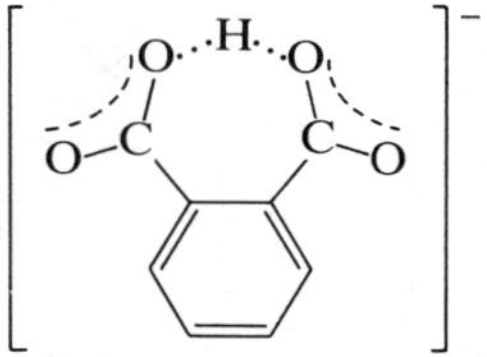 $[O—C(O)—C_6H_4—C(O)—O]^-$ (O···H···O)	CH_2Cl_2	21.0	−0.15	Altman *et al.*, 1978
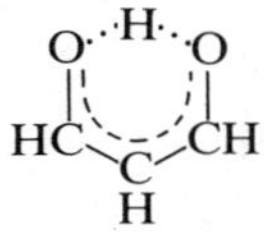 $O{=}HC—CH{=}CH—O$ (HC, C H, CH; O···H···O)	50%(v/v)CF_3Cl-CD_2Cl_2	16.4 14.0	 +0.42	Brown *et al.*, 1979 Altman *et al.*, 1978

spin-lattice relaxation times are used to determine deuterium quadrupole coupling constants. The results from these approaches will be described in detail. In the case of tropolone (Jackman *et al.*, 1980), the results were compatible with a double-minimum potential for the hydrogen bond and with a rapid equilibration of the tautomers in (37). Interestingly in the solid state, seven lines are observed in the ^{13}C spectrum of tropolone obtained by the cross polarisation/magic angle spinning technique, and this is taken to mean that equilibrium (37) is slow on the nmr timescale under these conditions (Szeverenyi *et al.*, 1982, 1983). In explaining this result, it was postulated that the rate of proton transfer was determined by a slow reorientation of the molecule in the crystal lattice.

In using the measurement of an isotope effect on the chemical shift of a hydrogen-bonded proton to distinguish between a single- and double-minimum potential (Gunnarsson *et al.*, 1976; Altman *et al.*, 1978), the chemical shifts of the protium, deuterium, and tritium compounds are measured. Since, in the vast majority of cases, equilibrium (36) is established rapidly for hydrogen-bonded species in solution, a single peak in the nmr spectrum is observed for the hydrogen-bonded proton in the absence of coupling with other protons. The signal is usually well downfield and often quite broad. Typical shifts for protons in inter- and intramolecular hydrogen bonds are given in Table 3. The signal for the proton in the intermolecular complex formed from 4-nitrophenol and 4-nitrophenolate ion occurs at δ 16.9 compared with δ 8.1 observed for 4-nitrophenol (Kreevoy and Liang, 1980). In the hydrogen bis(trifluoroacetate) complex δ 19.0 is observed for the hydrogen-bonded proton compared with δ 11.2 observed for trifluoroacetic acid under the same conditions (Kreevoy and Liang, 1980). The difference in chemical shift ($\Delta\delta = \delta^{H} - \delta^{D}$) of the protium and deuterium derivatives of various hydrogen-bonded complexes are given in Table 3 and the results vary from negative to positive values of $\Delta\delta$. The values of $\Delta\delta$ in Table 3 are probably uncertain to the extent of less than $\pm$ 0.03 ppm. Even more accurate methods for the measurement of isotope effects are now available (Evans, 1982; Saunders *et al.*, 1984). For compounds in which the proton is not hydrogen-bonded the $\Delta\delta$ values are usually close to zero; for benzyl alcohol $\Delta\delta = -0.02$ (Altman *et al.*, 1978) and for chloroform $\Delta\delta = +0.017$ (Evans, 1982). Three possible potential wells for a hydrogen-bonded species [23] are illustrated in Fig. 8, in which the abscissa (r) is the distance along a line between X and Y. Fig. 8*a* is thought to represent the situation for a weak hydrogen bond, Fig. 8*b* applies for shorter and stronger hydrogen bonds (where the atoms X and Y are closer together and the central barrier is

$$X \cdots H \cdots Y$$

[23]

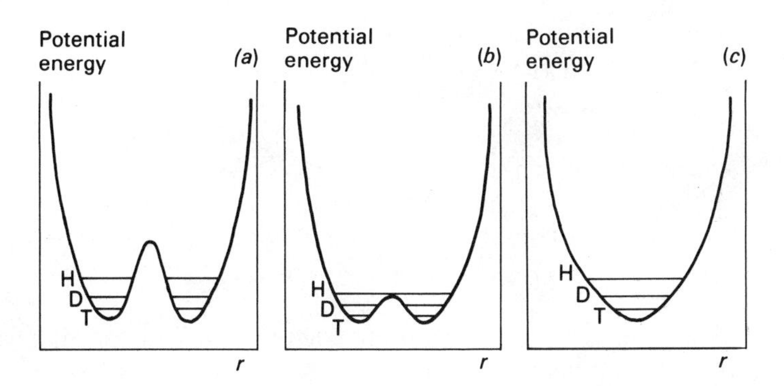

FIG. 8 Energy levels for hydrogen-bonded species X···H···Y, X···D···Y and X···T···Y

lower), and Fig. 8*c* represents the potential in a short and extremely strong hydrogen bond. In the case of Fig. 8*a*, the average position of the hydrogen isotopes in X···H···Y, X···D···Y, and X···T···Y will be similar, and similar chemical shifts for the isotopically different species would be expected ($\Delta\delta$ *ca* 0). In Fig. 8*b* the two halves of the potential function are strongly anharmonic and the light isotope will be closer to the mid-point between X and Y than are the heavier isotopes. The light isotope will therefore be most deshielded and the chemical shifts will be in the order $\delta^H > \delta^D > \delta^T$, so that $\Delta\delta > 0$. For a hydrogen bond for which the potential is a symmetrical single-minimum (Fig. 8*c*) the positions of the three isotopes are similar but because the amplitude of vibration of the light isotope will be largest, the proton will be least deshielded compared with the other hydrogen isotopes and $\Delta\delta$ will be close to zero or negative. Thus negative or near zero values of $\Delta\delta$ for strongly hydrogen-bonded species are taken as being diagnostic of single-minimum potential functions.

Of the complexes in Table 3 for which $\Delta\delta$ values have been determined, the hydrogen bonds in the bifluoride ion and in the phthalate ion are considered to have single-minimum potentials whereas the complexes giving large and positive $\Delta\delta$ values [hydrogen diformate ion, protonated 1,8-bis(dimethyl-amino)naphthalene, 2-hydroxypent-2-ene-4-one, and malondialdehyde] are probably best described by double-minimum potentials. Studies of the ^1H and ^{19}F nmr spectra of solid HF_2^- salts have led to the conclusion that the hydrogen bond (bond length 232.2 pm) is best represented either by a single-minimum potential or by a double-minimum with the minima separated by a low central barrier and less than 18 pm apart (Ludman *et al.*, 1977; Cousseau *et al.*, 1977). The majority of the evidence appears to favour a single-minimum potential for bifluoride ion (Emsley, 1980).

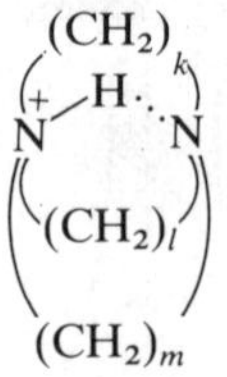

[24]

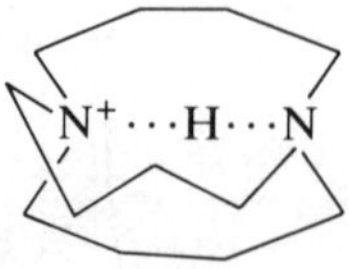

[25]

An interesting series of eleven protonated bicyclic amines [24] have been prepared with various ring sizes ($k = 4$ to 6, $l = 3$ to 5, $m = 2$ to 4) in which the geometry of the intramolecular N···H···N hydrogen bond is varied (Alder 1983; Alder *et al.*, 1983a,b). The nature of the hydrogen bonds was investigated by X-ray diffraction and by nmr and ir spectroscopy. The nmr spectra of the protonated amines in $CDCl_3$ showed a singlet for the hydrogen-bonded proton with chemical shift in the range δ 10.7 to 17.5 (Alder *et al.*, 1983a). The isotope effect on the chemical shift of eight of the protonated amines was measured and with one exception the values came in the range $\Delta\delta = +0.43$ to $+0.92$ as expected for moderately strong hydrogen bonds described by double-minimum potentials with relatively shallow minima. The exceptional value $\Delta\delta = +0.06$ was observed for in-protonated 1,6-diazabicyclo[4.4.4]tetradecane [25]. On the basis of this result in conjunction with the infra-red spectrum and crystal structure it was suggested that in this case the hydrogen bond was of the single-minimum type. Crystal structures of 1,6-diazabicyclo[4.4.4]tetradecane and the protonated amine show that protonation reduces the nitrogen to nitrogen distance from 281 to 253 pm and the N···H···N hydrogen bond length in the protonated amine is the shortest known (Alder *et al.*, 1983b; Schaefer and Marsh, 1984). The acid-base properties of these bicyclic amines will be discussed more fully in Section 5.

The Δδ value for pentane-2,4-dione in Table 3 is compatible with a double-minimum potential for the intramolecular hydrogen bond in the enol. This was also the conclusion reached from the microwave and photoelectron spectra. For unsymmetrically substituted diketones, for example 1,1,1-trifluoropentane-2,4-dione, measurement of ^{13}C chemical shifts has been used to estimate values for the equilibrium constant $K = [R]/[L]$ in (38)

$$[L] \rightleftharpoons [R] \qquad (38)$$

(Lazaar and Bauer, 1983). A single averaged ^{13}C signal was observed for the carbonyl carbon atoms and from the value of the chemical shift of this signal and the chemical shifts estimated for C═O and ═C—O—H from model compounds, the result $K = 0.42$ was calculated. It was also concluded that the residence time in the L or R forms was less than 10^{-3}s. For similar compounds microwave data (Baughcum *et al.*, 1981, 1984) indicate that the residence times are of the order of 10^{-12}s and this is of a similar magnitude to the result obtained from an analysis of line broadening in the infra-red and Raman spectra of the enol of pentane-2,4-dione (Cohen and Weiss, 1984).

The magnitude of deuteron quadrupole-coupling constants (DQCC) have also been used to determine whether the proton in an intramolecular hydrogen bond is best described by a single- or double-minimum potential. For pentane-2,4-dione in the pure liquid a double-minimum potential was deduced (Egan *et al.*, 1977), in agreement with the conclusion reached on the basis of the isotope effect on the chemical shift of the hydrogen-bonded proton (deuteron, triton) (Altman *et al.*, 1978) and with other data. The technique has been applied to the intramolecularly hydrogen-bonded species [26] to [33] (Jackman *et al.*, 1980). The ^{1}H and ^{13}C nmr spectra of these

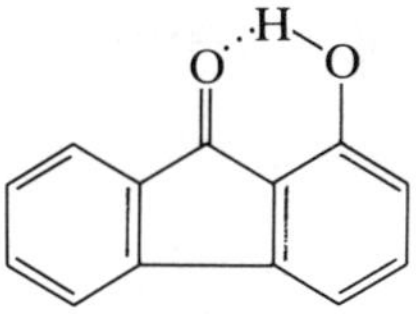

[26] 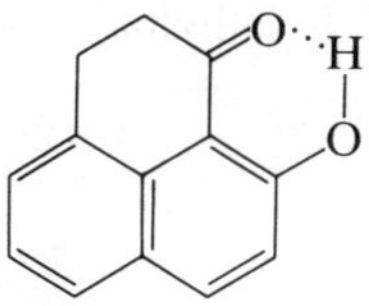 [27] 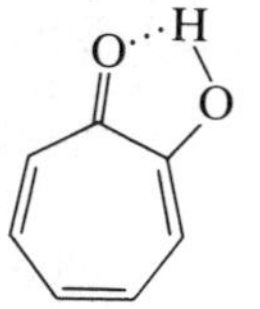 [28]

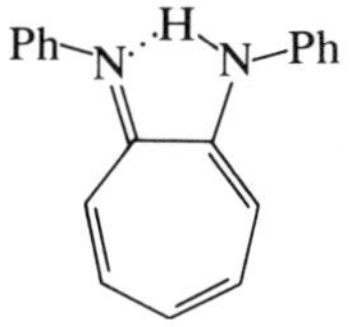

[29] 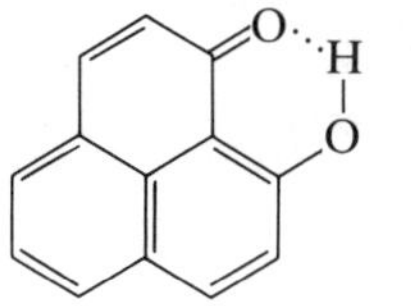 [30] 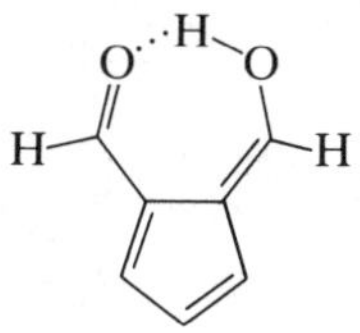 [31]

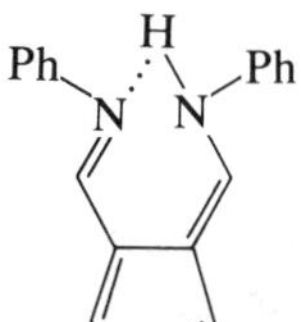

[32] 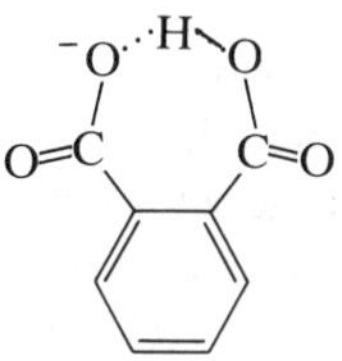 [33]

species in organic solvents are compatible with structures with C_{2v} symmetry or with rapidly equilibrating tautomers with C_s symmetry. In all cases, except for the phthalate ion [33] the magnitude of the DQCC provided evidence for a double-minimum potential. A similar conclusion was reached previously for tropolone [28] and for 6-hydroxy-2-formylfulvene [31] on the basis of the photoelectron spectra (Brown *et al.*, 1979). For the phthalate ion [33] the magnitude of the DQCC was interpreted in terms of a single-minimum potential which was also the conclusion reached from the negative value of the isotope effect ($\Delta\delta$) on the chemical shift of the hydrogen-bonded proton (deuteron) (Altman *et al.*, 1978).

In the majority of cases, interconversion of tautomers involving movement of a proton along a hydrogen bond is fast on the nmr timescale. However, in unusual cases slow interconversion has been found. For example, the ^1H nmr spectrum of 2-methylnaphthazarin [34] in methylene chloride-d_2 shows two hydroxyl signals (δ 12.41, 12.53) and a singlet for the ring protons at C(6) and C(7) (δ 7.19) and a quartet at δ 6.89 for the proton on C(3). The singlet and quartet were found to have areas in the ratio of 2 : 1. The signal for the methyl group appears as a doublet at δ 2.21. The spectrum is consistent (de la Vega *et al.*, 1982) with structure [34] and with relatively slow interconversion to [35]. For 2,6- and 2,7-dimethylnaphthazarins one signal is observed at

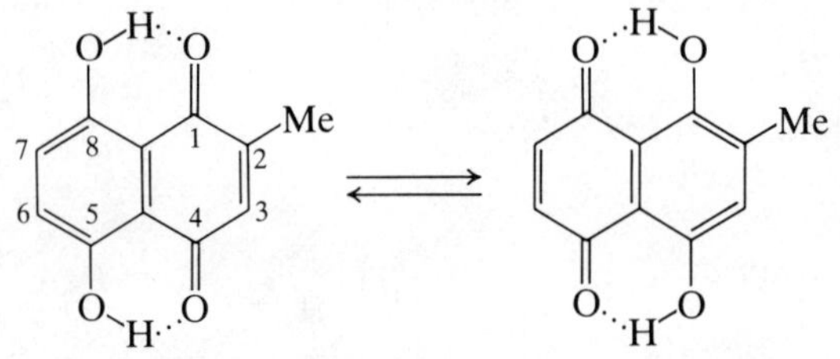

[34] ⇌ [35]

δ 6.96 for the ring protons and the methyl signals are singlets at δ 2.22 (for 2,6-dimethylnaphthazarin) and at 2.20 (for 2,7-dimethylnaphthazarin). In these cases rapid interconversion of the tautomers (39) occurs. It is thought that, for the symmetrical dimethylnaphthazarins and for naphthazarin, the proton can tunnel through the central barrier between the two minima of the potential well but lack of symmetry prevents tunnelling in the monomethyl derivative (de la Vega *et al.*, 1982; Hameka and de la Vega, 1984). Tunnelling would also be reduced if the energy of [35] were higher than the energy of [34] (Bell, 1973, 1981).

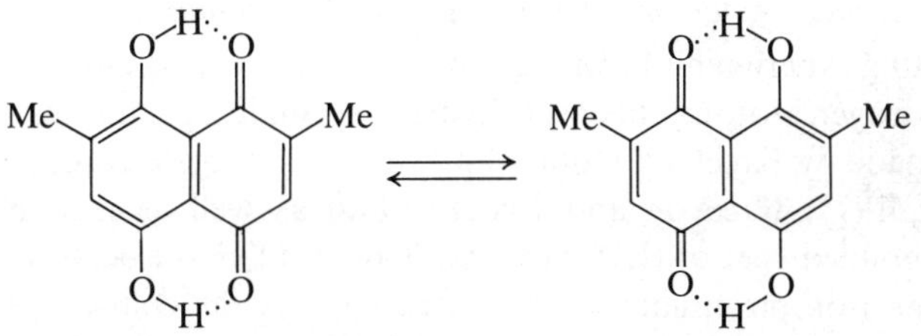

O H O Me Me ⇌ Me Me O H O (39)

Various nmr techniques have been used to investigate the intramolecular double proton transfer which occurs in the tautomerisation of meso-tetraphenylporphyrin (40) (Limbach *et al.*, 1982). The reaction has been studied (Storm and Teklu, 1972) by observation of the nmr signals due to the protons

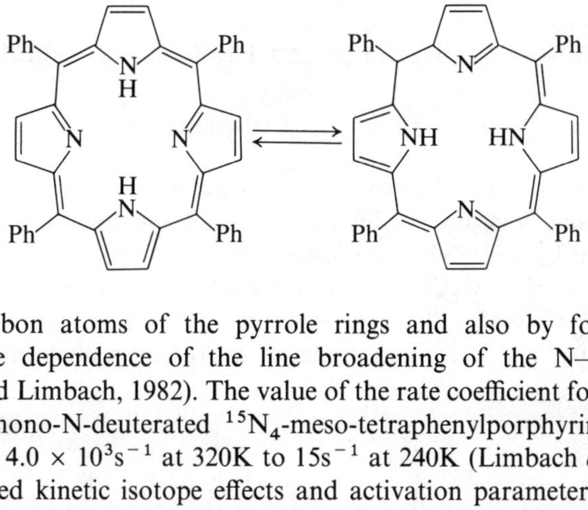

Ph Ph N H N N H N Ph Ph ⇌ Ph Ph N NH HN N Ph Ph (40)

on the carbon atoms of the pyrrole rings and also by following the temperature dependence of the line broadening of the N—H protons (Hennig and Limbach, 1982). The value of the rate coefficient for tautomerisation of mono-N-deuterated $^{15}N_4$-meso-tetraphenylporphyrin in $CDCl_3$ varies from $4.0 \times 10^3 s^{-1}$ at 320K to $15 s^{-1}$ at 240K (Limbach *et al.*, 1982). The observed kinetic isotope effects and activation parameters have been explained by a proton-tunnelling mechanism (Hennig and Limbach, 1984). The reaction has also been studied in the solid state (Limbach *et al.*, 1984). Other related reactions which can be studied by nmr spectroscopy include the tautomerisation of 2,5-dianilinobenzoquinone-1,4-dianil (41), for which a tunnelling mechanism is also thought to occur (Limbach *et al.*, 1982).

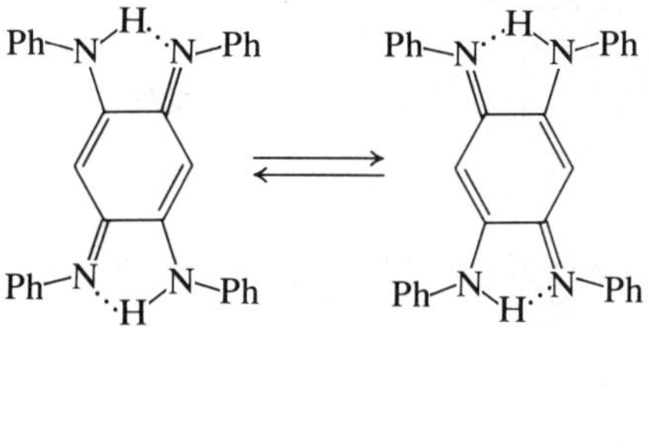

Ph—N H N—Ph Ph—N H N—Ph ⇌ Ph—N H N—Ph Ph—N H N—Ph (41)

Isotopic fractionation factors of hydrogen-bonded protons

An interesting experimental and theoretical study of the equilibrium distribution of hydrogen isotopes between hydrogen-bonded complexes and water has been made by Kreevoy. Homo- and hetero-conjugate complexes (Kreevoy *et al.*, 1977; Kreevoy and Liang, 1980) as well as intramolecularly hydrogen-bonded species (Kreevoy and Ridl, 1981) have been studied. The value of the isotopic fractionation factor ($\varphi_{A_1HA_2^-}$) as defined by (42) and (43) was obtained by combining values of the equilibrium constants obtained spectrophotometrically at 25°C for reactions (44), (45), and (46). The value

$$A_1HA_2^- + D(L_2O) \rightleftharpoons A_1DA_2^- + H(L_2O) \qquad (L = H \text{ or } D) \tag{42}$$

$$\Phi_{A_1HA_2^-} = ([A_1DA_2^-]/[A_1HA_2^-])\,(H/D)_{L_2O} \tag{43}$$

$$A_1HA_2^-(\text{soln}) + Ph_3COD(\text{soln}) \overset{CH_3CN}{\rightleftharpoons} A_1DA_2^-(\text{soln}) + Ph_3COH(\text{soln}) \tag{44}$$

$$Ph_3COD(\text{solid}) + Ph_3COH(\text{soln}) \overset{CH_3CN}{\rightleftharpoons} Ph_3COH(\text{solid}) + Ph_3COD(\text{soln}) \tag{45}$$

$$Ph_3COH(\text{solid}) + D(L_2O) \rightleftharpoons Ph_3COD(\text{solid}) + H(L_2O) \tag{46}$$

predicted fractionation factors at the same temperature for hydrogen-bonded protons described by different potential functions are given in Fig. 9 (Kreevoy and Liang, 1980). Homoconjugate biscarboxylate complexes such as $[CF_3CO_2 \cdots H \cdots O_2CCF_3]^-$, which was found experimentally to have a value ($\varphi_{A_1HA_2^-}$) = 0.42 ± 0.05, were assigned potential functions similar to that in Fig. 9*d*. The fractionation factor for hydrogen bis(formate) ion has

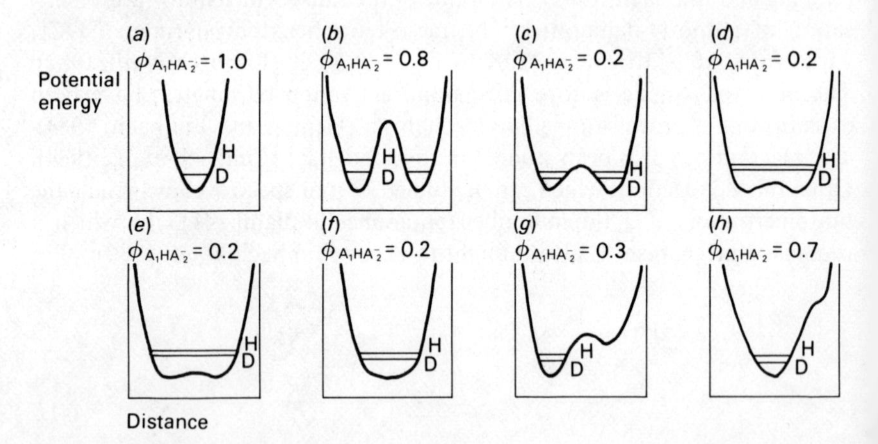

FIG. 9 Predicted fractionation factors ($\varphi_{A_1HA_2^-}$) for protons in different potential wells

also been measured in concentrated aqueous solutions of sodium formate (Fenn and Spinner, 1984). For bisphenolate complexes such as $[NO_2C_6H_4O \cdots H \cdots OC_6H_4NO_2]^-$ with $(\varphi_{A_1HA_2^-}) = 0.31 \pm 0.03$, the potential function in Fig. 9*c* was considered appropriate (Kreevoy and Liang, 1980), whereas Fig. 9*g* was thought to be applicable to the heteroconjugate complex $[3,5\text{-}(NO_2)_2C_6H_3 \quad O \cdots H \cdots Cl]^-$ for which the value $(\varphi_{A_1HA_2^-}) = 0.47 \pm 0.06$ was obtained. No evidence for a single-minimum function was found (Fig. 9*a*). For the enol form of 1,1,1,5,5,5-hexafluoropentane-2,4-dione, a value of $(\varphi_{A_1HA_2^-}) = 0.6 \pm 0.1$ was determined for the fractionation factor of the intramolecularly hydrogen-bonded proton (deuteron) (Kreevoy and Ridl, 1981). This value was accounted for by assuming that the compound exists as a pair of rapidly

$$F_3C{-}C({=}O\cdots H{-}O){-}CH{=}C{-}CF_3 \rightleftharpoons F_3C{-}C(O{-}H\cdots O{=}){=}CH{-}C{-}CF_3 \qquad (47)$$

equilibrating tautomers (47) with the proton oscillating between two identical sites. The double-minimum potential which represents this situation is shown in Fig. 10. The barrier height is *ca* 34 kJ mol^{-1} and the separation of the minima is 56 pm in a bent hydrogen bond of length 255 pm (Andreassen *et al.*, 1971).

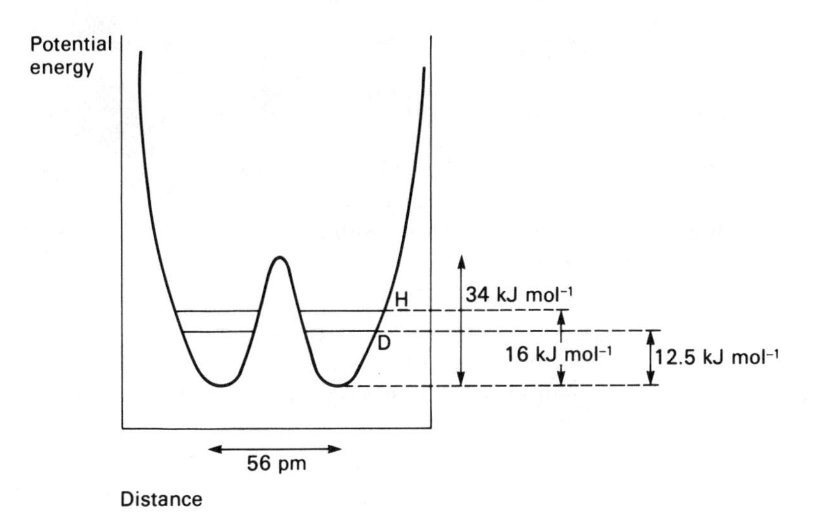

FIG. 10 Proposed potential function for the enolic proton in 1,1,1,5,5,5-hexafluoropentane-2,4-dione

Proton transfer in excited states

To conclude our description of techniques, the use of nanosecond and picosecond spectroscopy which has been applied to excited state intramolecular proton transfer (ESIPT) will be mentioned briefly (Beens *et al.*, 1965; Huppert *et al.*, 1981; Hilinski and Rentzepis, 1983). A large number of inter- and intramolecular proton transfers have been studied using these methods (Ireland and Wyatt, 1976) but in the case of processes which are thought to involve simple proton transfer along an intramolecular hydrogen bond it is usually only possible to estimate a lower limit for the rate coefficient.

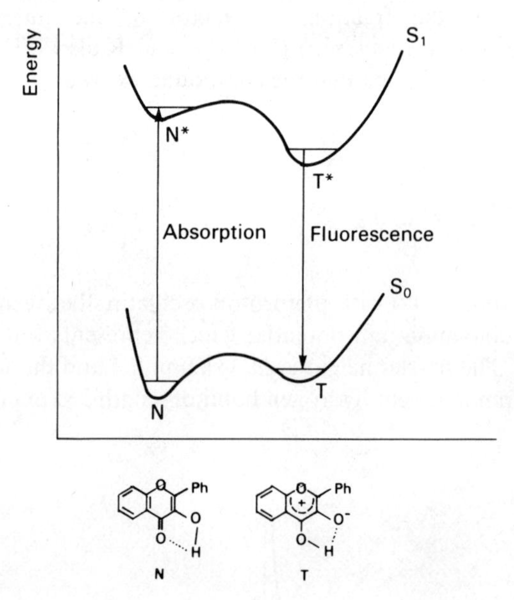

FIG. 11 Excited state proton transfer of 3-hydroxyflavone

The fluorescence observed when 3-hydroxyflavone is excited by the absorption of light is explained (Sengupta and Kasha, 1979) by Fig. 11. Measurement of the rise-time of fluorescence is thought to provide an estimate of the rate coefficient for proton transfer within the excited state (48). However, the value of the rate coefficient $k \sim 10^8 s^{-1}$ (Strandjord *et al.*, 1983) is much lower than previous estimates for similar processes in other

(48)

excited states (Woolf and Thistlethwaite, 1980, 1981, Itoh *et al.*, 1982). These reactions are considered to occur in less than 5 ps. The presence of impurities may be a complicating factor (McMorrow and Kasha, 1983, 1984). A more normal value is observed for proton transfer within the excited state of methyl salicylate (49), for which a lower limit of $k > 1 \times 10^{11} s^{-1}$ is estimated (Smith and Kaufman, 1978; Goodman and Brus, 1978; Huppert *et al.*, 1981).

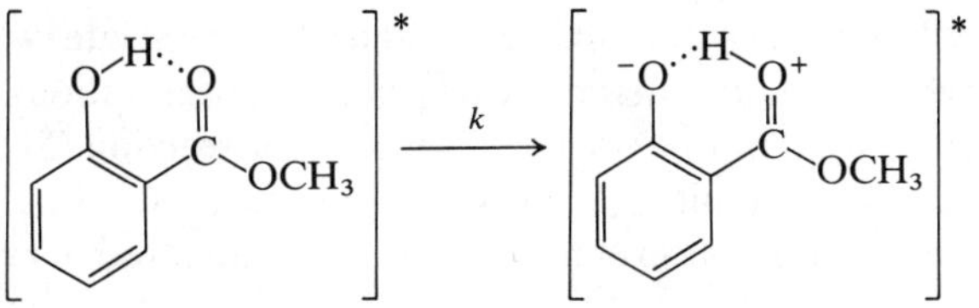

(49)

Rate coefficients in the expected range have been observed in excited state intramolecular proton transfers involving 2-(2-hydroxyphenyl)benzothiazole (Barbara *et al.*, 1980a) and salicylideneaniline (Barbara *et al.*, 1980b). In the former, the observation of fluorescence was explained by Scheme 2 and the

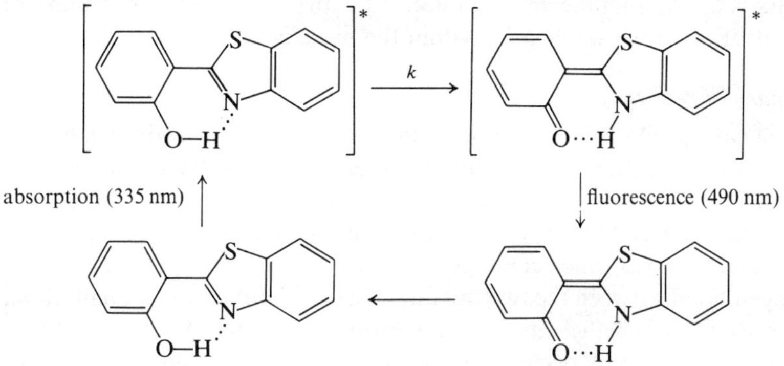

absorption (335 nm)

fluorescence (490 nm)

Scheme 2

measurement of the rise-time of fluorescence which occurred more rapidly than the resolution time of the detector (*ca* 5 ps) gave an estimate of $k > 2 \times 10^{11} s^{-1}$ for the excited state proton transfer. A similar result was obtained for the fluorescence rise-time in the enol of salicylideneaniline (50).

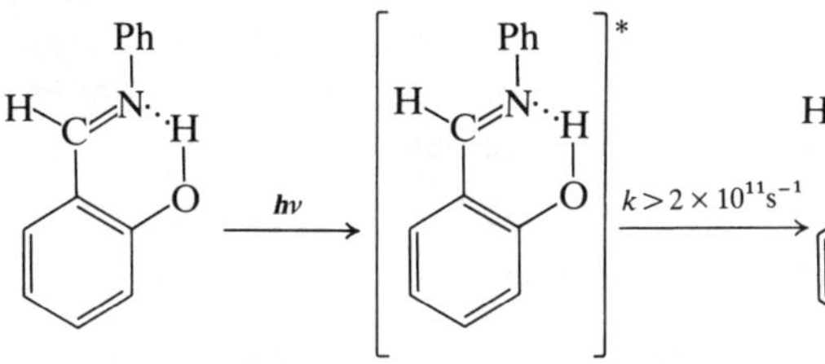

$h\nu$, $k > 2 \times 10^{11} s^{-1}$

(50)

The rates of intermolecular proton-transfer reactions can be studied using similar techniques (Escabi-Perez and Fendler, 1978; Huppert *et al.*, 1981; Hagopian and Singer, 1983; Webb *et al.*, 1984). In an interesting example involving nanosecond time-resolved fluorescence, the rate of protonation of the excited state of pyrene-1-carboxylate ($PyCO_2^-$) was compared in aqueous solution and in a reversed micelle of 0.1 mol dm^{-3} dodecylammonium propionate with 0.55 mol dm^{-3} water in benzene (Escabi-Perez and Fendler, 1978). The fluorescence life-time of pyrene-1-carboxylate was measured under these conditions at different hydrogen-ion concentrations and the data permitted calculation of the rate coefficient (k) for reaction (51). In aqueous solution at 25°C the result $k = 7.7 \times 10^{10}\ dm^3mol^{-1}s^{-1}$ was obtained. In the reverse micelle it is not certain which species is involved in protonation of

$$[PyCO_2{}^-] \xrightarrow{H^+} [PyCO_2H]^* \qquad (51)$$

the excited state of pyrene-1-carboxylate but if the acidic species is assumed to be hydronium ion, a rate coefficient $k = 2.0 \times 10^{12}\ dm^3mol^{-1}s^{-1}$ is calculated by dividing the observed first-order rate coefficient obtained from the fluorescence lifetime in the micelle by the estimated hydronium-ion concentration in the water pool within the micelle.

Summary of results

In most cases the location of a proton in an intramolecular hydrogen bond is best represented by a double-minimum potential, although two examples are known, phthalate ion and in-protonated 1,6-diazabicyclo[4.4.4]tetradecane, where the evidence favours a single-minimum potential. Interconversion of the chemical species which corresponds to movement of the proton along the hydrogen bond between the two minima of a double-minimum potential well occurs with rate coefficients that are probably greater than $1 \times 10^{12}s^{-1}$. Proton movement may occur by a tunnelling mechanism. Slow tautomerisation has been found for the double proton transfers of 2-methylnaphthazarin and meso-tetraphenylporphyrin. In intermolecular hydrogen-bonded complexes, the proton potential is usually of the double-minimum type although for bifluoride ion evidence for a single minimum potential has been obtained. It is likely that the central barrier to proton motion in an intermolecular hydrogen bond is low. Most of the data concerned with intermolecular hydrogen bonds are obtained from studies in non-aqueous solvents and the barrier to proton motion along an intermolecular hydrogen bond in aqueous solution may be higher than that in a non-aqueous solvent.

4 Proton removal from intramolecular hydrogen bonds

The reactivity of protons in chemical and biological systems is modified if the proton is held within an intramolecular hydrogen bond. In this section, the kinetics of proton transfer from an intramolecular hydrogen bond to an external base will be examined.

MECHANISM

Over the past 20 years, with the availability of fast reaction techniques (Eigen and de Maeyer, 1963; Hammes, 1974; Bernasconi, 1976), numerous kinetic studies have been made of the reactivity of hydrogen-bonded protons towards an external base (52). The majority of such studies have been made with hydroxide ion as the external base. Some examples of proton transfer to

$$\overline{\text{X—H}\cdots{}^{-}\text{Y}} + \text{B} \rightleftharpoons \overline{\text{X}^{-} \qquad {}^{-}\text{Y}} + \text{BH}^{+} \tag{52}$$

hydroxide are given in Table 4. They have been chosen to illustrate almost all of the possible types of hydrogen bond between oxygen and nitrogen acids and bases:

—O—H···⁻O— [36], [37], [38]

—O—H···N$\lessgtr$ [40], [41]; $\gtrless\overset{+}{\text{N}}$—H···N$\lessgtr$ [42]; $\gtrless\overset{+}{\text{N}}$—H···⁻O— [43];

$\gtrless\overset{+}{\text{N}}$—H···O$\lessgtr$ [39], [44]

All these reactions are thermodynamically favourable in the direction of proton transfer to hydroxide ion but the rate coefficients are somewhat below the diffusion-limited values. In broad terms, the typical effect of an intramolecular hydrogen bond on the rate coefficient for proton removal is to reduce the rate coefficient by a factor of up to *ca* 10^5 below the diffusion limit. Correspondingly the value of the dissociation constant of the acid is usually decreased by a somewhat smaller factor from that of a non-hydrogen-bonded acid. There are exceptions, however.

To understand the reasons for the effect of intramolecular hydrogen bonding on the rate of proton transfer it is first necessary to consider the mechanism of proton removal from an intramolecular hydrogen bond. The two mechanisms that have been suggested are shown in Schemes 3 and 4. In Scheme 3, which was first proposed by Eigen (Eigen and Kruse, 1963; Eigen, 1964; Eigen *et al.*, 1964), proton transfer to external base (B) occurs in two steps. The first step involves breakage of the hydrogen bond to give an open form of the acid which is externally hydrogen-bonded to solvent. The second step consists of proton removal by base from the open form present in low concentration. In Scheme 4, which was suggested by Eyring, proton transfer together with the accompanying solvation change is considered to occur in a

TABLE 4

Proton transfer from intramolecularly hydrogen-bonded nitrogen and oxygen acids to hydroxide ion[a]

X—H···⁻Y	pK_a	k_{OH^-}/dm^3 mol^{-1}s^{-1}	Solvent	Ionic strength/ mol dm^{-3}	Temp/°C	Reference
[36]	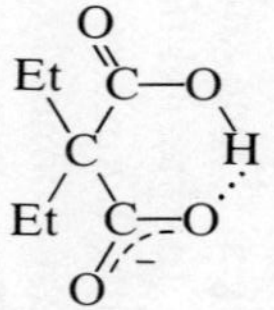 7.05	2.8×10^8	H_2O	0.1	25	Miles *et al.*, 1965
[37] R = —N=N—(3-nitrophenyl)	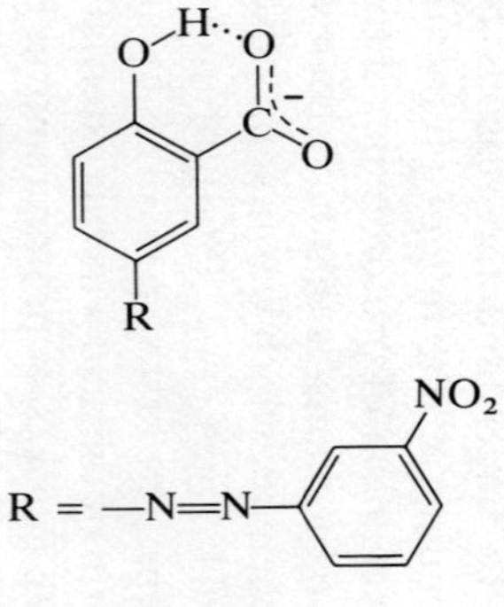 11.16	2.2×10^7	H_2O	0.5	6.5	Hibbert, 1981

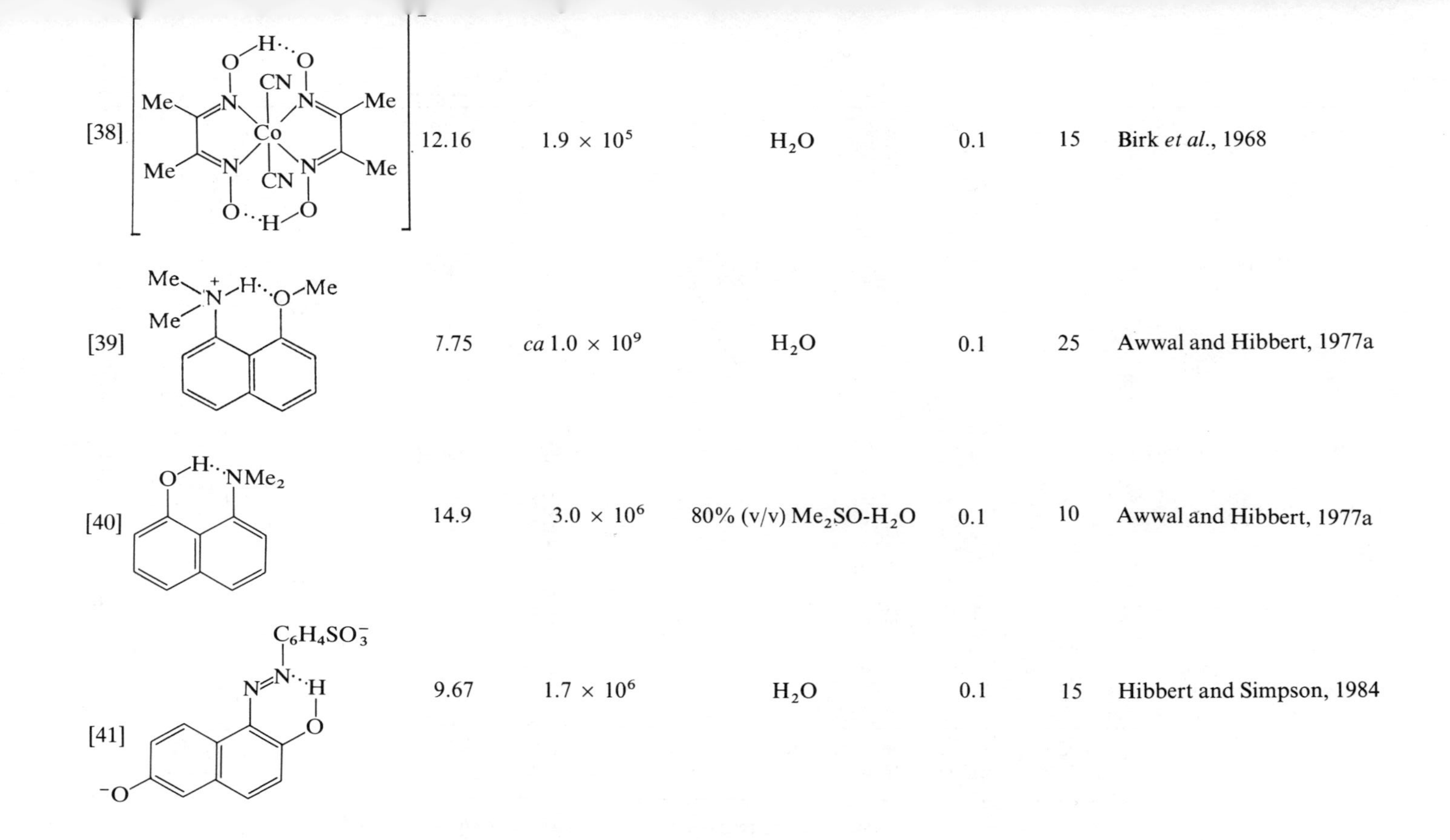

[38]		12.16	1.9×10^5	H_2O	0.1	15	Birk *et al.*, 1968
[39]		7.75	*ca* 1.0×10^9	H_2O	0.1	25	Awwal and Hibbert, 1977a
[40]		14.9	3.0×10^6	80% (v/v) Me_2SO-H_2O	0.1	10	Awwal and Hibbert, 1977a
[41]		9.67	1.7×10^6	H_2O	0.1	15	Hibbert and Simpson, 1984

Table 4 *continued*

Proton transfer from intramolecularly hydrogen-bonded nitrogen and oxygen acids to hydroxide ion[a]

X—H···⁻Y	pK_a	k_{OH^-}/dm^3 mol^{-1}s^{-1}	Solvent	Ionic strength/ mol dm^{-3}	Temp/°C	Reference
[42]	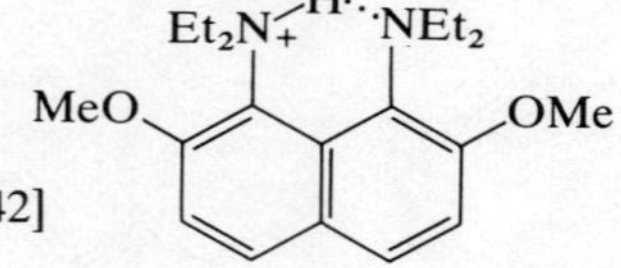 16.3	0.18	50% (v/v) Me_2SO-H_2O	0.1	25	Hibbert and Hunte, 1983
[43]	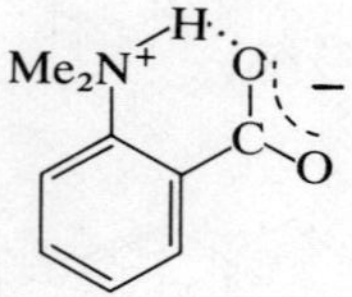 8.59	1.54×10^8	H_2O	0.1	25	Haslam and Eyring, 1967
[44]	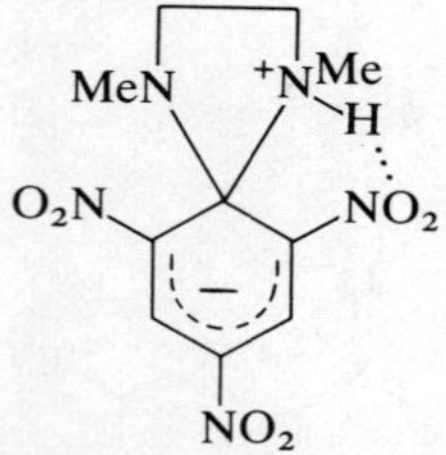 6.64	5.2×10^9	H_2O	0.5	25	Bernasconi and Terrier, 1975a,b

[a] $X{-}H\cdots{}^-Y + OH^- \xrightarrow{k_{OH^-}} X^- \; {}^-Y + H_2O$

$$\left[\begin{matrix}\text{X}\\ |\\ \text{H}\\ \vdots\\ \text{Y}^-\end{matrix}\right. \underset{k_{-1}}{\overset{k_1}{\rightleftharpoons}} \left[\begin{matrix}\text{X—H}\\ \\ \text{Y}^-\end{matrix}\right.$$

$$\left[\begin{matrix}\text{X—H}\\ \\ \text{Y}^-\end{matrix}\right. + \text{B} \xrightarrow{k_2} \left[\begin{matrix}\text{X}^-\\ \\ \text{Y}^-\end{matrix}\right. + \text{BH}^+$$

Scheme 3

$$\left[\begin{matrix}\text{X}\\ |\\ \text{H}\\ \vdots\\ \text{Y}^-\end{matrix}\right. + \text{B} \longrightarrow \left[\left[\begin{matrix}\text{X}\\ \\ \text{Y}\end{matrix}\right\rangle \text{H---B}\right]^{\ddagger} \longrightarrow \left[\begin{matrix}\text{X}^-\\ \\ \text{Y}^-\end{matrix}\right. + \text{BH}^+$$

Scheme 4

single step (Miles *et al.*, 1965; Haslam and Eyring, 1967). The literature in this field has been briefly reviewed (Kresge, 1973; Kresge 1975a,b) and it was considered at that time that no definite evidence to distinguish between Schemes 3 and 4 was available. Experiments to distinguish between the two mechanisms have been carried out in the last 10 years (Hibbert, 1984).

For a reaction occurring by the mechanism in Scheme 4, a first-order dependence of the rate on base concentration [B] would be observed. For Scheme 3, with the base present in excess, the rate expression (53) is obtained and the observed first-order rate coefficient (k_{obs}) for ionisation of the hydrogen-bonded acid is given by (54). Equation (54) reduces to (55) if the condition $k_{-1} \ll k_2$ [B] applies and (56) is obtained if $k_{-1} \ll k_2$ [B]. For all

$$-\mathrm{d}[\overline{\text{X—H}\cdots{}^-\text{Y}}]/\mathrm{d}t = k_1 k_2[\overline{\text{X—H}\cdots{}^-\text{Y}}][\text{B}]/(k_{-1} + k_2[\text{B}]) \tag{53}$$

$$k_{obs} = k_1 k_2[\text{B}]/(k_{-1} + k_2[\text{B}]) \tag{54}$$

$$k_{obs} = (k_1/k_{-1})k_2[\text{B}] \tag{55}$$

$$k_{obs} = k_1 \tag{56}$$

intramolecularly hydrogen-bonded acids that have been studied, except for derivatives of phenylazoresorcinol (see later), proton transfer is found to be first order in base. Scheme 4 and Scheme 3 (with $k_{-1} \gg k_2$ [B]) are compatible with this observation. A reduced rate of proton removal from an intramolecularly hydrogen-bonded acid can be understood in terms of Schemes 3 and 4. For Scheme 4, proton removal is unfavourable compared with proton removal from a normal acid because the base must attack the proton directly, with no intervening solvent molecule, and because the groups between which the proton transfer occurs cannot take up the most favourable in-line orientation in the transition state. In addition, in the transition state, the hydrogen bond is partially broken and this will also contribute to the activation energy. In Scheme 3, proton removal occurs through a low concentration intermediate (the non-hydrogen-bonded form), so that the rate of reaction will be reduced from that for a normal proton transfer by a factor corresponding to the equilibrium constant for the unfavourable formation of this intermediate open form (55).

Following the first studies by Eigen (Eigen and Kruse, 1963; Eigen, 1964; Eigen *et al.*, 1964), much of the early kinetic work in this area was due to E. M. Eyring. Various approaches were used to investigate the mechanism and transition-state structure for proton removal from intramolecularly hydrogen-bonded acids. The acids were mainly monoanions of dicarboxylic acids and the external base was hydroxide ion. In one approach, activation parameters for the reaction of a series of substituted malonic acids with hydroxide ion were measured (Miles *et al.*, 1965). Kinetic solvent isotope effects which refer to the ratio of rate coefficients ($k_{OH^-}^{H_2O}/k_{OD^-}^{D_2O}$) for the reactions in (57) and (58) were also determined (Miles *et al.*, 1966) and the

$$R^1R^2C(CO_2^-)(CO_2H)\ [\text{intramolecular } C{=}O\cdots H{-}O] + OH^- \xrightarrow{k_{OH^-}^{H_2O}} R^1R^2C(CO_2^-)_2 + H_2O \qquad (57)$$

$$R^1R^2C(CO_2^-)(CO_2D)\ [\text{intramolecular } C{=}O\cdots D{-}O] + OD^- \xrightarrow{k_{OD^-}^{D_2O}} R^1R^2C(CO_2^-)_2 + D_2O \qquad (58)$$

data are shown in Table 5. Activation parameters (Haslam *et al.*, 1965a) and solvent-isotope effects (Haslam *et al.*, 1965b; Eyring and Haslam, 1966; Haslam and Eyring, 1967) on related reactions have been measured. The data in Table 5 show an inverse variation of the ratio of the first and second dissociation constants (K_{a1}/K_{a2}) of the substituted malonic acids with the rate coefficient for hydroxide ion catalysed proton removal from the mono-anions ($k_{OH^-}^{H_2O}$). It was argued that an increase in the strength of the

TABLE 5

Proton transfer from substituted malonic acids[a,b]

	R^1	R^2	pK_{a1}	pK_{a2}	$10^{-5}K_{a1}/K_{a2}$	$10^{-7}k_{OH^-}^{H_2O}$ $/dm^3mol^{-1}s^{-1}$	$\Delta H^{\ddagger}$ $/kJ\ mol^{-1}$	$-T\Delta S^{\ddagger}$ $/J\ mol^{-1}K^{-1}$	$k_{OH^-}^{H_2O}/k_{OD^-}^{D_2O}$
[45]	C_2H_5	C_2H_5	2.15	7.05	0.8	27.8	19	21	2.0
[46]	C_2H_5	i-C_5H_{11}	2.15	7.31	1.45	16.0	23	12.5	1.9
[47]	C_2H_5	Ph	1.90	7.12	1.60	14.0	21	17	2.0
[48]	n-C_3H_7	n-C_3H_7	2.15	7.34	1.55	12.7	25	4	—
[49]	C_2H_5	i-C_3H_7	2.03	8.10	11.7	5.5	29	0	1.6
[50]	i-C_3H_7	i-C_3H_7	2.18	8.60	26.3	4.5	31	−8.5	1.5

[a] Conditions: aqueous solution, 25°C, ionic strength, 0.1 mol dm^{-3}
[b] Data taken from Miles *et al.*, 1965, 1966

intramolecular hydrogen bond in the monoanion has the effect of stabilising the monoanion; this means that, in the absence of other effects, the values of $(k_{OH^-}^{H_2O})$ and K_{a2} decrease as the value of K_{a1} increases. The single step mechanism in Scheme 6 was preferred over the two step mechanism of Scheme 5 in explaining the decrease in the value of $(k_{OH^-}^{H_2O})$ with increasing strength of the hydrogen bond in the monoanion. Activation parameters for the reactions were also explained in terms of this mechanism (Miles *et al.*, 1965). An increase in the enthalpy of activation along the series of compounds [45] to [50] was thought to be accounted for by the increase in strength of the intramolecular hydrogen bond. However, the activation parameters can be explained equally well by the two step mechanism for proton removal in Scheme 5. In this case, temperature is assumed to have an effect on the value of the equilibrium constant between open and closed forms of the malonate monoanions as well as a small effect on the rate coefficient for the proton removal.

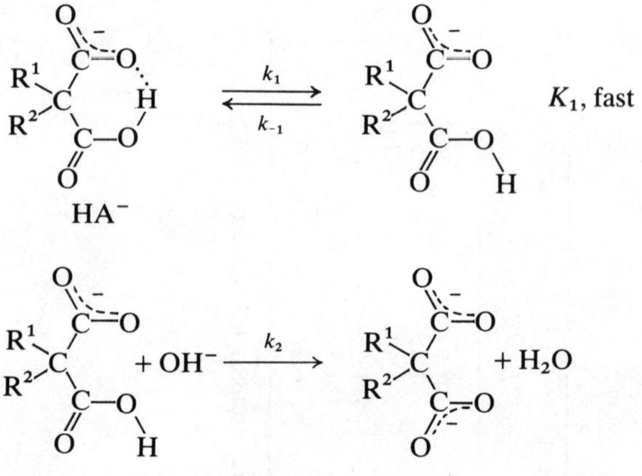

HA^- $\underset{k_{-1}}{\overset{k_1}{\rightleftharpoons}}$ K_1, fast

$+ OH^- \xrightarrow{k_2}$ $+ H_2O$

$$-d[HA^-]/dt = K_1k_2[HA^-][OH^-] = k_{OH^-}^{H_2O}[HA^-][OH^-]$$

Scheme 5

The kinetic solvent-isotope effects on these reactions are made up of primary and secondary kinetic isotope effects as well as a medium effect, and for either scheme it is difficult to estimate the size of these individual contributions. This means that the value of the isotope effect does not provide evidence for a choice between the two schemes (Kresge, 1973). The effect of gradual changes in solvent from an aqueous medium to 80% (v/v) Me_2SO—H_2O on the rate coefficient for hydroxide ion catalysed proton removal from the monoanions of several dicarboxylic acids was interpreted in terms of Scheme 6 (Jensen *et al.*, 1966) but an equally reasonable explanation is provided by Scheme 5.

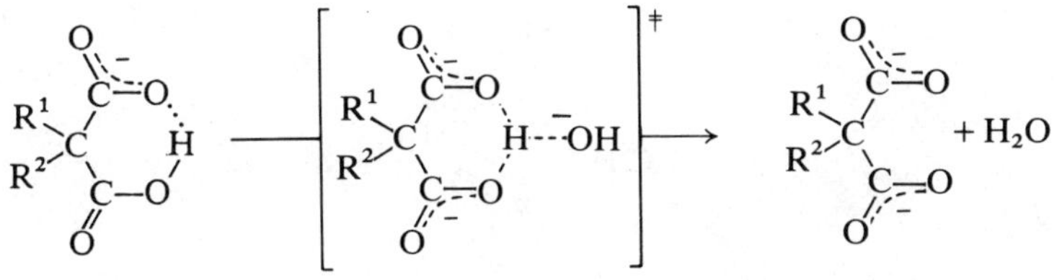

HA^-

$$-\mathrm{d}[HA^-]/\mathrm{d}t = k_{OH^-}^{H_2O}[HA^-][OH^-]$$

Scheme 6

The data in Table 5 were used (Miles *et al.*, 1966) to construct a Brønsted plot of the variation of the rate coefficient for proton removal with acidity along the series of substituted malonate monoanions; the plot is reproduced in Fig. 12. The value of the gradient of the best line (α = *ca* 0.5) was interpreted (Miles *et al.*, 1966) as indicating that proton removal by hydroxide ion occurs in a single step through a transition state in which the

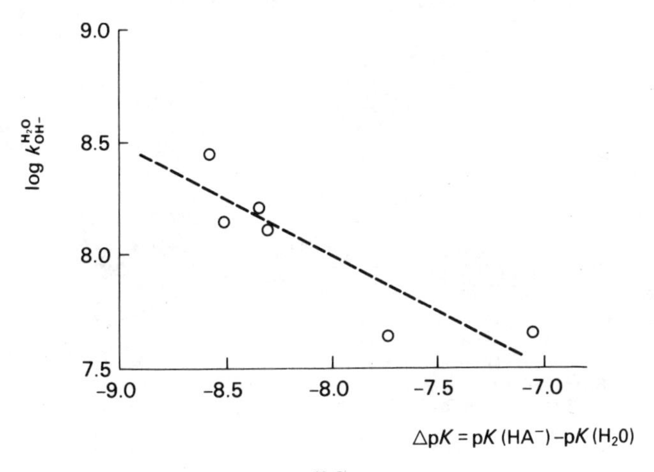

FIG. 12 Variation of rate coefficients ($k_{OH^-}^{H_2O}$) with ΔpK for proton transfer from substituted malonate monoanions to hydroxide ion

proton is roughly half-transferred to hydroxide ion. A rather better correlation of rate with acidity over a much smaller range has been obtained (Fueno *et al.*, 1973) for the reaction between hydroxide ion and a series of 1-ethyl-1-arylmalonate ions [51] with various aromatic substituents (X). The results were plotted in the form of a Hammett relationship using Taft σ^0-values (Taft and Lewis, 1959; Taft *et al.*, 1959), and the conclusion was

O
C—O
Et H
C
XC_6H_4
C—O
O

[51]

reached that the data could be explained by either the mechanism in Scheme 5 or that in Scheme 6. The results are given as a Brønsted plot in Fig. 13 and a Brønsted exponent of $\alpha = 0.39$ is obtained from the slope. Kinetic results for the reaction of various intramolecularly hydrogen-bonded acids with hydroxide ion were used (Rose and Stuehr, 1968, 1971, 1972) to construct a family of parallel Brønsted plots with slopes α *ca* 1.0. However, included in these correlations, were data for phenylazoresorcinols for which recent experiments have uncovered a complex dependence of rate on hydroxide ion concentration (see p. 177).

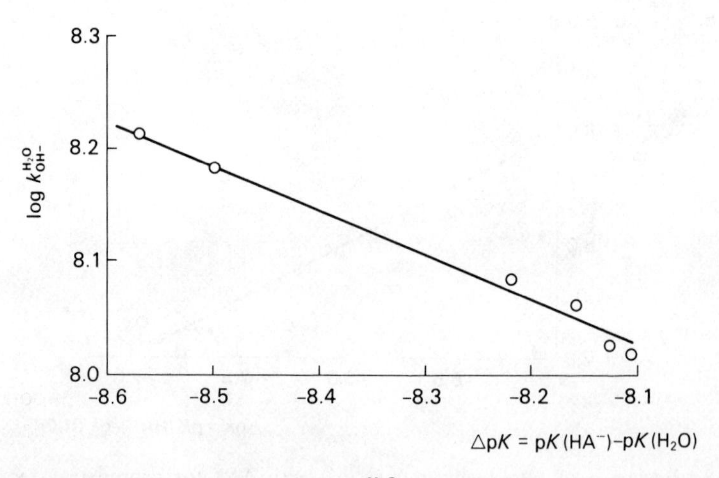

FIG. 13 Variation of rate coefficients ($k_{OH^-}^{H_2O}$) with ΔpK for proton transfer from 1-ethyl-arylmalonate monoanions to hydroxide ion

It is of interest to consider the form of the Brønsted plot or Eigen plot to be expected for reaction of a series of related intramolecularly hydrogen-bonded acids with hydroxide ion by the mechanisms in Schemes 5 and 6. The effect of a substituent on the value of the dissociation constant of an intramolecularly hydrogen-bonded acid ($\delta \log K$) will be two-fold. The stability of the undissociated acid will be modified because of a substituent effect on the strength of the intramolecular hydrogen bond. The substituent will also affect the stability of the dissociation product relative to that of the undissociated acid through an interaction with the developed charge. These effects can be thought of in terms of changes in the values of the equilibrium constants (K_1 and K_2) for the two steps in Scheme 5, and the overall change in acidity as a result of substitution will be given by (59) in which K refers to

$$\delta \log K = \delta \log K_1 + \delta \log K_2 \tag{59}$$

the equilibrium constant for the reaction between the hydrogen-bonded acid and hydroxide ion, $K = [A^{2-}]/[HA^-][OH^-]$. If proton transfer is considered to occur by the mechanism in Scheme 5 and the rate coefficient is given by (60), the change in rate coefficient ($\delta \log k_{OH^-}$) as a result of substitution will be given by (61). Since the proton-transfer step in this mechanism is the thermodynamically favourable ionisation of a normal acid, the value of k_2 will be diffusion-limited and the substituent effect on k_2 will be zero ($\delta \log k_2 = 0$). Equation (62) is then obtained, and it follows that the slope of a Brønsted plot for reaction of a series of substituted acids is given by (63). If the substituent effects $\delta \log K_1$ and $\delta \log K_2$ are of the same sign, α will lie

$$k_{OH^-} = K_1 k_2 \tag{60}$$

$$\delta \log k_{OH^-} = \delta \log K_1 + \delta \log k_2 \tag{61}$$

$$\delta \log k_{OH^-} = \delta \log K_1 \tag{62}$$

$$\alpha = \delta \log k_{OH^-}/\delta \log K = \delta \log K_1/(\delta \log K_1 + \delta \log K_2) \tag{63}$$

between zero and unity. If $\delta \log K_1$ and $\delta \log K_2$ are of opposite sign, results $\alpha < 0$ or $\alpha > 1$ may be obtained. Thus a wide range of values of α is compatible with the mechanism in Scheme 5. Miles *et al.* (1966) have argued that, for the mechanism in Scheme 6, a transition state for the reaction with partial rupture of the intramolecular hydrogen bond and partial proton transfer to hydroxide ion would be compatible with a Brønsted exponent $\alpha \sim 0.5$. Indeed any value of α between zero and unity could be explained in terms of different degrees of proton transfer in the transition state. This analysis shows that the value of the Brønsted exponent determined for the reaction of a series of intramolecularly hydrogen-bonded acids with hydroxide ion will not permit a choice between the mechanisms in Schemes 5 and 6. However, as shown below, results for proton transfer from a single intramolecularly hydrogen-bonded acid to a series of bases of varying base strength do provide a definite answer, and data of this type will now be considered.

The reaction of an intramolecularly hydrogen-bonded salicylate ion with a base B is shown in (64) and the two-step mechanism for the reaction is written in Scheme 7. We can show how, according to this mechanism, the

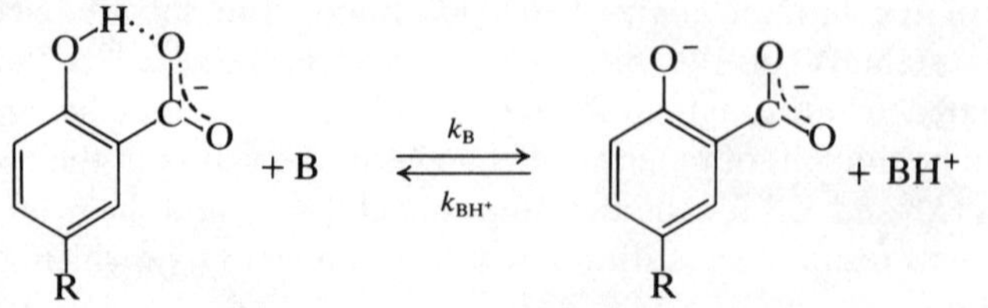

$$\text{(salicylate HA}^-\text{)} + B \underset{k_{BH^+}}{\overset{k_B}{\rightleftharpoons}} \text{(open dianion)} + BH^+ \qquad (64)$$

rate coefficients k_B and k_{BH^+} should vary as B is changed from strongly to more weakly basic species. A typical value of 1.0×10^{-3} will be assumed for K_1, the equilibrium constant between the hydrogen-bonded salicylate ion (HA^-) and its open form. For proton transfer to strong bases (B), the equilibrium in (64) will lie towards the right, corresponding to a value of the equilibrium constant $K > 1$, i.e. $-\log_{10}K = pK(HA^-) - pK(BH^+) = \Delta pK < 0$.

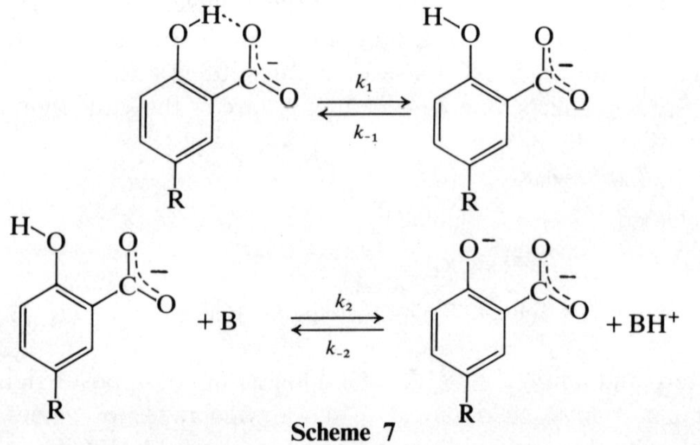

$$\text{(H-bonded form)} \underset{k_{-1}}{\overset{k_1}{\rightleftharpoons}} \text{(open form)}$$

$$\text{(open form)} + B \underset{k_{-2}}{\overset{k_2}{\rightleftharpoons}} \text{(dianion)} + BH^+$$

Scheme 7

In this region, the equilibrium constant for the proton-transfer step in Scheme 7 has a value $K_2 > 1$ and the proton transfer step is strongly favourable thermodynamically in the forward direction. This reaction step is a normal proton transfer between an oxygen acid which does not possess an intramolecular hydrogen bond and a base (B) and will therefore be diffusion-limited with a rate coefficient k_2 in the range 1×10^9 to $1 \times 10^{10} \text{dm}^3\text{mol}^{-1}\text{s}^{-1}$. It follows from (65) that k_B will have a value which is

$$k_B = K_1 k_2 \qquad (65)$$

$$k_{BH^+} = k_{-2} \qquad (66)$$

a factor of 10^3 below the diffusion limit and which is the same for proton transfer to bases of different strength. Also in this region ($\Delta pK < 0$) the value of k_{BH^+} is given by (66) and will vary inversely with K_2 and with K and will

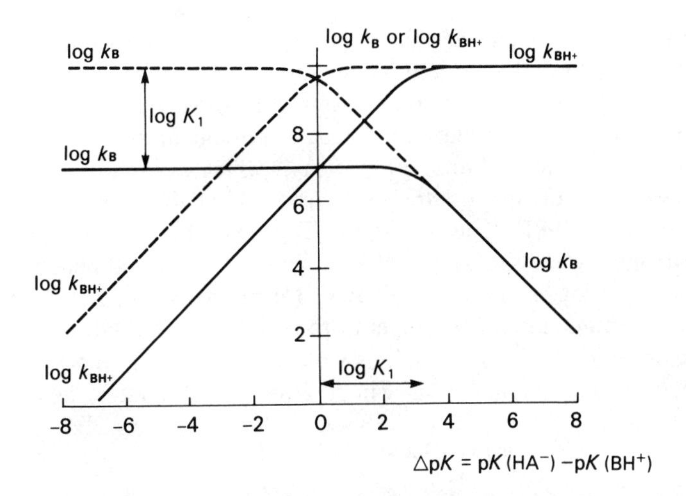

FIG. 14 Predicted variation in forward (k_B) and reverse (k_{BH^+}) rate coefficients for proton transfer from an intramolecularly hydrogen-bonded acid (HA^-) to bases (B) (solid line) and comparison with normal proton-transfer behaviour (dashed line)

vary directly with $K(BH^+)$. For reaction of salicylate ion with bases B, having basicity such that $\Delta pK = pK(HA^-) - pK(BH^+) = 0$, the proton-transfer step in Scheme 7 will still be strongly favourable thermodynamically even though for the overall reaction the equilibrium constant K has a value of unity. Hence at $\Delta pK = 0$, the value of k_2 will be diffusion-limited and the value of k_B will be a factor of 10^3 below the diffusion limit and independent of the base strength of B. This will apply up to the point where the proton-transfer step becomes thermodynamically unfavourable, which occurs for proton transfer to bases with $pK(BH^+)$ such that $\Delta pK > -\log K_1$ or $+3$. In this range the value of k_2 and the value of k_B will be inversely proportional to K_2 and K since proton transfer in the reverse direction will now be diffusion-limited (k_{-2} 1×10^9 to $1 \times 10^{10} \mathrm{dm^3 mol^{-1} s^{-1}}$). The conclusions from this analysis are given in Fig. 14 in comparison with results for proton transfer involving a normal oxygen acid. It is interesting that when proton transfer from the intramolecularly hydrogen-bonded acid to base is strongly thermodynamically unfavourable ($\Delta pK > 3$) the predicted rate coefficients in the forward and reverse directions are identical with those which would be observed for proton removal from a normal acid (Fig. 14). These predictions apply for proton transfer occurring by the mechanism in Scheme 7. If the reaction in (64) occurred by single-step attack by base on the hydrogen-bonded proton, it would be expected that, when the salicylate ion and BH^+ are of similar acidity, the proton in the transition state would be roughly half-transferred and that both k_B and k_{BH^+} would vary with the basicity of B in this region. It might therefore be expected that slopes of the Eigen plot would be *ca* 0.5 in this case. It follows that data in the region of $\Delta pK \sim 0$ would permit the two possible mechanisms for proton transfer to be distinguished.

The first experimental data for a reaction involving proton transfer from a hydrogen-bonded acid to a series of bases which were chosen to give ΔpK-values each side of $\Delta pK = 0$ are given in Fig. 15 (Hibbert and Awwal, 1976, 1978; Hibbert, 1981). The results were obtained for proton transfer from 4-(3-nitrophenylazo)salicylate ion to a series of tertiary aliphatic amines in aqueous solution, as in (64) with R = 3-nitrophenylazo. Kinetic measurements were made using the temperature-jump technique with spectrophotometric detection to follow reactions with half-lives down to 5×10^{-6}s. The reciprocal relaxation time (τ^{-1}), which is the time constant of the exponential

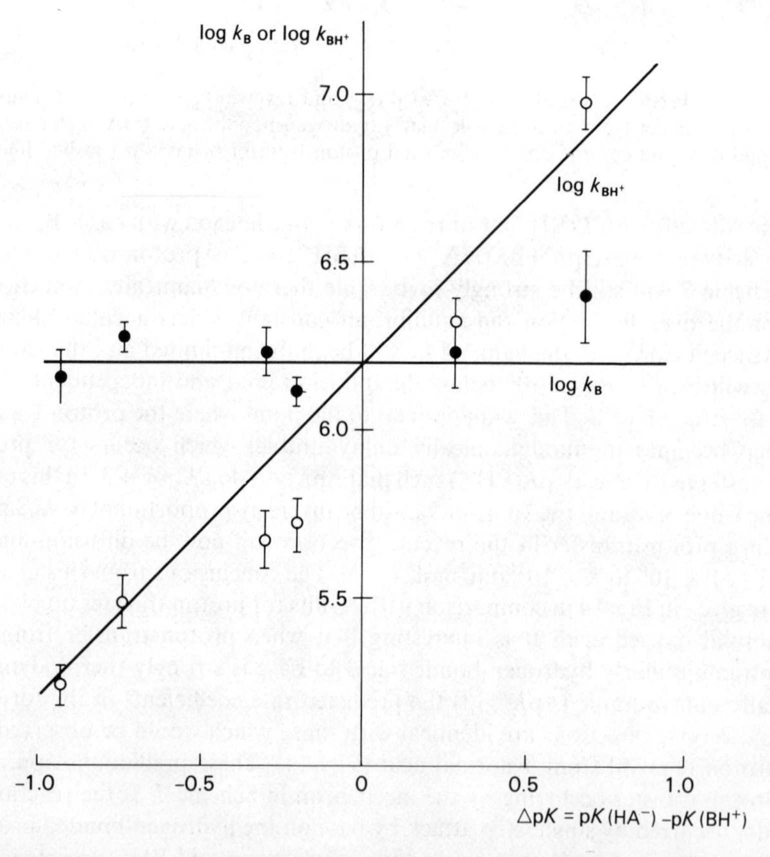

FIG. 15 Observed variation with ΔpK of forward (k_B) and reverse (k_{BH^+}) rate coefficients for proton transfer from 4-(3-nitrophenylazo)salicylate ion (HA^-) to bases (B)

approach to a new equilibrium position following a rapid temperature jump applied to a solution of 4-(3-nitrophenylazo)salicylate ion, is given by (67). The amine and protonated amine are present in excess and act as buffer. In (67), k_B and k_{BH^+} are the forward and reverse rate coefficients for proton transfer to the amine as in (64), and k_{OH^-} and k_{H_2O} are the rate coefficients for proton transfer to hydroxide ion (64) with $B = OH^-$ and r is the buffer ratio ($[BH^+]/[B]$). Values of k_B and k_{BH^+} for six amines were determined

$$\tau^{-1} = k_{OH^-}[OH^-] + k_{H_2O} + [B](k_B + k_{BH^+} r) \qquad (67)$$

from linear plots of τ^{-1} against buffer concentration at fixed buffer ratio and values of k_{OH^-} ($2.4 \times 10^7 dm^3 mol^{-1} s^{-1}$ at 6.5°C and ionic strength 0.5 mol dm^{-3}) and of k_{H_2O} were obtained from separate measurements in dilute aqueous hydroxide solutions in the absence of buffer. Often, in reactions of this type, the term involving catalysis by hydroxide ion predominates over the term for buffer catalysis and the rate coefficients for buffer catalysis cannot be determined accurately. Although the data in Fig. 15 show some scatter, the Eigen plot is closely similar to that given in Fig. 14 which was predicted by the two step mechanism in Scheme 7. The lines in Fig. 15 are drawn with slopes of 0.0 ($\log_{10} k_B$) and 1.0 ($\log_{10} k_{BH^+}$). It would have been of great interest to extend the data to higher ΔpK-values where, according to the predictions, curvature in the Eigen plot should be observed. However for 4-(3-nitrophenylazo)salicylate ion the approach to equilibrium in the presence of weakly basic catalysts, for which $\Delta pK > 1$, becomes too rapid because of the increase in the value of k_{BH^+}. A hydrogen-bonded acid which has been studied in the range $\Delta pK > 0$ will be discussed in the section which deals with proton transfer of hindered naphthalene diamines (p. 165).

More precise results than those shown in Fig. 15 were obtained for the reaction (68) studied in 70% (v/v) Me_2SO—H_2O (Bernasconi and Terrier, 1975a,b). For this reaction with phenolate ions as buffers (B), catalysis by

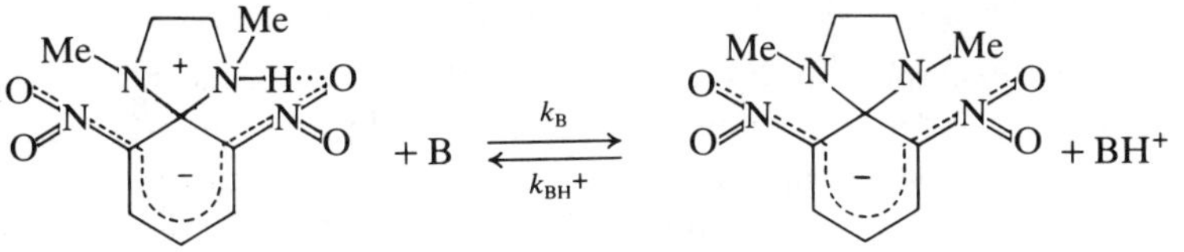

(68)

hydroxide ion does not contribute greatly in comparison with catalysis by buffer species under the conditions of the experiment. The data are given in Fig. 16 and, although the results do not extend to $\Delta pK = 0$, it is clear that the variation of $\log_{10} k_B$ and $\log_{10} k_{BH^+}$ with ΔpK is closely similar to the variation predicted in Fig. 14. The slopes of the lines through the data for $\log_{10} k_B$ and $\log_{10} k_{BH^+}$ in Fig. 16 have values of 0.0 and 1.0 respectively.

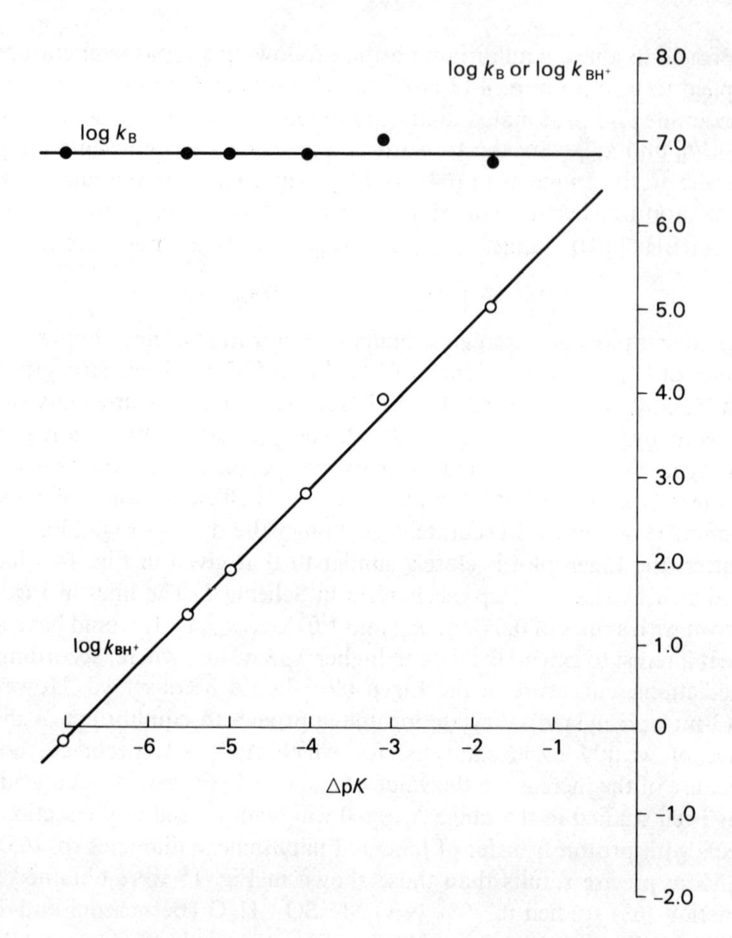

FIG. 16 Observed variation with ΔpK of forward (k_B) and reverse (k_{BH^+}) rate coefficients for proton transfer (68)

In the two examples of buffer catalysis of proton transfer from an intramolecularly hydrogen-bonded acid which have been discussed, it seems reasonably certain that the mechanism in Scheme 7 applies. The reactions are of the first order with respect to the catalyst B and it therefore follows that proton removal from the non-hydrogen-bonded species is rate-limiting; $k_{-1} > k_2$[B]. If this step consists of diffusion-controlled proton removal from a low concentration intermediate, the value $k_2 \sim 1 \times 10^9 \text{dm}^3 \text{mol}^{-1} \text{s}^{-1}$ will apply for proton transfer to an amine. In the case of proton removal by hydroxide ion from 4-(3-nitrophenylazo)salicylate ion, the reaction was found to be of the first order in hydroxide ion up to the highest concentrations which could be studied (0.003 mol dm^{-3}) with a rate coefficient $k_{OH^-} = 2.4 \times 10^7\,dm^3\,mol^{-1}\,s^{-1}$. The rate coefficient for proton transfer from the open form to hydroxide ion can be estimated as $k_2 \sim 1 \times 10^{10}\,dm^3\,mol^{-1}\,s^{-1}$, and in order to satisfy the condition $k_{-1} > k_2[OH^-]$, k_{-1} must have a value greater than *ca* $3 \times 10^7\ s^{-1}$. The result $K_1 \sim 2.4 \times 10^{-3}$ can be deduced from (65) using $k_2 \sim 1 \times 10^{10}\,dm^3\,mol^{-1}\,s^{-1}$ and $k_{OH^-} = 2.4 \times 10^7 dm^3\,mol^{-1}\,s^{-1}$; hence the rate coefficient for unimolecular opening of the intramolecular hydrogen bond is calculated to have a value $k_1 > 7 \times 10^4\,s^{-1}$. The minimum values deduced for k_1 and k_{-1} seem compatible with those expected for opening and closing of an intramolecular hydrogen bond.

In many ways, the behaviour of intramolecularly hydrogen-bonded salicylate ions towards proton removal by external base is fairly typical of what we expect to find for hydrogen-bonded acids. Reaction occurs through a non-hydrogen-bonded species present in low concentration, and the overall rate coefficient in the forward direction is about three to four orders of magnitude below the diffusion limit. The p*K*-value of the acid is *ca* 2 to 3 units higher than the p*K*-value of related acids that are unable to form intramolecular hydrogen bonds. Two classes of hydrogen-bonded acids which have received detailed study in recent years and which deviate in different ways from this pattern of behaviour will be considered in the next part of this review.

PROTON TRANSFER REACTIONS OF HINDERED DIAMINONAPHTHALENES

Since the discovery of the exceptional basicity of 1,8-bis(dimethylamino)-naphthalene (Alder *et al.*, 1968) and its unusual kinetic behaviour (Hibbert, 1973, 1974, 1975) there has been considerable interest in the acid-base properties of hindered diaminonaphthalenes. 1,8-Bis(dimethylamino)naphthalene (pK = 12.1 at 25°C and ionic strength 0.1 mol dm^{-3}) (Alder *et al.*, 1968; Hibbert, 1974; Chiang *et al.*, 1980) is more basic than most aliphatic amines and the p*K*-values of the partially methylated diamines [52] illustrate the dramatic effect of introducing the fourth methyl group (Alder *et al.*, 1968). Reaction of protonated 1,8-bis(dimethylamino)naphthalene with

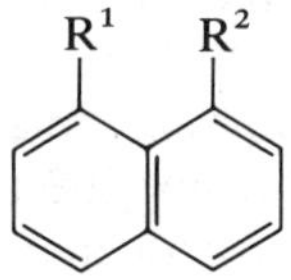

[52] (1,8-disubstituted naphthalene with substituents R^1 and R^2)

R^1	R^2	pK
NH_2	NH_2	4.61
NHMe	NHMe	5.61
NMe_2	NHMe	6.43
NMe_2	NMe_2	12.1

$$pK = -\log_{10}([amine][H_3O^+]/[amine\ H^+])$$

hydroxide ion in aqueous solution, although thermodynamically favourable, occurs with a rate coefficient ($k = 1.9 \times 10^5 \text{ dm}^3 \text{ mol}^{-1} \text{s}^{-1}$) which is five orders of magnitude below the diffusion-limited value expected for reaction between a protonated amine and hydroxide ion (Hibbert, 1973, 1974, 1975). Studies of some related diaminonaphthalenes have uncovered even more extreme behaviour. For 1,8-bis(diethylamino)-2,7-dimethoxynaphthalene (Alder *et al.*, 1978, 1981; Hibbert and Hunte, 1983) reaction of the protonated amine with hydroxide ion in aqueous solution is thermodynamically unfavourable and the amine remains in the protonated form in the presence of 1 mol dm^{-3} sodium hydroxide. Deprotonation occurs under the more basic conditions of $Me_2SO—H_2O$ solvent containing hydroxide ion (Dolman and Stewart, 1967; Rochester, 1970) and, in 50% (v/v) $Me_2SO—H_2O$, dissociation is half-complete in the presence of 0.1 mol dm^{-3} sodium hydroxide. A pK-value of 16.3 has been measured for this amine (Hibbert and Hunte, 1983). Even more surprisingly, it is found that the approach to equilibrium when 0.1 mol dm^{-3} sodium hydroxide is introduced into 50% (v/v) $Me_2SO—H_2O$ containing the protonated amine occurs with a half-life of 22 s, so that the increase in absorbance at 350 nm corresponding to formation of the free amine can be followed using a conventional spectrophotometer.

In 1,8-bis(dimethylamino)naphthalene steric and lone pair interactions are minimised in a structure (Einspahr *et al.*, 1973) having the amino groups twisted and displaced above and below the plane of the naphthalene ring, as in [53] in which the almost planar naphthalene ring is shown as a dashed line. In 1,8-bis(dimethylamino)-2,7-dimethoxynaphthalene, steric interactions involving the methoxy substituents bring about a further distortion as in [54] (Woolf, 1980).

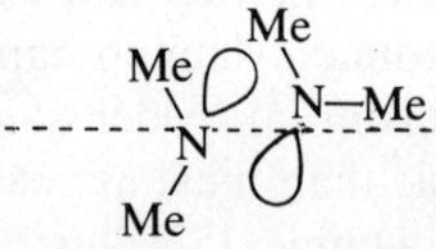

[53] 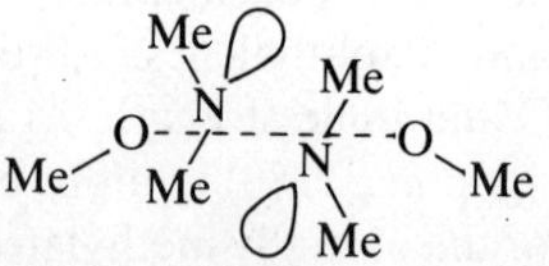 [54] 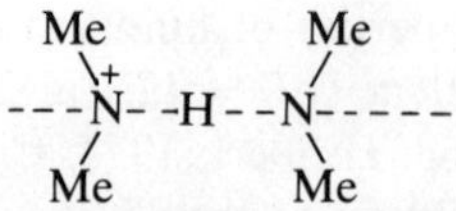 [55]

A considerable amount of the strain in 1,8-bis(dimethylamino)naphthalene is relieved by protonation and the N—H $\cdots$ N bond length (260 pm) in the protonated amine shows that the molecule is able to adopt a conformation [55] with an intramolecular hydrogen bond (Truter and Vickery, 1972). The infra-red spectrum of protonated 1,8-bis(dimethylamino)naphthalene and the chemical shift (δ 19.5) of the acidic proton in the nmr spectrum confirm the presence of an intramolecular hydrogen bond (Alder *et al.*, 1968). The magnitude of the isotope effect on the chemical shift (Altman *et al.*, 1978) and the appearance of two N_{1s} peaks in the photoelectron spectrum of the protonated amine (Haselbach *et al.*, 1972) are compatible with a double-minimum potential function for the proton in the intramolecular hydrogen bond. The fractionation factors of the acidic protons in protonated 1,8-bis(dimethylamino)naphthalene (Chiang *et al.*, 1980) and in protonated 1,8-bis(dimethylamino)-2,7-dimethoxynaphthalene (Hibbert and Robbins, 1980) have similar values ($\varphi \sim 0.9$) to those observed for other intramolecularly hydrogen-bonded acids (Kreevoy and Liang, 1980). The unusual basic properties of 1,8-bis(dimethylamino)naphthalene are probably the result of strain in the free amine and the presence of the intramolecular hydrogen bond in the protonated amine. In 1-methylamino-8-dimethylaminonaphthalene [52: $R^1 = NMe_2$, $R^2 = NHMe$] an intramolecular hydrogen bond is present in the free amine. This means that the relief of strain on protonation is much less than in the case of 1,8-bis(dimethylamino)naphthalene, and the p*K*-value of 1-methylamino-8-dimethylaminonaphthalene is only slightly higher than that expected for an aromatic amine.

The possibility arises that even more strongly basic amines with higher p*K*-values can be obtained if greater strain can be introduced and some of our results for substituted diaminonaphthalenes are given in Tables 6 and 7. The p*K*-value (12.1 ± 0.1) for 1,8-bis(dimethylamino)naphthalene [56] was determined in aqueous solution. The other amines listed in Table 6 are too strongly basic to permit studies in aqueous solution and the p*K*-values were obtained by comparing their acid-base equilibria with that of 1,8-bis-(dimethylamino)naphthalene in the more basic medium $Me_2SO—H_2O$ containing hydroxide ion. For example, in 35% (v/v) $Me_2SO—H_2O$, the equilibrium constants (*K*) for deprotonation of protonated 1,8-bis(dimethyl-amino)naphthalene [56] and protonated 1,8-bis(dimethylamino)-2,7-dimeth-oxynaphthalene [58] by hydroxide ion have values 4100 and 0.40 respectively, which means that 1,8-bis(dimethylamino)-2,7-dimethoxynaphthalene is more basic by a factor of 10^4 in this solvent. Assuming that the same factor applies in aqueous solution, the result $pK = 16.1$ is estimated for 1,8-bis(dimethylamino)-2,7-dimethoxynaphthalene. The p*K*-value of 16.3 found for 1,8-bis(diethylamino)-2,7-dimethoxynaphthalene [59] makes this amine the strongest uncharged base whose acid-base equilibrium has been directly studied. Bicyclic diamines and cryptands described in Section 5 are thought to have $pK > 17$, but the proton-transfer equilibria cannot be studied directly. The basicity of 1,8-bis(diethylamino)-2,7-dimethoxynaph-thalene may be the highest which can be achieved for molecules of this type. We have prepared 1,8-bis(diethylamino)-2,7-diethoxynaphthalene (Hibbert and Simpson, 1984) but the basicity is slightly lower than that of 1,8-bis-(diethylamino)-2,7-dimethoxynaphthalene. Further examples of amino-naphthalenes which have been prepared are given in Table 7, but these amines are all weaker bases than 1,8-bis(diethylamino)-2,7-dimethoxynaph-

TABLE 6

Kinetic and equilibrium acid-base properties of diaminonaphthalenes[a,b]

B		H_2O[c]	20[c]	30[d]	35[e]	50[f]	60[f]	70[f]	80[f]	90[g]
		Solvent: vol% $Me_2SO—H_2O$								
[56] Me_2N NMe_2 (1,8-naphthalene)	$K/dm^3\ mol^{-1}$	52	660	2800	4100					
pK 12.1 ± 0.1	$k_{OH^-}/dm^3\ mol^{-1}\ s^{-1}$	1.9×10^5	2.9×10^5	6.1×10^5						
[57] Et_2N NEt_2 (1,8-naphthalene)	K			380						
pK 13.0	k_{OH^-}			1.6×10^4						

Compound								
[58] (MeO, Me_2N, NMe_2, OMe)		K	0.4		252			
	pK 16.1	k_{OH^-}	110		440			
[59] (MeO, Et_2N, NEt_2, OMe)		K	0.26	8	116	3000	$>2 \times 10^4$	$>2 \times 10^4$
	pK 16.3	k_{OH^-}		0.18	3.3	10.4	45	*ca* 250

[a] $BH^+ + OH^- \underset{}{\overset{k_{OH^-}}{\rightleftharpoons}} B + H_2O$ $K = [B]/[BH^+][OH^-]$
[b] Data taken from Awwal and Hibbert, 1977b; Hibbert and Hunte, 1983
[c] Conditions: 25°C, ionic strength, 0.1 mol dm^{-3}
[d] Conditions: 35°C, ionic strength, 0.1 mol dm^{-3}
[e] Conditions: 25°C, ionic strength, 0.4 mol dm^{-3}
[f] Conditions: 20.1°C, ionic strength, 0.1 mol dm^{-3}
[g] Conditions: 20.1°C, ionic strength, 0.08 mol dm^{-3}

Table 7

Kinetic and equilibrium acid-base properties of diaminonaphthalenes[a]

BH$^+$	pK_{BH^+}	k_{OH^-}/dm^3 mol^{-1} s^{-1}	Solvent	Reference
[60][e] 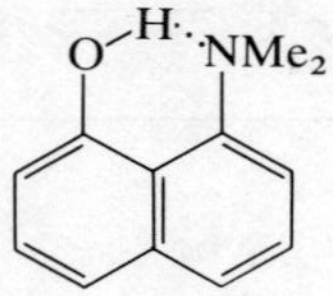	14.9	3.9×10^6	80%(v/v)Me_2SO-H_2O	*b*
[61][f] 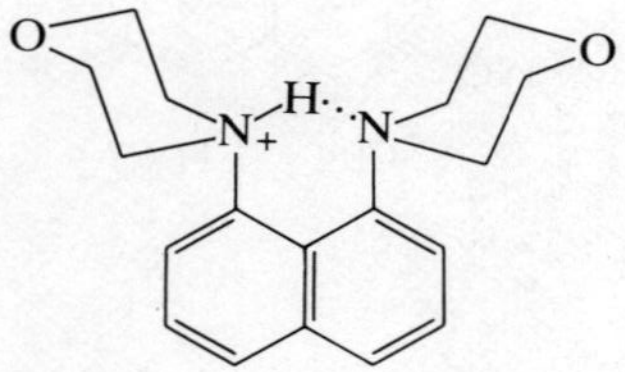	7.5	5.4×10^4	aq.	*c,d*
[62][f] 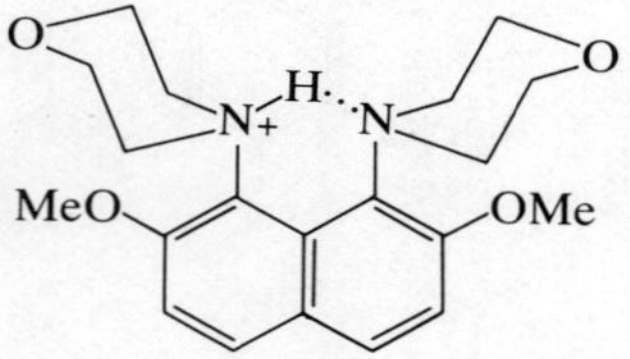	13.0	99.8	aq.	*c*

[63]	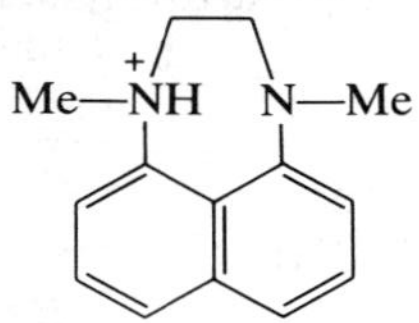 Me—$\overset{+}{N}$H N—Me	4.8		aq.	*c*
[64]	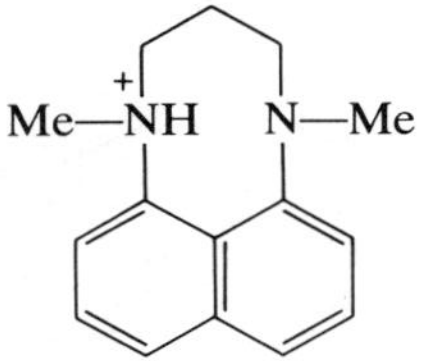 Me—$\overset{+}{N}$H N—Me	10.3		aq.	*c*
[65]	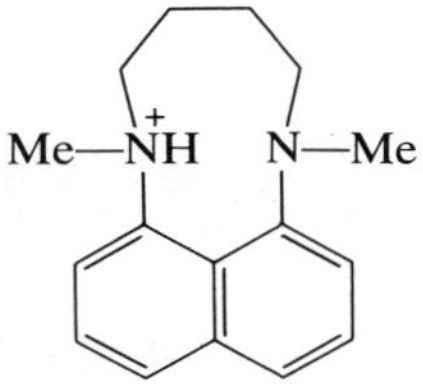 Me—$\overset{+}{N}$H N—Me	13.6	4.7×10^5	30%(v/v)Me_2SO-H_2O	*c*
[66]	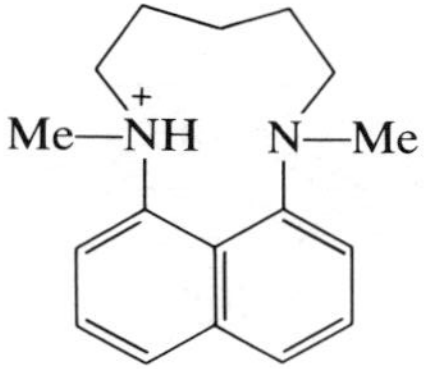 Me—$\overset{+}{N}$H N—Me	13.0	5.2×10^4	30%(v/v)Me_2SO-H_2O	*c*

TABLE 7 (continued)

BH^+	pK_{BH^+}	$k_{OH^-}/dm^3\ mol^{-1}\ s^{-1}$	Solvent	Reference
[67] 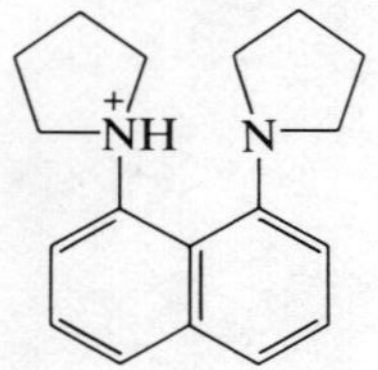	10.0	7.3×10^4	30%(v/v)Me_2SO-H_2O	*c*
[68] 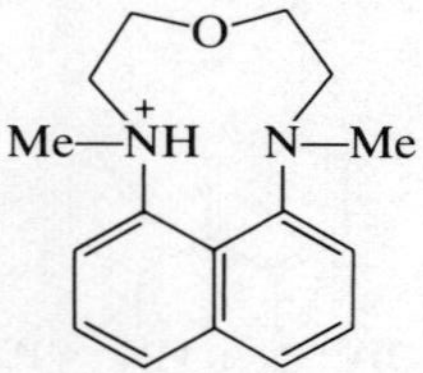	12.9	6.2×10^3	30%(v/v)Me_2SO-H_2O	*d*

[a] $BH^+ + OH^- \underset{}{\overset{k_{OH^-}}{\rightleftharpoons}} B + H_2O$
$pK_{BH^+} = -\log_{10}[B][H_3O^+]/[BH^+]$

[b] Conditions: 25°C, ionic strength, 0.1 mol dm^{-3} except where otherwise indicated. Awwal and Hibbert, 1977a

[c] Hibbert and Simpson, 1984

[d] Hibbert and Hunte, 1981

[e] Temperature: 10°C

[f] Ionic strength: 0.5 mol dm^{-3}

thalene. The relative pK-values of the diamines [63], [64], [65], and [66] illustrate the effect on basicity of the orientation of the nitrogen orbitals containing lone pair electrons. In [63], the lone pairs do not interact unfavourably and consequently there is negligible relief of strain on protonation; the basicity of [63] is similar to that of other aromatic amines. In [65] and [66] the interaction between the nitrogen lone pairs introduces strain into the amines which is reduced by protonation and the pK-values are therefore much higher. In addition, the protonated amines [65] and [66] are able to adopt a conformation in which the proton is held in an intramolecular hydrogen bond and this provides additional stabilisation of the protonated amines.

The rate coefficients for proton removal by hydroxide ion from the protonated amines in Tables 6 and 7 are exceptionally low, particularly in the case of 1,8-bis(diethylamino)-2,7-dimethoxynaphthalene [59] for which half-lives in the range of minutes are observed. As strain increases along the series of amines [56] to [59] in Table 6, the basicity increases and the value of the rate coefficient for proton removal by hydroxide ion from the protonated amines decreases. However, for a given change in substituent in the amine, the change in basicity, as reflected in the value of the equilibrium constant (K) between amine and protonated amine in the presence of hydroxide ion, is smaller than the change in rate coefficient for proton transfer (k_{OH^-}). For example, in going from 1,8-bis(dimethylamino)naphthalene [56] to 1,8-bis(diethylamino)naphthalene [57] the value of K decreases by a factor of 7.4 but the value of k_{OH^-} decreases 38-fold. Replacement of the N-methyl substituents in 1,8-bis(dimethylamino)-2,7-dimethoxynaphthalene [58] by N-ethyl substituents as in [59] has the effect of halving K, but the rate coefficient (k_{OH^-}) decreases by a factor of over 100. To understand these results in detail and to explain the exceptionally low rates of proton transfer, studies were carried out to identify the mechanism of reaction (Hibbert and Robbins, 1978, 1980; Barnett and Hibbert, 1984).

The kinetics of proton transfer from protonated 1,8-bis(dimethylamino)-2,7-dimethoxynaphthalene to substituted phenolate ions (69) were studied in 70% (v/v) $Me_2SO—H_2O$ using the temperature-jump technique with spectrophotometric detection to follow reactions with half-lives in the range 1–100 ms (Hibbert and Robbins, 1978). A limited

$$\text{[protonated 1,8-bis(dimethylamino)-2,7-dimethoxynaphthalene, } Me_2N^+\text{–H}\cdots NMe_2\text{, MeO, OMe]} + XC_6H_4O^- \underset{k_r}{\overset{k_f}{\rightleftharpoons}} \text{[1,8-bis(dimethylamino)-2,7-dimethoxynaphthalene, } Me_2N\text{, } NMe_2\text{, MeO, OMe]} + XC_6H_4OH \quad (69)$$

number of data were also obtained for 1,8-bis(diethylamino)-2,7-dimethoxynaphthalene under the same conditions. These much slower reactions with half-lives in the range of seconds to minutes were followed using a conventional spectrophotometer (Barnett and Hibbert, 1984). For reactions of both amines, suitable phenolate ions were chosen as base catalysts so that the condition $\Delta pK = pK(\text{protonated diamine}) - pK(XC_6H_4OH) \simeq 0$ was satisfied, and the results for the variation in $\log k_f$ and $\log k_r$ with ΔpK are given in Fig. 17. In the figure the dashed lines are drawn with slopes of zero

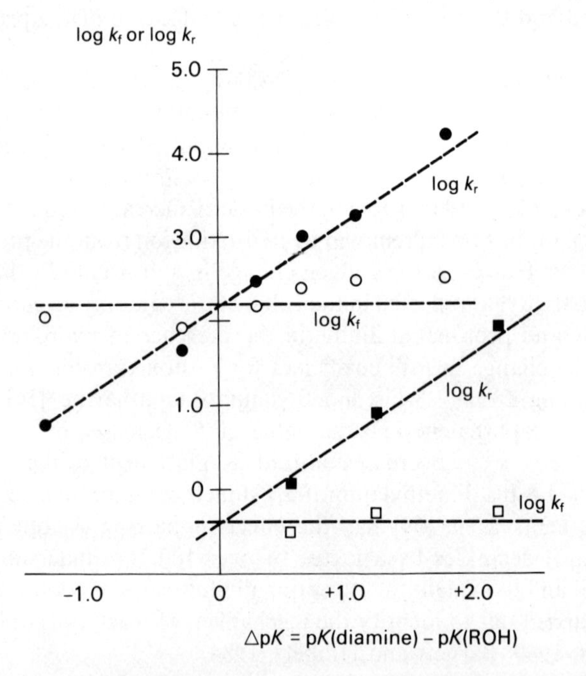

FIG. 17 Forward (k_f) and reverse (k_r) rate coefficients for the deprotonation of protonated 1,8-bis(dimethylamino)-2,7-dimethoxynaphthalene (○ and ●) and protonated 1,8-bis(diethylamino)-2,7-dimethoxynaphthalene (□ and ■) by bases

and unity and the behaviour is very closely similar to the results obtained for proton transfer involving 4-(nitrophenylazo)salicylate ion (Fig. 15). Actually the lines obtained by linear regression analysis for the data in Fig. 17 have slopes of 0.11 ($\log k_f$) and 1.04 ($\log k_r$) for 1,8-bis(dimethylamino)-2,7-dimethoxynaphthalene and 0.11 ($\log k_f$) and 1.14 ($\log k_r$) for 1,8-bis(diethylamino)-2,7-dimethoxynaphthalene. By the same arguments which were used for the reaction involving 4-(3-nitrophenylazo)salicylate ion, it can be concluded that the mechanism for proton removal from the protonated diaminonaphthalenes consists of unimolecular opening of the intramolecular hydrogen bond followed by proton removal from the open form of the protonated amine present in low concentration (Scheme 8; R = Me or Et). For this mechanism, it is predicted in Fig. 14 that curvature in the Eigen plot will occur at $\Delta pK = -\log K_1$ where K_1 is the equilibrium constant for opening of the intramolecular hydrogen bond (Scheme 8). This

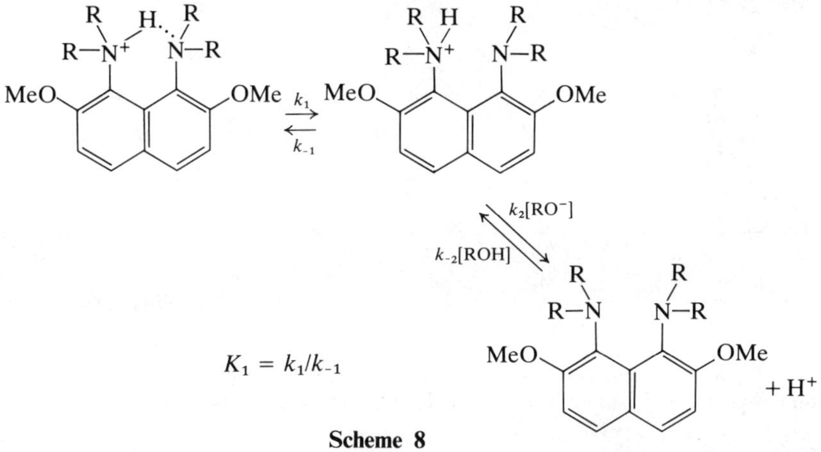

$$K_1 = k_1/k_{-1}$$

Scheme 8

prediction could not be tested for the reaction of 4-(3-nitrophenylazo)salicylate ion because the acid-base equilibrium with weak bases is established extremely rapidly when $\Delta pK > 1$. However, data in this region have been obtained for proton transfer from protonated 1,8-bis(dimethylamino)-2,7-dimethoxynaphthalene to phenolate ions and amines by use of an isotope-exchange method. Data which refer to reaction (69) in aqueous solution (Kresge and Powell, 1981) are given in Fig. 18. In constructing this figure from the original results the pK-value 16.1 for 1,8-bis(dimethylamino)-2,7-dimethoxynaphthalene was used. The behaviour in Fig. 18 supports the predictions in Fig. 14, according to which the point at which curvature in the Eigen plot occurs corresponds to $\Delta pK = -\log_{10} K_1$. Hence, from the data in Fig. 18, a K_1-value of about 10^{-6} is estimated for the equilibrium between open and closed forms of protonated 1,8-bis(dimethylamino)-2,7-dimethoxynaphthalene in aqueous solution. From (70) and the value $k_f \sim$

$$k_f = K_1 k_2; \quad k_r = k_{-2} \qquad (70)$$

$1.0\ dm^3\ mol^{-1}\ s^{-1}$, it follows that the proton transfer in the second step of Scheme 8 occurs with a rate coefficient $k_2 \sim 1 \times 10^6\ dm^3\ mol^{-1}\ s^{-1}$ when this proton transfer is thermodynamically favourable. Thus the overall

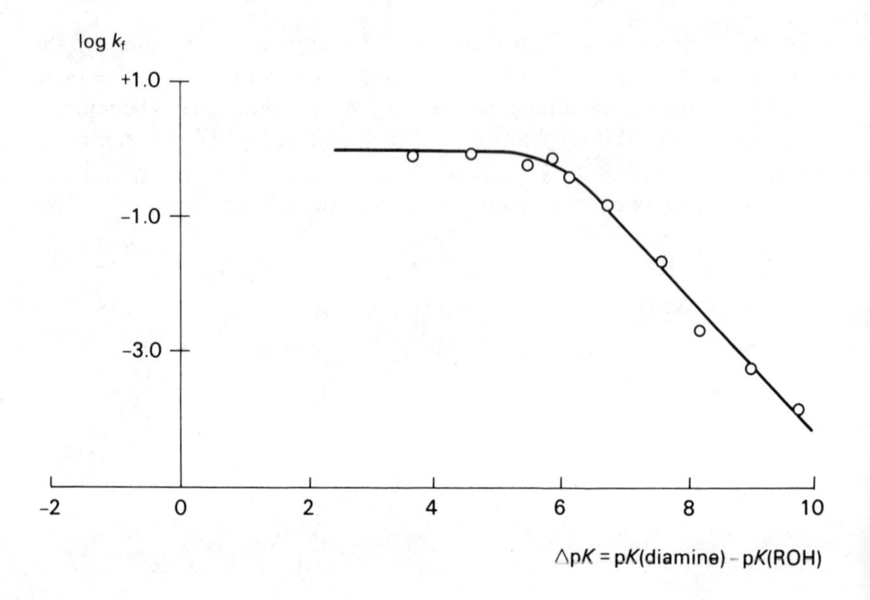

FIG. 18 Rate coefficients (k_f) for the deprotonation of protonated 1,8-bis-(dimethylamino)-2,7-dimethoxynaphthalene by phenolate ions in aqueous solution

reaction in (69) occurs slowly because (*a*) reaction occurs through an intermediate present to the extent of 1 part in 10^6 of the concentration of the protonated amine and (*b*) proton transfer from the open form occurs with a rate coefficient which is 10^3 to 10^4-fold below the diffusion-limited value. The equilibrium to give the open form of the protonated amine is very unfavourable partly because of the strong intramolecular hydrogen bond in the protonated amine and partly because the conformational change to give the open form is sterically hindered. A steric effect may also be responsible for the relatively low value of the rate coefficient for proton transfer from the open form to phenolate ions. A steric effect on proton transfer from the open form to hydroxide ion was used to explain the substituent effects on the equilibrium constant and rate coefficients for proton removal from the protonated amines in Table 6 (Awwal and Hibbert, 1977b; Hibbert and Hunte, 1983). Thus the rate coefficients for proton removal from the protonated amines [58] and [59] differ by a factor of 133 but the equilibrium constants for the same reactions differ only by a factor of 2.2. This may be due to a large steric effect to attack of hydroxide ion on the open form of protonated 1,8-bis(diethylamino)-2,7-dimethoxynaphthalene. For simple proton transfers with oxygen and nitrogen acids in aqueous solution, steric effects are usually quite unimportant. However, a modest steric effect has

$$Me_3C\text{-}C_5H_3N\text{-}CMe_3 + H_3O^+ \xrightarrow{k\,=\,3.7\times10^8\,dm^3\,mol^{-1}\,s^{-1}} Me_3C\text{-}C_5H_3\overset{+}{N}H\text{-}CMe_3 + H_2O \qquad (71)$$

been proposed to account for the magnitude of the rate coefficient for protonation of 2,6-di-t-butylpyridine in 20% (v/v) dioxan-water (71), which is about a fiftieth of the diffusion-limited value (Bernasconi and Carré, 1979a). A larger steric effect is observed (Bernasconi and Carré, 1979b) in reaction (72). Steric effects have been observed in proton transfer from carbon (Gold, 1962).

$$Ph_2C(\overset{+}{H}NR_1R_2)\text{-}C(NO_2)_2^- + \text{morpholine (NH)} \xrightarrow{k\,=\,2\times10^3\,dm^3\,mol^{-1}\,s^{-1}} Ph_2C(NR_1R_2)\text{-}C(NO_2)_2^- + H_2\overset{+}{N}\text{(morpholinium)} \qquad (72)$$

PROTON TRANSFER FROM HYDROGEN-BONDED PHENYLAZO-RESORCINOLS

Results which have an important bearing on the mechanism of proton removal from intramolecular hydrogen bonds have been obtained quite recently for the second ionisation of phenylazoresorcinols (73). Kinetic measurements were made using the temperature-jump technique (Perlmutter-Hayman and Shinar, 1975; Perlmutter-Hayman *et al.*, 1976).

$$\text{O=C}_6\text{H}_3(\text{O–H}\cdots\text{N(Ph)=N–}) + OH^- \underset{k_{H_2O}}{\overset{k_{OH^-}}{\rightleftharpoons}} \text{O=C}_6\text{H}_3(\text{O}^-)(\text{–N=N–Ph}) + H_2O \qquad (73)$$

For a reaction of the type shown in (74) with hydroxide ion in excess, the expected variation of the time constant (τ^{-1}) for the first-order approach to equilibrium after a temperature perturbation is given by (75). Thus a plot of reciprocal relaxation time (τ^{-1}) against hydroxide ion concentration is

$$ROH + OH^- \underset{k_r}{\overset{k_f}{\rightleftharpoons}} RO^- + H_2O \qquad (74)$$

$$\tau^{-1} = k_f[OH^-] + k_r \qquad (75)$$

expected to be linear and the ratio of the gradient (k_f) to the intercept (k_r) of the plot should be in agreement with the value of the separately measured equilibrium constant of the reaction. Data which appeared to behave in this way were obtained for different substituted phenylazoresorcinols in several laboratories (Eigen and Kruse, 1963; Inskeep *et al.*, 1968; Rose and Stuehr, 1971). However, using many of the same phenylazoresorcinols in extremely careful work, Perlmutter-Hayman showed that the dependence of relaxation time on hydroxide ion concentration was complex and τ^{-1} actually passed through a minimum value (Perlmutter-Hayman and Shinar, 1975, Perlmutter-Hayman *et al.*, 1976). In some cases, the minimum in τ^{-1} occurs at quite low hydroxide-ion concentrations and it is easy to understand why in the earlier experiments the dependence of τ^{-1} on hydroxide-ion concentration was thought to be linear.

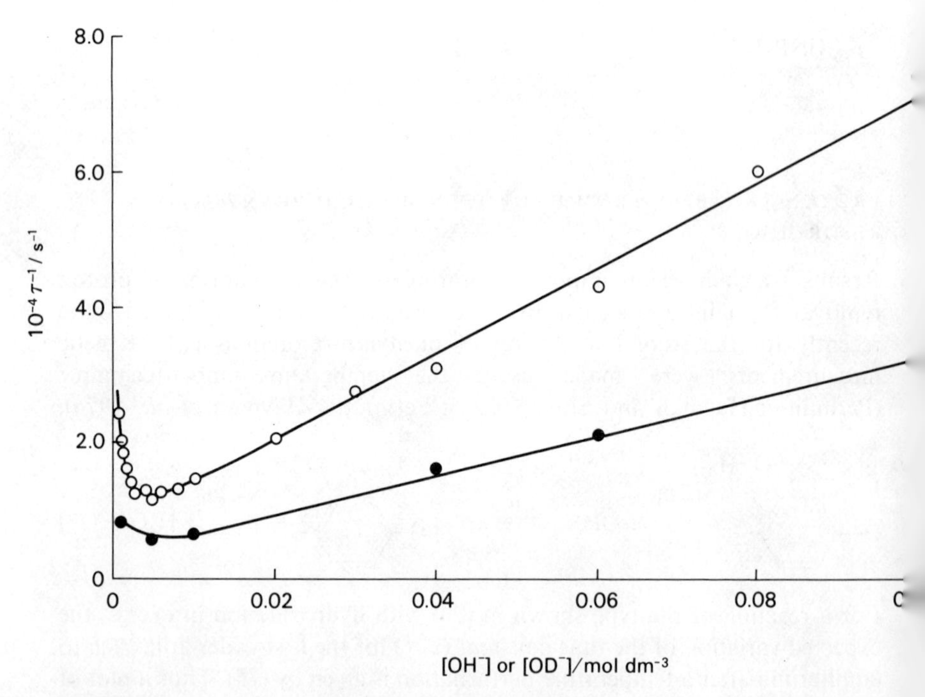

FIG. 19 Reciprocal relaxation time for the equilibration of 4-(4-sulphonatophenyl-azo)resorcinol with hydroxide ion in H_2O (○) and D_2O (●)

The first published observation of a minimum in τ^{-1} with increasing hydroxide-ion concentration refer to the reaction (76). These measurements have been repeated and the results are given in Fig. 19 (Hibbert and Simpson,

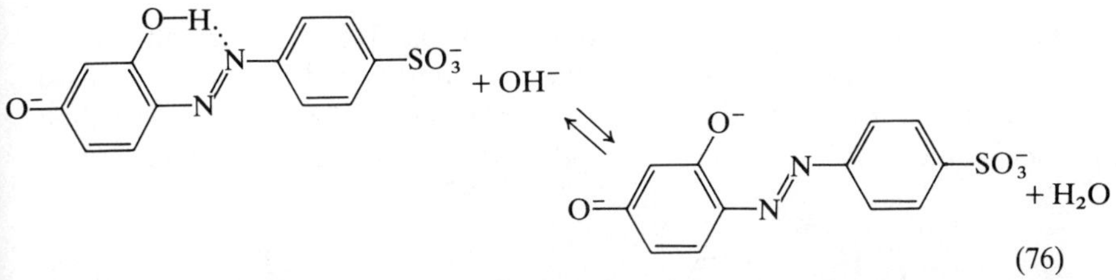

(76)

1984). At low hydroxide-ion concentrations, the rate of approach to equilibrium after a temperature jump decreases as the hydroxide-ion concentration increases. At higher concentrations the reaction becomes first order in hydroxide ion. The value of the kinetic solvent deuterium isotope effect on the reaction shows little variation over the range of hydroxide-ion concentrations studied as shown in Fig. 19. The ratio $\tau^{-1}(H_2O)/\tau^{-1}(D_2O)$ at a particular concentration of OL^- (L = H or D) remains within the range 2.0 to 3.0 for OL^- concentrations of 0.001 to 0.100 mol dm^{-3} and provides little mechanistic information. Similar results were obtained in the original work (Perlmutter-Hayman and Shinar, 1978).

A possible mechanism which explains the unusual kinetic behaviour is given in Scheme 9. According to this two routes from reactant to product are

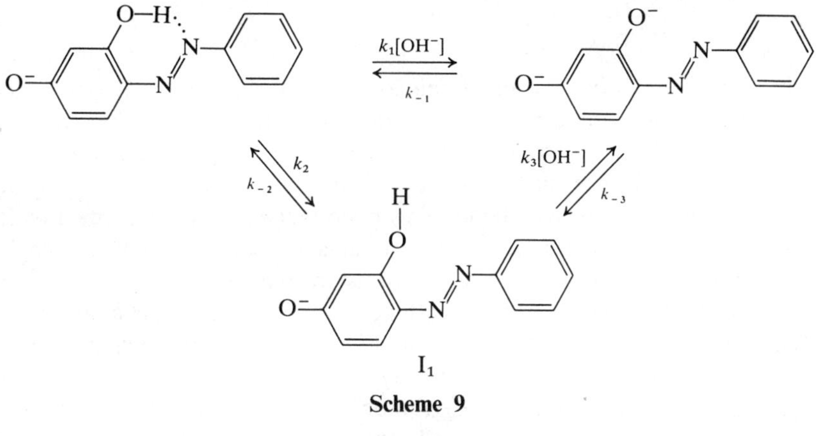

I_1

Scheme 9

important, direct single-step attack by hydroxide ion on the hydrogen-bonded proton (the upper route of Scheme 9) and two-step proton removal through an open form (I_1) of phenylazoresorcinol monoanion (the lower route). In order to satisfy the observed kinetic behaviour it is necessary to assume that the rate of proton removal from the open form (I_1) is much greater than the rate at which the open form reverts to the hydrogen-bonded species. If it can be shown that proton transfer occurs by the mechanism in Scheme 9, it would mean that phenylazoresorcinol monoanion behaves quite differently from other hydrogen-bonded acids. However, an alternative mechanism which will account for the minimum in τ^{-1} has been proposed (Yoshida and Fujimoto, 1977) and is given in Scheme 10. In this Scheme, the upper route, which is first order in hydroxide ion, can consist of direct attack

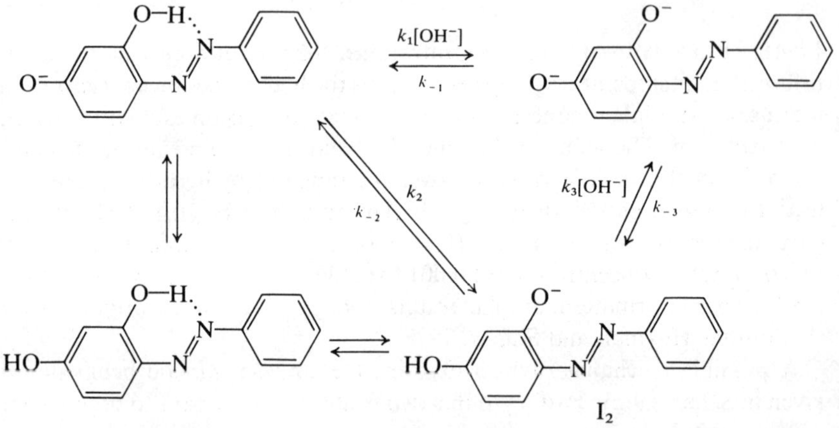

Scheme 10

by hydroxide ion or involve an intermediate step in which the hydrogen bond opens, whereafter a proton is removed in a rate-limiting step. The intermediate (I_2) on the lower route is an isomeric form of phenylazoresorcinol monoanion in which the proton is at a site where it is unable to participate in an intramolecular hydrogen bond. This intermediate could be formed by a protonation-deprotonation sequence through a low concentration of phenylazoresorcinol, as shown. The mechanisms in Schemes 9 and 10 predict the same dependence of τ^{-1} on hydroxide-ion concentration as given in (77), although the rate coefficients k_1 and k_2 refer to different reaction steps in the two mechanisms. In deriving (77) it is necessary to assume that $k_3[OH^-] \gg k_{-2}$. In (77), K is the equilibrium constant between the monoanion and dianion of phenylazoresorcinol in the presence of hydroxide ion;

$$\tau^{-1} = (k_2 + k_1[OH^-])(1 + 1/K[OH^-]) \tag{77}$$

K = [dianion]/[monoanion] $[OH^-]$. The line through the experimental data for the reaction of 4-(4-sulphonatophenylazo)resorcinol in H_2O as given in Fig. 19 is a best-fit of (77) using the values $k_1 = 6.4 \times 10^5\ dm^3\ mol^{-1}\ s^{-1}$ and $k_2 = 1.2 \times 10^3\ s^{-1}$ and the value $K = 112\ dm^3\ mol^{-1}$ obtained from separate equilibrium measurements. Thus both Schemes are compatible with the observed kinetic behaviour.

To provide information which may help the choice between the mechanisms in Schemes 9 and 10, the reciprocal relaxation time for the equilibration between the bisphenylazoresorcinol monoanions [69], [70], and [71] and the corresponding dianions was measured over a range of hydroxide-ion concentrations (Hibbert and Simpson, 1983, 1985). In each case a minimum

[69] [70] [71]

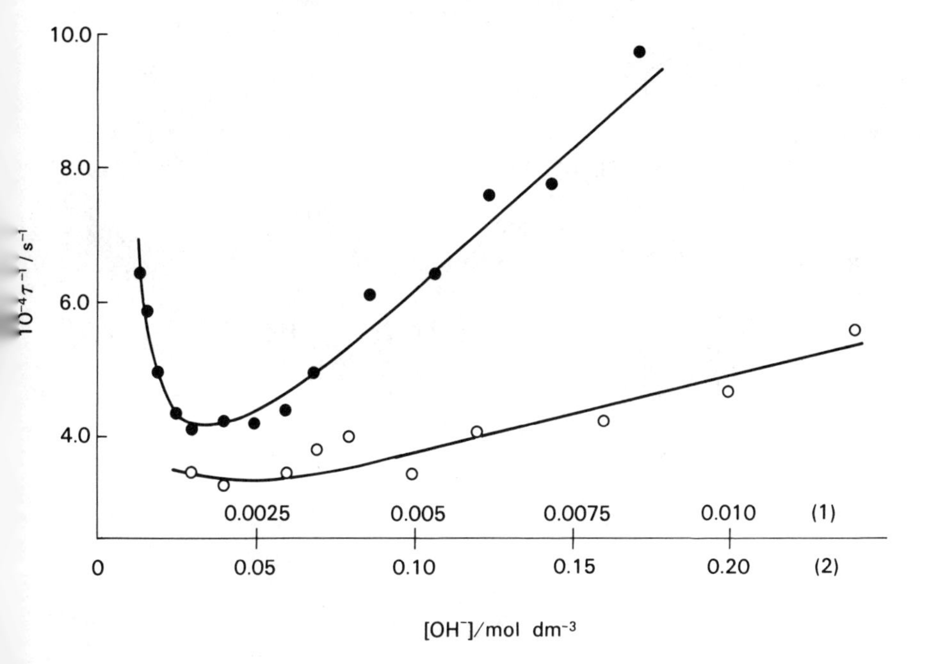

FIG. 20 Reciprocal relaxation time for the equilibration of the monoanions and dianions of 2,4-bis(phenylazo)resorcinol (●) and 4,6-bis(phenylazo)resorcinol (○) with hydroxide ion

τ^{-1} was observed although for [69] the minimum was indistinct. The results for [69] in 20% (v/v) dioxan-water at 5°C and ionic strength 0.1 mol dm^{-3} and for [70] in aqueous solution at 5°C and ionic strength 0.2 mol dm^{-3} are given in Fig. 20. If the mechanism in Scheme 10 is considered for [69] and [71], it is seen that the intermediate I_2 becomes identical to the intermediate I_1 in Scheme 9. For [69], [70], and [71] it is not possible to draw tautomeric forms in which a hydroxyl group is unable to participate in an intramolecular hydrogen bond. For 4-phenylazoresorcinol monoanion the intermediate I_2 in Scheme 10 is a tautomeric form of the hydrogen-bonded monoanion in which the proton is relocated at a site where it is not hydrogen-bonded. Thus for [69], [70], and [71] this modification means that Schemes 9 and 10 are reduced to Scheme 11. In Scheme 11 two routes to the open form of the monoanion (I_3) are considered, but it is likely that the route involving single-step opening of the intramolecular hydrogen bond predominates. The data for [69], [70], and [71] require that the upper route in Scheme 11 involve

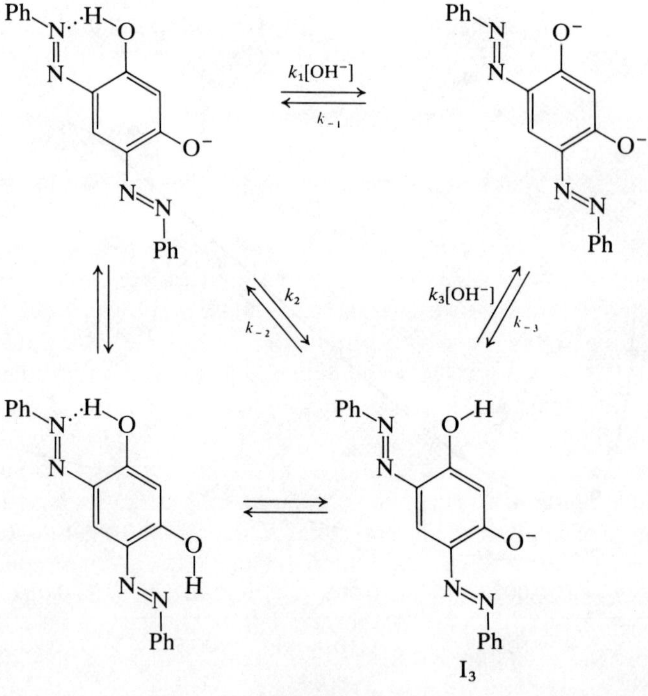

Scheme 11

direct single-step attack by hydroxide ion on the hydrogen-bonded proton (Hibbert and Simpson, 1983, 1985). The solid lines in Fig. 20 are constructed using best-fit values for the rate coefficients k_1 and k_2 in (77) and separately measured values for K. For [69] in 20% dioxan-water at 5°C and ionic strength 0.1 mol dm^{-3} the values were $k_1 = 2.5 \times 10^6 \text{dm}^3\text{mol}^{-1}\text{s}^{-1}$ and $k_2 = 2.3 \times 10^4 \text{s}^{-1}$ with $K = 2300$ dm^{3}mol^{-1}. For [70] in aqueous solution at 5°C and ionic strength 0.2 mol dm^{-3}, the results $k_1 = 4.1 \times 10^5 \text{dm}^3\text{mol}^{-1}\text{s}^{-1}$, $k_2 = 7.4 \times 10^3 \text{s}^{-1}$, and $K = 32.0$ dm^{3}mol^{-1} were obtained, whereas for [71] in 20% dioxan-water at 5°C and ionic strength 0.1 mol dm^{-3} the results were $k_1 = 4.5 \times 10^5$ dm^{3}mol^{-1}s^{-1}, $k_2 = 1.2 \times 10^4 \text{s}^{-1}$, and $K = 209$ dm^{3}mol^{-1}. In deriving equation (77) it is necessary to assume that proton removal from the open form of the phenylazoresorcinol monoanions occurs more rapidly than conversion of the open form to the hydrogen-bonded monoanion. Proton transfer from the open form is likely to be diffusion-controlled with k_3 *ca* $1 \times 10^{10} \text{dm}^3\text{mol}^{-1}\text{s}^{-1}$. Thus over the range of hydroxide-ion concentrations for which [70] was studied, it follows that $k_{-2} \ll k_3[\text{OH}^-]$ or $k_{-2} <$ *ca* $5.0 \times 10^7 \text{dm}^3\text{mol}^{-1}\text{s}^{-1}$. The result $k_2 = 7.4 \times 10^3 \text{s}^{-1}$ was deduced from the best-fit of (77) to the experimental data. Hence, combining this value with the maximum value $k_{-2} <$ *ca* $5.0 \times 10^7 \text{dm}^3\text{mol}^{-1}\text{s}^{-1}$ gives the result $K >$ *ca* 1.5×10^{-4} for the intramolecular hydrogen bond in 2,4-bis(phenylazo)resorcinol monoanion.

There is a further feature of the behaviour of phenylazoresorcinol monoanions which is of interest. It would be predicted that reaction by the upper route of Scheme 11 would be susceptible to general base catalysis, but reaction by the lower route would not be catalysed. Hence, catalysis of proton transfer by general bases will consist of direct attack by base on the hydrogen-bonded proton. In earlier parts of this section, data for base catalysis of the two-step proton transfer from a substituted salicylate ion and a protonated diaminonaphthalene were presented, and it was predicted that different results would be observed for direct single-step attack by general bases. Thus data for direct attack by general base on the hydrogen-bonded proton in phenylazoresorcinol monoanions are of great interest. Such data are available (Perlmutter-Hayman and Shinar, 1977), but the variation with base strength of catalytic coefficients for proton removal from 4-(4-sulphonatophenylazo)resorcinol monoanion shows scatter and no definite conclusions can be reached. This point is under further study in our laboratory.

The conclusions reached about proton transfer from phenylazoresorcinol monoanions are quite different from the behaviour which has been described for other hydrogen-bonded acids. For phenylazoresorcinol monoanions, it appears that direct attack by base on the hydrogen-bonded proton is an important process and can compete with two-step proton removal. For two-step proton transfer through an open form of the phenylazoresorcinol monoanion it is found that the rate of proton transfer from the open form is greater than the rate at which the hydrogen bond closes and the kinetic results show a complex dependence on the concentration of base. The reasons for the unusual behaviour of phenylazoresorcinols are obscure and studies of these interesting hydrogen-bonded species will continue.

5 Hindered proton transfer from molecular cavities

In Section 4 data were described that show that removal of the intramolecularly hydrogen-bonded proton by a general base from protonated 1,8-bis(dimethylamino)- and protonated 1,8-bis(diethylamino)-2,7-dimethoxynaphthalene occurs by a two-step process involving an open non-hydrogen-bonded form of the protonated amine. The reactions occur slowly because the hydrogen bonds are moderately strong and the open forms are present in low concentration and because attack of base on the non-hydrogen-bonded forms is sterically hindered. It was estimated that the effect of steric hindrance reduced the rate coefficient for proton removal by general bases from the open form of protonated 1,8-bis-(dimethylamino)-2,7-dimethoxynaphthalene by roughly four orders of magnitude below the diffusion limit. For 1,8-bis(diethylamino)-2,7-dimethoxy-naphthalene the rate reduction may be larger by a further two orders of magnitude (Barnett and Hibbert, 1984). Except in extreme cases such as this, steric effects on simple proton transfers involving oxygen and nitrogen acids in aqueous solution are usually quite small. It is thought that such proton transfer can occur through intervening solvent water bridges so that the reactants need not approach closely. In non-aqueous solvents larger steric effects are often observed. For example, proton exchange between trimethyl-piperidinium ion and trimethylpiperidine (78) occurs with a rate coefficient $k = 1.13 \times 10^8 \mathrm{dm^3 mol^{-1} s^{-1}}$ at 25°C in aqueous solution, whereas in

[Scheme: trimethylpiperidinium ion (H, Me, N^+, Me, Me) + trimethylpiperidine (Me, N, Me, Me) $\underset{}{\overset{k}{\rightleftharpoons}}$ trimethylpiperidine (Me, N, Me, Me) + trimethylpiperidinium ion (H, Me, N^+, Me, Me)] (78)

dimethyl sulphoxide in which proton transfer occurs directly between the acid and base without an intervening solvent molecule, the rate coefficient has a value of $2.58 \times 10^4 \mathrm{dm^3 mol^{-1} s^{-1}}$ (Bianchin *et al.*, 1975). For the proton exchange between ammonia and ammonium ion, for which close approach of the reactants is not sterically hindered, the value of the rate coefficient is $1.17 \times 10^9 \mathrm{dm^3 mol^{-1} s^{-1}}$ in aqueous solution and $1.21 \times 10^9 \mathrm{dm^3 mol^{-1} s^{-1}}$ in dimethyl sulphoxide. For roughly equal concentrations of 2,6-di-t-butylpyridine and the conjugate acid in methylene chloride at $-23°C$, distinct nmr signals are observed for each species, showing that proton exchange is slow on the nmr timescale and in fact is thought to proceed by intermediate proton transfer to the counter ion (Gold and Lee, 1984). The exchange becomes rapid, on the nmr timescale, at room temperature.

The most dramatic rate retardations of proton transfers have been observed when the acidic or basic site is contained within a molecular cavity. The first kinetic and equilibrium studies of the protonation of such a basic site were made with large ring bicyclic diamines [72] (Simmons and Park, 1968; Park and Simmons, 1968a). It was also observed (Park and Simmons, 1968b) that chloride ion could be trapped inside the diprotonated amines. The binding of metal ions and small molecules by macrocyclic compounds is now a well-known phenomenon (Pedersen, 1967, 1978; Lehn, 1978). In the first studies of proton encapsulation, equilibrium and kinetic measurements were made with several macrobicyclic diamines [72] using an nmr technique.

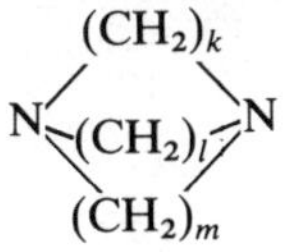

$(CH_2)_k$, $(CH_2)_l$, $(CH_2)_m$ bridging N and N

$k = l = m = 7, 8, 9, 10$ or
$k.l.m. = 6.6.8, 6.8.10, 8.8.10$

[72]

The diamines and their mono- and di-protonated ions can exist in various conformations, in which the nitrogen lone pairs and the protons on nitrogen are directed in (i or i^+) or directed out (o or o^+) from the molecular cavity. Diprotonated 1,10-diazabicyclo[8.8.8]hexacosane, for example, may exist in either of the three forms in equation (79). When the out-out isomer of 1,10-diazabicyclo[8.8.8]hexacosane bis hydrochloride ($o^+ o^+$) is dissolved in aqueous acidic solution, isomerisation to the in-in isomer ($i^+ i^+$) occurs so

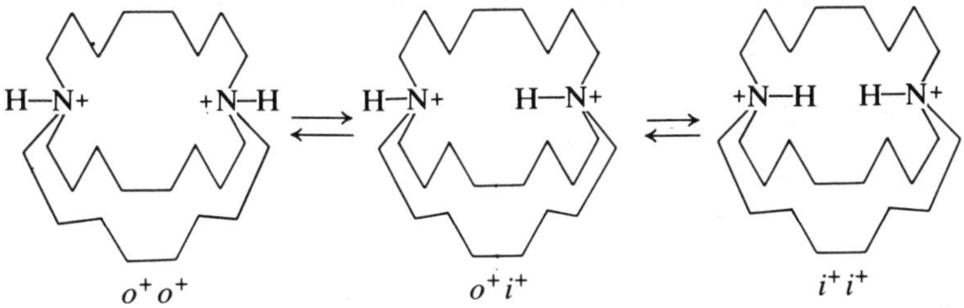

$$o^+o^+ \rightleftarrows o^+i^+ \rightleftarrows i^+i^+ \quad (79)$$

that at equilibrium a negligible concentration of the out-out isomer remains (Simmons and Park, 1968; Park and Simmons, 1968b). For some of the bicyclic diamines, for example $k.l.m. = 6.8.10$ in [72], almost equal concentrations of the out-out and in-in isomers are present at equilibrium (Simmons and Park, 1968). For out-out diprotonated 1,10-diazabicyclo[8.8.8]hexacosane, slow isomerisation to the in-in isomer was followed by nmr spectroscopy and the process was found to be first order in the diprotonated diamine. In the presence of low concentrations of acid (0.05–0.2 mol dm^3) the value of the first-order rate coefficient was independent of acid concentration and the isomerisation occurred with a half-life of 61 min at 23°C. At higher acid concentrations (0.2–1.0 mol dm^{-3}) the rate coefficient was found to be inversely proportional to acid concentration. An eight-stage mechanism involving monoprotonated diamines in various conformations as intermediates was proposed to explain the results.

Medium-ring bicyclic diamines have now been prepared and their proton transfer behaviour has been investigated (Alder, 1983). The diamine 1,6-diazabicyclo[4.4.4]tetradecane [73] is reluctant to undergo outside-protonation, presumably because outward pyramidalisation of the nitrogen atoms introduces strain; the diamine exists in the in-in conformation. The strain involved

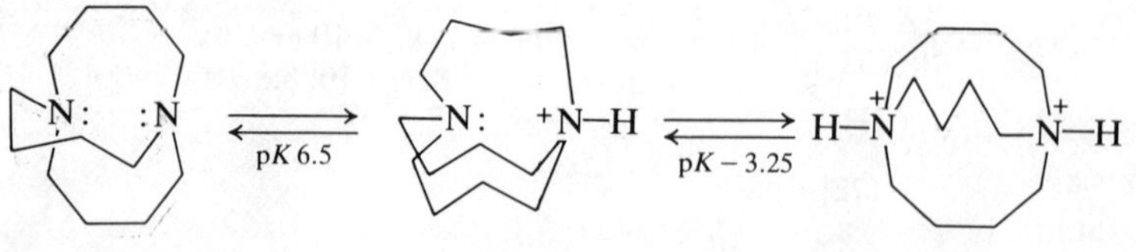

pK 6.5 pK − 3.25

[73] [74] [75]

in outward pyramidalisation means that the outside-monoprotonated species [74] (pK = 6.5) is four units more acidic than a monoprotonated acyclic diamine and the outside diprotonated species [75] (pK = −3.25) is ten units more acidic than a diprotonated acyclic diamine (Alder and Sessions, 1979; Alder *et al.*, 1979, 1982). Inside-monoprotonated 1,6-diazabicyclo[4.4.4]-tetradecane [76] could not be prepared by a simple proton-transfer reaction, but could be prepared and isolated under conditions favourable to electron

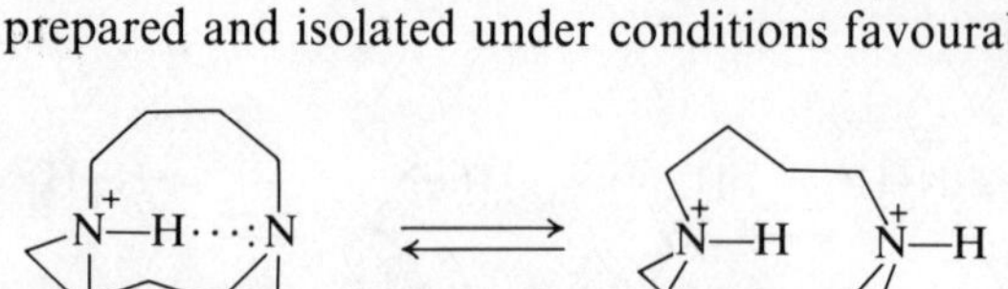

i^+i [76] i^+o^+ [77]

transfer. Under these conditions it was shown that the proton in the isolated product actually originated from a methylene group in a position α to nitrogen (Alder *et al.*, 1979). The proton in [76] is held in a strong intramolecular hydrogen bond with a $^+$N—H···N distance of 253 pm

(Alder *et al.*, 1983b). In the nmr spectrum the proton resonance occurs at δ 17.4 and the infra-red spectrum shows a broad absorbance due to $^{+}N{-}H\cdots N$ at *ca* 1400 cm^{-1}. The monoprotonated ion [76] is extremely stable towards deprotonation and is recovered unchanged from 1 mol dm^{-3} NaOD in D_2O and from solutions of sodium amide in liquid ammonia. It is not known whether the reaction is kinetically unfavourable or thermodynamically unfavourable or both (Alder *et al.*, 1979). When treated with 1 : 1 HSO_3F/SbF_5, [76] forms an in-out diprotonated ion [77] and the hydrogen bond which was present in [76] is no longer present in [77] (Alder *et al.*, 1979).

The formation of inside-protonated monocations from medium-ring bicyclic diamines has been examined for a number of ring sizes (Alder *et al.*, 1983a). The larger ring species (*k.l.m* in [72] of 5.4.2, 6.4.2, 6.3.3, 5.5.2, 6.4.3, 5.5.3, 6.5.3, and 5.5.4) form inside-monoprotonated species (i^{+} i) by simple proton transfer in $CDCl_3$ containing CF_3CO_2H. However, the reactions are slow and in some cases, for example 5.5.4, take several days to reach completion. The smaller ring species (4.4.4, 5.4.3, and 5.4.4) require electron-transfer conditions for formation of the inside-monoprotonated ions. For several bicyclic diamines in which the rings are even smaller, for example 3.3.2, formation of the inside-monoprotonated species could not be detected under any conditions.

There is considerable interest in the properties of macrobicyclic cryptands, for example [78] to [81], and particularly in their ability to complex protons, metal ions, and small molecules (Lehn, 1978). In the proton cryptates there exists the possibility of intramolecular $^{+}N{-}H\cdots N$ hydrogen bonding as well as interaction of the proton with the oxygen atoms, and the properties are also strongly influenced by the size of the molecular cavity. In the [1.1.1]-cryptand [78] the molecular cavity is small (Cheney *et al.*, 1978) and

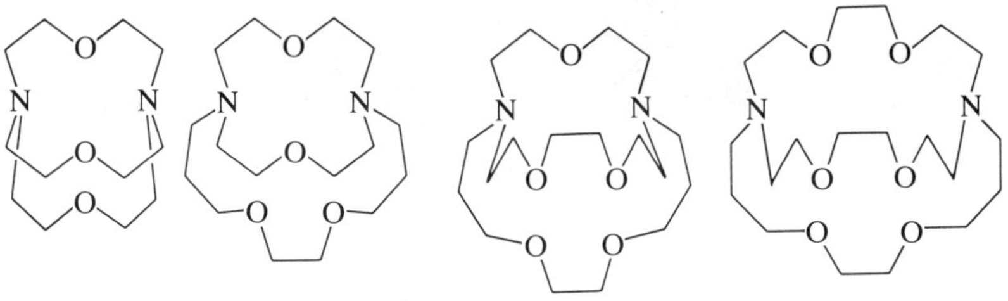

[78] [79] [80] [81]

the proton transfer behaviour shows several remarkable features (Cheney and Lehn, 1972; Smith *et al.*, 1981). The five monoprotonated or diprotonated species in Scheme 12 have been observed by nmr spectroscopy under various conditions (Smith *et al.*, 1981). The p*K*-values (7.1 and *ca* 1) corresponding to dissociation of the two outside-monoprotonated ions (io^{+} and

oo^+, respectively) were obtained from nmr studies in aqueous solution. The i^+i species is about half-protonated to give the i^+o^+ species in 1 mol dm^{-3} hydrochloric acid and this suggests p$K \sim 0$ for dissociation of i^+o^+ into i^+i. The equilibria between ii and i^+i and between i^+i and i^+i^+ could not be studied directly and values of pK *ca* 8 (for i^+i^+) and p$K >$ *ca* 18 (for i^+i) were estimated from kinetic data (Smith *et al.*, 1981).

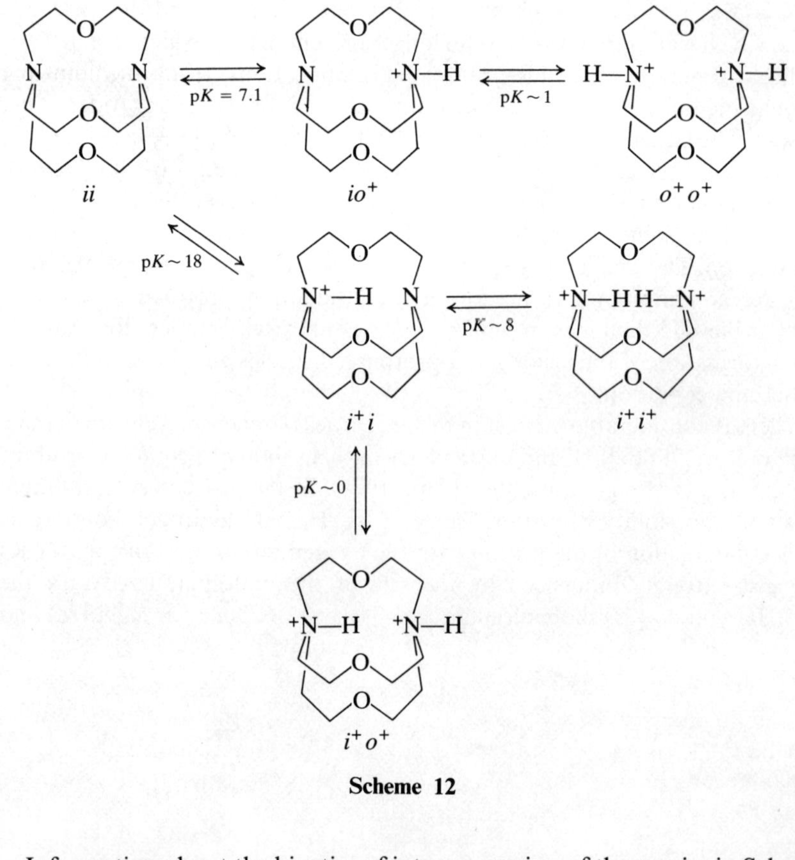

Scheme 12

Information about the kinetics of interconversion of the species in Scheme 12 has been obtained (Smith *et al.*, 1981). The values of the rate coefficients for external protonation of ii to give io^+ and o^+o^+ are probably close to the diffusion-controlled limit. However, the rate of internal monoprotonation of ii to ii^+ is quite low and the reaction can be followed by observing the change in nmr signals with time. At pH 1 and 25°C the half-life is 7 min. Under these conditions, insertion of the second proton into the cavity takes several weeks to reach completion, but can be observed in convenient times at higher temperatures. Removal of the first proton from i^+i^+ to give i^+i takes weeks in $5\,\text{mol dm}^{-3}$ aqueous potassium hydroxide but at 90°C occurs with a half-life of 1.38 hours. The pK-value for dissociation of i^+i^+ to give i^+i is similar to that of a diprotonated acyclic diamine. The apparent stability of i^+i^+ towards proton removal arises because the reaction is kinetically difficult and not because it is thermodynamically unfavourable. For ii^+, removal of the proton cannot be accomplished in concentrated potassium hydroxide solutions at high temperatures and attempts to use more extreme conditions result in decomposition of the cryptate. The [1.1.1]-cryptand is an extremely strong base towards internal protonation (p$K > 18$), but the process occurs quite slowly indicating that the cryptand is a poor base kinetically.

For cryptands in which the molecular cavity is larger than in the case of the [1.1.1]-species [78], proton transfer in and out of the cavity can be observed more conveniently. Proton transfer from the inside-monoprotonated cryptands [2.1.1] [79], [2.2.1] [80], and [2.2.2] [81] to hydroxide ion in aqueous solution has been studied by the pressure-jump technique, using the conductance change accompanying the shift in equilibrium position after a pressure jump to follow the reaction (Cox *et al.*, 1978). The temperature-jump technique has also been used to study the reactions. If an equilibrium, such as that given in equation (80), can be coupled with the faster acid-base equilibrium of an indicator, then proton transfer from the proton cryptate to hydroxide ion

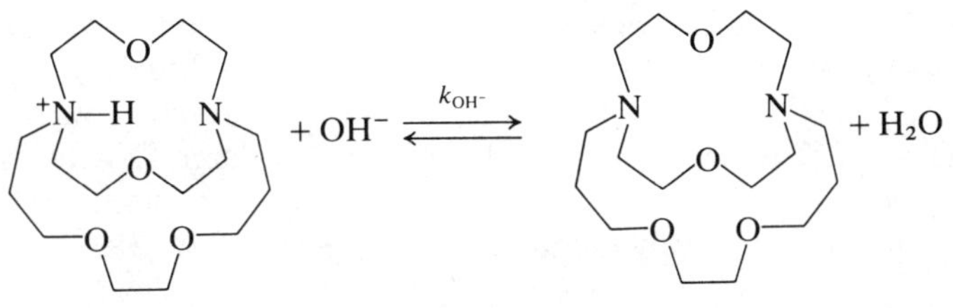

$$\text{(proton cryptate, }{}^{+}\text{N—H)} + OH^- \underset{}{\overset{k_{OH^-}}{\rightleftharpoons}} \text{(cryptand)} + H_2O \qquad (80)$$

can be followed spectrophotometrically (Pizer, 1978). The cryptands [79], [80], and [81] are quite strong bases and the proton cryptates have pK-values of 11.17, 11.78, and 9.86 respectively. Proton transfer to hydroxide ion (80) occurs quite slowly and $k_{OH^-} = 1.1 \times 10^3\,\text{dm}^3\text{mol}^{-1}\text{s}^{-1}$ at 25°C for the [2.1.1]-cryptand (Cox *et al.*, 1978). For the [2.2.2]-species with a somewhat larger cavity, $k_{OH^-} = 1.0 \times 10^7\,\text{dm}^3\text{mol}^{-1}\text{s}^{-1}$. The observed first-order dependence of the rate on hydroxide ion concentration is compatible with direct single-step attack by hydroxide ion on these proton cryptates. The rate coefficients have values which are well below the diffusion-controlled limit and this reflects the hindrance to attack of hydroxide ion as a result of the protection of the proton within the molecular cavity and because of the intramolecular hydrogen bond. For the [2.2.1]-cryptand, the kinetic data were not compatible with a single-step attack by hydroxide ion because the rate of reaction was found to approach a limiting value at high hydroxide-ion concentrations. Possible explanations involving conformational changes of the cryptand or cryptate before and after proton removal have been considered (Cox *et al.*, 1978).

Monoprotonation of the [2.1.1]-cryptand occurs rapidly but protonation of the monoprotonated species by hydronium ion and other acids can be followed kinetically in various solvents (Cox *et al.*, 1982, 1983). In methanol, protonation of ii^+ species by substituted acetic and benzoic acids to give i^+i^+ has been studied using the stopped flow technique with conductance detection. The values of the rate coefficients (k_{HA}) for protonation (81) vary with the acidity of the donor acid from $k_{HA} = 563\ dm^3mol^{-1}s^{-1}$ (for 4-hydroxybenzoic acid) to $k_{HA} = 2.3 \times 10^5\ dm^3mol^{-1}s^{-1}$ (for dichloroacetic acid).

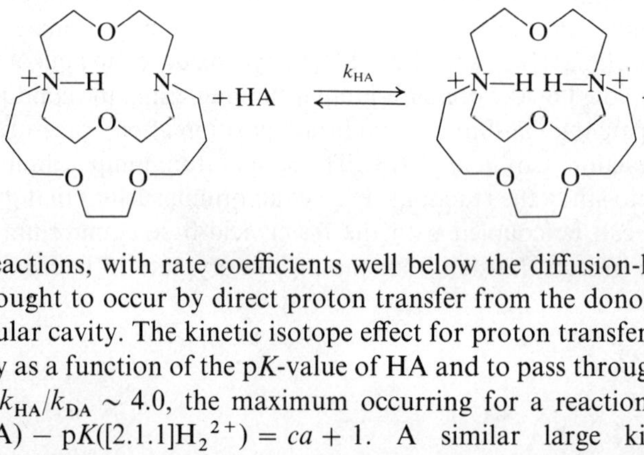

$$\overset{+}{\mathrm{N}}\text{—H} \quad \mathrm{N} \;+\; \mathrm{HA} \;\overset{k_{HA}}{\rightleftharpoons}\; \overset{+}{\mathrm{N}}\text{—H H—}\overset{+}{\mathrm{N}} \;+\; \mathrm{A}^- \qquad (81)$$

The reactions, with rate coefficients well below the diffusion-limited values, are thought to occur by direct proton transfer from the donor acid into the molecular cavity. The kinetic isotope effect for proton transfer was observed to vary as a function of the p*K*-value of HA and to pass through a maximum value $k_{HA}/k_{DA} \sim 4.0$, the maximum occurring for a reaction with $\Delta pK = pK(HA) - pK([2.1.1]H_2{}^{2+}) = ca + 1$. A similar large kinetic isotope effect $k_{HA}/k_{DA} = 3.9$ was observed for protonation of the cryptand by H_2O and D_2O in the isotopically different solvents (Kjaer *et al.*, 1979).

6 Multiple proton transfers

Many of the proton transfers which are postulated to occur as steps in chemical reactions are more complex than the processes which have been considered in Sections 1 to 5 and they frequently involve simultaneous or stepwise transfer of several protons. For example, in the charge-relay mechanism of hydrolyses catalysed by chymotrypsin and other serine protease enzymes (Blow, 1976), it is thought that a proton is transferred from a serine hydroxyl group (Ser-195) to aspartate (Asp-102) through an intervening histidine group (Hist-57). The proposed mechanism of the chymotrypsin-catalysed hydrolysis of 4-nitrophenyl trimethylacetate given in (82) to (86) involves formation of trimethylacetylchymotrypsin,

$$(CH_3)_3C.CO.OC_6H_4NO_2 + EOH \overset{K_m}{\rightleftharpoons} (CH_3)_3C.CO.OC_6H_4NO_2,EOH \quad (82)$$

$$(CH_3)_3C.CO.OC_6H_4NO_2,EOH \xrightarrow{k_2} (CH_3)_3C.CO.OE + HOC_6H_4NO_2 \quad (83)$$

$$(CH_3)_3C.CO.OE + H_2O \xrightarrow{k_3} (CH_3)_3C.CO.OH + EOH \quad (84)$$

$$d[HOC_6H_4NO_2]/dt = (k_2/K_m)[(CH_3)_3C.CO.OC_6H_4NO_2][EOH] \quad (85)$$

$$d[(CH_3)_3C.CO.OH]/dt = k_3[(CH_3)_3C.CO.OE] \quad (86)$$

$(CH_3)_3C.CO.OE$, by attack of the serine hydroxyl group on the carbonyl carbon atom of 4-nitrophenyl trimethylacetate. This attack is assisted by proton removal from the hydroxyl group by the charge-relay mechanism, Scheme 13. It is also considered that breakdown of trimethylacetylchymotrypsin may be assisted by the charge-relay mechanism. In this case, a proton

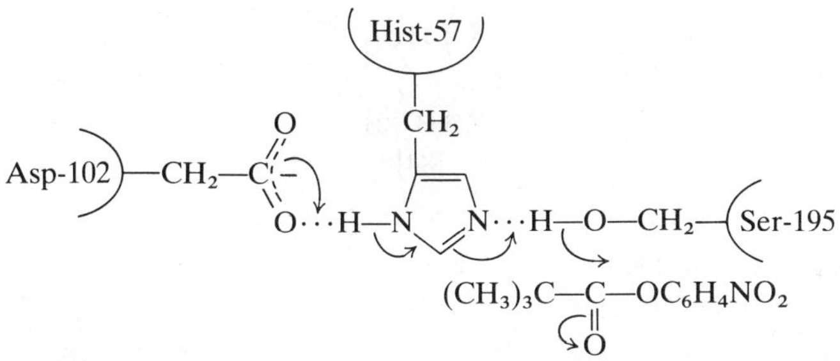

Hist-57

CH_2

Asp-102 — CH_2—C(=O)—O$^-$ $\cdots$ H—N ⟨imidazole⟩ N $\cdots$ H—O—CH_2— Ser-195

$(CH_3)_3C$—C(=O)—$OC_6H_4NO_2$

Scheme 13

is removed from the water molecule which attacks the carbonyl group of the acylated enzyme (Blow, 1976). The rate-pH dependencies of the acylation and deacylation steps shown in Fig. 21 are compatible with participation of a group on the enzyme with p$K \sim 7.0$ that is active in the basic form (Bender *et al.*, 1962; Bender, 1971). Assuming this to be the carboxylate group of aspartate-102 (Blow, 1976), the inactive and active forms of chymotrypsin may correspond to the species [82] and [83] respectively. Theoretical calculations (Gandour *et al.*, 1974) of potential energy surfaces for models of the charge-relay system and for models of bifunctional catalysis (Jencks, 1969; Bender, 1971; Kirby, 1980) as for example in [84], have suggested that stepwise transfer of one proton followed by the other, or concerted proton transfer in which the protons are transferred simultaneously are both plausible mechanisms. Chemical models of the charge-relay system have also been devised (Mallick *et al.*, 1984).

Hist-57
CH₂ → CH_2
Asp-102—CH_2—C(=O)—O—H···N⟨imidazole⟩N—H···O(H)—CH_2—Ser-195

[82] inactive

Hist-57
CH_2
Asp-102—CH_2—C(O)(O)⁻ O···H—N⟨imidazole⟩N···H—O—CH_2—Ser-195

[83] active

X ↑ H—A
Y M
Z—H :B

[84]

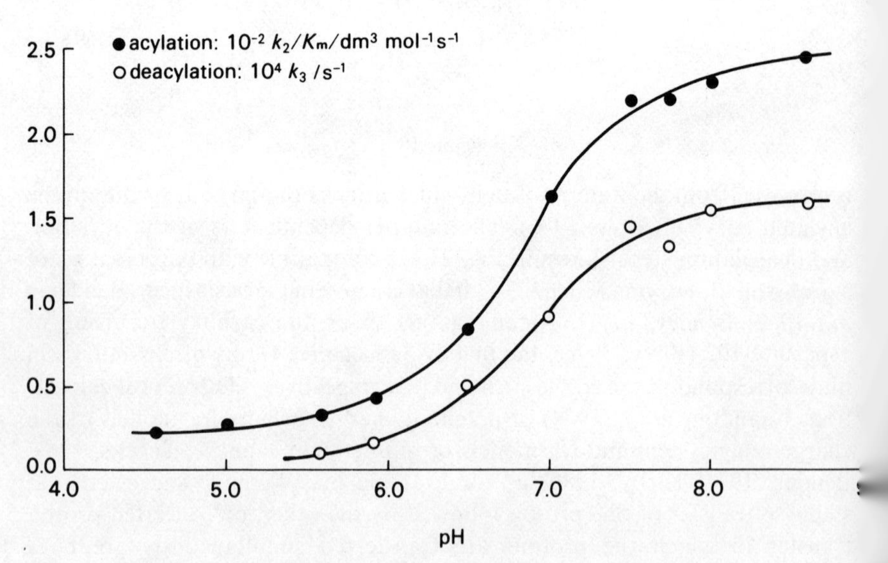

FIG. 21 Rate-pH dependence of the acylation and deacylation steps in the chymotrypsin-catalysed hydrolysis of 4-nitrophenyl trimethylacetate

In carbonyl addition reactions, a commonly occurring and important mechanistic step is the transfer of a proton from one site to another in a reactive intermediate (proton switch). If the proton switch occurs sufficiently rapidly compared with the rate of collapse of the intermediate to reactants, the overall reaction may be facilitated by trapping of the unstable intermediate by the proton switch (Jencks, 1976). For example, in the formation of oximes from the reaction of benzaldehyde with O-methylhydroxylamine shown in (87)–(89) (Sayer and Jencks, 1973; Rosenberg *et al.*, 1974), the first unstable intermediate (I_1) on the reaction pathway is converted by a proton switch (88) to the intermediate (I_2) which has less tendency than I_1 to

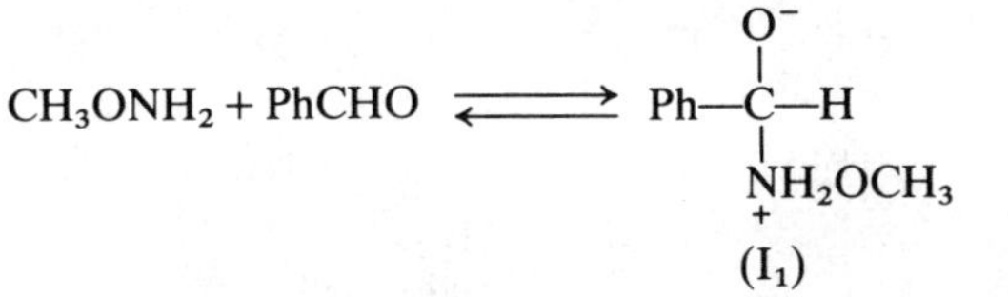

$$CH_3ONH_2 + PhCHO \rightleftharpoons Ph{-}C(O^-)(H){-}\overset{+}{N}H_2OCH_3 \quad (I_1) \tag{87}$$

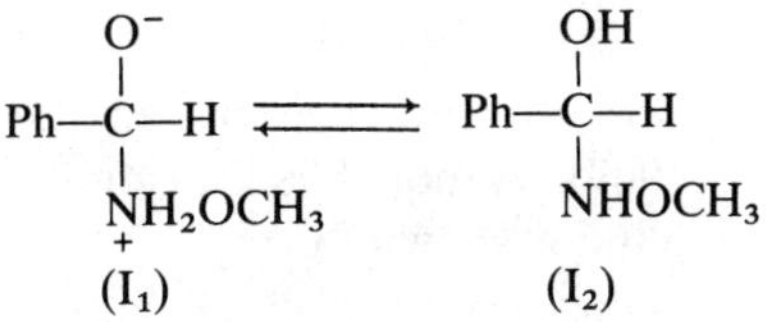

$$Ph{-}C(O^-)(H){-}\overset{+}{N}H_2OCH_3 \; (I_1) \rightleftharpoons Ph{-}C(OH)(H){-}NHOCH_3 \; (I_2) \tag{88}$$

$$Ph{-}C(OH)(H){-}NHOCH_3 \longrightarrow PhCH{=}N{-}OCH_3 \tag{89}$$

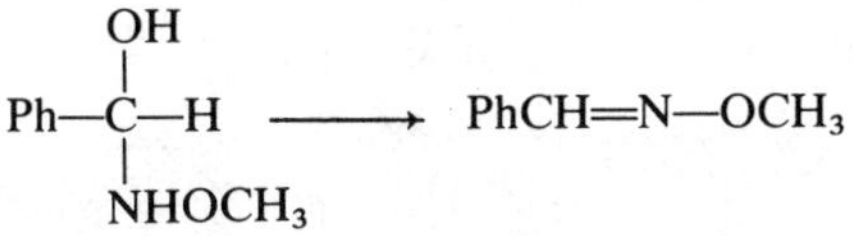

collapse to reactants. It has been suggested that the proton switch occurs through two bridging solvent molecules between which protons are transferred in a concerted pathway. In the transition state for this concerted transfer, it is thought that proton transfer to oxygen is more advanced than proton removal from nitrogen, as in [85]. In some cases bifunctional catalysts

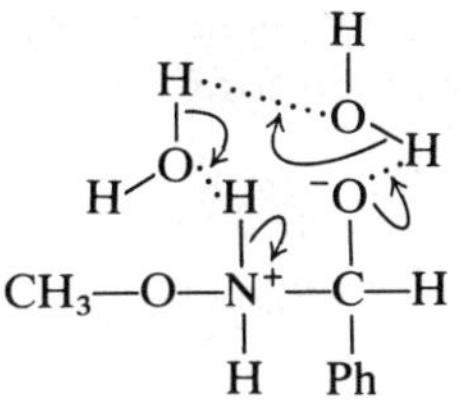

$$CH_3{-}O{-}\overset{+}{N}(H)(\cdots H\cdots O(H)H\cdots O(H)H\cdots{}^-O){-}C(H)(Ph)$$

[85]

may take part in proton-switch mechanisms. An example occurs in the intramolecular aminolysis of S-acetylmercaptoethylamine (90)–(92), in which the catalytic coefficient observed for bicarbonate ion is 30-fold larger

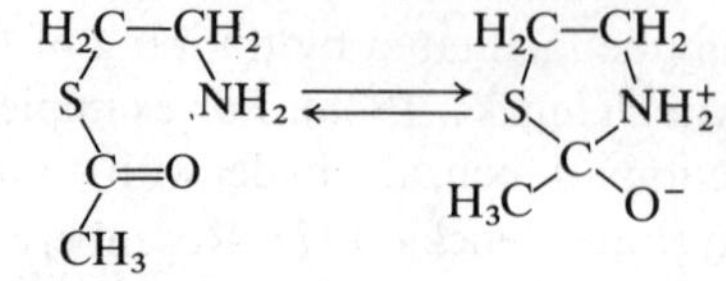

(90)

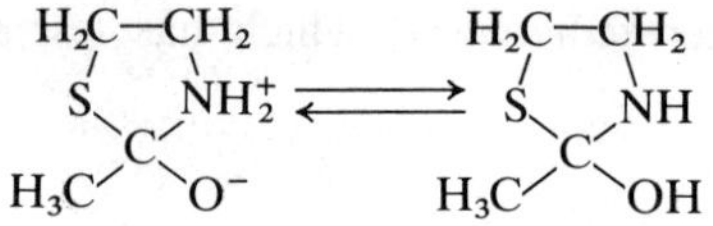

(91)

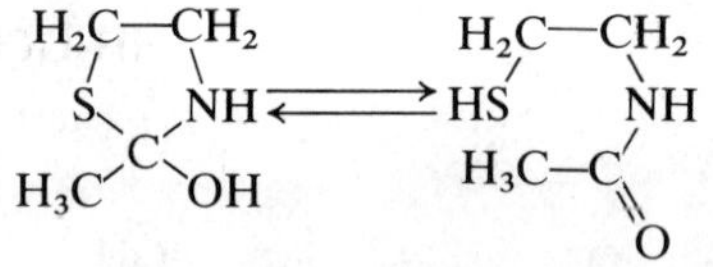

(92)

than that observed for monofunctional general acid catalysts of the same pK (Barnett and Jencks, 1969). A possible transition state for bifunctional catalysis by bicarbonate ion is shown in [86].

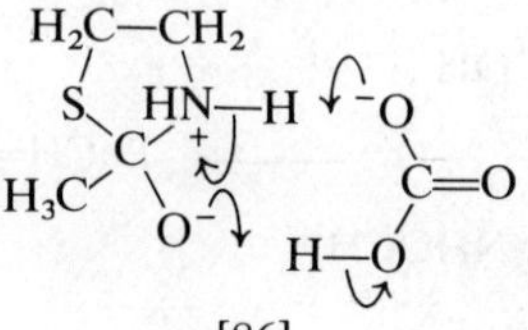

[86]

The tautomerisation of the purine bases adenine and guanine and of the pyrimidine bases thymine, cytosine, and uracil has important implications in molecular biology, and the occurrence of rare tautomeric forms of these bases has been suggested as a possible cause of spontaneous mutagenesis (Löwdin, 1965; Pullman and Pullman, 1971; Kwiatowski and Pullman, 1975). Three of the most likely tautomers for cytosine are shown in [87]–[89], together with the less likely imino forms [90] and [91] (Scanlan and Hillier,

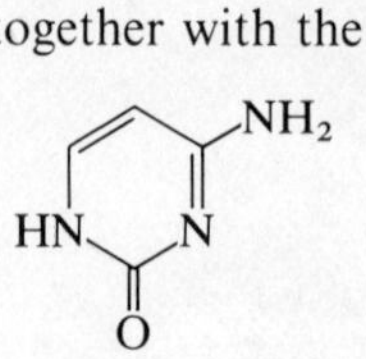

[87] 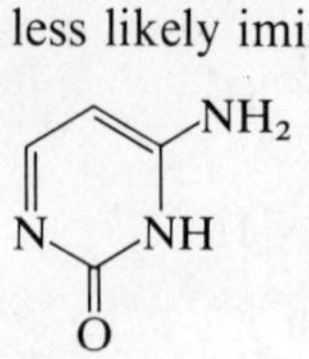 [88] 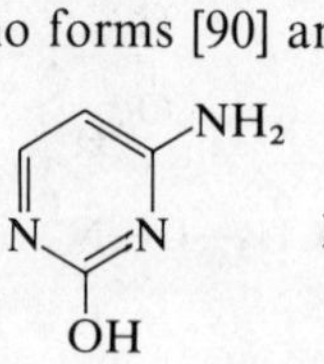 [89] 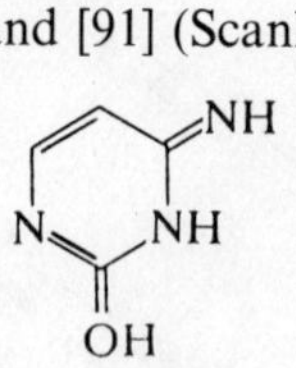 [90] 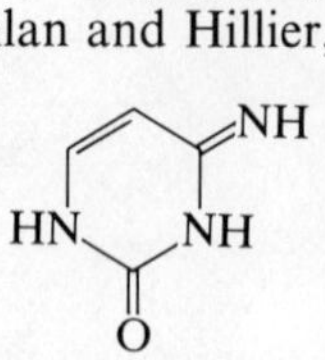 [91]

1984). In DNA, base pairing occurs between guanine and cytosine in its most stable form [87], as in [92]. However, cytosine in its imino form [91] is able to form a hydrogen bond to adenine as in [93], and such a base pairing could lead to the formation of mutations if it continued in DNA replication. The kinetics of the tautomerisation of purine and pyrimidine bases has been

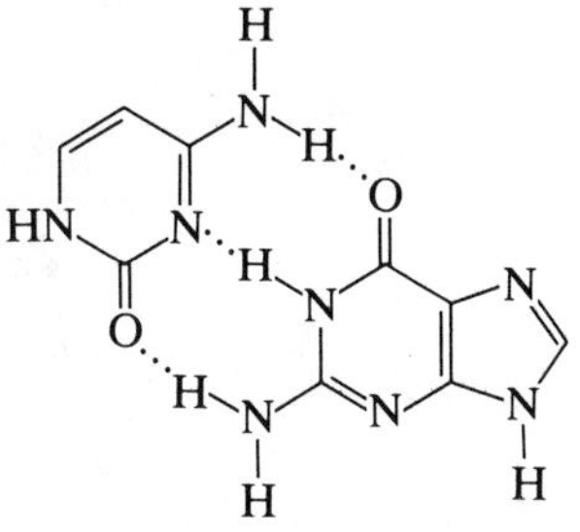

[92]

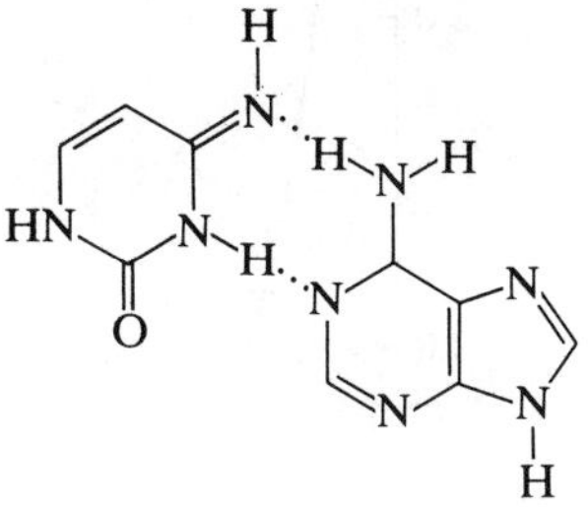

[93]

studied by the research groups of Grunwald and Dubois as described in the following paragraph. Studies of the kinetics of the proton-transfer steps in these reactions and in other tautomerisations involving oxygen and nitrogen centres are considered in this section.

TAUTOMERISATION OF OXYGEN AND NITROGEN ACIDS AND BASES

The tautomerisation of the N(9)H- and N(7)H-isomers of adenine (93) has been studied in aqueous solution using the temperature-jump technique with Joule heating (Dreyfus *et al.*, 1975). The equilibrium constant $K =$ [N(7)H]/[N(9)H] has a value of 0.28 at 20°C. The reciprocal relaxation time

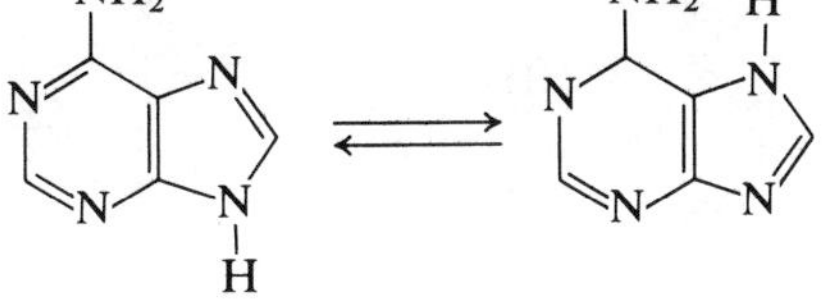

[N(9)H] ⇌ [N(7)H] (93)

$$\tau^{-1} = k_0 + k[H_3O^+] + k'[OH^-] + k''[OH^-][\text{adenine}] \quad (94)$$

for the equilibration between the isomers following a temperature-jump shows a U-shaped dependence on pH as in Fig. 22. The data are fitted by (94) in which k_0 has the value 210 s^{-1} and is thought to refer to water-catalysed interconversion; this may involve water acting as an acid catalyst (95). The second term in (94) involves catalysis by hydronium ion according to (96).

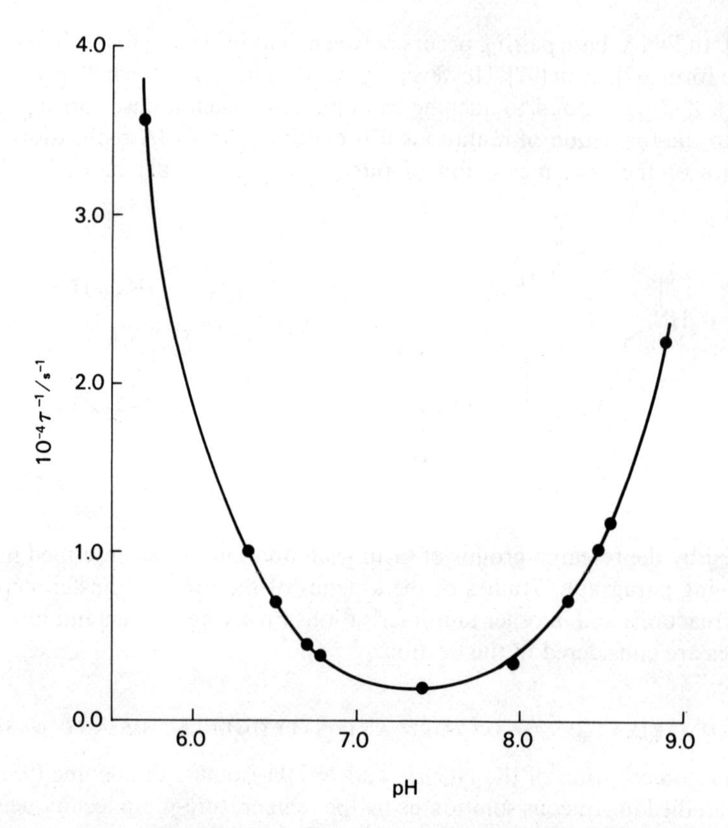

FIG. 22 Variation with pH of the reciprocal relaxation time for the equilibration between N(7)H and N(9)H isomers of cytosine

$$\mathrm{N(9)H + H_2O} \rightleftarrows [\mathrm{NH_2},\ \mathrm{N},\ \mathrm{NH},\ +,\ \mathrm{NH},\ \mathrm{N}] + \mathrm{OH^-} \rightleftarrows \mathrm{N(7)H + H_2O} \qquad (95)$$

For this mechanism, values of $k_1 = k_{-2} = 1.3 \times 10^{10}\,\mathrm{dm^3 mol^{-1} s^{-1}}$ are calculated from the experimental value for k in (94), and this means that the proton-transfer steps in (96) are diffusion-controlled in the thermodynamically favourable directions. The hydroxide ion catalysed tautomerisation

$$\mathrm{N(9)H + H_3O^+} \underset{k_{-1}}{\overset{k_1}{\rightleftarrows}} [\mathrm{NH_2},\ \mathrm{N},\ \mathrm{NH},\ +,\ \mathrm{NH},\ \mathrm{N}] + \mathrm{H_2O} \underset{k_{-2}}{\overset{k_2}{\rightleftarrows}} \mathrm{N(7)H + H_3O^+} \qquad (96)$$

occurs according to (97) and the proton-transfer steps are again diffusion-controlled with values of the rate coefficients $k_3 = k_{-4} = 0.85 \times 10^{10} dm^3 mol^{-1} s^{-1}$. The final term in (94) arises from a

$$\mathrm{N(9)H + OH^-} \underset{k_{-3}}{\overset{k_3}{\rightleftharpoons}} [\text{adeninate anion}] + \mathrm{H_2O} \underset{k_{-4}}{\overset{k_4}{\rightleftharpoons}} \mathrm{N(7)H + OH^-} \qquad (97)$$

mechanism for tautomerisation catalysed by adeninate ion, (98), and the values $k_5 = 0.75 \times 10^8 dm^3 mol^{-1} s^{-1}$ and $k_{=5} = 2.7 \times 10^8 dm^3 mol^{-1} s^{-1}$ are found for the forward and reverse rate coefficients. These results are

$$\mathrm{N(9)H} + [\text{adeninate anion}] \underset{k_{-5}}{\overset{k_5}{\rightleftharpoons}} \mathrm{N(7)H} + [\text{adeninate anion}] \qquad (98)$$

similar to the value obtained (Marshall and Grunwald, 1969) in nmr studies of the identity reaction (99) between purine and its anion for which a rate coefficient $k_6 = 1.0 \times 10^8 dm^3 mol^{-1} s^{-1}$ was found. Thus the kinetics of the interconversion of the N(7)H and N(9)H isomers of adenine can be explained by stepwise proton transfers through cationic or anionic intermediates which are present in low concentrations under most conditions. Such pathways are

$$[\text{purine}] + [\text{purine anion}] \overset{k_6}{\rightleftharpoons} [\text{purine anion}] + [\text{purine}] \qquad (99)$$

described as dissociative. Similar results have been obtained (Dreyfus *et al.*, 1976) for the tautomerisation of the 1(H)- and 3(H)-amino-oxo forms of cytosine (100), and for the equilibration of the amino-oxo and imino-oxo-tautomers of 3-methylcytosine (101). In both cases a U-shaped dependence

$$\text{1(H)} \rightleftharpoons \text{3(H)} \qquad (100)$$

$$K = [3(H)]/[1(H)] = 2.5 \times 10^{-3}$$

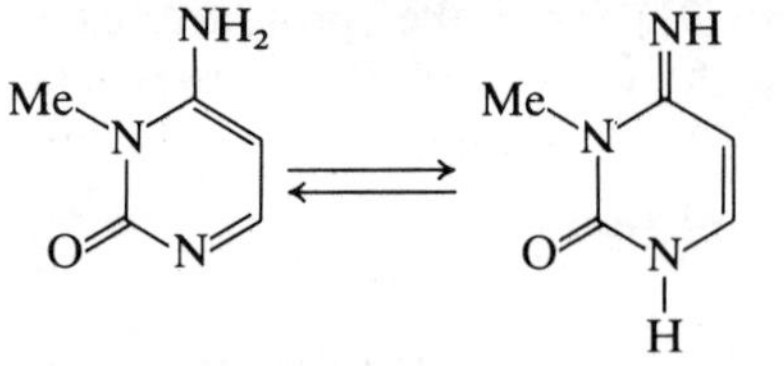

$$K = [\text{imino-oxo}]/[\text{amino-oxo}] = 0.03 \qquad (101)$$

of τ^{-1} on pH was observed. The reactions were studied using the temperature-jump technique with Joule heating and with spectrophotometric detection. The kinetics of the interconversions (100) and (101) were explained by assuming dissociative pathways through cationic and anionic intermediates which were formed or reacted at diffusion-controlled rates. The reactions occurred measurably slowly with relaxation times τ *ca* 10^{-4} s because the intermediates are present in low concentration. For cytosine, a pH-independent term in the rate expression was explained by the occurrence of a pathway, (102) and (103), involving acid catalysis by water. A similar route was found in the tautomerisation of 3-methylcytosine.

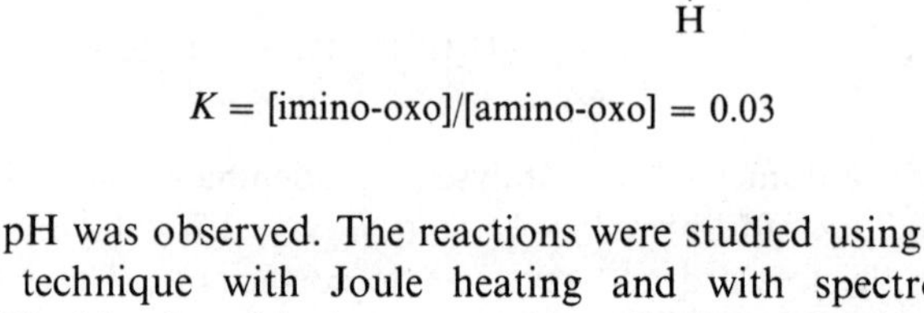

$$+ H_2O \rightleftharpoons + OH^- \qquad (102)$$

$$+ OH^- \rightleftharpoons + H_2O \qquad (103)$$

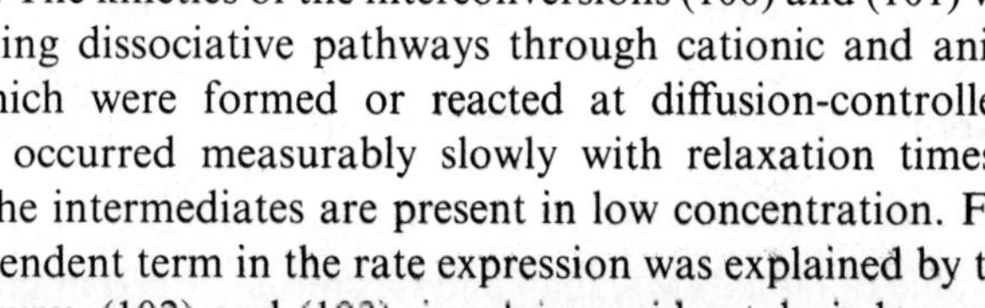

There is interest in the possibility of tautomerisation occurring by a non-dissociative pathway. The pH-independent terms in the rate expressions for the tautomerisations of adenine, cytosine, and 3-methylcytosine were explained by dissociative pathways with water acting as an acid catalyst. However, evidence from nmr studies (Grunwald *et al.*, 1963; Grunwald and Meiboom, 1963; Luz and Meiboom, 1963; Grunwald and Eustace, 1975) indicates that proton transfer between the oxygen atoms of carboxylic acids in aqueous solution and in methanol may occur through intervening solvent molecules in a non-dissociative pathway (104). The rate coefficient for

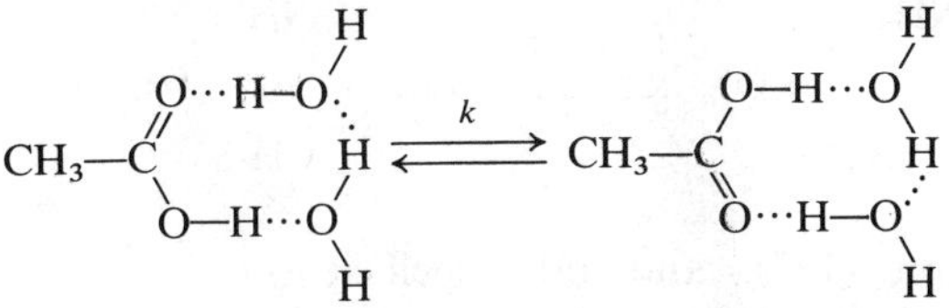

$$CH_3C(=O\cdots H{-}O(H){\cdots}H{-}O(H)\cdots H{-})O \;\underset{}{\overset{k}{\rightleftharpoons}}\; CH_3C(O{-}H\cdots O(H){-}H\cdots O(H){-}H\cdots)=O \qquad (104)$$

process (104) was found to have a value $k = 4.8 \times 10^7 s^{-1}$. Quite similar results have been obtained from ultrasonic absorption measurements for the intramolecular proton transfers occurring between the thiol and amino groups of cysteine [94], homocysteine [95], penicillamine [96], cysteamine [97], and glutathione [98] (Maass and Peters, 1972). The reciprocal relaxation time (τ^{-1}) for the intramolecular proton transfer (105) is identical with the

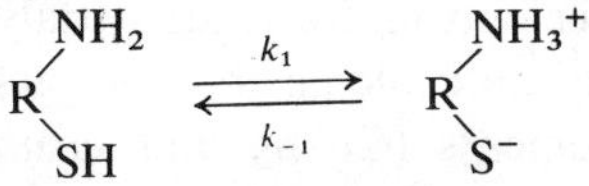

$$R(NH_2)(SH) \underset{k_{-1}}{\overset{k_1}{\rightleftharpoons}} R(NH_3^+)(S^-) \qquad (105)$$

sum of the forward and reverse rate coefficients and the values of τ^{-1} for each of the compounds [94] to [98] are given below the formulae. These results may refer to direct proton transfer without the participation of water

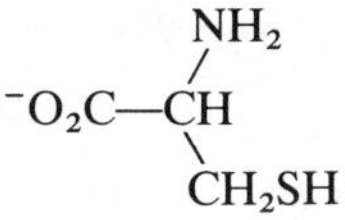

$^-O_2C{-}CH(NH_2)CH_2SH$ [94] $3.6 \times 10^8 s^{-1}$

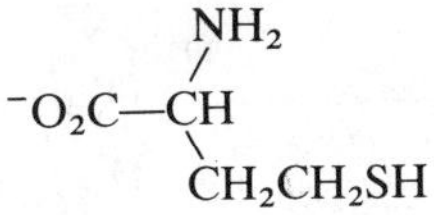

$^-O_2C{-}CH(NH_2)CH_2CH_2SH$ [95] $2.3 \times 10^7 s^{-1}$

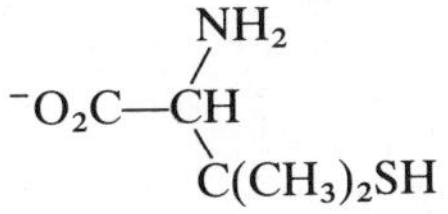

$^-O_2C{-}CH(NH_2)C(CH_3)_2SH$ [96] $2.0 \times 10^7 s^{-1}$

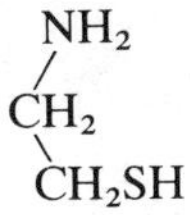

$NH_2CH_2CH_2SH$ [97] $3.0 \times 10^7 s^{-1}$

$^-O_2C{-}CH(NH_2)(CH_2)_2CONHCH(CONHCH_2CO_2^-)CH_2SH$ [98] $6.3 \times 10^7 s^{-1}$

molecules between the amino and thiol groups. Studies using nmr spectroscopy to follow the reaction of cysteine [94] give a result which is different from that which is obtained by the ultrasonic technique. The nmr technique is thought to give rate coefficients for the reaction shown in (106) and values of $k_f \sim k_r \sim 2 \times 10^6 s^{-1}$ were obtained (Grunwald *et al.*, 1976). In the same work, proton transfer between the amino groups of lysine was

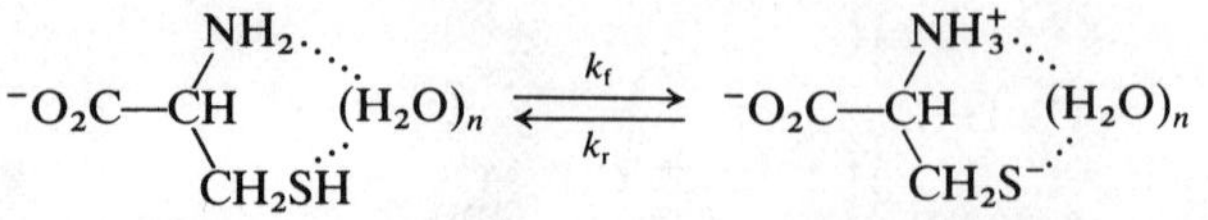

$$^{-}O_2C{-}CH(NH_2\cdots)(CH_2SH)\ (H_2O)_n \underset{k_r}{\overset{k_f}{\rightleftharpoons}} {}^{-}O_2C{-}CH(NH_3^+\cdots)(CH_2S^-\cdots)\ (H_2O)_n \qquad (106)$$

studied, equation (107), and rate coefficients $k_f = 4.7 \times 10^6 s^{-1}$ and $k_r = 5.2 \times 10^7 s^{-1}$ were obtained. The average value $n = 1.6 \pm 0.5$ was found for the number of water molecules participating in the proton transfer.

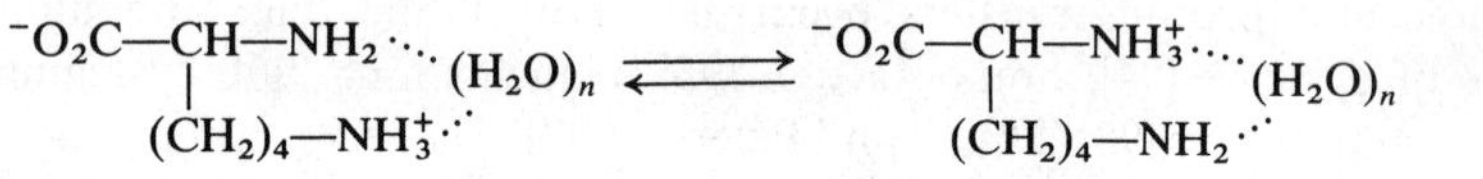

$$^{-}O_2C{-}CH(NH_2\cdots)(CH_2)_4{-}NH_3^+\cdots\ (H_2O)_n \rightleftharpoons {}^{-}O_2C{-}CH(NH_3^+\cdots)(CH_2)_4{-}NH_2\cdots\ (H_2O)_n \qquad (107)$$

A non-dissociative mechanism for water-catalysed tautomerisation of uracil monoanion and of imidazole has been suggested on the basis of nmr line broadening measurements (Chang and Grunwald, 1976). However, doubts have been expressed about these conclusions (Bensaude *et al.*, 1978). The mechanisms deduced from nmr measurements are shown in (108) and (109) and involve water behaving as a bifunctional catalyst. Rate coefficients

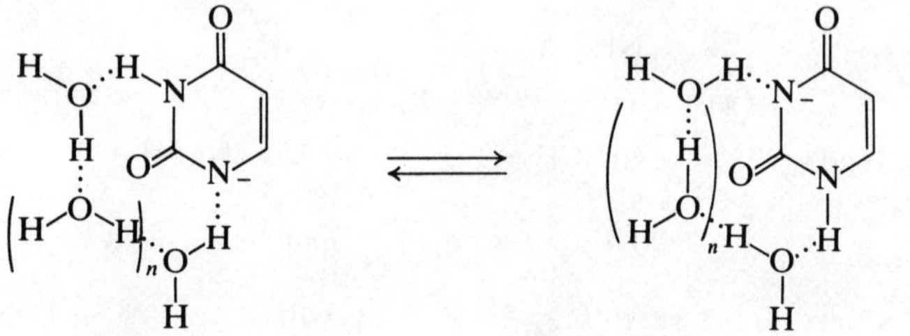

(108)

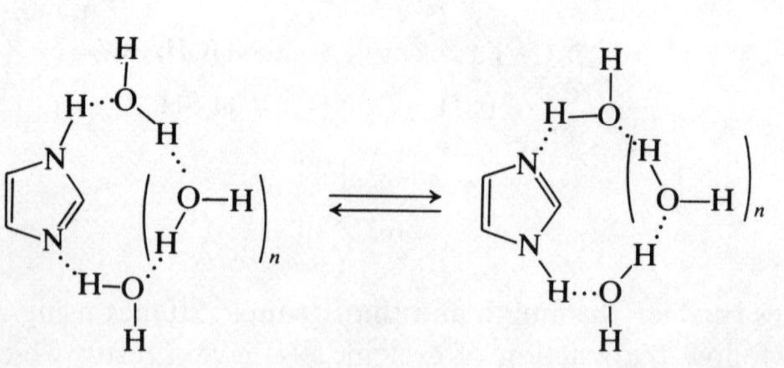

(109)

for the reactions in (108) and (109) were calculated to be $1.18 \times 10^8 s^{-1}$ and $1.5 \times 10^6 s^{-1}$ respectively. However, the reactions of uracil and 5-fluorouracil monoanions have been studied by laser- and Joule-heating temperature-jump techniques and quite different conclusions have been reached (Bensaude *et al.*, 1978). No relaxation in the time range consistent with the nmr measurements was found. However, chemical relaxations thought to correspond to the tautomerisation of uracil monoanions and of 5-fluorouracil monoanions were observed and the pH-dependence of the reciprocal relaxation times was fitted by equation (110) and explained by the mechanism in Scheme 14. The first term in (110) is independent of pH and is attributed to

$$\tau^{-1} = k_0 + k_{OH^-}[OH^-] + k_{UH_2}[UH_2] \tag{110}$$

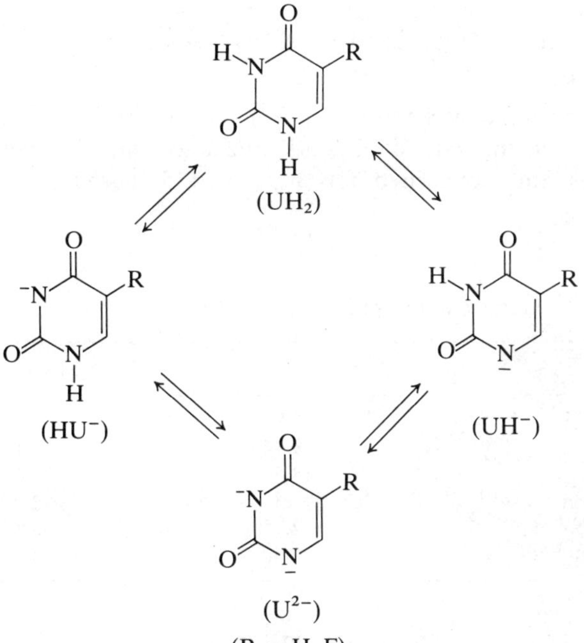

(UH₂)

(HU⁻)

(UH⁻)

(U^{2-})

(R = H, F)

Scheme 14

reaction by the mechanism in (111) and (112). From the magnitude of this term for uracil ($k_0 = 3.65 \times 10^5 s^{-1}$) and for 5-fluorouracil ($k_0 = 8.5 \times 10^3 s^{-1}$), the values $k_r = 2.3 \times 10^{10}$ and $2.0 \times 10^{10} dm^3 mol^{-1} s^{-1}$ were calculated for uracil and 5-fluorouracil respectively. Thus the pH-independent terms are explained by dissociative mechanisms in which

$$HU^- + H_2O \underset{k_r}{\overset{k_f}{\rightleftharpoons}} UH_2 + OH^- \tag{111}$$

$$UH_2 + OH^- \underset{k_f}{\overset{k_r}{\rightleftharpoons}} UH^- + H_2O \quad (112)$$

water behaves as an acid catalyst and the intermediates are formed or collapse at diffusion-limited rates. Similarly, the second and third terms in

$$HU^- + OH^- \overset{k_{OH^-}}{\rightleftharpoons} U^{2-} + H_2O \quad (113)$$

$$U^{2-} + H_2O \underset{k_{OH^-}}{\rightleftharpoons} UH^- + OH^- \quad (114)$$

$$HU^- + UH_2O \underset{k_{UH_2}}{\overset{k_{UH_2}}{\rightleftharpoons}} UH^- + UH_2 \quad (115)$$

(110) arise from the mechanisms in (113) + (114) and in (115) respectively. The values of the rate coefficients calculated for the steps in these mechanisms are summarised for uracil and 5-fluorouracil in Table 8.

TABLE 8

Kinetic data for tautomerisation of uracil monoanions

	Uracil	5-Fluorouracil
$K = [UH^-]/[HU^-]$	0.85	0.51
pK_1	9.74	8.13
pK_2	13.5	12.6
$10^{-10}k_r/dm^3mol^{-1}s^{-1}$	2.3	2.0
$10^{-9}k_{OH^-}/dm^3mol^{-1}s^{-1}$	2.25	2.15
$10^{-8}k_{UH_2}/dm^3mol^{-1}s^{-1}$	0.39	3.9

The values appear reasonable and, in contrast to the interpretation of the nmr data, (which may possibly relate to a different process) the results can be explained by a dissociative mechanism.

A non-dissociative mechanism for interconversion of the tautomers of 6-substituted-2-pyridones (116) has been found (Bensaude *et al.*, 1977). In

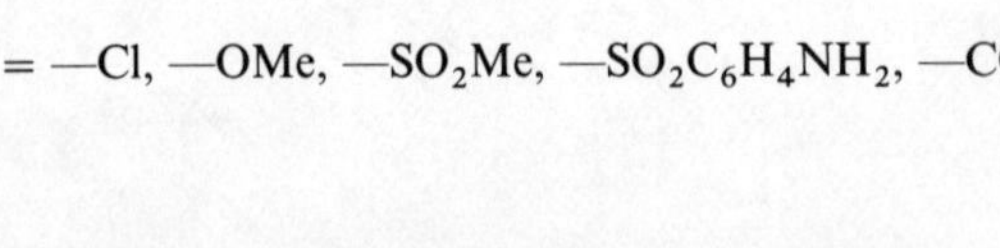

(116)

(X = —Cl, —OMe, —SO_2Me, —$SO_2C_6H_4NH_2$, —$COCCl_3$)

$$\tau^{-1} = k_0 + k_{H_3O^+}[H_3O^+] + k_{OH^-}[OH^-] + k_{A^-}[A^-] \tag{117}$$

these systems, the dependence on pH of the reciprocal relaxation time for equilibration of the tautomers is given by (117) in which A^- is the deprotonated 2-pyridone. The final three terms in (117) are the contributions made by dissociative pathways in which tautomerisation occurs by intermediate formation of protonated or deprotonated forms of 2-pyridone as in

X, N, OH $+ H_3O^+ \rightleftharpoons$ X, H, N, +, OH $+ H_2O$ (118)

X, H, N, +, OH $+ H_2O \rightleftharpoons$ X, H, N, O $+ H_3O^+$ (119)

(118) + (119) and (120) + (121). However, the first term in (117) is pH-independent but cannot be explained as being due to acid or base catalysis

X, N, OH $\xrightleftharpoons{OH^-, A^-}$ X, N, O^- (120)

X, N, O^- $\xrightleftharpoons{H_2O, HA}$ X, H, N, O (121)

by water, as in (122) + (123) or (124) + (125); because the magnitude of

X, N, OH $+ H_2O \rightleftharpoons$ X, H, N, +, OH $+ OH^-$ (122)

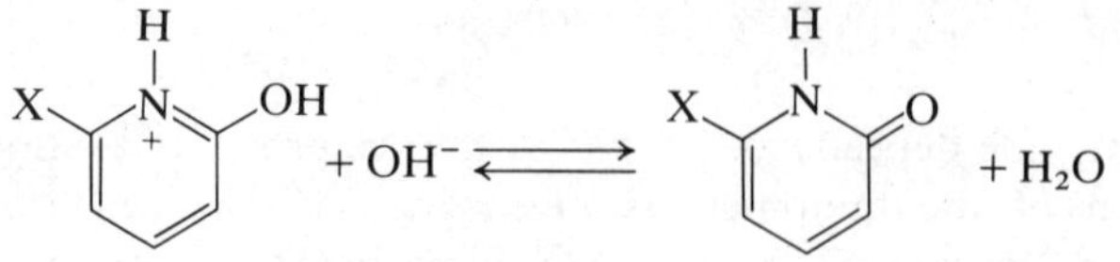

$$+ OH^- \rightleftharpoons + H_2O \qquad (123)$$

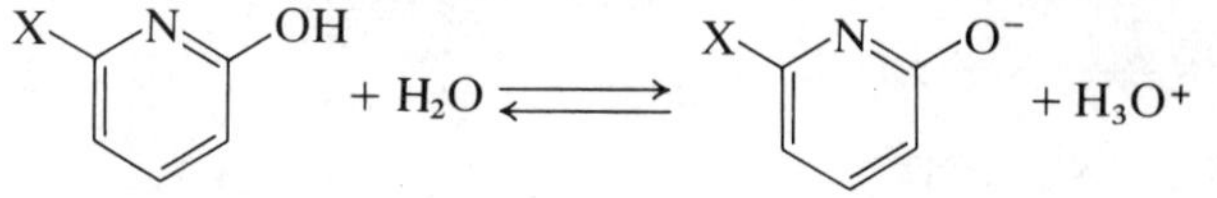

$$+ H_2O \rightleftharpoons + H_3O^+ \qquad (124)$$

$$+ H_3O^+ \rightleftharpoons + H_2O \qquad (125)$$

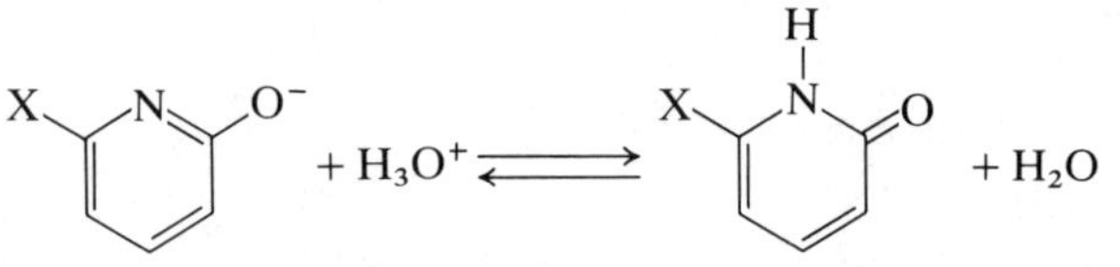

k_0 is such that the required values of the rate coefficients in these equations would exceed the diffusion-limited values. It was therefore concluded that a non-dissociative process in which water behaves as a bifunctional catalyst was needed to explain the magnitude of the pH-independent term in (117) and a reasonable possibility is shown in (126). Activation parameters

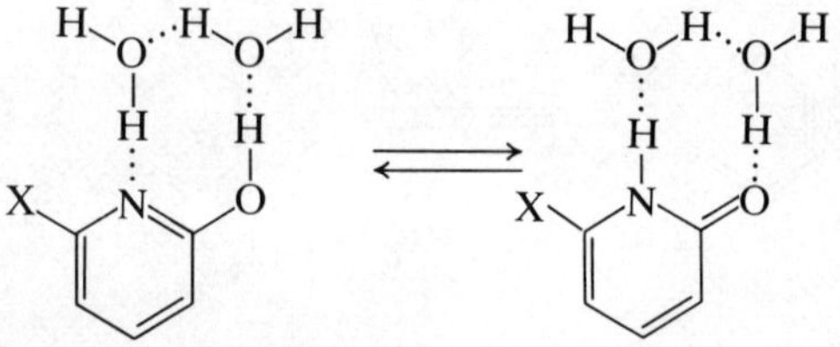

$$\rightleftharpoons \qquad (126)$$

(Chevrier *et al.*, 1983), solvent effects (Bensaude *et al.*, 1979), and the effect of added salt on the rate of reaction (Bensaude *et al.*, 1978) have been studied to provide information about this process. Molecular-orbital calculations confirm that a suitable transition state for the reaction is one involving bridging water molecules (Field *et al.*, 1984).

An attempt has been made to predict the general requirements for the operation of a non-dissociative mechanism of tautomerisation between oxygen and nitrogen centres (Bensaude *et al.*, 1977). The proximity of the two centres between which the proton is transferred is obviously of critical importance but more experimental results are required before a clear picture will emerge.

Future work

Although the mechanisms of proton transfers between oxygen and nitrogen acids and bases are quite well understood, it is apparent that there remain some areas where knowledge is incomplete. Important developments can be expected over the next few years. These may include further progress in assessing the importance, to intermolecular acid-base reactions, of proton transfer within the hydrogen-bonded complex between the acid and the base. A comparison with the less well-known behaviour of proton transfer to and from sulphur atoms would be useful. It is likely that further examples of slow proton transfer along intramolecular hydrogen bonds of the type illustrated by meso-tetraphenylporphyrin and 2-methylnaphthazarin will be uncovered and will lead to a better understanding of these reactions. More information is also needed about the rates of opening and closing of intramolecular hydrogen bonds in aqueous solution and nmr methods may be applied to this problem. The discovery of more examples of strongly basic amines is anticipated and perhaps the most promising possibilities are those with a basic centre inside a molecular cavity. Strongly basic amines which exhibit moderately high rates of proton transfer would be important as catalysts in organic reactions. Further work will be directed towards an understanding of the mechanisms of tautomerisation at oxygen and nitrogen atoms and towards the complex mechanisms involved in enzyme catalysis. It may be possible to study directly the proton transfers within model enzyme-substrate complexes.

References

Ahrens, M. L. and Maass, G. (1968). *Angew. Chem. Int. Ed.* **7**, 818
Albery, W. J. (1980). *Ann. Rev. Phys. Chem.* **31**, 227
Albery, W. J. and Kreevoy, M. M. (1978). *Adv. Phys. Org. Chem.* **16**, 87
Albery, W. J., Campbell-Crawford, A. N. and Curran, J. S. (1972). *J. Chem. Soc. Perkin Trans. 2*, 2206
Alder, R. W. (1983). *Acc. Chem. Res.* **16**, 321
Alder, R. W. and Sessions, R. B. (1979). *J. Am. Chem. Soc.* **101**, 3651
Alder, R. W., Bowman, P. S., Steele, W. R. S. and Winterman, D. R. (1968). *Chem. Commun.*, 723
Alder, R. W., Goode, N. C., Miller, N., Hibbert, F., Hunte, K. P. P. and Robbins, H. J. (1978). *J. Chem. Soc. Chem. Commun.* 79
Alder, R. W., Casson, A. and Sessions, R. B. (1979). *J. Am. Chem. Soc.* **101**, 3652
Alder, R. W., Bryce, M. R., Goode, N. C., Miller, N. and Owen, J. (1981). *J. Chem. Soc. Perkin. Trans. 1*, 2840
Alder, R. W. Sessions, R. B., Bennet, A. J. and Moss, R. E. (1982). *J. Chem. Soc. Perkin Trans. 1*, 603
Alder, R. W., Moss, R. E. and Sessions, R. B. (1983a). *J. Chem. Soc. Chem. Commun.* 997, 1000

Alder, R. W., Orpen, A. G. and Sessions, R. B. (1983b). *J. Chem. Soc. Chem. Commun.* 999

Altman, L. J., Laungani, D., Gunnarsson, G. Wennerström, H. and Forsén, S. (1978). *J. Am. Chem. Soc.* **100**, 8264

Andreassen, A. L. and Bauer, S. H. (1972). *J. Mol. Struct.* **12**, 381

Andreassen, A. L., Zebelman, D. and Bauer, S. H. (1971). *J. Am. Chem. Soc.* **93**, 1148

Awwal, A. and Hibbert, F. (1977a). *J. Chem. Soc. Perkin Trans. 2,* 152

Awwal, A. and Hibbert, F. (1977b). *J. Chem. Soc. Perkin Trans. 2,* 1589

Bacon, G. E., Walker, C. R. and Speakman, J. C. (1977). *J. Chem. Soc. Perkin Trans. 2,* 979

Bannister, J. J., Gormally, J., Holzwarth, J. F. and King, T. A. (1984). *Chem. Brit. 20,* 227

Barbara, P. F., Rentzepis, P. M. and Brus, L. E. (1980a). *J. Am. Chem. Soc.* **102**, 563

Barbara, P. F., Brus, L. E. and Rentzepis, P. M. (1980b). *J. Am. Chem. Soc.* **102**, 2786

Barnett, G. H. and Hibbert, F. (1984). *J. Am. Chem. Soc.* **106**, 2080

Barnett, R. E. and Jencks, W. P. (1969). *J. Am. Chem. Soc.* **91**, 2358

Barrow, M. J., Currie, M., Muir, K. W., Speakman, J. C. and White, D. N. J. (1975). *J. Chem. Soc. Perkin Trans. 2,* 15

Baughcum, S. L., Duerst, R. W., Rowe, W. F., Smith, Z. and Wilson, E. B. (1981). *J. Am. Chem. Soc.* **103**, 6296

Baughcum, S. L., Smith, Z., Wilson, E. B. and Duerst, R. W. (1984). *J. Am. Chem. Soc.* **106**, 2260

Beens, H., Grellman, K. H., Gurr, M. and Weller, A. H. (1965). *Disc. Faraday Soc.* **39**, 183

Bell, R. P. (1973). "The Proton in Chemistry", 2nd edn. Cornell University Press, Ithaca, New York

Bell, R. P. (1981). "Proton Tunnelling in Chemistry", Chapman and Hall, London

Bender, M. L. (1971). "Mechanisms of Homogeneous Catalysis from Protons to Proteins". Wiley Interscience, New York

Bender, M. L., Schonbaum, G. R. and Zerner, B. (1962). *J. Am. Chem. Soc.* **84**, 2562

Bensaude, O., Dreyfus, M., Dodin, G. and Dubois, J. E. (1977). *J. Am. Chem. Soc.* **99**, 4438

Bensaude, O., Aubard, J., Dreyfus, M., Dodin, G. and Dubois, J. E. (1978). *J. Am. Chem. Soc.* **100**, 2823

Bensaude, O., Chevrier, M. and Dubois, J. E. (1978). *J. Am. Chem. Soc.* **100**, 7055

Bensaude, O., Chevrier, M. and Dubois, J. E. (1979). *J. Am. Chem. Soc.* **101**, 2423

Bergman, N-Å., Chiang, Y. and Kresge, A. J. (1978). *J. Am. Chem. Soc.* **100**, 5954

Bernasconi, C. F. (1976). "Relaxation Kinetics". Academic Press, London

Bernasconi, C. F. and Carré, D. J. (1979a). *J. Am. Chem. Soc.* **101**, 2707

Bernasconi, C. F. and Carré, D. J. (1979b). *J. Am. Chem. Soc.* **101**, 2698

Bernasconi, C. F. and Terrier, F. (1975a). *J. Am. Chem. Soc.* **97**, 7458

Bernasconi, C. F. and Terrier, F. (1975b). *In* Proceedings of the NATO Advanced Study Institute "Chemical and Biological Applications of Relaxation Spectrometry" (E. Wyn-Jones, ed.) p. 379. D. Reidel Publishing, Dordrecht, Holland

Bianchin, B., Chrisment, J., Delpuech, J. J., Deschamps, M. N., Nicole, D. and Serratrice, G. (1975). *In* Proceedings of the NATO Advanced Study Institute "Chemical and Biological Applications of Relaxation Spectrometry" (E. Wyn-Jones, ed.) p. 365. D. Reidel Publishing, Dordrecht, Holland

Bicerano, J., Schaeffer, H. F. and Miller, W. H. (1983). *J. Am. Chem. Soc.* **105,** 2550

Birk, J. P., Chock, P. B. and Halpern, J. (1968). *J. Am. Chem. Soc.* **90**, 6959

Blow, D. M. (1976). *Acc. Chem. Res.* **9**, 145
Bouma, W. J., Vincent, M. A. and Radom, L. (1978). *Int. J. Quantum Chem.* **14**, 767
Brown, R. S., Tse, A., Nakashima, T. and Haddon, R. C. (1979). *J. Am. Chem. Soc.* **101**, 3157
Caldin, E. (1975). *Chem. Brit.* **11**, 4
Camerman, A., Mastropaolo, D. and Camerman, N. (1983). *J. Am. Chem. Soc.* **105**, 1584
Chang, K-C. and Grunwald, E. (1976). *J. Am. Chem. Soc.* **98**, 3737
Cheney, J. and Lehn, J-M (1972). *J. Chem. Soc. Chem. Commun.* 487
Cheney, J., Kintzinger, J. P. and Lehn, J-M. (1978). *Nouv. J. Chim.* **2**, 411
Chevrier, M., Guillerez, J. and Dubois, J. E. (1983). *J. Chem. Soc. Perkin Trans. 2,* 979
Chiang, Y., Kresge, A. J. and More O'Ferrall, R. A. (1980). *J. Chem. Soc. Perkin. Trans. 2,* 1832
Chiang, Y., Kresge, A. J. and Holzwarth, J. F. (1982). *J. Chem. Soc. Chem. Commun.* 1203
Clements, R., Dean, R. L., Singh, T. R. and Wood, J. L. (1971a). *J. Chem. Soc. Chem. Commun.* 1125
Clements, R., Masri, F. N. and Wood, J. L. (1971b). *J. Chem. Soc. Chem. Commun.* 1530
Clements, R., Dean, R. L. and Wood, J. L. (1971c). *J. Chem. Soc. Chem. Commun.,* 1127
Cohen, A. O. and Marcus, R. A. (1968). *J. Phys. Chem.* **72**, 4249
Cohen, B. and Weiss, S. (1984). *J. Phys. Chem.* **88**, 3159
Cousseau, J., Gouin, L., Pang, E. K. C. and Smith, J. A. S. (1977). *J. Chem. Soc. Faraday Trans. 2,* **73**, 1015
Cox, B. G., Knop, D. and Schneider, H. (1978). *J. Am. Chem. Soc.* **100**, 6002
Cox, B. G., Murray-Rust, J., Murray-Rust, P., Schneider, H. and van Truong, Ng. (1982). *J. Chem. Soc. Chem. Commun.* 377
Cox, B. G., van Truong, Ng. and Schneider, H. (1983). *J. Chem. Soc., Perkin Trans. 2,* 515
Cox, M. M. and Jencks, W. P. (1978). *J. Am. Chem. Soc.* **100**, 5956
Currie, M. (1972). *J. Chem. Soc. Perkin Trans. 2,* 832
de la Vega, J. R. (1982). *Acc. Chem. Res.* **15**, 185
de la Vega, J. R., Busch, J. H., Schauble, J. H., Kunze, K. L. and Haggert, B. E. (1982). *J. Am. Chem. Soc.* **104**, 3295
Dolman, D. and Stewart, R. (1967). *Can. J. Chem.* **45**, 911
Dreyfus, M., Dodin, G., Bensaude, O. and Dubois, J. E. (1975). *J. Am. Chem. Soc.* **97**, 2369
Dreyfus, M., Bensaude, O., Dodin, G. and Dubois, J. E. (1976). *J. Am. Chem. Soc.* **98**, 6338
Egan, W., Gunnarsson, G., Bull, T. E. and Forsén, S. (1977). *J. Am. Chem. Soc.* **99**, 4568
Eggers, F. (1957). *Acustica* **19**, 323
Eigen, M. (1963). *Pure Appl. Chem.* **6**, 97
Eigen, M. (1964). *Angew. Chem. Int. Ed.* **3**, 1
Eigen, M. (1967). *In* "Fast Reactions and Primary Processes in Chemical Kinetics" (S. Claesson, ed.) Nobel Symposium 5, p. 333. Interscience, New York
Eigen, M. and de Maeyer, L. (1963). *In* "Technique of Organic Chemistry" (S. L. Friess, E. S. Lewis and A. Weissberger, eds) Vol. VIIIB, p. 895. Interscience, New York

Eigen, M. and Kruse, W. (1963). *Z. Naturforsch.* **B18**, 857
Eigen, M., Kruse, W., Maass, G. and de Maeyer, L. (1964). *Prog. Reaction Kinetics* **2**, 285
Einspahr, H., Robert, J-B., Marsh, R. E. and Roberts, J. D. (1973). *Acta Cryst.* **29B**, 1611
Emsley, J. (1980). *Chem. Soc. Rev.* **9**, 91
Emsley, J. (1984). *Structure and Bonding,* **57**, 147
Engebretson, G. R. and Rundle, R. E. (1964). *J. Am. Chem. Soc.* **86**, 574
Escabi-Perez, J. R. and Fendler, J. H. (1978). *J. Am. Chem. Soc.* **100**, 2234
Evans, D. F. (1982). *J. Chem. Soc. Chem. Commun.* 1226
Eyring, E. M. and Haslam, J. L. (1966). *J. Phys. Chem.* **70**, 293
Fenn, M. D. and Spinner, E. (1984). *J. Phys. Chem.* **88**, 3993
Field, M. J., Hillier, I. H. and Guest, M. F. (1984). *J. Chem. Soc. Chem. Commun.* 1310
Fischer, H., DeCandis, F. X., Ogden, S. D. and Jencks, W. P. (1980). *J. Am. Chem. Soc.* **102**, 1340
Friess, S. L., Lewis, E. S. and Weissberger, A. (eds.) (1963). "Technique of Organic Chemistry", Vol. VIIIB. Interscience, New York
Fueno, T., Kajimoto, O., Nishigaki, Y. and Yoshioka, T. (1973). *J. Chem. Soc. Perkin Trans. 2,* 738
Fuess, H. and Lindner, H. J. (1975). *Chem. Ber.* **108**, 3096
Fujiwara, F. Y. and Martin, J. S. (1974). *J. Am. Chem. Soc.* **96**, 7625
Gandour, R. D., Maggiora, G. M. and Schowen, R. L. (1974). *J. Am. Chem. Soc.* **96**, 6967
German, E. D., Dogonadze, R. R., Kuznetsov, A. M., Levich, V. G. and Kharkats, Y. I. (1971). *J. Res. Inst. Catalysis,* **99**, 115
Gold, V. (1962). *In* "Progress in Stereochemistry" (P. B. D. de la Mare and W. Klyne, eds.) Ch. 5. Butterworths, London
Gold, V. and Lee, R. A. (1984). *J. Chem. Soc. Chem. Commun.* 1032
Golic, L. and Speakman, J. C. (1965). *J. Chem. Soc.* 2530
Goodman, J. and Brus, L. E. (1978). *J. Am. Chem. Soc.* **100**, 7472
Grunwald, E. and Eustace, D. (1975). *In* "Proton Transfer Reactions" (E. F. Caldin and V. Gold eds) Ch. 4, p. 103. Chapman and Hall, London
Grunwald, E. and Ku, A. Y. (1968). *J. Am. Chem. Soc.* **90**, 29
Grunwald, E. and Meiboom, S. (1963). *J. Am. Chem. Soc.* **85**, 2047
Grunwald, E. and Ralph, E. K. (1971). *Acc. Chem. Res.* **4**, 107
Grunwald, E., Jumper, C. F. and Meiboom, S. (1963). *J. Am. Chem. Soc.* **85**, 522
Grunwald, E., Chang, K-C., Skipper, P. L. and Anderson, V. K. (1976). *J. Phys. Chem.* **80**, 1425
Gunnarsson, G., Wennerström, H., Egan, W. and Forsén, S. (1976). *Chem. Phys. Lett,* **38**, 96
Hafner, K., Kramer, H. E. A., Musso, H., Ploss, G. and Schulz, G. (1964). *Chem. Ber.* 2066
Hagopian, S. and Singer, L. A. (1983). *J. Am. Chem. Soc.* **105**, 6760
Hameka, H. F. and de la Vega, J. R. (1984). *J. Am. Chem. Soc.* **106**, 7703
Hammes, G. G., ed. (1974). "Investigation of Rates and Mechanisms of Reactions", 3rd edn, Part II. John Wiley, New York
Hammond, G. S. (1955). *J. Am. Chem. Soc.* **77**, 334
Haselbach, E., Henriksson, A., Jachimowicz, F. and Wirz, J. (1972). *Helv. Chim. Acta,* **55**, 1757
Haslam, J. L. and Eyring, E. M. (1967). *J. Phys. Chem.* **71**, 4471

Haslam, J. L., Eyring, E. M., Epstein, W. W., Christiansen, G. A. and Miles, M. H. (1965a). *J. Am. Chem. Soc.* **87**, 1
Haslam, J. L., Eyring, E. M., Epstein, W. W., Jensen, R. P. and Jaget, C. W. (1965b). *J. Am. Chem. Soc.* **87**, 4247
Hennig, J. and Limbach, H-H. (1982). *J. Magn. Res.* **49**, 322
Hennig, J. and Limbach, H-H. (1984). *J. Am. Chem. Soc.* **106**, 292
Hibbert, F. (1973). *J. Chem. Soc. Chem. Commun.* 463
Hibbert, F. (1974). *J. Chem. Soc. Perkin Trans. 2,* 1862
Hibbert, F. (1975). *In* Proceedings of the NATO Advanced Study Institute "Chemical and Biological Applications of Relaxation Spectrometry" (E. Wyn-Jones, ed.) p. 387. D. Reidel Publishing, Dordrecht, Holland
Hibbert, F. (1977). *In* "Comprehensive Chemical Kinetics" (C. H. Bamford and C. F. H. Tipper, eds) Vol. 8, Ch. 2, p. 97. Elsevier, Amsterdam
Hibbert, F. (1981). *J. Chem. Soc. Perkin Trans. 2,* 1304
Hibbert, F. (1984). *Acc. Chem. Res.* **17**, 115
Hibbert, F. and Awwal, A. (1976). *J. Chem. Soc. Chem. Commun.* 995
Hibbert, F. and Awwal, A. (1978). *J. Chem. Soc. Perkin Trans. 2,* 939
Hibbert, F. and Hunte, K. P. P. (1981). *J. Chem. Soc. Perkin Trans. 2,* 1562
Hibbert, F. and Hunte, K. P. P. (1983). *J. Chem. Soc. Perkin Trans. 2,* 1895
Hibbert, F. and Robbins, H. J. (1978). *J. Am. Chem. Soc.* **100**, 8239
Hibbert, F. and Robbins, H. J. (1980). *J. Chem. Soc. Chem. Commun.* 141
Hibbert, F. and Simpson, G. R. (1983). *J. Am. Chem. Soc.* **105**, 1063
Hibbert, F. and Simpson, G. R. (1984). Unpublished work
Hibbert, F. and Simpson, G. R. (1985). *J. Chem. Soc. Perkin Trans. 2,* 1247
Hilinski, E. F. and Rentzepis, P. M. (1983). *Acc. Chem. Res.* **16**, 224
Hollander, F. J., Templeton, D. H. and Zalkin, A. (1973). *Acta Cryst.* **B29**, 1552
Huppert, D., Gutman, M. and Kaufman, K. J. (1981). *Adv. Chem. Phys.* **XLVII**, Part 2, p. 643.
Inskeep, W. H., Jones, D. L., Silfvast, W. T. and Eyring, E. M. (1968). *Proc. Nat. Acad. Sci.* **59**, 1027
Ireland, J. F. and Wyatt, P. A. H. (1976). *Adv. Phys. Org. Chem.* **12**, 131
Itoh, M., Tokumura, K., Tanimoto, Y., Okada, Y., Takeuchi, H., Obi, K. and Tanaka, I. (1982). *J. Am. Chem. Soc.* **104**, 4146
Jackman, L. M., Trewella, J. C. and Haddon, R. C. (1980). *J. Am. Chem. Soc.* **102**, 2519
Jencks, W. P. (1969). "Catalysis in Chemistry and Enzymology". McGraw-Hill, New York
Jencks, W. P. (1976). *Acc. Chem. Res.* **9**, 425
Jensen, R. P., Eyring, E. M., Walsh, W. M. (1966). *J. Phys. Chem.* **70**, 2264
Joesten, M. D. and Schaad, L. J. (1974). "Hydrogen Bonding". Marcel Dekker, New York
Jones, R. D. G. (1976). *Acta Cryst.* **B32**, 1807
Karlström, G., Jönsson, B., Roos, B. and Wennerström, H. (1976). *J. Am. Chem. Soc.* **98**, 6851
Kirby, A. J. (1980). *Adv. Phys. Org. Chem.* **17**, 183
Kjaer, A. M., Sorensen, P. E. and Ulstrup, J. (1979). *J. Chem. Soc. Chem. Commun.* 965
Koeppl, G. W. and Kresge, A. J. (1973). *J. Chem. Soc. Chem. Commun.* 371
Kreevoy, M. M. and Konasewich, D. E. (1971). *Adv. Chem. Phys.* **21**, 243
Kreevoy, M. M. and Liang, T. M. (1980). *J. Am. Chem. Soc.* **102**, 3315
Kreevoy, M. M. and Oh, S. (1973). *J. Am. Chem. Soc.* **95**, 4805

Kreevoy, M. M. and Ridl, B. A. (1981). *J. Phys. Chem.* **85**, 914
Kreevoy, M. M. and Liang, T. M. and Chang, K.-C. (1977). *J. Am. Chem. Soc.* **99**, 5207
Kresge, A. J. (1973). *Chem. Soc. Rev.* **2**, 475
Kresge, A. J. (1975a). *In* "Proton Transfer Reactions" (E. F. Caldin and V. Gold, eds) p. 179. Chapman and Hall, London
Kresge, A. J. (1975b). *Acc. Chem. Res.* **8**, 354
Kresge, A. J. and Powell, M. F. (1981). *J. Am. Chem. Soc.* **103**, 972
Kurz, J. L. and Kurz, L. C. (1972). *J. Am. Chem. Soc.* **94**, 4451
Kwiatkowski, J. S. and Pullman, B. (1975). *Adv. Heterocyclic Chem.* **18**, 199
Laane, J. (1970). *Appl. Spectroscopy* **24**, 73
Lazaar, K. I. and Bauer, S. H. (1983). *J. Phys. Chem.* **87**, 2411
Leffler, J. E. (1953). *Science* **117**, 340
Leffler, J. E. and Grunwald, E. (1963). "Rates and Equilibria of Organic Reactions", p. 157. John Wiley, New York
Lehn, J-M. (1978). *Acc. Chem. Res.* **11**, 49
Lewis, E. S. and Hu, D. D. (1984). *J. Am. Chem. Soc.* **106**, 3292
Limbach, H.-H., Hennig, J., Gerritzen, D. and Rumpel, H. (1982). *J. Chem. Soc. Faraday. Disc.* **74**, 229
Limbach, H.-H., Hennig, J., Kendrick, R. and Yannoni, C. S. (1984). *J. Am. Chem. Soc.* **106**, 4059
Lindemann, R. and Zundel, G. (1977). *J. Chem. Soc. Faraday Trans. 2,* **73**, 788
Löwdin, P-O. (1965). *Adv. Quantum Chem.* **2**, 213
Lowrey, A. H., George, C., D'Antonio, P. and Karle, J. (1971). *J. Am. Chem. Soc.* **93**, 6399
Ludman, C. J., Waddington, T. C., Pang, E. K. C. and Smith, J. A. S. (1977). *J. Chem. Soc. Faraday Trans. 2,* **73**, 1003
Luz, Z. and Meiboom, S. (1963). *J. Am. Chem. Soc.* **85**, 3923
Maass, G. and Peters, F. (1972). *Angew. Chem. Int. Ed.* **11**, 428
Macdonald, A. L., Speakman, J. C. and Hadzi, D. (1972). *J. Chem. Soc. Perkin Trans. 2,* 825
Mallick, I. M., D'Souza, V. T., Yamaguchi, M., Lee, J., Chalabi, P., Gadwood, R. C. and Bender, M. L. (1984). *J. Am. Chem. Soc.* **106**, 7252
Mandel, G. S. and Marsh, R. E. (1975). *Acta Cryst.* **B31**, 2862
Marcus, R. A. (1964). *Ann. Rev. Phys. Chem.* **15**, 155
Marcus, R. A. (1968). *J. Phys. Chem.* **72**, 891
Marcus, R. A. (1969). *J. Am. Chem. Soc.* **91**, 7224
Marcus, R. A. (1975). *Faraday Symp. Chem. Soc.* **10**, 60
Marshall, T. H. and Grunwald, E. (1969). *J. Am. Chem. Soc.* **91**, 4541
McMorrow, D. and Kasha, M. (1983). *J. Am. Chem. Soc.* **105**, 5133
McMorrow, D. and Kasha, M. (1984). *J. Phys. Chem.* **88**, 2235
Miles, M. H., Eyring, E. M., Epstein, W. W. and Ostlund, R. E. (1965). *J. Phys. Chem.* **69**, 467
Miles, M. H., Eyring, E. M., Epstein, W. W. and Anderson, M. T. (1966). *J. Phys. Chem.* **70**, 3490
More O'Ferrall, R. A. (1975). *In* "Proton Transfer Reactions" (E. F. Caldin and V. Gold, eds) Ch. 8, p. 201. Wiley, New York
Muir, K. W. and Speakman, J. C. (1979). *J. Chem. Res. (S),* 277
Murdoch, J. R. (1972). *J. Am. Chem. Soc.* **94**, 4410
Murdoch, J. R. (1980). *J. Am. Chem. Soc.* **102**, 71
Murdoch, J. R. (1983a). *J. Am. Chem. Soc.* **105**, 2660

Murdoch, J. R. (1983b). *J. Am. Chem. Soc.* **105**, 2667
Novak, A. (1974). *Structure and Bonding,* **18**, 177
Park, C. H. and Simmons, H. E. (1968a). *J. Am. Chem. Soc.* **90**, 2429
Park, C. H. and Simmons, H. E. (1968b). *J. Am. Chem. Soc.* **90**, 2431
Pawlak, Z., Tusk, M., Kuna, S., Strohbusch, F. and Fox, M. F. (1984). *J. Chem. Soc. Faraday. Trans. 2.* **80**, 1757
Pedersen, C. J. (1967). *J. Am. Chem. Soc.* **89**, 2495
Pedersen, C. J. (1978). "Synthetic and Multidentate Macrocyclic Compounds" (R. M. Izatt, J. J. Christensen, eds). Academic Press, New York
Pellerite, M. J. and Brauman, J. I. (1983). *J. Am. Chem. Soc.* **105**, 2672
Perlmutter-Hayman, B. and Shinar, R. (1975). *Int. J. Chem. Kin.* **7**, 453
Perlmutter-Hayman, B. and Shinar, R. (1977). *Int. J. Chem. Kin.* **9**, 1
Perlmutter-Hayman, B. and Shinar, R. (1978). *Int. J. Chem. Kin.* **10**, 407
Perlmutter-Hayman, B., Sarfaty, R. and Shinar, R. (1976). *Int. J. Chem. Kin.* **8**, 741
Pickett, H. M. (1973). *J. Am. Chem. Soc.* **95**, 1770
Pizer, R. (1978). *J. Am. Chem. Soc.* **100**, 4239
Pullman, B. and Pullman A. (1971). *Adv. Heterocyclic Chem.* **13**, 77
Ratajczak, H. and Orville-Thomas, W. J. (eds) (1980). "Molecular Interactions", Vol. 1, John Wiley, New York
Ratajczak, H. and Orville-Thomas, W. J. (eds) (1981). "Molecular Interactions" Vol. 2. John Wiley, New York
Ratajczak, H. and Orville-Thomas, W. J. (eds) (1982). "Molecular Interactions", Vol. 3. John Wiley, New York
Rochester, C. H. (1970). "Acidity Functions". Academic Press, London
Rose, M. C. and Stuehr, J. E. (1968). *J. Am. Chem. Soc.* **90**, 7205
Rose, M. C. and Stuehr, J. E. (1971). *J. Am. Chem. Soc.* **93**, 4350
Rose, M. C. and Stuehr, J. E. (1972). *J. Am. Chem. Soc.* **94**, 5532
Rosenberg, S., Silver, S. M. Sayer, J. M. and Jencks, W. P. (1974). *J. Am. Chem. Soc.* **96**, 7986
Rozière, J., Belin, C. and Lehman, M. S. (1982). *J. Chem. Soc. Chem. Commun.* 388
Saunders, M., Saunders, S. and Johnson, C. A. (1984). *J. Am. Chem. Soc.* **106**, 3098
Sayer, J. M. and Jencks, W. P. (1973). *J. Am. Chem. Soc.* **95**, 5637
Scanlan, M. J. and Hillier, I. H. (1984). *J. Chem. Soc. Chem. Commun.* 102
Schaefer, W. P. and Marsh, R. E. (1984). *J. Chem. Soc. Chem. Commun.* 1555
Schuster, P., Zundel, G. and Sandorfy, C. (eds) (1976a). "The Hydrogen Bond", Vol. I. North Holland, Amsterdam
Schuster, P., Zundel, G. and Sandorfy, C. (eds) (1976b). "The Hydrogen Bond", Vol. II. North Holland, Amsterdam
Schuster, P., Zundel, G. and Sandorfy, C. (eds) (1976c). "The Hydrogen Bond", Vol. III. North Holland, Amsterdam
Sengupta, P. K. and Kasha, M. (1979). *Chem. Phys. Lett.* **68**, 382
Simmons, H. E. and Park, C. H. (1968). *J. Am. Chem. Soc.* **90**, 2428
Smith, P. B., Dye, J. L., Cheney, J. and Lehn, J.-M. (1981). *J. Am. Chem. Soc.* **103**, 6044
Smith, K. K. and Kaufman, K. J. (1978). *J. Phys. Chem.* **82**, 2286
Stevens, E. D., Lehmann, M. S. and Coppens, P. (1977). *J. Am. Chem. Soc.* **99**, 2829
Storm, C. B. and Teklu, Y. (1972). *J. Am. Chem. Soc.* **94**, 1745
Strandjord, A. J. G., Courtney, S. H., Friedrich, D. M. and Barbara, P. F. (1983). *J. Phys. Chem.* **87**, 1125
Szeverenyi, N. M., Sullivan, M. J. and Maciel, G. E. (1982). *J. Magn. Res.* **47**, 462
Szeverenyi, N. M., Bax, A. and Maciel, G. E. (1983). *J. Am. Chem. Soc.* **105**, 2579

Taft, R. W. and Lewis, I. C. (1959). *J. Am. Chem. Soc.* **81**, 5343
Taft, R. W., Ehrenson, S., Lewis, I. C. and Glick, R. E. (1959). *J. Am. Chem. Soc.* **81**, 5352
Toullec, J. (1982). *Progr. Phys. Org. Chem.* **18**, 1
Truter, M. R. and Vickery, B. L. (1972). *J. Chem. Soc. Dalton Trans.* 395
Webb, S. P., Yeh, S. W., Phillips, L. A., Tolbert, M. A. and Clark, J. H. (1984). *J. Am. Chem. Soc.* **106**, 7286
Williams, D. E. (1966). *Acta Cryst.* **21**, 340
Williams, D. E., Dunke, W. L. and Rundle, R. E. (1962). *Acta Cryst.* **15**, 627
Wood, J. L. (1973). *In* "Spectroscopy and Structure of Molecular Complexes" (J. Yarwood, ed.) p. 336. Plenum Press, New York
Woolf, A. A. (1980). University of Bath. Personal communication
Woolf, G. J. and Thistlethwaite, P. J. (1980). *J. Am. Chem. Soc.* **102**, 6917
Woolf, G. J. and Thistlethwaite, P. J. (1981). *J. Am. Chem. Soc.* **103**, 6916
Yoshida, N. and Fujimoto, M. (1977). *Chem. Lett.* 1301
Zundel, G. (1976). In "The Hydrogen Bond" (P. Schuster, G. Zundel and C. Sandorfy (eds) Vol. II, p. 683. North Holland, Amsterdam
Zundel, G. (1978). *J. Mol. Structure,* **45**, 55

Organic Reactivity in Aqueous Micelles and Similar Assemblies

CLIFFORD A. BUNTON[1] and GIANFRANCO SAVELLI[2]

[1] *Department of Chemistry, University of California, Santa Barbara, California, 93106, U.S.A.*
[2] *Department of Chemistry, University of Perugia, Perugia, Italy*

Symbols used in the text 214
1 Introduction 214
2 Micellar structure and ion binding 219
3 Quantitative treatments of rates and equilibria 222
The pseudophase ion-exchange model 228
Reactive-ion micelles 237
Validity of the ion-exchange and mass-action models 241
4 Spontaneous, unimolecular and water-catalysed reactions 244
Unimolecular reactions of anionic substrates 244
Water-catalysed, uni- and bimolecular reactions 245
5 The source of micellar rate enhancements 251
6 Functional micelles and comicelles 259
7 Micellar effects on acid-base equilibria 265
8 Reactions in non-micellar aggregates 268
Reactions in vesicles 268
Reactions in microemulsions 271
Submicellar self-assembly aggregates 273
9 Stereochemical effects 277
10 Applications in synthetic and analytical chemistry 279
Acknowledgements 281
Appendix 282
A survey of micellar effects on chemical and photochemical reactions 282
Quantitative treatment of micelle-assisted bimolecular reactions 295
References 299
Notes added in proof 309

ADVANCES IN PHYSICAL ORGANIC CHEMISTRY
VOLUME 22 ISBN 0 12 033522 0

Symbols used in the text

α	Degree of micellar ionization
β	Fraction of counterions bound to micelle, $\beta = 1 - \alpha$
cmc	Critical micelle concentration
CPyCl	Hexadecylpyridinium chloride, cetylpyridinium chloride
CTABr(OH)	n-Hexadecyltrimethylammonium bromide (hydroxide) or cetyltrimethylammonium bromide (hydroxide)
[D]	Stoichiometric concentration of surfactant (detergent)
$[D_n]$	Concentration of micellized surfactant: generally $[D_n] = [D]$-*cmc*
DDDACl(OH)	Didodecyldimethylammonium chloride (hydroxide)
DODACl	Dioctadecyldimethylammonium chloride
K_b^M	Base dissociation constant in a micelle based on concentration as a mole ratio
K_b^v	Base dissociation constant in a micelle based on molarity in the Stern layer
K_X^Y	Ion exchange constant for ions of like charge
K_X'	Mass action binding constant of ion X.
K_s	Binding constant of solute based on concentration of micellized surfactant
k_ψ	Observed first-order rate constant
k_W'	First-order rate constant (s^{-1}) in the aqueous pseudophase
k_W	Second-order rate constant ($M^{-1}s^{-1}$) in the aqueous pseudophase
k_M'	First-order rate constant in the micellar pseudophase (s^{-1})
k_M	Second-order rate constant in the micellar pseudophase (s^{-1}), with concentration expressed as mole ratio
k_2^m	Second-order rate constant in the micellar pseudophase ($M^{-1}s^{-1}$): $k_2^m = k_M V_M$
SDS	Sodium dodecylsulfate
V_M	Molar volume element of reaction in the micelle
m_x^s	Mole ratio of X in the micelle
[ ]	Molarity in terms of the total solution volume

1 Introduction

Surfactants, sometimes called surface active agents or detergents, are amphiphilic materials which contain both apolar, hydrophobic (lipophilic) and polar, hydrophilic (lipophobic) groups (Hartley, 1948, 1977; Fendler and Fendler, 1975; Fendler, 1982; Lindman and Wennerström, 1980: Sudhölter *et al.*, 1979; Brown, 1979; Kunitake and Shinkai, 1980). In solvents which have a strong three-dimensional structure, for example water, hydrazine, 1,2-diols (Ray, 1971; Ray and Nemethy, 1971; Ionescu and Fung, 1981; Evans and Ninham, 1983) or sulfuric acid (Menger and Jerkunica, 1979) this dual character of the amphiphile leads to self-association or micellization. In the small colloidal particles, or micelles, which result, the apolar groups tend to pack together away from the polar solvent, and the polar or ionic head groups tend to be at the surface of the micelle where they interact with the solvent. Water is the preferred solvent for study of this phenomenon and all the results discussed in this chapter relate to experiments in water or in solvents of high water content unless otherwise specified. Micellization is a manifestation of the strong self-association of water and similar solvents, and is an example of the hydrophobic or solvophobic effect[1] which forces self-association of apolar materials (Tanford, 1980).

Micelles are small, relative to the wavelength of light. Their solutions are therefore transparent, but they scatter light, and this property provided compelling evidence for the formation of discrete micelles (Debye, 1949; Debye and Anacker, 1951).

Most studies of micellar systems have been carried out on synthetic surfactants where the polar or ionic head group may be cationic, e.g. an ammonium or pyridinium ion, anionic, e.g. a carboxylate, sulfate or sulfonate ion, non-ionic, e.g. hydroxy-compound, or zwitterionic, e.g. an amine oxide or a carboxylate or sulfonate betaine. Surfactants are often given trivial or trade names, and abbreviations based on either trivial or systematic names are freely used (Fendler and Fendler, 1975). Many commercial surfactants are mixtures so that purity can be a major problem. In addition, some surfactants, e.g. monoalkyl sulfates, decompose slowly in aqueous solution. Some examples of surfactants are given in Table 1, together with values of the critical micelle concentration, cmc. This is the surfactant concentration at the onset of micellization (Mukerjee and Mysels, 1970) and can therefore be taken to be the maximum concentration of monomeric surfactant in a solution (Menger and Portnoy, 1967). Its value is related to the change of free energy on micellization (Fendler and Fendler, 1975; Lindman and Wennerström, 1980).

Micellization depends upon a balance of forces and the cmc decreases with increasing hydrophobicity of the apolar groups, and for ionic amphiphiles also depends on the nature and concentration of counterions in solution. Added electrolytes decrease the cmc, and the effect increases with decreasing charge density of the counterion. Divalent counterions, however, lead to

[1] The term "solvophobic" applies to interactions in a variety of associated solvents, whereas "hydrophobic" applies to interactions in highly aqueous solvents.

TABLE 1
Synthetic surfactants

	Surfactant	cmc/M
$C_{11}H_{23}CO_2^- Na^+$	Sodium dodecanoate	2.6×10^{-3}
$C_{12}H_{25}OSO_3^- Na^+$	Sodium dodecylsulfate, SDS Sodium laurylsulfate, NaLS	8×10^{-3}
$C_{14}H_{29}OSO_3^- Na^+$	Sodium tetradecylsulfate	2×10^{-3}
$C_{14}H_{29}\overset{+}{N}Me_3\ Br^-$	Tetradecyltrimethylammonium bromide Myristyryltrimethylammonium bromide, MTABr	3.5×10^{-3}
$C_{16}H_{33}\overset{+}{N}Me_3\ Br^-$	Hexadecyltrimethylammonium bromide, HTABr Cetyltrimethylammonium bromide, CTABr	9×10^{-4}
$C_{12}H_{25}\overset{+}{N}C_5H_5\ Cl^-$	Dodecylpyridinium chloride	1.5×10^{-2}
$C_{12}H_{25}\overset{+}{N}Me_2O^-$	Dodecyldimethylamine oxide	2×10^{-3}
$C_{12}H_{25}\overset{+}{N}Me_2(CH_2)_3SO_3^-$	12-3 Zwittergent, Sulfobetaine	
$RO(CH_2CH_2O)_n(CH_2)_2OH$	Brij, Igepal, Triton	

lower values of the cmc than do univalent ions because ion binding, of itself, leads to a decrease in entropy (Gunnarsson *et al.*, 1980).

Non-ionic and zwitterionic amphiphiles typically have lower values of the cmc than otherwise similar amphiphiles, because there is no formal coulombic repulsion between the head groups.

Scheme 1 gives a representation of an approximately spherical micelle in water, with ionic head groups at the surface and counterions clustered around the micelle partially neutralizing the charges. Counterions which are closely associated with the micelle can be assumed to be located in a shell, the so-called Stern layer, the thickness of which should be similar to the size of the micellar head groups. Monomeric co-ions will be repelled by the ionic head groups. The hydrophobic alkyl groups pack randomly and parts of the chains are exposed to water at the surface (Section 2).

The cmc is a key property, because it is related to the free energy difference between monomer and micelles. The onset of micellization is detected by marked changes in such properties as surface tension, refractive index and

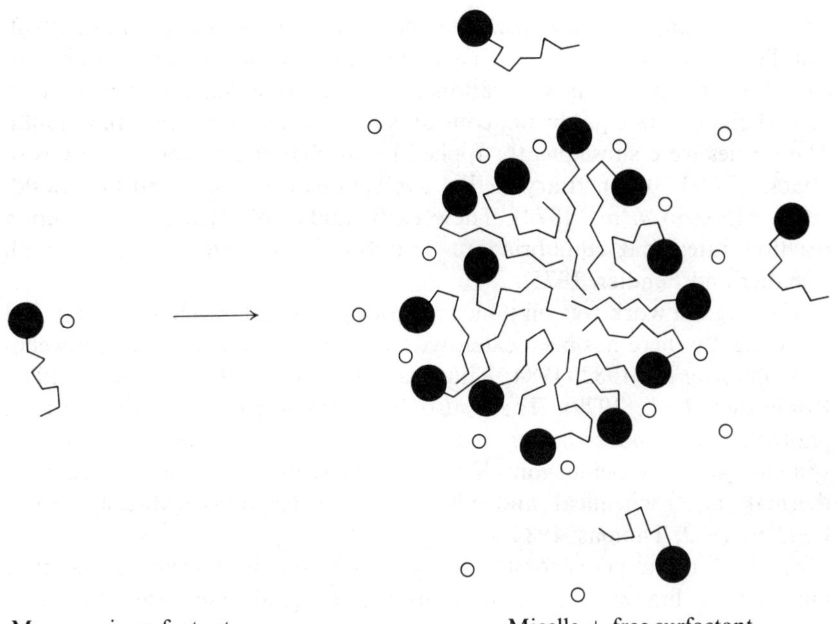

Scheme 1

conductivity (for ionic micelles); light scattering also increases sharply on micellization, as does solubilization of hydrophobic solutes. To a first approximation the solution can be assumed to contain monomeric amphiphile and fully formed micelles with sub-micellar particles playing a minor role.

It has long been known that aqueous micelles can influence chemical rates and equilibria (Hartley, 1948; Duynstee and Grunwald, 1959; Kurz, 1962; Motsavage and Kostenbauder, 1963), and there are a number of related self-assembling colloids which share this ability. Microemulsions generally contain water, an oil, a surfactant and a cosurfactant which is generally a medium chain length alcohol, amine or similar polar organic molecule (Bellocq *et al.*, 1984; Mackay, 1981). Oil in water (o/w) microemulsions are formed when water is the bulk solvent. The droplets are larger than normal micelles in water (Zana *et al.*, 1981; Bellocq *et al.*, 1984) but the two structures have the common feature that the polar or ionic head groups are in contact with water. Water in oil (w/o) microemulsions are formed when oil is the bulk solvent. They are akin to the reverse micelles which form when surfactants, usually with small amounts of water, are dissolved in an apolar organic solvent (Fendler and Fendler, 1975; Danielsson and Lindman, 1981). The interiors of these droplets contain water and the apolar regions of the amphiphiles are in contact with the apolar solvent (O'Connor *et al.*, 1984). Micelles in water have long been known to influence acid-base indicator equilibria and the effects were rationalized in terms of Hartley's Rules which related changes in equilibrium constants to micellar charge (Hartley, 1948). These rules were subsequently applied to micellar effects upon the rates of attack of OH^- upon triarylmethyl dye cations (Duynstee and Grunwald, 1959; Albrizzio *et al.*, 1972). These early studies of micellar effects upon reaction rates and equilibria are described in an extensive monograph (Fendler and Fendler, 1975).

The original work was on ionic reactions in normal micelles in water, but subsequently there has been extensive work on reactions in reverse micelles (O'Connor *et al.*, 1982, 1984; Kitahara, 1980; O. A. El Seoud *et al.*, 1977; Robinson, *et al.*, 1979). There also has been a great deal of work on photochemical and radiation induced reactions in a variety of colloidal systems, and microemulsions have been used as media for a variety of thermal, electrochemical and photochemical reactions (Mackay, 1981; Fendler, 1982; Thomas, 1984).

Much of the impetus for the study of reactions in micelles is that they model, to a limited extent, reactions in biological assemblies. Synthetic vesicles and cyclodextrins are other model reaction media and the term "Biomimetic Chemistry" has been coined to describe this general area of study. Work in this area is reviewed in recent publications (Kunitake and Shinkai, 1980; Fendler, 1982).

This chapter is restricted to discussion of thermal reactions in aqueous colloidal systems. Our aim is to analyse the reactions from a mechanistic point of view, and therefore our discussion is focused on organic reactions in aqueous micelles. We do not consider inorganic reactions, although their rates are influenced by micelles, and the factors which control rates of organic reactions are at work here (Balekar and Engberts, 1978; Robinson *et al.* 1979; Pelizzetti and Pramauro, 1984; Pelizzetti *et al.*, 1984). The structure of micelles or other colloidal droplets is considered only to the extent needed to understand reactivity. A key feature of micelles and similar colloidal aggregates is that they can incorporate solutes. For example an ionic micelle can bind a non-ionic solute and also, by virtue of its charge, attract counterions (Scheme 1). Therefore it may affect reaction rates and equilibria by bringing reactants together or keeping them apart, but because the micelle can exert a medium effect it is necessary to separate the "medium" and "concentration" effects of the micelle (Bunton, 1973a,b, 1979a,b). Our choice of topics is dictated partly by our own interests, but also by space requirements. The overall subject of reaction in submicroscopic aggregates has expanded so rapidly and in so many different directions that one has to focus one's interest on specific areas.

The general principles which govern the effects of normal, aqueous, micelles upon reaction rates and equilibria are considered first, and then we discuss some specific reactions and the relation of micellar effects to mechanism. We also briefly consider some non-micellar species generated by amphiphiles which can also mediate reactivity.

2 Micellar structure and ion binding

Micellization depends upon a balance between hydrophobic and van der Waals interactions which bring monomers together and repulsions between the polar or ionic head groups (Hartley, 1948; Tanford, 1980). But repulsions between ionic head groups will be offset by attraction of the counterions to the micellar surface (Stigter, 1964, 1978).

The cmc decreases with increasing chain length of the apolar groups, and is higher for ionic than for non-ionic or zwitterionic micelles. For ionic micelles it is reduced by addition of electrolytes, especially those having low charge density counterions (Mukerjee and Mysels, 1970). Added solutes or cosolvents which disrupt the three-dimensional structure of water break up micelles, unless the solute is sufficiently apolar to be micellar bound (Ionescu *et al.*, 1984).

Onset of micellization is detected by sharp changes in such properties as surface tension, refractivity or conductivity (of ionic micelles). To a first approximation the solution is assumed to contain monomeric amphiphiles, whose concentration is given by the cmc, and fully formed micelles, with submicellar aggregates playing a minor role.

Formation of micelles in water seems to require a minimum length of the hydrophobic alkyl group which is eight to ten methylene (and methyl) groups. Aggregation numbers are generally greater than 50, and increase with increasing hydrophobicity of the alkyl group. Any factor which increases the balance of hydrophobicity over hydrophilicity stabilizes the micelle as evidenced by decrease of the cmc and increase of the aggregation number. Normal aqueous micelles are generally formed from single-chain surfactants and chain-branching inhibits micellization.

Micellar structure is a controversial topic, although there seems to be a consensus on some points (Israelachvili *et al.*, 1976; Menger, 1979a,b; Wennerström and Lindman, 1979; Dill and Flory, 1980, 1981; Dill, 1984; Dill *et al.*, 1984; Fromherz, 1981; Gruen, 1984; Butcher and Lamb, 1984). At surfactant concentrations not markedly above the cmc, micelles are approximately spherical and their radius is similar to that of the extended chain length of the surfactant (Missel *et al.*, 1980; Corti and Degiorgio, 1981; Briggs *et al.*, 1982; Dorshow *et al.*, 1982, 1983). Micellar head groups and associated counterions are fully hydrated (Stigter, 1964, 1978), and one widely accepted micellar model, the so-called Hartley micelle, involved a hydrocarbon-like interior surrounded by polar or ionic head groups. There are questions as to the way in which hydrophobic alkyl groups pack together in the micellar interior. For example, Menger (1979a) used space-filling models to demonstrate that there will be large voids in the micelle if eclipsing is to be avoided in the alkyl groups, and he concluded that these voids will accommodate water molecules which penetrate deeply into the micelle. Fromherz (1981) and Butcher and Lamb (1984) postulated a different packing of monomers which also leads to partial exposure of alkyl groups to water, and Dill and Flory (1980, 1981) showed how a micelle could be constructed with eclipsing of some of the alkyl groups and contact between methylene groups and water; Scheme 1 illustrates this principle.

A modified Dill-Flory model has been reported to be consistent with results of small angle neutron scattering which exclude penetration of water into the micellar interior (Dill *et al.*, 1984). However, the structure of the micelle is so dynamic that every segment of the hydrophobic alkyl groups is exposed to water at some time, and in a small micelle over half the chain segments will be at the micellar surface at any given time. These theoretical predictions agree with proton and fluorine n.m.r. spectroscopy (Dill, 1984; Gruen, 1984; Cabane, 1981; Muller and Simsohn, 1971; Cabane *et al.*, 1984; Jonsson *et al.*, 1984).

Although micelles at surfactant concentrations close to the cmc seem to be approximately spherical, they grow with increasing [amphiphile] and added salt, and the counterion of the added salt has a specific effect on the growth. Micellar growth of necessity involves a change of shape from spherical to ellipsoidal (Missel *et al.*, 1980; Corti and Degiorgio, 1981; Tanford, 1978, 1980; Dorshow *et al.*, 1982, 1983; Hoffmann *et al.*, 1984). Even relatively hydrophobic micellar-bound solutes tend to be exposed to water and generally bind close to the head groups at the micelle–water interface (Menger, 1979a; Mukerjee *et al.*, 1977; Russell *et al.*, 1981; Russell and Whitten, 1982; Whitten *et al.*, 1984).

Monomeric surfactant and added solutes rapidly diffuse into the micelles, so that transfer of material between water and micelle is much faster than most activated thermal chemical reactions. This generalization cannot be applied to photochemical reactions where some steps of the reaction may be very rapid and therefore faster than solute transfer (Fendler, 1982; Thomas, 1984).

The following discussion of chemical reactivity and mechanism will be based on the premise that for most thermal reactions equilibrium is maintained between water and the micelles, which can be regarded as distinct reaction media, and most kinetic treatments are based on this so-called pseudophase model (Cordes and Gitler, 1973; Bunton, 1973b). Reaction typically occurs at the micellar surface in the Stern layer (for ionic micelles) which contains specifically bound counterions. The Stern layer is highly aqueous and contains a high concentration of ionic head groups and counterions. The fraction of ionic head groups neutralized by counterions is generally designated as α, which is the fractional micellar charge. For most ionic micelles α is in the range 0.1–0.3 and is insensitive to the overall concentration of counterion; in other words the surface of an ionic micelle is treated as if it is saturated with counterions (Romsted, 1977, 1984; Evans and Wightman, 1982). The remaining counterions are distributed in the diffuse, Gouy–Chapman layer, and their distribution is governed by their non-specific electrostatic interactions with the micelle which can be regarded as a bulky macroion (Bell and Dunning, 1970; Mille and Vanderkooi, 1977; Gunnarsson *et al.*, 1980).

Kinetic treatments are usually based on the assumption that reaction does not occur across the micelle-water interface. In other words a bimolecular reaction occurs between reactants in the Stern layer, or in the bulk aqueous medium. Thus the properties of the Stern layer are of key importance to the kineticist, and various probes have been devised for their study. Unfortunately, many of the probes are themselves kinetic, so it is hard to avoid circular arguments. However, the charge transfer and fluorescence spectra of micellar-bound indicators suggest that the micellar surface is less polar than water (Cordes and Gitler, 1973; Fernandez and Fromherz, 1977; Ramachandran *et al.*, 1982).

There are unanswered questions regarding micellar size and shape. The size of an approximately spherical micelle is geometrically constrained, but based on a variety of measurements it appears that micelles grow with increasing concentration of amphiphile or added electrolyte, and therefore become ellipsoidal. Growth of an ionic micelle depends very much upon the properties of the counterion; for example for a C_{16} quaternary ammonium ion ($C_{16}H_{33}N^{+}Me_{3}$) there is extensive growth with an increase in concentration of surfactant and counterion for the bromide, much less growth for the chloride, and very little growth for the hydroxide (Dorshow *et al.*, 1983; Athanassakis *et al.*, 1985). However, some of the evidence cited for micellar growth is highly suspect because of the neglect of intermicellar interactions, but fortunately for the kineticist it seems that micellar size, *per se*, is not a dominant factor in determining chemical reactivity, probably because it does not markedly affect the nature of the micelle–water interface.

Studies of chemical reactivity in self-assembling colloids were initially based on reactions in aqueous micelles, but recently reactivity has been examined in other colloidal systems such as microemulsions and synthetic vesicles (Mackay, 1981; Fendler, 1982; O'Connor *et al.*, 1982, 1984; Cuccovia *et al.* 1982b). Some hydrophobic trialkylammonium salts, which are phase-transfer catalysts, will also speed reactions in water. Although they do not micellize their effect on reaction rate is similar to that of cationic micelles (Okahata *et al.*, 1977; Kunitake *et al.*, 1980; Bunton *et al.*, 1981b; Biresaw *et al.*, 1984; Bunton and Quan, 1984, 1985).

3 Quantitative treatments of rates and equilibria

Micellar effects upon reaction rates and equilibria have generally been discussed in terms of a pseudophase model, and this approach will be followed here.

Provided that equilibrium is maintained between the aqueous and micellar pseudophases (designated by subscripts W and M) the overall reaction rate will be the sum of rates in water and the micelles and will therefore depend upon the distribution of reactants between each pseudophase and the appropriate rate constants in the two pseudophases. Early studies of reactivity in aqueous micelles showed the importance of substrate hydrophobicity in determining the extent of substrate binding to micelles; for example, reactions of a very hydrophilic substrate could be essentially unaffected by added surfactant, whereas large effects were observed with chemically similar, but hydrophobic substrates (Menger and Portnoy, 1967; Cordes and Gitler, 1973; Fendler and Fendler, 1975).

Ester saponifaction was a favoured reaction for this type of study, because the hydrophobicity of the acyl moiety could easily be controlled by increasing the length of an n-alkyl group, and saponification of *p*-nitrophenyl n-alkanoates could be followed with very dilute substrate. Substrate concentration is an important factor, because provided that it is kept low it is reasonable to assume that the micelle structure is relatively unperturbed.

Menger and Portnoy (1967) developed a quantitative treatment which adequately described inhibition of ester saponification by anionic micelles. Micelles bound hydrophobic esters, and anionic micelles excluded hydroxide ion, and so inhibited the reaction, whereas cationic micelles speeded saponification by attracting hydroxide ion (Menger, 1979b).

Provided that only substrate distribution has to be considered, which is the situation for micelle-inhibited bimolecular, or spontaneous unimolecular, reactions, Scheme 2 describes substrate distribution and reaction in each pseudophase (Bunton *et al.*, 1968).

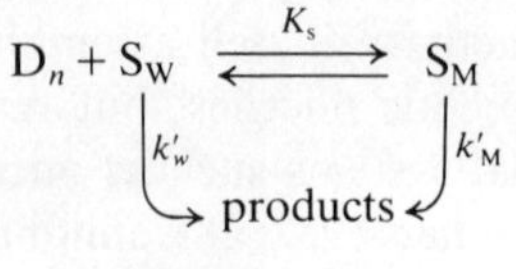

$$D_n + S_W \underset{}{\overset{K_s}{\rightleftharpoons}} S_M$$

$$S_W \xrightarrow{k'_w} \text{products} \xleftarrow{k'_M} S_M$$

Scheme 2

In Scheme 2 D_n denotes micellized surfactant detergent, S is substrate, subscripts W and M denote aqueous and micellar pseudophases respectively, and k'_W and k'_M are first-order rate constants. The binding constant, K_s, is written in terms of the molarity of micellized surfactant, but it could equally be written in terms of the molarity of micelles. The two constants differ in magnitude by the aggregation number of the micelles.

The concentration of micellized surfactant is that of total surfactant less that of monomer which is assumed to be given by the critical micelle concentration (cmc). The overall first-order rate constant k_ψ is then given by (1).

$$k_\psi = \frac{k'_W + k'_M\ K_S([D]-\text{cmc})}{1 + K_s([D]-\text{cmc})} \qquad (1)$$

This equation is formally similar to the Michaelis–Menten equation of enzyme kinetics, although the analogy is limited because most enzymic reactions are studied with substrate in large excess over enzyme. Equation (1) could be rearranged to give (2) which is formally similar to the Lineweaver–Burk equation, and which permits calculation of k'_M and K_s, provided that k'_W is known (Menger and Portnoy, 1967; Menger, 1979b).

$$\frac{1}{k_\psi - k'_W} = \frac{1}{k'_M - k'_W} + \frac{1}{(k'_M - k'_W)\, K_s([D]-\text{cmc})} \qquad (2)$$

Equations (1) and (2) depend on some major assumptions, in particular that the cmc gives the concentration of monomeric surfacant and that rate and binding constants in the micellar pseudophase are unaffected by reactants and products. These equations have been used very extensively and provided the basis for quantitative analysis of micellar rate effects. Much of this work has been reviewed in a comprehensive monograph which gives extensive compilations of data up to 1974 (Fendler and Fendler, 1975).

Two other general ways of treating micellar kinetic data should be noted. Piszkiewicz (1977) used equations similar to the Hill equation of enzyme kinetics to fit variations of rate constants and surfactant concentration. This treatment differs from that of Menger and Portnoy (1967) in that it emphasizes cooperative effects due to substrate–micelle interactions. These interactions are probably very important at surfactant concentrations close to the cmc because solutes may promote micellization or bind to submicellar aggregates. Thus, eqn (1) and others like it do not fit the data for dilute surfactant, especially when reactants are hydrophobic and can promote micellization.

Katiyar and coworkers have also developed equations which attempt to take into account the way in which reactant interactions affect reactivity in micelles (Srivastava and Katiyar, 1980). However, most workers have followed the original formalism despite its approximations.

As applied to micelle-inhibited reactions, (1) or (2) showed that micelles could bind hydrophobic solutes very strongly, consistent with the results of direct measurements. The binding constants of non-ionic solutes to micelles can be measured directly by gel-permeation chromatography, by ultrafiltration or by solubilization (Cordes and Gitler, 1973; Fendler and Fendler, 1975; Armstrong, 1981; Almgren and Rydholm, 1979). In favorable cases spectral shifts on incorporation of solutes into micelles are large enough to allow measurement of K_s (Sepulveda, 1974). Direct measurement is not feasible with very labile solutes but there is impressive agreement between direct and kinetic measurements of substrate binding, which justifies use of (1).

Equation (1) is generally used to estimate the rate constant, k_M', in the micellar pseudophase, but for inhibited bimolecular reactions it provides an indirect method for estimation of otherwise inaccessible rate constants in water. Oxidation of a ferrocene to the corresponding ferricinium ion by Fe^{3+} is speeded by anionic micelles of SDS and inhibited by cationic micelles of cetyltrimethylammonium bromide or nitrate (Bunton and Cerichelli, 1980). The variation of the rate constants with [surfactant] fits the quantitative treatment described on p. 225. Oxidation of ferrocene by ferricyanide ion in water is too fast to be easily followed kinetically, but the reaction is strongly inhibited by anionic micelles of SDS which bind ferrocene, but exclude ferricyanide ion. Thus reaction occurs essentially quantitatively in the aqueous pseudophase, and the overall rate depends upon the rate constant in water and the distribution of ferrocene between water and the micelles. It is easy therefore to calculate the rate constant in water from this micellar inhibition.

Equations (1) and (2) generally fail for bimolecular, micelle-assisted reactions. Equation (1) predicts that the first-order rate coefficients should reach a constant, limiting value at high surfactant concentration when the substrate is fully micellar bound, but rate maxima are observed for the corresponding nonsolvolytic bimolecular reactions. The rate-surfactant concentration profiles can be treated quantitatively by taking into account the distribution of *both* reactants between water and micelles. This can be done by extending (1), and a simple formalism involves writing the first-order rate constants k_W' and k_M' in terms of second-order rate constants in water and micelles, and reactant concentrations in each pseudophase (Martinek *et al.*, 1977; Romsted, 1977; Balekar and Engberts, 1978; Bunton and Wolfe, 1973; Cuccovia *et al.*, 1978; Bunton and Romsted, 1979). However, one immediately runs into the problem of defining concentration in the micellar pseudophase. One approach is to write concentration in terms of moles of reagent per litre of micelles, or to assume some volume of the micellar pseudophase, V_M, in which reaction takes place. The problem is similar to that of comparing second-order rate constants, written conventionally in the dimensions $M^{-1} s^{-1}$, for a variety of solvents. The comparisons will be completely different if concentrations are written in terms of molarities or mole fractions.

Another approach is to define concentration in the micellar pseudophase in terms of a mole ratio. Concentration is then defined unambiguously, and the equations take a simple form (Bunton, 1979a,b; Romsted, 1984). However, this approach does not allow direct comparison of second-order rate constants in aqueous and micellar pseudophases and by evading one problem one faces another.

The first-order rate constants are written in (3) and (4) as second-order rate constants, k_W and k_M, for reaction of a reagent, Y, where the mole ratio, $m_Y^s = [Y_M]/([D]-\text{cmc})$. (Here, and elsewhere, the quantities in square brackets denote molarity in terms of total solution volume, which is approximately that of the aqueous pseudophase.)

$$k'_W = k_W[Y_W] \tag{3}$$

$$k'_M = k_M m_Y^s \tag{4}$$

$$k_\psi = \frac{k_W[Y_W] + k_M K_s m_Y^s([D]-\text{cmc})}{1 + K_s([D]-\text{cmc})} \tag{5}$$

$$= \frac{k_W[Y_W] + k_M K_s[Y_M]}{1 + K_s([D]-\text{cmc})} \tag{6}$$

By combining (1), (3) and (4), expressions (5) and (6) are obtained. These, or similar, equations readily explain why first-order rate constants of micelle-assisted bimolecular reactions typically go through maxima with increasing surfactant concentration if the overall reactant concentration is kept constant. Addition of surfactant leads to binding of both reactants to micelles, and this increased concentration increases the reaction rate. Eventually, however, increase in surfactant concentration dilutes the reactants in the micellar pseudophase and the rate falls. This behavior supports the original assumption that substrate in one micelle does not react with reactant in another, and that equilibrium is maintained between aqueous and micellar pseudophases.

Equations (5) or (6) and others which are essentially identical but are written in different ways, can be applied to bimolecular micelle-assisted reactions provided that the distribution of both reactants can be determined.

In some cases the problem is relatively simple. For example the binding of organic, nucleophilic anions, e.g. aryloxides or areneimidazolide anions, can be estimated from the spectral shifts which occur when the ion is transferred from water to micelles (Sepulveda, 1974; Bunton and Sepulveda, 1979;

Bunton *et al.*, 1981b). Estimation of the extent of micellar binding becomes a non-problem if the organic ion is very hydrophobic, because then it is completely micellar bound under essentially all conditions (Martinek *et al.*, 1977). Perhaps for this reason, there are many examples of good fits between experimental rate constant–surfactant profiles and those calculated using (5), (6) or equivalent expressions.

An example of the application of the pseudophase model to imidazolide ion dephosphorylation is shown in Fig 1. The areneimidazole is a weak acid,

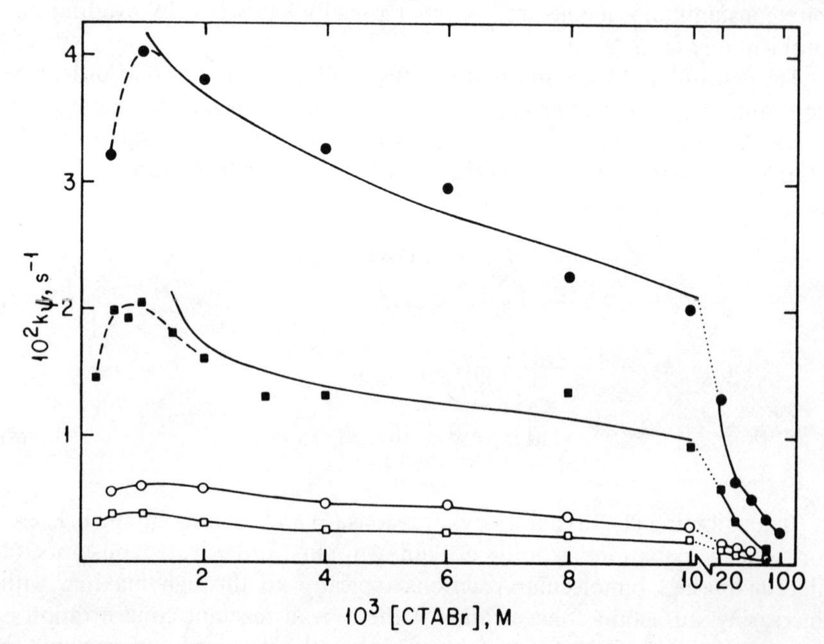

FIG. 1 Micellar effects upon reaction of *p*-nitrophenyl diphenyl phosphate with benzimidazolide ion (solid points); open points are for reaction in the absence of benzimidazole: ■, 10^{-4} M benzimidazole, pH 10.7; ●, 1.2×10^{-4} M benzimidazole, pH 11: □, ○, pH 10.7, and 11 respectively. The solid lines are theoretical. (Reprinted by permission of the American Chemical Society)

but at high pH it gives the imidazolide ion which is an effective dephosphorylating agent displacing *p*-nitrophenoxide ion from *p*-nitrophenyl diphenyl phosphate. These reactions were run at high pH so there is a concomitant reaction of OH^- which has to be taken into account. Concentrations of areneimidazolide ion and phosphate ester in the aqueous and micellar pseudophases can be estimated. The solid lines are calculated from the distribution of the reagents between aqueous and micellar pseudophases, and values of rates and equilibrium parameters are discussed in Sections 5–7. Table 2 gives examples of reactions of anionic or non-ionic nucleophiles

TABLE 2

Micellar effects upon reactions of organic nucleophiles

Reaction	Surfactant	K_s/M^{-1}	k_{rel}	k_2^m/k_W	Ref.
$C_6H_{13}CO_2C_6H_4NO_2(4)$ + imidazolide ion	CTABr	3600		*ca* 10	Martinek *et al.*, 1977
$C_6H_{13}CO_2C_6H_4NO_2(4)$ + imidazole	CTABr	3600		*ca* 10^{-2}	Martinek *et al.*, 1977
$MeCO_2C_6H_4NO_2(4) + PhS^-$	CTABr	27	69	0.42	Cuccovia *et al.*, 1978
$MeCO_2C_6H_4NO_2(4) + MePhS^-$	CTABr	27	55	0.32	Cuccovia *et al.*, 1978
$C_7H_{15}CO_2C_6H_4NO_2(4) + C_7H_{15}S^-$	DODAC[a]	4000	*ca* 10^7	*ca* 15	Cuccovia *et al.*, 1982b
$(PhO)_2PO_2C_6H_4NO_2(4) + PhO^-$	CTABr	16 000	3000	0.53	Bunton and Sepulveda, 1979
$(PhO)_2PO_2C_6H_4NO_2(4) + MeC_6H_4O^-$	CTABr	16 000	4000	0.50	Bunton and Sepulveda, 1979
$2,4\text{-}(NO_2)_2C_6H_3F + PhNH_2$	CTABr	54	3–8	0.17	Bunton *et al.*, 1979
$2,4\text{-}(NO_2)_2C_6H_3F + PhNH_2$	SDS		*ca* 3	0.12	Bunton *et al.*, 1979
$(PhO)_2PO_2C_6H_4NO_2(4) + C_7H_5N_2^-$ (benzimidazolide)	CTABr	10^4		*ca* 1	Bunton *et al.*, 1981b

[a] Vesicles of dioctadecylammonium chloride

for which second-order rate constants are compared in the aqueous and micellar pseudophases.

The problem is more difficult for bimolecular reactions of hydrophilic ions. In favorable cases the concentration of ionic reagent in the aqueous pseudophase can be measured electrochemically, because specific-ion electrodes do not respond to micellar bound ions (Larsen and Tepley, 1974); so, neglecting activity effects, the concentration of bound ion is calculated by difference. Alternatively interionic exchange can be followed using an organic anion as an indicator (Bartet *et al.*, 1980; Gamboa *et al.*, 1981), and fluorescence quenching can be used with some counterions, e.g. Br^- (Abuin *et al.*, 1983a,b; Lissi *et al.*, 1984).

Specific-ion electrodes are expensive, temperamental and seem to have a depressingly short life when exposed to aqueous surfactants. They are also not sensitive to some mechanistically interesting ions. Other methods do not have these shortcomings, but they too are not applicable to all ions. Most workers have followed the approach developed by Romsted who noted that counterions bind specifically to ionic micelles, and that qualitatively the binding parallels that to ion exchange resins (Romsted 1977, 1984). In considering the development of Romsted's ideas it will be useful to note that many micellar reactions involving hydrophilic ions are carried out in solutions which contain a mixture of anions; for example, there will be the chemically inert counterion of the surfactant plus the added reactive ion. Competition between these ions for the micelle is of key importance and merits detailed consideration. In some cases the solution also contains buffers and the effect of buffer ions has to be considered (Quina *et al.*, 1980).

THE PSEUDOPHASE ION-EXCHANGE MODEL

The competition of reactive and inert ions for the micellar surface was recognized early in the study of micellar rate effects (Bunton and Robinson, 1968, 1969; Bunton, 1973a; Cordes and Gitler, 1973) and it soon became evident that ions behaved specifically in micellar binding. A high charge density ion such as hydroxide or fluoride was readily displaced from a cationic micelle by a low charge density ion such as bromide, nitrate or especially tosylate. Thus these ions sharply inhibited reactions of hydroxide or fluoride ion mediated by cationic micelles. The problem was to quantify the results. Romsted pointed out that the charge per head group, α, of an ionic micelle seemed to depend little upon the nature or concentration of the counterion, and was generally close to 0.3 (Romsted, 1977, 1984). In other words the micellar surface was, to a first approximation, saturated by counterions. The counterions were regarded as being bound in close proximity to the micellar ionic head groups in the Stern layer, and the remaining (*ca* 30%) of the counterions were distributed non-specifically in the diffuse Gouy–Chapman layer and apparently did not react with micellar-bound substrate. [This assumption may be invalid for some reactions of hydrophilic anions (see pp. 239, 241).]

With these assumptions, ion exchange between a reactive anion, Y^-, and an inert anion, X^-, for example, was written in terms of (7).[2] It then was relatively straightforward to write the concentration of reactive ion in the micelle in terms of an assumed constancy of fractional micellar charge, α, and the ion exchange parameter, K_X^Y, and to analyse rates in terms of these parameters, the binding constants of the substrate, K_s, and the second-order rate constants, k_W and k_M (Romsted, 1977, 1984; Quina and Chaimovich, 1979; Bunton and Romsted, 1979).

$$K_X^Y = [Y_W^-][X_M^-]/([Y_M^-][X_W^-]) \tag{7}$$

This approach has been used successfully by a number of research groups for a variety of reactions of OH^-, some involving substrates which were strongly micellar bound and some occurring wholly in the aqueous pseudophase (Almgren and Rydholm, 1979; Chaimovich *et al.*, 1979; Funasaki and Murata, 1980; Al-Lohedan *et al.*, 1981). Different values of the ion-exchange parameters, K_X^{OH}, have been used, and there are significant differences between the values adopted by different groups. However, for the most widely-studied system, which is hydroxide ion with a bromide ion micelle, most values of K_{Br}^{OH} are between 10 and 20, and are not very different from values estimated by direct competition measurements (Romsted, 1984). The major source of uncertainty is that the kinetic data can often be fitted by combinations of values of the various parameters (Al-Lohedan *et al.*, 1982a), and to add to the confusion some workers have included in their kinetic equations a term for the surface potential of the micelle (Almgren and Rydholm, 1979). Nonetheless, the ion-exchange model has been applied to reactions of a variety of hydrophilic ions, although OH^- is the most widely studied anion. Some examples are shown in Tables 3 and 4. Data for reactions in vesicles derived from twin-tailed cationic surfactants are also included.

Figure 2 illustrates the variation of the first-order rate constants, k_ψ, with [CTABr] for reactions of benzoic anhydride with 0.01 M NaOH and with 0.02 M sodium formate. The lines are calculated using the ion-exchange treatment with the following parameters: $K_s = 650\ M^{-1}$, $\beta = 0.75$, $K_{Br}^X = 10$ for $X = OH^-$ and HCO_2^- and $k_M = 200$ and $0.06\ s^{-1}$ for OH^- and HCO_2^- respectively. Similar values of the rate constants were used in fitting the data

[2] Equation (7) is sometimes written in the inverse form.

TABLE 3

Micellar effects upon reactions of hydroxide ion

Substrate	Surfactant	K_s/M^{-1}	K_X^{OH}	k_{rel}	k_2^m/k_W	Ref.
$MeCO_2C_6H_4NO_2(4)$	CTABr	54	12.5		0.14	Quina *et al.*, 1980
$C_7H_{15}CO_2C_6H_4NO_2(4)$	CTABr	15 000	12.5		0.11	
$C_3H_7CO_2C_6H_4NO_2(4)$	CTABr	530	10	*ca* 3	0.13	Funasaki 1979
4-CN-pyridinium, N^+-$C_{10}H_{21}$	CTABr		12.5		3.4	Bonilha *et al.*, 1984
4-CN-pyridinium, N^+-$C_{10}H_{21}$	CPyCl[a]		5.6		3.4	
$(PhCO)_2O$	CTABr	650	10	7	0.06	Al-Lohedan and Bunton, 1982
$C_{12}H_{25}\overset{+}{N}Me_2CH_2CO_2Me$	$C_{14}H_{29}NMe_3Cl$	22	4	5	0.28	Al-Lohedan *et al.*, 1981
O_2N-C_6H_4-CO_2Et	CTABr	240	10		0.034	Funasaki and Murata, 1980

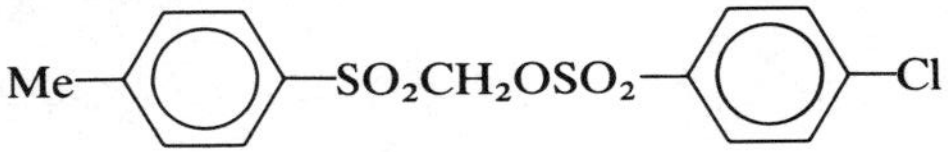	CTABr	4000	25	*ca* 20	0.14	van der Langkruis and Engberts, 1984
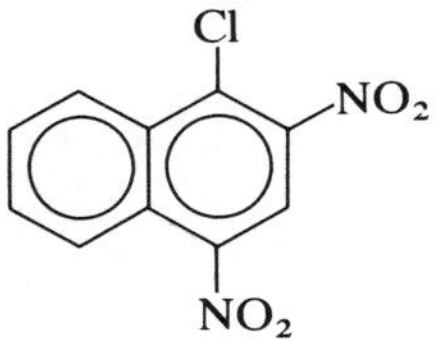	DDDACl	1600	4		1.6	Cipiciani *et al.*, 1984
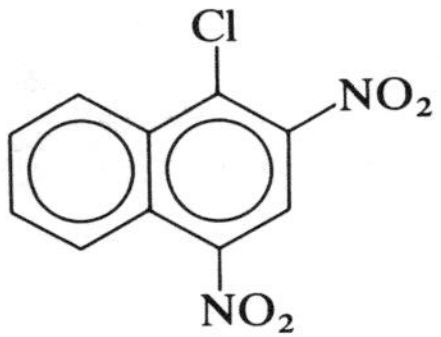	DDDAOH	1600			1.5	
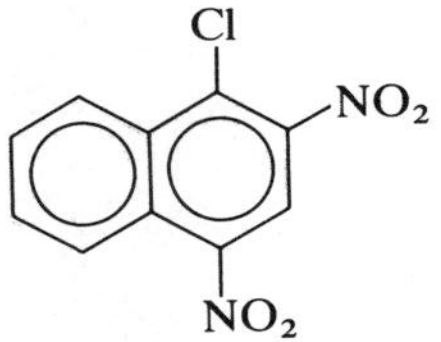	CTAOH	1600			2.2	Bunton *et al.*, 1981a
$(PhO)_2PO_2C_6H_4NO_2(4) + OH^-$	CTAOH	10^4			*ca* 0.2	

[a] Cetylpyridinium chloride

TABLE 4

Micellar effects upon reactions of hydrophilic anions

Reaction	Surfactant	K_s	K_X^{N} [a]	k_2^m/k_W	Ref.
N-dodecylpyridinium ($C_5H_5N^+$–$C_{12}H_{25}$) + CN^-	CTABr	70	1	2.3	Bunton *et al.*, 1980a
N-hexadecylpyridinium ($C_5H_5N^+$–$C_{16}H_{33}$) + CN^-	CTABr	3500	1	2.5	Bunton *et al.*, 1980a
N-dodecylpyridinium ($C_5H_5N^+$–$C_{12}H_{25}$) + CN^-	CTACN	70	1	2.3	Bunton *et al.*, 1980a
N-hexadecylpyridinium ($C_5H_5N^+$–$C_{16}H_{33}$) + CN^-	CTACN	3500	1	2.6	Bunton *et al.*, 1980a
$(PhCO)_2O + HCO_2^-$	CTABr	650	10	0.21	Al-Lohedan and Bunton, 1982
$(PhCO)_2O + HCO_2^-$	$CTAHCO_2$	650		0.21	Al-Lohedan and Bunton, 1982

Reaction	Surfactant				Reference
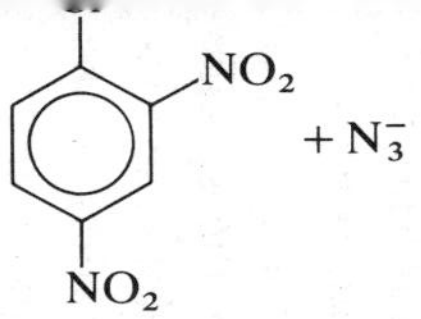 NO_2, NO_2 (ring) $+ N_3^-$	CTACl	82	1.3	52	Bunton *et al.*, 1982b
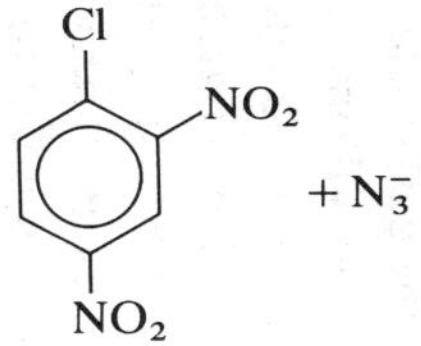 Cl, NO_2, NO_2 (ring) $+ N_3^-$	$CTAN_3$	115		28	Bunton *et al.*, 1982b
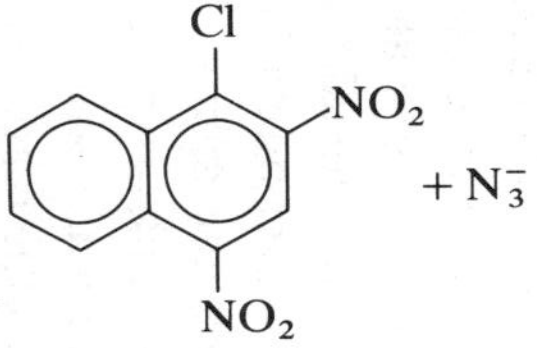 Cl, NO_2, NO_2 (naphthalene) $+ N_3^-$	CTABr	600	2	400	Bunton *et al.*, 1982b
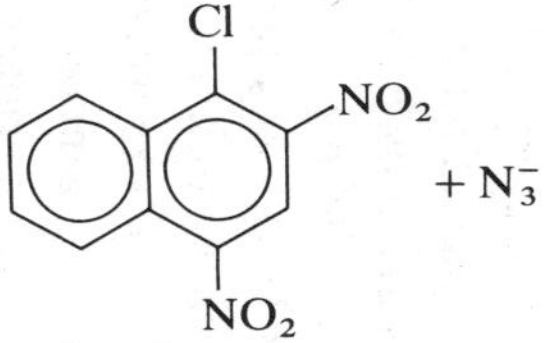 Cl, NO_2, NO_2 (naphthalene) $+ N_3^-$	$CTAN_3$	*ca* 600		200	Bunton *et al.*, 1982b
$PhSO_3Me + N_3^-$	CTAOMes	55	1.1	0.7	Bunton *et al.*, 1982b
n-BuBr $+ S_2O_3^{2-}$	CTABr	450		*ca* 8	Cuccovia *et al.*, 1982b

[a] N denotes the reactive ion and X the inert ion

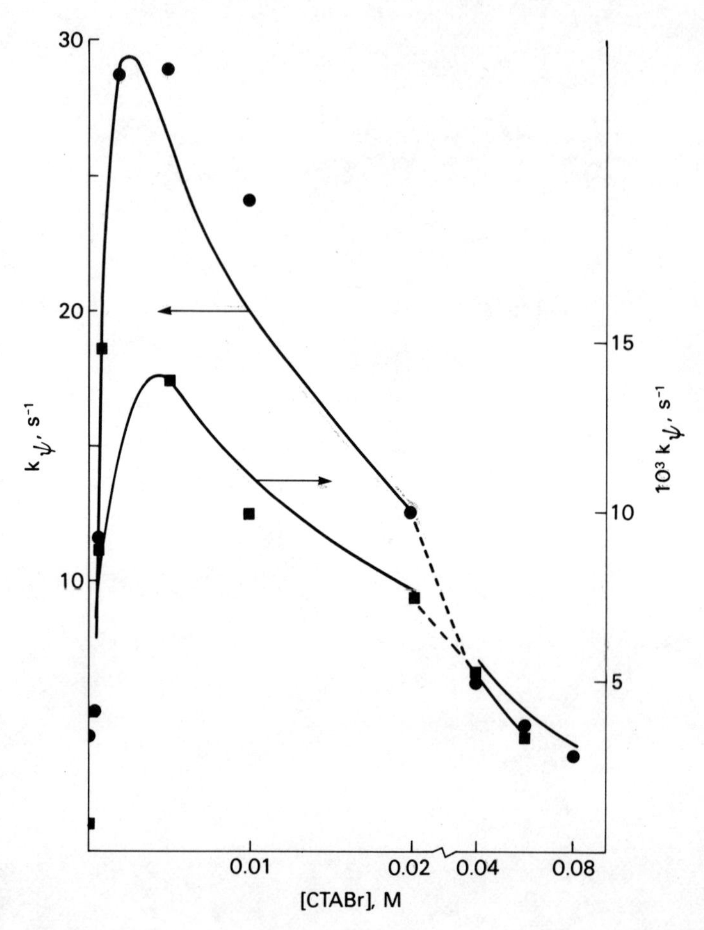

FIG. 2 Reactions of benzoic anhydride in CTABr: ●, 0.01 M NaOH; ■, 0.02 M HCO_2Na. The lines are calculated using the ion-exchange model. (Reprinted with permission of the American Chemical Society)

in mixtures of CTABr and NaBr, and here too the fits of experiment and theory are satisfactory (Al-Lohedan and Bunton, 1982).

Some examples of micellar rate enhancements of bimolecular reactions of electrophiles are shown in Table 5. Generally the surfactant was SDS with added electrophile, e.g. H_3O^+ or a metal ion, but sulfonic acids were also used so that H_3O^+ was the counterion and there was no interionic competition. The maximum rate enhancements, k_{rel}, depend upon the specific conditions of the experiment, and, as predicted by the pseudophase ion-exchange model, generally decrease with increasing concentration of the electrophilic ion. In some cases the reactions were too fast for measurement

TABLE 5

Micellar effects upon reactions of electrophiles[a]

Reaction	K_s/M^{-1}	k_{rel}	k_2^m/k_W	Ref.
$4\text{-}O_2NC_6H_4CH(OEt)_2 + HCl$	100	20	0.05	Bunton *et al.*, 1978b
$4\text{-}O_2NC_6H_4CH(OEt)_2 + C_{14}H_{29}SO_3H$	73	—	0.02	Bunton *et al.*, 1979
$4\text{-}O_2NC_6H_4CH(OEt)_2 + 4\text{-}C_{12}H_{25}C_6H_4SO_3H$	91	—	0.06	Bunton *et al.*, 1979
PhNH—NHPh + 2HCl	220	2×10^3	0.013	Bunton *et al.*, 1978b
N-(CF_3CO)pyrrole + HCl	420	4	0.027	Cipiciani *et al.*, 1981
$(C_5H_5)_2Fe + Fe^{3+}$	340		0.4	Bunton and Cerichelli, 1980

[a] The surfactant is SDS except for reactions in sulfonic acid micelles.

of the maximum rate constants, but values of k_2^m could be calculated from the rate-surfactant concentration profiles. Values of k_{rel} in micellized sulfonic acids are not given because both $[H_3O^+]$ and [surfactant] are changing simultaneously.

The original ion-exchange treatment was developed for competition between reactive and inert monoanions, but Chaimovich, Quina and their coworkers have extended it to competition between mono and dianions (Cuccovia *et al.*, 1982a; Abuin *et al.*, 1983a). The ion-exchange constant for exchange between thiosulfate dianion and bromide monoanion is not dimensionless as in (7) but depends on salt concentration, and the formalism was developed for analysing micellar effects upon reaction of dianionic nucleophiles, e.g. thiosulfate ion.

The ion-exchange model also nicely explains some results from experiments on the effects of anionic micelles on rates of reactions of hydrophilic anions. Anionic micelles strongly inhibit bimolecular attack of anions upon non-ionic substrates, and one might expect reaction to stop when the substrate is fully micellar bound. However, in the early saponification work of Menger and Portnoy (1967), there was a small but measurable reaction even at very high [surfactant]. These experiments were made in buffered solutions with added salt to maintain the ionic strength, and Chaimovich, Quina and their coworkers pointed out that the high concentration of added Na^+ would increase the pH at the micellar surface by driving out hydronium ions. This is equivalent to saying that Na^+ increases the concentration of reactive OH^- in the anionic micelle (Quina *et al.*, 1982). Added NaCl speeded reactions of OH^- with substrates bound to SDS micelles as predicted.

Added salts accelerate the (micelle-inhibited) reaction of CN^- with a triphenylmethyl dye cation. The salt order is $Cs^+ > K^+ > Na^+ > Li^+$ and here too the cation displaces hydronium ion from the anionic micelle and so increases the concentration of bound CN^- (Srivastava and Katiyar 1980).

A somewhat similar example of mediated co-ion binding to micelles was the observation of anomalous salt effects on the reaction of Malachite Green with OH^- or BH_4^- in cationic micelles of CTABr. Typically inert counterions inhibit reactions of hydrophilic ions by excluding them from micelles, but in those nucleophilic additions to a carbocation in a cationic micelle added inert anions could inhibit reaction by excluding OH^- or BH_4^- from the micelle or assist it by increasing the micellar binding of Malachite Green. Depending upon the conditions of the experiment and the nature of the added salt either effect could prevail (Bunton *et al.*, 1978a).

The pseudophase model is often applied to reactions of hydrophobic ionic substrates, e.g. pyridinium ion in solutions of cationic micelles (Tables 3 and 4) with the hydrophobic attraction between micelle and substrate overcoming the coulombic repulsion. However, the model can also be applied to reactions for which a cationic substrate is so hydrophilic that it does not bind to a cationic micelle. In this situation reaction is assumed to take place wholly in the aqueous pseudophase, and the rate surfactant-profiles can be analysed on this basis. This approach has been applied to reactions of OH^- with 4-cyano-N-methylpyridinium ion (Chaimovich *et al.*, 1979, 1983; Quina *et al.*, 1980) and with betaine esters (Al-Lohedan *et al.*, 1981).

The acid hydrolysis of micellized alkyl sulfates (Kurz, 1962; Motsavage and Kostenbauder, 1963) has recently been very carefully reinvestigated (Garnett *et al.*, 1983). For relatively dilute micellized alkyl sulfate, salt inhibition follows the predictions of the pseudophase ion-exchange model, with the expected salt order. But this order is not followed with more concentrated alkyl sulfate, and these results are a very interesting deviation from the widely observed pattern of micellar salt effects.

The ion-exchange model has also been successfully applied to reactions of hydrophilic anions in microemulsions or alcohol-swollen micelles (Mackay, 1982; Bunton and de Buzzaccarini, 1982; Athanassakis *et al.*, 1982).

REACTIVE-ION MICELLES

The pseudophase ion-exchange model accounts very nicely for the variation of rate constant with [surfactant] for reactions of hydrophilic ions. However, it depends upon some assumptions which are difficult to verify: for example β may change as the mixture of counterions is changed, and the ion exchange constant (7), may also vary as the counterion concentrations change. An alternative approach is to use micelles of an ionic surfactant whose counter-ion is the reactant. If the micellar surface is saturated with counterions the concentration of ionic reagent at that surface should be constant, and the mole ratio of reactive ion to micellar head group will be given by $\beta = 1 - \alpha$, Then the overall rate constant will depend only upon the distribution of substrate between water and micelles, i.e. it will follow the simple distribution equation (1).

This hypothesis is satisfactory for nucleophilic reactions of cyanide and bromide ion in cationic micelles (Bunton *et al.*, 1980a; Bunton and Romsted, 1982) and of the hydronium ion in anionic micelles (Bunton *et al.*, 1979). As predicted, the overall rate constant follows the uptake of the organic substrate and becomes constant once all the substrate is fully bound. Addition of the ionic reagent also has little effect upon the overall reaction rate, again as predicted. Under these conditions of complete substrate binding the first-order rate constant is given by (8), and, where comparisons have been made for reaction in a reactive-ion micelle and in solutions

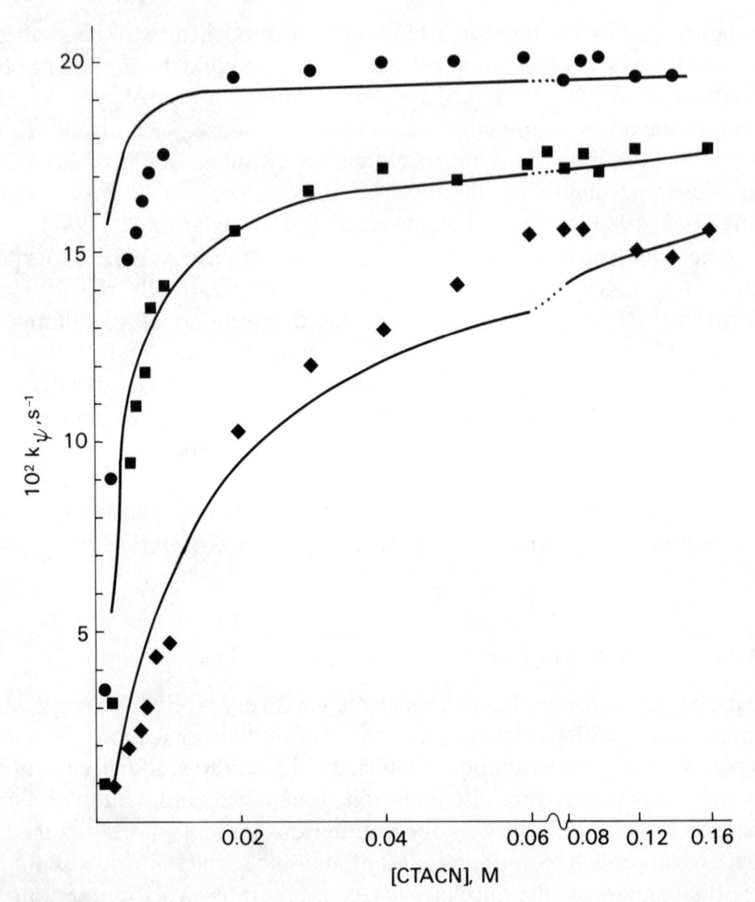

FIG. 3 Reaction of N-alkyl-2-carbamoylpyridinium ion in CTACN: ◆, ■, ●, alkyl = n-$C_{12}H_{25}$, n-$C_{14}H_{29}$, n-$C_{16}H_{33}$ respectively. (Reprinted with permission of the American Chemical Society)

containing both inert and reactive ions, there has been reasonable agreement between values of k_M in the two sets of conditions.

$$k_\psi = \beta k_M \tag{8}$$

The application of the concept of a reactive-ion micelle is illustrated by addition of cyanide ion to N-alkylpyridinium ions (Fig. 3) and rate constants are compared in Table 4. Second-order rate constants are essentially independent of substrate hydrophobicity and are only slightly affected by added inert salts. They are also very similar to second-order rate constants in water.

This treatment seems to be satisfactory for reactions of anions such as cyanide and bromide (Bunton *et al.*, 1980a; Al-Lohedan *et al.*, 1983) which should bind strongly and specifically to cationic micelles, and to reactions of the hydronium ion, which is very hydrophilic, but should bind strongly and specifically by hydrogen-bonding to the sulfate or sulfonate head group of an anionic micelle.

The situation is different for reactions of very hydrophilic ions, e.g. hydroxide and fluoride, because here overall rate constants increase with increasing concentration of the reactive anion even though the substrate is fully micellar bound (Bunton *et al.*, 1979, 1980b, 1981a). The behavior is similar for equilibria involving OH^- (Cipiciani *et al.*, 1983a, 1985; Gan, 1985). In these systems the micellar surface does not appear to be saturated with counterions. The kinetic data can be treated on the assumption that the distribution between water and micelles of reactive anion, e.g. Y^-, follows a mass-action equation (9) (Bunton *et al.*, 1981a).

$$K'_Y = [Y^-_M]/[Y^-_W]([D]\text{-cmc-}[Y^-_M]) \quad (9)$$

The application of this mass-law treatment to the reaction of benzoic anhydride in CTAOH is illustrated in Fig. 4 (Al-Lohedan and Bunton, 1982). The lines are calculated taking: $K_s = 650\ M^{-1}$, $K'_{OH} = 55\ M^{-1}$ and, $k_M = 180\ s^{-1}$. The value of K_s is that estimated in CTABr and values of k_M are almost the same in CTAOH and CTABr + NaOH.

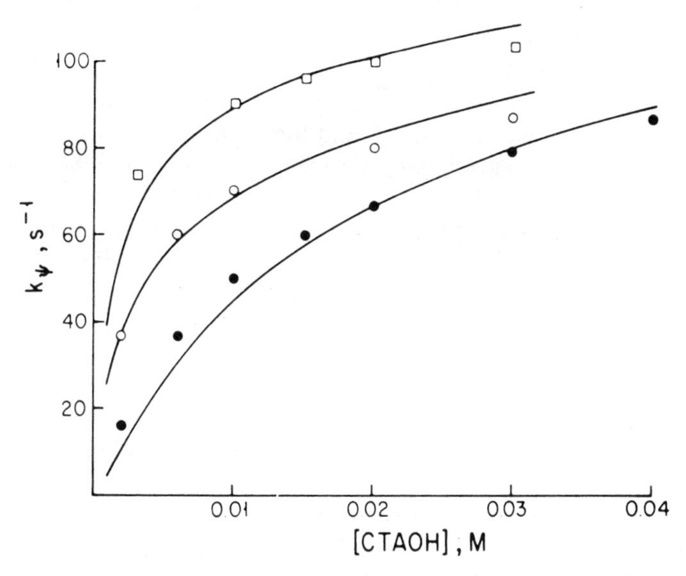

FIG. 4 Reaction of benzoic anhydride in CTAOH: ●, no added NaOH; ○, 0.01 M added NaOH; □, 0.02 M added NaOH. The lines are calculated using the mass-action model. (Reprinted with permission of the American Chemical Society)

Although kinetic data can be fitted to the mass-law model its physical significance is uncertain. For example, ion binding is assumed to follow a site model, with the micelle having a number of binding sites whose occupancy depends on an affinity parameter, K_Y', and on the concentration of ions in the aqueous pseudophase. In other words, α and β are not constant: but they will be approximately constant if K_Y' is large, as it should be for ions such as cyanide or bromide. Alternatively, we could suppose that micelles of a hydroxide ion surfactant, for example, have a large size dispersion, and that micelles, regardless of their size, can bind non-ionic substrates, but that the small micelles are relatively ineffective at binding hydrophilic ions. On this basis, the ionic distribution represented by (9) may be due to an increase in the average size of the micelle on addition of the counterion. Another possibility is that the micellar reaction is not restricted to reactants in the Stern layer (see p. 244).

There is not a great deal of physical evidence on these questions. For example, CTAOH forms micelles which have small aggregation numbers and larger fractional charge, α, than micelles of CTACl or CTABr, but the evidence available at present is not sufficient to show that the properties of CTAOH micelles are constant over a range of conditions (Bunton *et al.*, 1981a; Lianos and Zana 1983; Athanassakis *et al.*, 1985).

Despite these uncertainties values of k_M for reactions of hydroxide ion in CTAOH and mixtures of CTABr or CTACl with NaOH calculated using the ion-exchange or mass-action models agree reasonably well, and some examples are given in Table 3.

The ion-exchange constants, K_X^Y, in (7) should be related to the individual mass action constants, K_X' and K_Y', and this seems to be correct, based on limited evidence and some major approximations (Bunton *et al.*, 1983a). A recent and more detailed data analysis shows that the value of the ion-exchange constant is consistent with values of the individual constants (Rodenas and Vera, 1985).

The concept of counterion binding to micelles in a well-defined Stern layer provides a very convenient way of describing micellar effects on reaction rates and equilibria, but there are serious questions as to its physical significance. The distribution of counterions around a colloidal macroion can be calculated in terms of coulombic interactions between the macroion and the counterions which are treated as point charges (Stigter, 1964, 1978; Bell and Dunning, 1970; Mille and Vanderkooi, 1977). This treatment neglects specific interaction between a macroion, e.g. a micelle and counter-ions, and therefore does not explain the apparent differences in affinities of various counterions for micelles (Romsted 1977, 1984). However, it gives values for α for SDS which are in reasonable agreement with experiment, although specific binding of counterions in the micellar Stern layer is not taken into account (Gunnarsson *et al.*, 1980).

It seems possible that a very hydrophilic anion such as OH^- might not in fact penetrate the micellar surface (Scheme 1) so that its interaction with a cationic micelle would be non-specific, and it would exist in the diffuse, Gouy–Chapman layer adjacent to the micelle. In other words, OH^- would not be bound in the Stern layer, although other less hydrophilic anions such as Br^-, CN^- or N_3^- probably would bind specifically in this layer. In fact the distinction between micellar and aqueous pseudophases is partially lost for reactions of very hydrophilic anions. The distinction is, however, appropriate for micellar reactions of less hydrophilic ions.

The question then becomes that of the significance of the ion-exchange and mass-law equations which successfully account for the dependence of micellar rate constants upon the concentrations of surfactant and reactive and inert counterions. It seems reasonable to continue to use these descriptions at the present time, despite uncertainties as to the location of hydrophilic counterions at the micellar surface.

VALIDITY OF THE ION-EXCHANGE AND MASS-ACTION MODELS

A very careful analysis of the pseudophase ion-exchange model has been given by Romsted who reviewed the evidence up to 1982 and considered the limitations of the treatment (Romsted, 1984).

Key questions in these treatments are the constancy of α (or β) and the nature of the reaction site at the micellar surface. Other questions are less troubling; for example the equations include a term for the concentration of monomeric surfactant which is assumed to be given by the cmc, but cmc values depend on added solutes and so will be affected by the reactants. In addition submicellar aggregates may form at surfactant concentrations near the cmc and may affect the reaction rate. But these uncertainties become less important when [surfactant] $\gg$ *cmc* and kinetic analyses can be made under these conditions. In addition, perturbation of the micelle by substrate can be reduced by keeping surfactant in large excess over substrate.

The problems of the constancy of α and the site of reaction are closely linked. It is very convenient to assume that the charge on the micellar head groups is extensively neutralized by counterions which bind specifically to the micellar surface. In this way micellar stability is associated with a balance between hydrophobic attractions between apolar groups and coulombic repulsions of the ionic head groups which will be reduced by favorable interactions with the counterions in both the Stern and the diffuse Gouy-Chapman layers. It is the behavior of the counterions which is important in considerations of their chemical reactivity.

The treatments discussed thus far are based on the assumption that reaction occurs in the micellar Stern layer or in the aqueous medium with no reaction across the boundary between them. The concentration of counterions is considered to be uniform within the Stern layer (Stigter, 1964, 1978; Romsted, 1977) but it will decrease with increasing distance into the water from the micelle–water interface, following the Gouy–Chapman equation with the interactions being non-specific and governed by coulombic forces.

But this static picture is clearly inadequate, because solutes and surfactant monomers move rapidly from water to micelles, and the surfactant head groups will oscillate about some mean position at the micelle surface (Aniansson, 1978). Non-ionic substrates are not localized within the micelle or its Stern layer and there is no reason to believe that they are distributed uniformly within the Stern layer.

The breakdown of the simple ion-exchange model with constant α can be associated with the hydrophilicity of small ions of high charge density such as hydroxide and fluoride. They are strongly hydrated and therefore less likely to bind specifically in the Stern layer than less hydrophilic ions such as bromide. Therefore micelles of CTAOH, for example, may attain stability by keeping the head groups further apart, but this keeps surfactant monomers further apart and reduces hydrophobic interactions, so that these micelles tend to have smaller aggregation numbers than micelles which have other counterions (Lianos and Zana, 1983; Athanassakis *et al.*, 1985). Addition of hydroxide ion to a solution of CTAOH should screen the charge of the micellar head groups, regardless of the location of the hydroxide ion. An additional point is the artificiality of a sharp distinction between the Stern layer and the solvent, and the fact that reactants bound to the micelle will be both oscillating about some mean position and translating within the micelle. It is reasonable to believe that such motions of a non-ionic substrate will bring it into contact with a reactive counterion which is close to, but not within, the Stern layer, assuming that the latter is the region encompassing the micellar ionic head groups. Hicks and Reinsborough (1982) have pointed out that the volume in which micellar reaction takes place is probably not constant, and this may provide a partial explanation of ion specificity. In particular this volume may be very dependent upon the hydrophilicity of the anion and be larger for ions such as OH^- or F^- than for less hydrophilic ions such as Br^- or CN^-.

We should not regard eqns (7) and (9) based on the ion-exchange or mass-action models as anything more than useful vehicles for describing and comparing kinetic and equilibrium data. The ion-exchange parameter, K_X^Y in (7), may not in fact be describing the relative affinities of two counterions for the micellar ionic head groups, especially if one ion is very hydrophilic, e.g. OH^-, and the other much less so, e.g. Br^-. Instead the equation may be providing a reasonable approximation to the probability of the reactive ion being close enough to react with micellar-bound substrate. Similarly the parameter K_Y' in (9) may be describing the proximity of a hydrophilic, nucleophilic anion, e.g. OH^-, to micellar bound substrates. Recent experimental observations and theoretical calculations illustrate some of the limitations of the simple ion-exchange model. Ion-exchange constants estimated by bromide ion fluorescence quenching are in reasonable agreement with those estimated electrochemically, kinetically or spectrophotometrically, except for OH^-. The quenching experiments show that OH^- competes much less weakly than expected. The value of K^{OH}_{Br} is *ca* 40, which is greater than values estimated by other methods by factors of 2–4 (Abuin *et al.*, 1983b), and similar discrepancies were found with F^-. Thus kinetics and several physical methods overestimate the amount of OH^- bound intimately in the Stern layer.

Calculations based on non-specific coulombic interactions between the micelle and its counterions gave reasonable values of α, which were insensitive to the concentration of added salt (Gunnarsson *et al.*, 1980). Although these calculations do not explain the observed specificity of ion binding, they suggest that such hydrophilic ions as OH^- and F^- may not in fact enter the Stern layer, as is generally assumed. Instead they may cluster close to the micelle surface in the diffuse layer.

These uncertainties as to the location of ions such as OH^- or F^- cast doubt on the validity of the quantitative models which are used to treat micellar rate effects. The problem is less serious for reactions of less hydrophilic ions which bind strongly and specifically to micelles, and it should be relatively unimportant for bimolecular reactions of non-ionic reagents. It is probable also that the volume element of reaction decreases as the concentration of ionic reagent is increased, which would speed reaction.

Similar considerations apply to situations in which substrate and micelle carry like charges. If the ionic substrate carries highly apolar groups, it should be bound at the micellar surface, but if it is hydrophilic so that it does not bind in the Stern layer, it may, nonetheless, be distributed in the diffuse Gouy–Chapman layer close to the micellar surface. In this case the distinction between sharply defined reaction regions would be lost, and there would be some probability of reactions across the micelle–water interface.

Despite our reservations as to the validity of the various pseudophase models of micellar rate effects they provide a convenient mental scaffolding for discussion of the data and we use them for that purpose (Mortimer, 1982).

The above discussion has been based upon the assumption that aqueous and micellar pseudophases are distinct reaction media, i.e. a reactant in water does not attack a micellar-bound reactant. Recently this view has been challenged by Nome and coworkers who have shown that rate constants for attack of OH^- upon hydrophobic organic chlorides, such as DDT and related compounds (10), have the predicted dependence upon [surfactant] or

$[OH^-]$ only in dilute NaOH (Nome *et al.*, 1982; Stadler *et al.*, 1984; Ionescu and Nome, 1984). With $[OH^-] > 0.03$ M the reaction is faster than predicted and it was suggested that OH^- in the water could attack micellar-bound substrate. In other words there is an additional reaction path not taken into account in (5) or (6) or Scheme 2.

$$(4\text{-}ClC_6H_4)_2CH\text{—}CXYCl \xrightarrow{OH^-} (4\text{-}ClC_6H_4)_2C = CXY \qquad (10)$$
$$(X,Y = H, Cl)$$

The problem may be a semantic one because OH^- does not bind very strongly to cationic micelles (Romsted, 1984) and competes ineffectively with other ions for the Stern layer. But it will populate the diffuse Gouy–Chapman layer where interactions are assumed to be coulombic and non-specific, and be just as effective as other anions in this respect. Thus the reaction may involve OH^- which is in this diffuse layer but adjacent to substrate at the micellar surface. The concentration of OH^- in this region will increase with increasing total concentration. This question is considered further in Section 6.

4 Spontaneous, unimolecular and water-catalysed reactions

The quantitative treatment of micellar rate effects upon spontaneous reactions is simple in that the overall effect can be accounted for in terms of distribution of the substrate between water and the micelles and the first-order rate constants in each pseudophase (Scheme 2). The micelles behave as a submicroscopic solvent and to a large extent their effects can be related to known kinetic solvent effects upon spontaneous reactions. It will be convenient first to consider unimolecular reactions and to relate micellar effects to mechanism.

UNIMOLECULAR REACTIONS OF ANIONIC SUBSTRATES

Spontaneous decarboxylations of carboxylate ions and hydrolyses of aryl phosphate dianions and aryl sulfate monoanions are much faster in organic solvents than in water (Thomson, 1970; Kemp and Paul, 1970; Bunton *et al.*, 1967; Kirby and Varvoglis, 1967). This solvent effect is consistent with the Hughes–Ingold qualitative solvent theory because these reactions involve dispersion of charge in forming the transition state.

The micellar surface appears to be less polar than water, based largely on shifts in fluorescence or charge-transfer spectra (Section 1). Although it may not be reasonable to apply bulk solvent parameters such as Z or dielectric constant to submicroscopic species such as micelles, the spectral and kinetic evidence are self-consistent. An additional point is that these reactions have to be carried out in cationic, non-ionic or zwitterionic, micelles because anionic micelles do not take up these anionic substrates, and the positive charge in a cationic or zwitterionic micelle could interact favourably with the transition state with its delocalized charge.

The binding constants between the anionic substrates and cationic micelles are large because of the combination of coulombic and hydrophobic effects; so rate enhancements may be large even with dilute surfactant. There is binding with non-ionic and zwitterionic micelles despite the absence of coulombic attraction (Bunton *et al.*, 1975).

Some examples of rate and binding constants for these micelle-assisted reactions are in Table 6. There are very large differences in k'_M/k'_W for these reactions, but the rate effects on decarboxylation are large and depend upon the charge on the head group. Reaction of 2,4-dinitrophenyl phosphate is often written as generating intermediate metaphosphate ion, but this species is so short-lived that reaction follows an enforced association mechanism (Buchwald and Knowles, 1982).

WATER-CATALYSED, UNI- AND BIMOLECULAR REACTIONS

Spontaneous hydrolysis of many activated derivatives of carboxylic or carbonic acids involves nucleophilic addition of water to the carbonyl group, assisted by another water molecule which acts as a general base (Johnson, 1967; Menger and Venkatasubban, 1976). The tetrahedral intermediate then rapidly goes forward to products (Scheme 3).

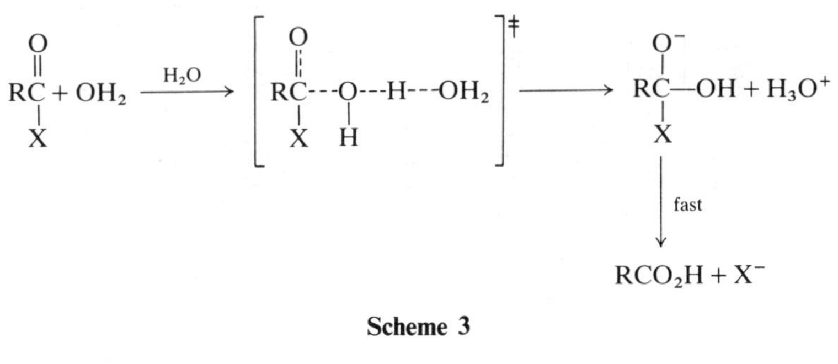

$$RC(=O)X + OH_2 \xrightarrow{H_2O} \left[RC(\overset{O}{\vdots})(X)\cdots O(H)\cdots H\cdots OH_2 \right]^{\ddagger} \longrightarrow RC(O^-)(X)\text{—}OH + H_3O^+ \xrightarrow{\text{fast}} RCO_2H + X^-$$

Scheme 3

These reactions are slowed by addition of chemically inert organic solvents to water and are generally slower in micelles than in water, although the effect is often small. The micellar inhibition has been discussed in terms of water activity, or more likely polarity, being somewhat lower at the micellar surface than in water (Menger *et al.*, 1981).

These rate effects are consistent with a "wet" micellar surface and the water molecules associated with the micelle seem to have reactivities which

Table 6

Spontaneous reactions of anionic substrates

Substrate	Surfactant	k_{rel}	Ref.
$2,6(NO_2)_2C_6H_3OPO_3^{2-}$	CTABr	25	Bunton *et al.*, 1968
$2,4(NO_2)_2C_6H_3OSO_3^-$	CTABr	3	Fendler *et al.*, 1970
$4\text{-}NO_2C_6H_4CO.OPO_3^{2-}$	$(C_{16}H_{33}\overset{+}{N}Me_2)_2(CH_2)_62Br^-$	6	Bunton and McAneny, 1977
CO_2^-, N, O, O_2N	CTABr	95	Bunton *et al.*, 1973
CO_2^-, N, O, O_2N	$C_{12}H_{25}\overset{+}{N}Me_2CH_2CO_2^-$	200	Bunton *et al.*, 1975
$PhCH(CN)CO_2^-$	CTABr	660	Bunton *et al.*, 1975
$PhCH(CN)CO_2^-$	$C_{12}H_{25}\overset{+}{N}Me_2CO_2^-$	770	Bunton *et al.*, 1972

are not very different from those of bulk water. For example the first-order rate constant for hydrolysis of acyltriazoles (11) in micelles is very similar to that in aqueous dioxane of high water content (Fadnavis and Engberts, 1982). Micellar inhibition of these hydrolyses is reduced by admixture with atactic poly(N-vinylpyrrolidone) (Fadnavis and Engberts, 1984).

$$\text{PhCO—N(triazole)} \xrightarrow{H_2O} PhCO_2H + \text{HN(triazole)} \qquad (11)$$

The rate of attack of water upon the tri-*p*-anisylmethyl cation is unaffected by binding of this cation to anionic micelles of sodium dodecyl sulfate (SDS) (Bunton and Huang, 1972) and equilibrium constants for aldehyde hydration are only slightly reduced by binding to micelles (Albrizzio and Cordes, 1979). These observations are also consistent with substrate binding at a "wet" micellar surface rather than in the interior of the micelle.

In the discussions of micellar effects thus far there has been essentially no discussion of the possible effect of micellar charge upon reactivity in the micellar pseudophase. This is an interesting point because in most of the original discussions of micellar rate effects it was assumed that rate constants in micelles were affected by the presence of polar or ionic head groups. It is impracticable to seek an answer to this question for spontaneous reactions of anionic substrates because they bind weakly if at all to anionic micelles (p. 245). The problem can be examined for spontaneous unimolecular and water-catalysed reactions of non-ionic substrates in cationic and anionic micelles, and there appears to be a significant relation between reaction mechanism and the effect of micellar charge upon the rate of the spontaneous hydrolysis of micellar-bound substrates.

Spontaneous hydrolyses of carboxylic anhydrides, diaryl carbonates and aryl chloroformates are faster in cationic than in anionic micelles, regardless of the nature of the counteranion in the cetyltrimethylammonium micelle (Al-Lohedan *et al.*, 1982b; Bunton *et al.*, 1984). This charge effect does not seem to be related to substrate hydrophobicity, although the extent of micellar inhibition (relative to reaction in water) is clearly dependent upon substrate hydrophobicity for anhydride hydrolyses.

Examples of this behaviour are shown in Table 7 where k^+ is related to reaction of substrate fully bound to a CTAX micelle and k^- to reaction in an anionic micelle of SDS. The ratio k^+/k^- is consistently larger than unity for hydrolyses of open chain anhydrides, diaryl carbonates and aryl chloroformates. In addition hydrolysis of 4-nitrophenyl chloroformate is slightly faster in cationic micelles than in water. Spontaneous hydrolyses of N-acyl triazoles are also inhibited less by cationic micelles of CTABr than anionic micelles of SDS (Fadnavis and Engberts, 1982).

TABLE 7

Spontaneous hydrolyses of acid chlorides and anhydrides[a]

Substrate	k_{rel} SDS	k_{rel} CTAX	k^+/k^-
$4\text{-}O_2NC_6H_4COCl$	0.16	1.9,X = Cl	12
$4\text{-}ClC_6H_4COCl$	0.011	0.05,X = Cl	5
PhCOCl	0.014	0.008,X = Br	0.6
$4\text{-}MeOC_6H_4COCl$	0.013	0.0014,X = Cl	0.1
$(4\text{-}O_2NC_6H_4CO)_2O$	0.06	0.2,X = Br	3.4
$(PhCO)_2O$	0.024	0.055,X = Br	2.2

[a] Al-Lohedan *et al.*, 1982b; Bunton *et al.*, 1984

Hydrolyses of alkyl halides and arenesulfonates have long been known to be micelle-inhibited (Gani *et al.*, 1973; Lapinte and Viout, 1973, 1979) but now $k^+/k^- < 1$, except for hydrolysis of methyl benzenesulfonate which involves extensive bond making in the transition state (Al-Lohedan *et al.*, 1982b; Bunton and Ljunggren, 1984). Thus values of $k^+/k^- < 1$ seem to be characteristic of hydrolyses in the S_N1–S_N2 mechanistic spectrum which involve considerable bond breaking in the transition state, and k^+/k^- is very low for hydrolyses of diphenylmethyl halides where the transition state has considerable carbocation character (Table 8).

TABLE 8

Spontaneous hydrolyses of alkyl halides and arenesulfonates[a]

Substrate	k_{rel} SDS	k_{rel} CTAX	k^+/k^-
MeO_3SPh	0.36	0.65,X = OMS	1.8
$PhCH_2Br$	0.11	0.074,X = Br	0.7
$Me_3CCHMeOTos$	0.006	0.0025,X = Br	0.4
2-AdamantylOBros	0.004	0.0007,X = Br	0.2
Ph_2CHCl	0.06	0.004,X = Cl	0.07

[a] Al-Lohedan *et al.*, 1982b; Bunton and Ljunggren, 1984; Bunton *et al.*, 1984

Zwitterionic micelles of the sulfobetaine $C_{16}H_{33}N^+Me_2(CH_2)_3SO_3^-$ have effects very similar to those of cationic micelles (Table 7). This result is understandable if the substrate binds close to the quaternary ammonium center and the anionic sulfate moiety extends into the aqueous region.

It is reasonable to relate the values of k^+/k^- at least qualitatively to the extents of bond making and breaking in the transition state. Bond making is all important in hydrolyses of carboxylic anhydrides, diaryl carbonates and methyl arenesulfonates. Bond breaking will be important in hydrolyses of alkyl halides and sulfonates, except for methyl derivatives, and especially so in water which can effectively solvate the leaving anion.

These considerations suggest that the charge effect of micelles on carbonyl-addition reactions can be rationalized in terms of favorable interactions between cationic head groups and the organic moiety with its partial negative charge, noting the role of a general base in deprotonating the attacking water molecule. The interactions in anionic micelle will be correspondingly unfavorable. This explanation of the micellar charge effect is consistent with rate enhancements of spontaneous bimolecular solvolyses of carboxylic anhydrides and diaryl carbonates by electron withdrawing substituents.

In S_N hydrolyses of alkyl halides and sulfonates where there is considerable bond breaking, there will be unfavorable interactions between the carbocation-like reaction center and the head groups of a cationic micelle, but favorable interactions with anionic head groups. The leaving anion will interact preferentially with water molecules, although we might expect there to be differences in this regard between halide and arenesulfonate ions. These ions also differ in their affinities for cationic micelles, with arenesulfonate ions being more strongly bound. This difference should make values of k^+/k^- larger for hydrolyses of arenesulfonates than of otherwise similar chlorides and there is some supporting evidence for this view (Bunton and Ljunggren, 1984).

The foregoing discussion of micellar charge effects has implicitly assumed that differences in water activity or substrate location in cationic and anionic micelles are not of major importance. If such differences were all important it would be difficult to explain the differences in k^+/k^- for carbonyl addition and S_N reactions, because increase of water content in an aqueous-organic solvent speeds all these reactions (Johnson, 1967; Ingold, 1969). As to substrate location, there is very extensive evidence that polar organic molecules bind close to the micelle–water interface in both anionic and cationic micelles, although the more hydrophobic the solute the more time it will spend in the less polar part of the micelle. Substrate hydrophobicity has a marked effect on the overall rate effects in both cationic and anionic micelles, but less so on values of k^+/k^-. It seems impossible to explain all these charge effects in terms of differences in the location of substrates in cationic and anionic micelles.

If micellar charge effects can be related to charge distribution in the transition state, as we suggest, they should be applicable to the elucidation of mechanisms of spontaneous hydrolyses, and we illustrate this approach by applying it to hydrolyses of benzoyl chlorides. Solvolyses of acyl halides and chloroformates are generally considered to involve nucleophilic participation, except where steric and electronic factors may strongly favor bond breaking (Johnson, 1967; Bentley *et al.*, 1984). However, the question of electronic distribution in the transition state is less readily answered. One extreme description is akin to the S_N2 model with concerted making of the new bond and breaking of the carbon–halogen bond. Alternatively reaction could follow a carbonyl-addition model with an addition step followed by loss of halide ion. The differences are essentially in the states of hybridization of the halide and the carbonyl oxygen in the transition state (Scheme 4). The two models are merely mechanistic extremes which may well merge under some conditions.

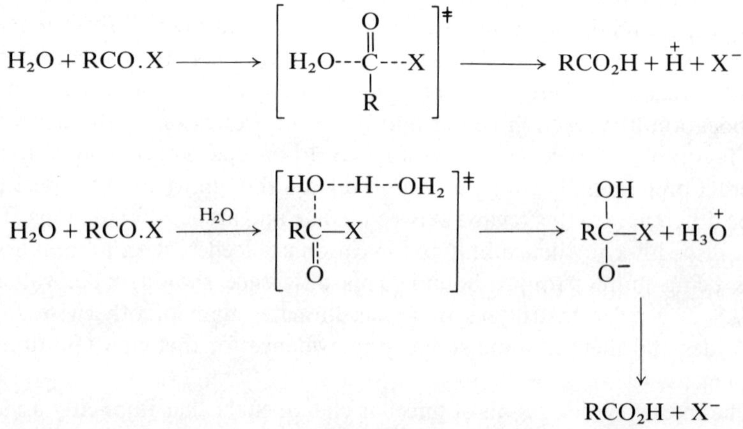

$$H_2O + RCO.X \longrightarrow \left[H_2O\text{---}\overset{\overset{\displaystyle O}{\|}}{\underset{\underset{\displaystyle R}{|}}{C}}\text{---}X \right]^{\ddagger} \longrightarrow RCO_2H + \overset{+}{H} + X^-$$

$$H_2O + RCO.X \xrightarrow{H_2O} \left[\begin{array}{l} HO\text{---}H\text{---}OH_2 \\ \;\vdots \\ RC\text{—}X \\ \;\,\vdots\| \\ \;O \end{array} \right]^{\ddagger} \longrightarrow \begin{array}{c} OH \\ | \\ RC\text{—}X + H_3\overset{+}{O} \\ | \\ O^- \end{array} \longrightarrow RCO_2H + X^-$$

Scheme 4

Electronic effects upon rates of solvolysis of benzoyl chlorides suggest that the mechanism, i.e. the extents of rehybridization of chloride and carbonyl oxygen atoms, depends upon the solvent. In aqueous acetone of low water content, a plot of $\log k$ against σ is linear and $\rho \sim 2$ for a limited range of substituents, whereas in moist formic acid the corresponding plot also is linear, but now $\rho \sim -4$, and in 50% aqueous acetone there is a pronounced minimum in the plot (Johnson, 1967).

For reaction of benzoyl chlorides in water, reactivity follows the electron-releasing ability of *p*-substituents, and there are striking differences in the micellar effects (Table 7). With electron-withdrawing substituents $k^+/k^- > 1$, but its value progressively decreases with increasing electron release from a *p*-substituent. Thus an increase in electron release changes k^+/k^- from values characteristic of a carbonyl addition reaction to values characteristic of S_N reactions at saturated carbon. This classification also includes chloroformates because aryloxy groups are strongly electron withdrawing.

There is no reason to believe that any of these reactions involve formation of an intermediate acylium ion (Bentley *et al.*, 1984) but the extent of C—Cl bond breaking probably increases markedly with increasing electron release from a substituent. Kinetic primary chloride isotope effects show that C—Cl bond breaking increases with increasing electron release in hydrolyses of 4-substituted benzoyl chlorides in aqueous acetone (Fry, 1970)

These micellar charge effects also seem to be present for reactions at heteroatoms. For example, in the spontaneous hydrolysis of a series of benzenesulfonyl chlorides (12), values of k^+/k^- for reactions in CTACl and SDS increase from 0.85 for X = OMe to 22 for $X = NO_2$, suggesting that there is a substituent effect upon the relative extents of bond-making and breaking (Bunton *et al.*, 1985).

$$X-C_6H_4-SO_2Cl + H_2O \xrightarrow{H_2O} X-C_6H_4-SO_3^- + H_3O^+ + Cl^- \quad (12)$$

The effect of micelles on these spontaneous hydrolyses is difficult to explain in terms of kinetic solvent effects on these reactions. Mukerjee and his coworkers have refined earlier methods for estimating apparent dielectric constants or effective polarities at micellar surfaces. For cationic and zwitterionic betaine sulfonate micelles D_{eff} is lower by *ca* 15 from the value in anionic dodecyl sulfate micelles (Ramachandran *et al.*, 1982). We do not know whether there is a direct connection between these differences in effective dielectric constant and the relation between reaction rates and micellar charge, but the possibility is intriguing.

5 The source of micellar rate enhancements

Micelles have effects on spontaneous reactions which can be related to mechanism and the properties of the micellar surface. Inhibition of bimolecular, micelle-inhibited reactions is also straightforward because micelles keep reactants apart.

For micelle-assisted bimolecular non-solvolytic reactions it is necessary to consider both medium and proximity effects.

Analysis of the variation of the overall rate constant of reaction with [surfactant] was discussed in Section 3 (p. 222) and the treatment allows calculation of the second-order rate constants of reaction in the micellar pseudophase. These rate constants can be compared with second-order rate constants in water provided that both constants are expressed in the same dimensions and typically the units are $M^{-1}\,s^{-1}$. Inevitably the comparison depends upon the assumed volume element of reaction; this may well differ from one type of micelle to another and probably also depends upon the structure of the reactants and especially upon the hydrophilicity of a reactive ion. Different investigators make different assumptions about values of this volume, but most estimates are within a factor of two (Romsted, 1977, 1984). In addition, some micelles grow with increasing concentration of surfactant and electrolyte, so these factors may influence the volume element of reaction. Fortunately, the uncertainties so introduced may be more apparent than real; for an approximately spherical micelle the volume of the Stern layer is approximately half the total volume of the micelle, and estimates of the volume element of reaction and of the second-order rate constants should therefore be within a factor of two, provided that the reactant composition is uniform within the Stern layer. The second-order rate constants, k_M(s^{-1}) in (4), are related to those for which concentration is expressed as molarity by (13). Here V_M is the molar volume of reaction, and estimates of its value vary from 0.14–0.3 litre; a variety of values within this range is used.

$$k_2^m = V_M k_M \tag{13}$$

Another problem is that different workers make their calculations of second-order rate constants in the micelle in different ways. For example, the surface potential of a micelle may be specifically included in the calculation in order to estimate ion binding, but there is a circularity in this argument because surface potentials are often estimated from micellar effects upon indicator acid-base equilibria which themselves depend upon ion-binding to micelles (Fernandez and Fromherz, 1977).

Despite all these uncertainties the picture is reasonably self-consistent, in that calculated second-order rate constants in micelles are generally similar over wide ranges of surfactant and reagent concentration and are often little affected by changes in the micellar counterion. Examples are given in Tables 2–5.

Several generalizations can be made:

(i) For most micelle-assisted reactions second-order rate constants in the micellar pseudophase are similar to, or smaller than, those in water.

(ii) Reactions of non-ionic nucleophiles are slower in the micellar pseudophase than in water, and second-order rate constants are similar in anionic and in cationic micelles.

(iii) Hydronium ion catalysed reactions are generally slower in anionic micelles than in water.

(iv) Reactant hydrophobicity has little effect upon second-order rate constants in the micellar pseudophase. However, the more hydrophobic the reagents the lower is the surfactant concentration at the rate maximum.

An impressive body of evidence supports these generalizations. This evidence has been reviewed (Romsted, 1984) and it does not seem necessary to discuss it in detail here, but some examples will be given and some exceptions to these generalizations will be mentioned. Some reactions of OH^- are shown in Table 3 for both inert and reactive ion surfactants, and Table 4 gives data for reactions of other hydrophilic ions. Reactions of hydrophobic nucleophiles are shown in Table 2. For all these reactions second-order rate constants in the micellar pseudophase are compared with those in water. For some reactions we also give values of k_{rel}, i.e. the rate constant relative to that in water. These values depend upon the reactant concentration and are included merely to provide an indication of the micellar rate effects. Other examples of micellar rate effects are given in the Appendix.

It is easy to understand the lower reactivity of non-ionic nucleophiles in micelles as compared with water. Micelles have a lower polarity than water and reactions of non-ionic nucleophiles are typically inhibited by solvents of low polarity. Thus, micelles behave as a submicroscopic solvent which has less ability than water, or a polar organic solvent, to interact with a polar transition state. Micellar medium effects on reaction rate, like kinetic solvent effects, depend on differences in free energy between initial and transition states, and a favorable distribution of reactants from water into a micellar pseudophase means that reactants have a lower free energy in micelles than in water. This factor, of itself, will inhibit reaction, but it may be offset by favorable interactions with the transition state and, for bimolecular reactions, by the concentration of reactants into the small volume of the micellar pseudophase.

The lower rate constants of reactions catalysed by hydronium ion in micelles as compared with water suggest that micellar binding decreases the acidity of this ion. Although the hydronium ion is very hydrophilic it binds strongly to anionic micelles; for example, based on values of the ion-exchange parameter in (7), H_3O^+ and Na^+ have similar affinities for an alkyl sulfate or an alkane sulfonate micelle (Bunton *et al.*, 1977). The situation is quite different for binding of anions to cationic micelles where very hydrophilic ions such as OH^- or F^- bind only weakly (Abuin *et al.*, 1983b; Romsted, 1984). The difference in the two situations is that H_3O^+ can bind specifically to a sulfate or sulfonate head group by hydrogen bonding, whereas there is no such specific interaction of an anion such as OH^- with a quaternary ammonium ion head group of a cationic micelle.

Alternatively one could suggest that a micellized sulfuric or sulfonic acid is not strong. For example, apparent acid dissociation constants of weak acids decrease when the acids are bound to anionic micelles (Hartley, 1948), and the rapid hydrolysis of micellized alkyl sulfates at low pH is consistent with formation of covalent alkyl sulfuric acid (Kurz, 1962; Motsavage and Kostenbauder, 1963; Garnett *et al.*, 1983).

Many reactions involve nucleophilic organic anions, e.g. aryloxide, thiolate, oximate or hydroxamate, and micellar effects upon their reactions have been extensively studied and analysed in terms of the pseudophase model (Martinek *et al.*, 1977; Cuccovia *et al.*, 1978; Bunton and Sepulveda, 1979). Sometimes the distribution of these ions between water and micelles can be determined spectrophotometrically by taking advantage of small spectral shifts on micellar binding. In other cases the extent of binding is estimated by indirect methods, such as changes in apparent dissociation constants of weak acids (Martinek *et al.*, 1977). Some of these methods are inherently suspect, but this is of little consequence provided that the organic ions are sufficiently hydrophobic to bind essentially quantitatively to micelles. Fortunately, or otherwise, this is generally the case.

Some examples of reactions of inorganic ions with organic substrates are given in Tables 3–5. It is important to note that different values are assumed for the volume element of reaction, i.e. the molar volume of the micellar pseudophase (V_M) and it should be noted that the second-order rate constant, k_2^m in the micellar pseudophase varies proportionately (13).

For several reactions, values of k_2^m/k_W are similar for micelles which have different structures of apolar or head group or different counterions: in addition some of the data relate to reaction in different concentrations of reactive ion. Comparison can also be made for a few reactions between values of k_2^m/k_W in the presence and absence of an inert counterion. The agreement here is reasonable, although, if anything, values are larger when inert counterion is present. This is consistent with there being a small, but significant, beneficial effect possibly due to shrinkage of the volume element of reaction owing to the presence of additional electrolyte in the solution. Alternatively one could assume that added electrolyte increases the micellar radius, and provided that the head-group area is constant the potential at the micellar surface will also increase (Evans and Ninham, 1983). Therefore more ions, reactive and inert, will be attracted to the micelle. This beneficial effect on rate will oppose the inhibition due to interionic competition.

Most values of k_2^m/k_W are not very different from unity (Tables 2–5) especially in view of the approximations involved in their estimation. The self-consistency of the data suggests that it is reasonable to use them to analyse the factors which control values of k_2^m/k_W, and the special case of reaction of the hydronium ion has already been noted. Aromatic nucleophilic substitution by azide ion is a conspicuous exception to these generalizations.

The high ionic concentration at the micellar surface may result in an ionic strength effect on reaction rate. Salt effects in water, however, are generally smaller for ion-molecule reactions than for reactions which involve an increase or decrease of charge, and they should be approximately zero in the absence of specific effects. There may well be specific effects upon ion–molecule reactions in the micellar pseudophase, but their significance may be masked by the fact that many cationic micelles have a quaternary ammonium ion as the head group and chloride or bromide as the counterion. The range of structural variations in anionic micelles is also limited, so that specific effects of the surfactant ions on these bimolecular reactions should be reasonably constant.

Considering first reactions of inorganic anions, values of k_2^m/k_W are less than unity for such substrates as carboxylic esters and related compounds and inorganic esters, but are greater than unity for aromatic nucleophilic substitution, especially for attack of azide ion (Bunton *et al.*, 1982b). The variation in k_2^m/k_W is generally not large in view of differences in the parameters, especially V_M, used in the calculations, but they seem to be significant.

The value of k_2^m/k_W is larger than unity for reaction of thiosulfate ion with n-butyl bromide. There is insufficient evidence to know whether this difference is characteristic of micelle-assisted reactions of dianionic nucleophiles (Cuccovia *et al.*, 1982a).

Hydrophilic anions will stay in the outer, water-rich regions of micelles and the more hydrophobic organic substrates may be located more deeply in the micelle. These effects, and those due to the preferred orientation of substrates in the micelle, do not seem to be of major importance in determining k_2^m/k_W because values are not very different for reactions of substrates which have the same reactive groups but different hydrophobicities. This conclusion is illustrated by comparing values of k_2^m/k_W for reactions of OH^- with 4-nitrophenyl acetate and octanoate or of CN^- with a series of N-alkylpyridinium ions (Tables 3 and 4). Despite large differences in hydrophobicities, as shown by variations in K_s, there is little change in k_2^m/k_W. (It is necessary to make these comparisons with data from the same laboratory because of differences in the parameters such as α, K_X^Y and V_M chosen by the various research groups.)

The mechanisms of reactions of anionic nucleophiles with carboxyl derivatives and phosphate esters are similar in that negative charge tends to be localized on oxygen in the transition states, as shown for deacylation in Scheme 5 (Bunton, 1977). The transition state here should be stabilized by

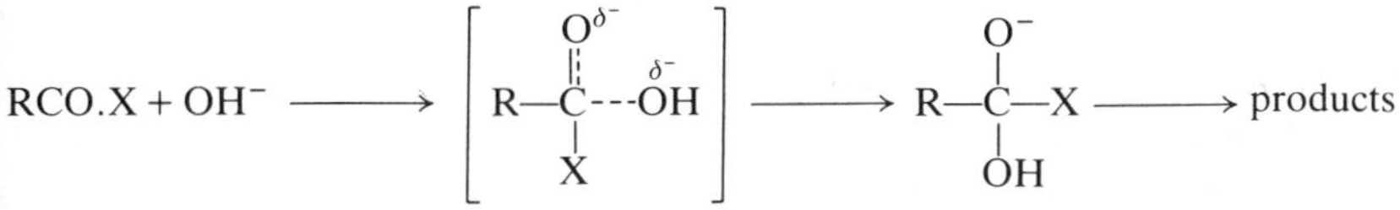

$$\mathrm{RCO.X + OH^-} \longrightarrow \left[\mathrm{R{-}\overset{\overset{\displaystyle O^{\delta-}}{\vdots}}{\underset{\underset{\displaystyle X}{|}}{C}}{-}{-}{-}\overset{\delta-}{O}H}\right] \longrightarrow \mathrm{R{-}\overset{\overset{\displaystyle O^{-}}{|}}{\underset{\underset{\displaystyle OH}{|}}{C}}{-}X} \longrightarrow \text{products}$$

$X = OC_6H_4NO_2;\ OCO.R'$

Scheme 5

hydrogen bonding to water molecules which may be less available at the surface of an ionic micelle than in bulk water.

The situation is different for aromatic nucleophilic substitution (14) where the transition state will be like a Meisenheimer complex with its negative

$$1\text{-X-2,4-}(NO_2)_2C_6H_3 + Y^- \longrightarrow [1\text{-X-1-Y-2,4-}(NO_2)_2C_6H_3]^- \longrightarrow 1\text{-Y-2,4-}(NO_2)_2C_6H_3 + X^- \quad (14)$$

charge delocalized over the nitroaromatic moiety (Miller, 1968). This bulky, low charge density anion should interact favorably with cationic micellar head groups and be stabilized by this interaction.

We can also describe the differences between these reaction types in terms of Pearson's hard–soft description (Pearson, 1966; Pearson and Songstad, 1967). Cationic micellar head groups interact best with soft bases, e.g. relatively large anions of low charge density such as bromide or arenesulfonate, or anionic transition states such as those for nucleophilic aromatic substitution. They interact less readily with hard bases, e.g. high charge density anions such as OH^-, or anionic transition states for deacylation.

We note in this context that k_2^m/k_W is greater than unity both for nucleophilic attack upon neutral aromatic substrates, e.g. 2,4-dinitrochlorobenzene and cationic N-alkylpyridinium ions (Tables 3 and 4).

It is more difficult to interpret micellar effects upon reactions of azide ion. The behavior is "normal", in the sense that $k_2^m/k_W \sim 1$, for deacylation, an S_N2 reaction, and addition to a carbocation (Table 4) (Cuenca, 1985). But the micellar reaction is much faster for nucleophilic aromatic substitution. Values of k_2^m/k_W depend upon the substrate and are slightly larger when both N_3^- and an inert counterion are present, but the trends are the same. We have no explanation for these results, although there seems to be a relation between the anomalous behavior of the azide ion in micellar reactions of aromatic substrates and its nucleophilicity in water and similar polar, hydroxylic solvents. Azide is a very powerful nucleophile towards carbocations, based on Ritchie's N^+ scale, but in water it is much less reactive towards 2,4-dinitrohalobenzenes than predicted, whereas the reactivity of other nucleophiles fits the N^+ scale (Ritchie and Sawada, 1977). Therefore the large values of k_2^m/k_W may reflect the fact that azide ion is unusually unreactive in aromatic nucleophilic substitution in water, rather than that it is abnormally reactive in micelles.

Very large micellar rate effects are observed for attack of azide ion upon N-alkylbromopyridinium ions (15) (Cuenca, 1985). However, even in water,

$$\text{2-bromo-1-R-pyridinium} + N_3^- \longrightarrow \text{2-azido-1-R-pyridinium} + Br^- \tag{15}$$

$R = C_{12}H_{25}, C_{14}H_{29}, C_{16}H_{33}$

rate constants increase sharply with increasing concentration of the bromopyridinium ion provided that the N-alkyl group is long (C_{12}—C_{16}) and the substrate sufficiently hydrophobic to self-associate giving aggregates which attract N_3^-. The effects of inert cationic surfactants, e.g. CTACl or CTABr, cannot be fitted to the pseudophase ion-exchange model, which suggests that there may be some specific interaction between N_3^- and the pyridinium ion. These bromopyridinium ions react with OH^-, and here too reactivity in the absence of added surfactant increases with increasing substrate hydrophobicity (Al-Lohedan *et al.*, 1982a). However, the effect of added surfactant follows the ion exchange model, perhaps because specific interactions are small between OH^- and the N-alkyl-2-bromopyridinium ions.

It is difficult to generalize about the relation between values of k_2^m/k_W and mechanism or structure for reactions of nucleophilic organic anions (Romsted, 1984) not least because the structural variations have been limited (Table 2). Second-order rate constants in the micellar pseudophase for deacylations by benzimidazole ion are little affected by substrate hydrophobicity (Martinek *et al.*, 1977). However, values of k_2^m/k_W are usually slightly less than unity, except for attack of phenoxide ion upon 2,4-dinitrochlorobenzene where it is 1.5. Overall rate enhancements are often larger than those found for reactions of inorganic anions, but this is simply because the organic anions are sufficiently hydrophobic to be micellar bound even with dilute surfactant, and their concentration is therefore high in the micellar pseudophase.

The similarity for many reactions of second-order rate constants in aqueous and micellar pseudophases, and the observation that substrate hydrophobicity usually affects binding and not inherent reactivity in the micelle, suggests that substrate location or orientation is relatively unimportant. This conclusion is strongly supported by a quantitative analysis of the effects of CTABr micelles on the reaction of OH^- and arylsulfonylalkyl arenesulfonates (16) (van der Langkruis and Engberts, 1984).

$$\begin{aligned} p\text{-}XC_6H_4SO_2CHROSO_2C_6H_4Y\text{-}p + OH^- &\longrightarrow \\ p\text{-}XC_6H_4SO_2^- + RCHO + p\text{-}YC_6H_4SO_3^- \end{aligned} \tag{16}$$

If orientation effects were important it would be possible to dissect rate and binding constants into contributions depending specifically upon groups

X and Y. This behavior was not observed, suggesting that the substrate had considerable freedom of motion at the micellar surface. However, product selectivity is observed in some reactions in aqueous surfactants, and may be related to orientation of reactants in micelles (Sections 9 and 10).

Micellar rate enhancements of bimolecular, non-solvolytic reactions are due largely to increased reactant concentrations at the micellar surface, and micelles should favor third- over second-order reactions. The benzidine rearrangement typically proceeds through a two-proton transition state (Shine, 1967; Banthorpe, 1979). The first step is a reversible pre-equilibrium and in the second step proton transfer may be concerted with N—N bond breaking (17) (Bunton and Rubin, 1976; Shine *et al.*, 1982). Electron-donating substituents permit incursion of a one-proton mechanism, probably involving a pre-equilibrium step.

Anionic micelles strongly favor the two-proton mechanism, because of the increased concentration of hydronium ions at the micellar surface (Bunton and Rubin, 1976; Bunton *et al.*, 1978b).

$$\mathrm{ArNH{-}NHAr} \underset{}{\overset{\mathrm{H^+}}{\rightleftharpoons}} \mathrm{Ar\overset{+}{N}H_2{-}NHAr} \xrightarrow{\mathrm{H^+}} \mathrm{H_2NAr{-}ArNH_2} \qquad (17)$$

However, micelles do not always favor reactions of higher order. In dilute OH^-, reaction of activated amides, for example (18), is typically second order in OH^-, but the order decreases to one with increasing $[OH^-]$ because the tetrahedral intermediate is converted rapidly into products (Menger and Donohue, 1973; Cipiciani *et al.*, 1979). These reactions are speeded by cationic micelles, but in the micelles they are always first order in OH^-, even when the total concentration of OH^- is low. This is simply because the micelles concentrate OH^-, so that the tetrahedral intermediate in (18) is

$$\text{(pyrrole-N-C(=O)R)} + \mathrm{OH^-} \rightleftharpoons \text{(pyrrole-N-C(O^-)(OH)R)} \xrightarrow{\mathrm{OH^-}} \mathrm{RCO_2^-} + \text{(pyrrole-NH)} \qquad (18)$$

readily deprotonated and rapidly goes to products. In this system micelles are changing the rate-limiting step of a reaction. (Cipiciani *et al.*, 1983a).

Quantitative fits of the rate constant to concentrations of surfactant or reagent are sometimes poor when [surfactant] is close to the cmc. Several factors can be involved here; (*i*) the "kinetic" cmc can be lower than that in water because the reagents promote micellization; (*ii*) reaction is promoted by submicellar aggregates; or (*iii*) the simplifying assumptions involved in the kinetic equations (2–6) may be invalid when concentrations of substrate and micellized surfactant are similar (Romsted, 1984).

6 Functional micelles and comicelles

A nucleophilic or basic group can be chemically bound to a surfactant which on aggregation generates a functional micelle, or comicelle if it is admixed with inert surfactant. Typical reactive groups are imidazole, oxime, amino, hydroxamic, thiol, hydroxy or hydroperoxy. Most of these groups have to be deprotonated to give the reactive anion (Tonellato, 1977, 1979; Kunitake and Shinkai, 1980; Pillersdorf and Katzhendler, 1979; Gobbo *et al.*, 1984). Very large rate enhancements, relative to reaction in water, can be obtained using functional micelles, but in most systems reaction of substrate with the functional group gives a relatively unreactive product so that the "turnover" step is very slow.

Several workers have attempted to draw analogies between reactions in functional micelles and in enzymes, but in general this seems to be unjustified because there is generally little substrate- or stereo-specificity in the micellar reactions and usually limited "turnover" rates.

A major question has been that of "bifunctional catalysis". For example, if a micelle contains both nucleophilic groups and groups which can transfer protons one might hope to achieve high rates of deacylation by having concerted nucleophilic attack and proton transfer (Scheme 6). Such concerted processes are well established in enzymic reactions, but evidence in

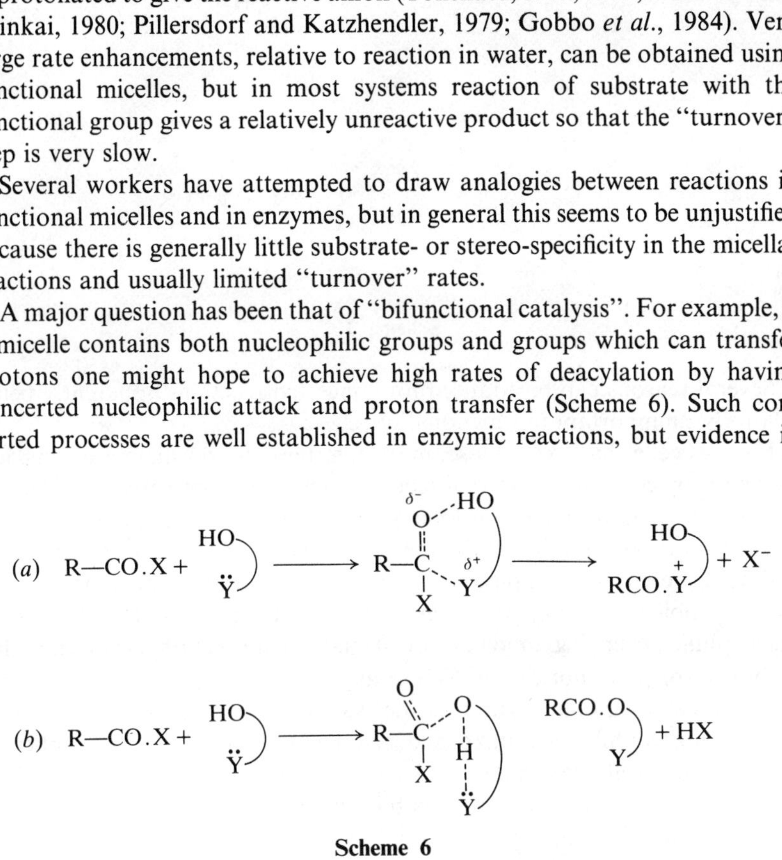

(*a*) R—CO.X + HO⌒Ÿ ⟶ R—C(=O$^{\delta-}$···HO)(X)···Y$^{\delta+}$ ⟶ HO⌒$\overset{+}{Y}$.RCO + X^-

(*b*) R—CO.X + HO⌒Ÿ ⟶ R—C(=O)(X)···O—H···Ÿ ⟶ RCO.O⌒Y + HX

Scheme 6

micellar systems is much less compelling. In one carefully studied case a concerted reaction was excluded, although the final O-acyl product could have been produced by a reaction analogous to Scheme 6*b* (Moss *et al.*, 1977; Tonellato, 1977).

The functional micellized surfactant contained both imidazole and hydroxyethyl groups and the final products of reaction with *p*-nitrophenyl alkanoates are acyl derivatives of the hydroxyethyl group. However, these O-acylated products were formed from an acylimidazole intermediate which was detected spectrophotometrically. Thus reaction was stepwise, rather than concerted, despite the proximity of the imidazole and hydroxyl groups (Scheme 7).

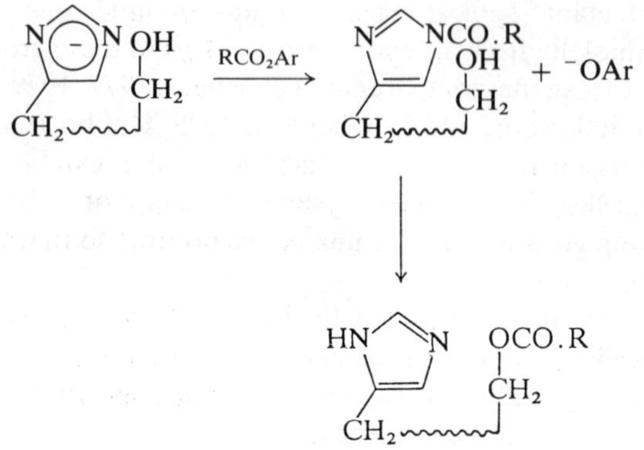

$$\text{Im}^{-}\text{-CH}_2\text{~~~CH}_2\text{OH} \xrightarrow{RCO_2Ar} \text{Im(NCO.R)-CH}_2\text{~~~CH}_2\text{OH} + {}^{-}OAr$$

$$\downarrow$$

$$\text{HN-Im-CH}_2\text{~~~CH}_2\text{OCO.R}$$

Scheme 7

Several reasons can be adduced for concerted reactions apparently being relatively unimportant in functional micelles.

First, micelles have very loose, mobile, structures and there are considerable entropy costs in a concerted reaction. These costs are much less serious in enzymic systems where conformation at the active site is "tailor-made" to fit the transition state. Secondly, the sites of micellar reactions are very wet and omnipresent water molecules are available to transfer protons.

The problem of slow "turnover" has been noted. Initial reaction of a nucleophilic group, e.g. imidazole or oximate, in a functional micelle, with a carboxylic or phosphoric ester, for example, gives an acylated or phosphorylated imidazole or oxime, and these derivatives hydrolyze slowly to regenerate the nucleophile. Kunitake and Shinkai (1980) discuss a number of reactions in micelles which contain both nucleophilic and basic groups which are potentially capable of acting as bifunctional reagents (Tonellato, 1979, Kunitake and Shinkai, 1980; Bunton, 1984)

The clearest example of rapid "turnover" of a functional micelle is a reaction of a micellized *gem*-diol. Attack of the anion of the *gem*-diol gives an acetylated or phosphorylated intermediate which rapidly breaks down regenerating the catalytic *gem*-diol (Scheme 8) (Menger and Whitesell, 1985).

Another example of rapid turnover in a micellar system is the cleavage of carboxylic and phosphate esters by *o*-iodosobenzoate in cationic micelles. This reaction was not studied with a functional micelle, but it is useful to note it in this context (Moss *et al.*, 1983, 1986).

The sources of rate enhancements in functional and non-functional

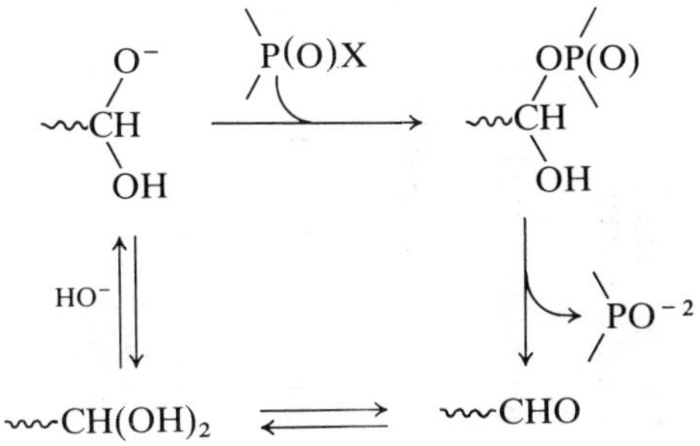

O^- / ~CH \ OH + \P(O)X/ ⟶ ~CH(OP(O)<)(OH)

HO$^-$ ⇅ ; ⟶ $\rangle PO^{-2}$

$\sim CH(OH)_2$ ⇌ $\sim CHO$

Scheme 8

micelles are similar, and the quantitative treatments applied to rate enhancements by non-functional micelles (Section 3) should be equally applicable to functional micelles (Bunton, 1979a,b; Fornasier and Tonellato, 1980).

In a functional micelle in which the reactive group is fully deprotonated there is a 1 : 1 relationship between the concentrations of reactive nucleophile and micellar head group in the micellar pseudophase. If under these conditions the substrate is fully micellar bound, (5) or (6) take the very simple form (19). This rate constant, k_M, can then be converted into the second-order rate constant, k_2^m in $M^{-1}s^{-1}$, estimating the volume element of reaction, V_M, which can be assumed to be that of the micelle or of its Stern layer, and these second-order rate constants can be compared with reaction in water of a chemically similar, non-micellized, nucleophile.

$$k_\psi = k_M \tag{19}$$

In practice it is usually necessary to take into account partial deprotonation of the potentially nucleophilic group, or partial micellar binding of the substrate, and if comicelles are used allowance has to be made for "dilution" of the nucleophile in the micellar pseudophase.

There are inevitably uncertainties in these treatments, due largely to changes in micellar charge as the result of deprotonation of nucleophiles at the micellar head groups. These changes in charge may lead to changes in the micellar surface and therefore changes in rate and equilibrium constants at the surface, but rather surprisingly these changes do not seem to affect the overall conclusions.

For a number of reactions in functional micelles and comicelles second-order rate constants are similar in micelles and in water. Except for aromatic nucleophilic substitution they are slightly smaller in the micelles than in water, and the pattern of behavior is exactly that found for reactions of organic nucleophilic anions in non-functional micelles. Some examples of these comparisons are in Table 9.

TABLE 9

Reactivities in functional micelles and comicelles[a]

Functional surfactant	Model	Substrate	k_2^m/k_W	Ref.
$C_{16}H_{33}\overset{+}{N}Me_2CH_2$–(imidazol-4-yl)[b,c]	$Et_2Me\overset{+}{N}CH_2$–(imidazol-4-yl)	$C_5H_{11}CO.OC_6H_4NO_2(4)$	1.9	Fornasier and Tonalleto, 1980
$C_{16}H_{33}\overset{+}{N}Me_2CH_2$–(imidazol-4-yl)	$Me_3\overset{+}{N}CH_2$–(imidazol-4-yl)	$(PhO)_2PO.OC_6H_4NO_2(4)$	0.5	Brown *et al.*, 1980
$C_{12}H_{25}\overset{+}{N}Me_2CH_2CPh{=}NO^-$[b]	$Et_3\overset{+}{N}CH_2CPh{=}NO^-$	$(PhO)_2PO.OC_6H_4NO(4)$	0.2–0.5	Bunton *et al.*, 1980d
$C_{16}H_{33}\overset{+}{N}Me_2CH_2CH_2O^-$	$Me_3\overset{+}{N}CH_2CH_2O^-$	$PhCO.OC_6H_4NO_2(4)$	2	Biresaw *et al.*, 1984
$C_{16}H_{33}\overset{+}{N}Me_2CH_2CH_2O^-$	$Me_3\overset{+}{N}CH_2CH_2O^-$	$(PhO)_2PO.OC_6H_4NO_2(4)$	0.25	

[a] Based on a molar volume of the Stern layer of 0.14 M^{-1} litre unless specified
[b] Comicelle with inert surfactant
[c] Micellar molar volume of 0.36 M^{-1} litre

It is easy to explain the large rate of enhancements which have been observed with some functional micelles, relative to reaction in water.

(i) The functional group of the micelle introduces a new reaction path, e.g. attack of imidazolide or hydroxamate ion upon an ester substrate.

(ii) If reaction is followed at a pH well below the pK_a-value of the potential nucleophile, there will be increased generation of nucleophilic anion due to increased deprotonation in the micelle.

(iii) The high concentration of both reactants in the micellar pseudophase will increase the reaction rate.

It is important to note that only the third effect is covered by the comparisons given in Table 9.

Most reactions in functional micelles involve nucleophilic attack by the reactive group, and covalent intermediates can often be detected by trapping or spectrophotometrically, or they may be observable as initial products (Tonellato 1977, 1979; Moss *et al.*, 1977; Schiffman *et al.*, 1977a,b). There is, however, no reason why a group in a functional micelle should not act as a general base catalyst for attack of water upon a substrate. Such catalysis has been postulated for dephosphorylation in imidazole-functionalized micelles at a pH such that there should be little nucleophilic imidazolide ion present (Brown *et al.*, 1974, 1980). The evidence was based largely on the magnitude of the kinetic solvent hydrogen isotope effect, $k_{H_2O}/k_{D_2O} = 2$–2.9 which was larger than expected for nucleophilic reaction, but this is not a rigorous proof. It is difficult in these systems to carry out the usual kinetic tests for general catalysis based on the Brønsted Catalysis Law because the pK_a-value of the reactive group may be affected by the micelle and cannot be varied widely.

Analysis of deacylation by histidinyl-functionalized micelles suggests that the histidinyl group can act both nucleophilically, generating an acylated histidine intermediate, and as a general base. These conclusions are consistent with the kinetic solvent hydrogen isotope effect (Murakami *et al.*, 1981).

Reactions in most functional micelles involve nucleophilic attack by an anionic moiety, e.g. oximate, hydroxamate, thiolate or alkoxide. Therefore it may be necessary to take into account the acid-base equilibrium which generates the anionic moiety. The simplest approach is to work at a pH such that deprotonation is essentially quantitative, but if this cannot be done the extent of deprotonation has to be measured directly or estimated.

$$C_{16}H_{33}N^+Me_2CH_2CH_2OH \rightleftharpoons C_{16}H_{33}N^+Me_2CH_2CH_2O^- + H^+ \qquad (20)$$

The hydroxyethyl quaternary ammonium ion in (20) is a weak acid: the pK_a-value of the non-micellizing model compound, choline, is 12.8 (Haberfield and Pessin, 1982). Reactions of the alkoxide zwitterion are therefore followed at relatively high pH, and from the variation of rate constant with pH one can estimate the apparent acid or base dissociation constants, pK_a or pK_b. The value of pK_a is ~ 12.3 so that the micellized derivative is more acidic than choline, as expected (Bunton and Ionescu, 1973; Bunton and Diaz, 1976; Biresaw *et al.*, 1984). Unexpectedly, the apparent pK_a-value varies very little with change in the concentration of surfactant or added OH^- (Biresaw *et al.*, 1985; Biresaw and Bunton, 1986). The ion-exchange model predicts that apparent acid (or base) dissociation constants should *not* be constant under these conditions (Romsted, 1984) and for deprotonation of 5-nitroindole bound to micelles of the hydroxyethyl surfactant (20) these constants vary in the direction predicted by the ion-exchange model (p. 266). Thus a model which applies to deprotonation of a micellar bound indicator fails when applied to deprotonation of an acidic function covalently bound to the micelle.

This paradoxical result suggests that the indicator, 5-nitroindole, is bound close to the quaternary ammonium ion, so that ion-exchange governs its deprotonation, at least qualitatively. On the other hand, the hydroxyl group probably extends away from the micellar Stern layer and into the diffuse layer. There will be interionic competition in the diffuse layer, as in the Stern layer, but it should be non-specific. Thus deprotonation of the hydroxyl group in the diffuse layer will be governed largely by the total amount of OH^- in solution. At the same time Br^-, which is the surfactant counterion, will be a much less effective competitor than it is for ion binding in the Stern layer.

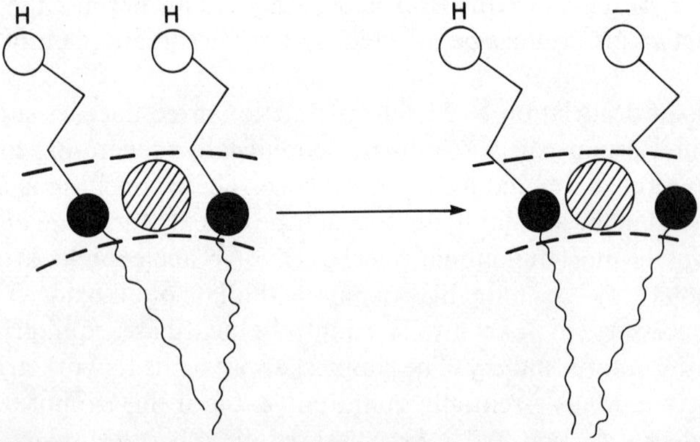

FIG. 5 Schematic representation of locations of a weakly acidic indicator, ⊘, adjacent to the cationic head groups, ●, and of a functional hydroxyl group in the water-rich region

This situation is illustrated qualitatively in Fig. 5 where an indicator, BH, is located close to quaternary ammonium ion head groups and will be deprotonated by hydroxide ions which are in their vicinity. But the functional hydroxyl groups are in the diffuse layer and their deprotonation will therefore depend on hydroxide ions which are close to, but outside the Stern layer. If this hypothesis is correct the volume element of reaction for functional micelles will be larger than that estimated for non-functional micelles (Bunton, 1979a,b, 1984; Fornasier and Tonellato, 1980; Romsted, 1984). The earlier use of the lower value of $V_M = 0.14$ (Section 5) will therefore have underestimated the true second-order rate constants at the surfaces of functional micelles and comicelles.

The effect of mixtures of surfactants and polyelectrolytes on spontaneous, water-catalysed hydrolysis (Fadnavis and Engberts, 1982) was mentioned in Section 4, but mixtures of functionalized polyelectrolytes and cationic surfacants are effective deacylating agents (Visser *et al.*, 1983). Polymerized isocyanides were functionalized with an imidazole group and the deacylation of 2,4-dinitrophenyl acetate in the polyelectrolyte was speeded by addition of single or twin chain quaternary ammonium ion surfactants, up to a plateau value. Anionic surfactants had essentially no effect. It is probable that the cationic surfactants accelerate the reaction by increasing the deprotonation of the imidazole groups.

7 Micellar effects on acid-base equilibria

Discussion of acid-base equilibria may seem out of place in a chapter devoted to reactivity, but micellar effects upon indicator equilibria played a key role in the development of ideas regarding micellar effects upon reactivity, so a brief discussion is in order (Hartley, 1948, Fendler and Fendler, 1975).

Hartley showed that micellar effects upon acid-base indicator equilibria could be related to the ability of anionic micelles to attract, and cationic micelles to repel, hydrogen ions. More recently attempts have been made to quantify these ideas in terms of the behavior of a micelle as a submicroscopic solvent, together with an effect due to its surface potential (Fernandez and Fromherz, 1977).

An alternative approach to this problem is to assume that deprotonation of a weak acid in the micellar pseudophase will be related to the concentration of bound OH^-, which should follow the ion-exchange model (Section 5). As a result much of the work has been based on the use of very weak acids which are deprotonated only at high pH.

The pH of the aqueous pseudophase, but not that at the micellar surface, can be controlled by buffers, although then it may be necessary to allow for exchange of buffer anions between water and micelles (Romsted, 1984). This approach has been used by some workers who have developed equations which include terms for exchange of all ionic species between water and micelles (Quina *et al.*, 1980).

A conceptually simpler approach is to avoid the use of buffers by using aqueous NaOH. The problem then becomes that of estimating the concentration of micellar bound OH^-, following the ion-exchange treatment, and of the acid and base forms of the indicator in the micelle. (The base form is generally a hydrophobic anion which binds essentially quantitatively to a cationic micelle.)

It is convenient to define the basicity constant, K_b^M, of the indicator, BH, in the micellar pseudophase as a dimensionless quantity in terms of the mole ratio of micellar bound OH^- (p. 225) using (21). The quantity $[OH_M^-]$ (or m_{OH}^s) can be calculated from the ion-exchange relation and the experimental data fitted, usually by computer simulation, following the approach discussed for treatment of rate constants (p. 229). This treatment fits micellar effects upon the deprotonation of benzimidazole for a variety of CTAX surfactants (X = Cl, Br, NO_3) over a range of concentrations of NaOH and of added salts (Bunton *et al.*, 1982a). A similar, but less general approach, was also applied to deprotonation of phenols and oximes (Bunton *et al.*, 1980c).

$$K_b^M = \frac{[BH_M]\, m_{OH}^s}{[B_M^-]} = \frac{[BH_M][OH_M^-]}{[B_M^-][D_n]} \qquad (21)$$

Deprotonation of benzimidazole can be studied only in dilute NaOH, but deprotonation of 5-nitroindole and 5-nitroindole-2-carboxylate ion can be followed in 0.01–0.1 M NaOH in both CTABr and CTAOH. For CTAOH the concentration of micellar bound OH^- was estimated from (19), and values of K_b^M were similar in both CTABr and CTAOH micelles (Cipiciani *et al.*, 1983b, 1985).

The basicity constants, K_b^M cannot be compared directly with the classical basicity constants in water, K_b, but comparison can be made by converting m_{OH}^s into moles per litre of micellar Stern layer using (22), based on the micellar volume applied to reaction rates (Section 5).

$$K_b^V = K_b^M/0.14 \qquad (22)$$

The basicity constants in water and micelles then have the same units (M^{-1}), and values of K_b^V and K_b are not very different for arenimidazoles and nitroindoles under a variety of conditions (Table 10). The comparisons suggest that inherent basicities are not very different in water and cationic micelles, but, as with second-order rate constants of bimolecular reactions (Section 5), there is a limited degree of specificity because K_b^V/K_b is slightly larger for the nitroindoles than for the arenimidazoles, almost certainly because of interactions between the cationic micellar head groups and the indicator anions.

These quantitative treatments were based on the assumption that the anionic forms of the indicators were sufficiently hydrophobic to bind

TABLE 10

Micellar effects on deprotonation of weak acids

Acid	Surfactant	K_s/M^{-1}	K_X^{OH}	K_b^V/K_b^W	Ref.
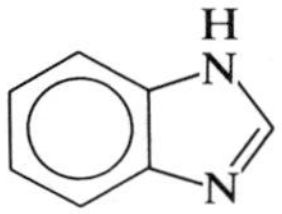 H N N	CTACl	43	4	0.6	Bunton *et al.*, 1982a
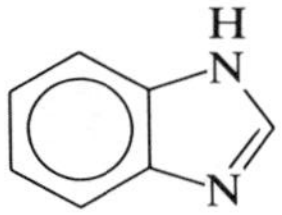 H N N	CTABr	36	12	0.6	
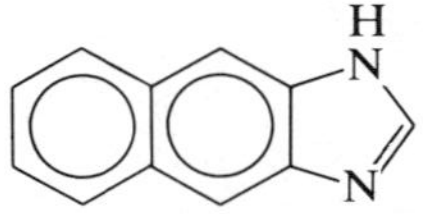 H N N	CTABr	1100	12	0.6	
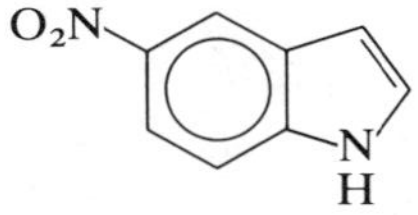 O_2N N H	CTABr	350	15	0.14	Cipiciani *et al.*, 1983b
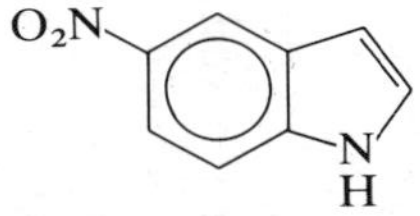 O_2N N H	CTAOH	*ca* 300		0.14	

quantitatively to the cationic micelles. This assumption seems to be satisfactory except when high concentrations of added salts are present. In these cases the salt anion tends to displace the indicator anion from the micelle, and a correction should be made for this displacement (Romsted, 1985).

8 Reactions in non-micellar aggregates

REACTIONS IN VESICLES

Surfactants which have two n-alkyl groups generally do not form simple micelles in water; dialkyldimethylammonium salts with apolar alkyl groups are sparingly soluble and tend to form liquid crystals. Vesicles are formed when dispersions of these salts are sonicated. Large vesicles can also be prepared by dissolving the surfactant in a volatile solvent and vaporizing it in warm water. Vesicles are single or multicompartment closed bilayer assemblies. Liposomes are vesicles made up of phospholipids and the assemblies formed from synthetic surfactants have been described as surfactant vesicles (Fendler, 1982).

The structure of vesicles formed from a given surfactant depends upon the extent of sonication, and over a period of time vesicles fuse and separation of phases occurs. The ease of fusion depends upon vesicular charge and the extent to which it is neutralized by added electrolyte.

Vesicles are permeable to apolar non-ionic solutes, and, if ionic, can bind counterions at the inner and outer surfaces. Permeability to ions depends critically upon vesicle structure.

Mechanistic studies of organic reactivities in vesicles have focused on two questions; the first is the application of the pseudophase model to reactions in vesicles and the second that of reaction at the inner and outer vesicular surfaces.

Fendler and Hinze compared the reactivity of OH^- with 5,5-dithiobis-(2-nitrobenzoate ion) (23) in micelles of CTABr and vesicles of dioctadecyldimethylammonium chloride (DODACl) (Fendler and Hinze 1981). The kinetic forms were similar for reaction in micelles and vesicles, with k_ψ going through maxima with increasing surfactant concentration. The data could be fitted reasonably well to the pseudophase model in both micelles and vesicles. Although overall rate enhancements were much greater in vesicles than in micelles this was simply a consequence of higher concentration of reactants in the vesicles; second-order rate constants in the micellar and vesicular pseudophases were not very different. Similar experiments have been carried out on dephosphorylation by OH^- in vesicles of $(C_{16}H_{33})_2N^+Me_2Br^-$ (Moss *et al.*, 1984b).

$$\mathrm{RS{-}SR} \xrightarrow{\mathrm{OH^-}} \mathrm{RS^-} + \mathrm{RSOH}$$

$$2\mathrm{RSOH} \xrightarrow[\mathrm{OH^-}]{\text{fast}} \mathrm{RS^-} + \mathrm{RSO_2^-} \qquad (23)$$

R = 3-carboxylato-4-nitrophenyl ($-C_6H_3(NO_2)(CO_2^-)$)

Chaimovich and coworkers have prepared large unilamellar vesicles of DODACl by a vaporization technique which gives vesicles of *ca* 0.5 μm diameter. These vesicles are much larger than those prepared by sonication, where the mean diameter is 30 nm, and their effects on chemical reactivity are very interesting. The reaction of *p*-nitrophenyl octanoate by thiolate ions is accelerated by a factor of almost 10^7 by DODACl vesicles (Table 2), but this unusually large effect is due almost completely to increased concentration of the very hydrophobic reactants in the small region of the vesicular surface and an increased extent of deprotonation of the thiol. There is uncertainty as to the volume element of reaction in these vesicles, but it seems that second-order rate constants at the vesicular surface are similar to those in cationic micelles or in water (Cuccovia *et al.*, 1982b; Chaimovich *et al.*, 1984).

A major question is the possibility of separating exo- and endo-vesicular reactions. If either (or both) reactants move slowly through the vesicle wall, it may be possible to observe both slow and fast components of a reaction. This approach has been used extensively by Moss and coworkers who have shown that in some cases an initial spectral change due to a relatively fast reaction is followed by a much slower change. This slow second process could correspond to an endo-vesicular reaction, or to slow reagent transfer. However, for several reactions such slow spectral changes have been shown to be due to some physical change, rather than to a chemical reaction (Moss and Bizzigotti, 1981; Moss *et al.*, 1982; Moss and Ihara, 1983; Moss and Shin, 1983; Moss and Schreck, 1983).

A major problem in all these studies is that the colloidal aggregates scatter light. Thus, any change in their size or shape which materially affects the light scattering may appear to be a change in absorbance due to a chemical reaction. There is this problem in all spectrophotometric studies of reactions in colloidal systems, but intensity of light scattering increases very sharply with increasing size of the colloidal particle. Thus, the artifacts are more likely to appear in vesicular reactions, where the particles are large, than in micellar reactions, and controls are needed in elucidation of endo- and exo-vesicular processes (Moss and Hui, 1983; Moss and Schreck, 1983;

Mizutani and Whitten, 1985). Vesicles prepared by sonication or vaporization are metastable, and it is necessary to use standard conditions to obtain reproducible kinetic data.

In sharp contrast to the behavior of such vesicle-forming surfactants as DODACl, didodecyldimethylammonium hydroxide (DDDAOH) is reported to form vesicles spontaneously with no tendency for vesicle fusion, and similar vesicles were formed with other very hydrophilic counterions, e.g. formate (Talmon *et al.*, 1983; Ninham *et al.*, 1983; Kachar *et al.*, 1984, 1985; Brady *et al.*, 1984). This behavior is related to the high α for aggregates with very hydrophilic counterions, e.g. for CTAOH, because the high charge keeps the aggregates apart (Bunton *et al.*, 1981a; Lianos and Zana, 1983; Hashimoto *et al.*, 1983; Athanassakis *et al.*, 1985).

The relation between head-group area and extended chain length and volume of the apolar groups favors formation of vesicles or bilayers from these twin-tailed surfactants. However, it seems that DDDAOH forms vesicles only in dilute solution, because the solution becomes viscous with increasing [surfactant] which is incompatible with the presence of approximately spherical vesicles. It appears that the solution then contains long, non-vesicular aggregates (Ninham *et al.*, 1983).

Reactions of 2,4-dinitrochloro-benzene and -naphthalene are speeded by DDDAOH and the corresponding chloride + NaOH (Cipiciani *et al.*, 1984). The rate/surfactant concentration profiles and the rate constants are very similar to those for reactions in solutions of the corresponding C_{16} single chain surfactants which form normal micelles. The spontaneous hydrolysis of 2,4-dinitrophenyl phosphate dianion is also speeded by DDDACl and rates reach plateau values in very dilute surfactant (Savelli and Si, 1985).

This limited amount of kinetic evidence suggests that the kinetic models developed for reactivity in aqueous micelles are directly applicable to reactions in vesicles, and that the rate enchancements have similar origins. There is uncertainty as to the appropriate volume element of reaction, especially if the vesicular wall is sufficiently permeable for reaction to occur on both the inner and outer surfaces, because these surfaces will have different radii of curvature and one will be concave and the other convex. Thus binding, exchange and rate constants may be different at the two surfaces.

Several workers have introduced polymerizable groups into twin-tailed amphiphiles and formed vesicles by sonication. They then link the amphiphiles by initiating polymerization, either chemically or photochemically. The polymerized vesicles which are so generated show little tendency to fuse, and are much more stable than the vesicles formed by sonication or vaporization. They therefore have considerable potential for compartmentalizing reagents, although as with "normal" vesicles there is always the problem of diffusion through the vesicle wall, in part because not all the amphiphiles are cross linked (Fendler, 1982).

REACTION IN MICROEMULSIONS

Large amounts of water and oil can be admixed in microemulsions which find wide application in household and pharmaceutical products and have potential for enhanced oil recovery (Mackay, 1981; Bellocq *et al.*, 1984). Microemulsions have been reported to form in the absence of surfactant over a very limited range of concentration, but there is a question as to the nature of these dispersions (Barden and Holt, 1979).

It is convenient to differentiate between oil-in-water (o/w) microemulsions and water-in-oil (w/o) microemulsions in which water and oil are the respective major components. It is reasonable to regard (o/w) microemulsions as akin to swollen normal micelles and w/o microemulsions as reverse micelles (Section 1).

These microdroplets can act as a reaction medium, as do micelles or vesicles. They affect indicator equilibria and can change overall rates of chemical reactions, and the cosurfactant may react nucleophilically with substrate in a microemulsion droplet. Mixtures of surfactants and cosurfactants, e.g. medium chain length alcohols or amines, are similar to o/w microemulsions in that they have ionic head groups and cosurfactant at their surface in contact with water. They are probably best described as swollen micelles, but it is convenient to consider their effects upon reaction rates as being similar to those of microemulsions (Athanassakis *et al.*, 1982).

A microemulsion droplet is a multicomponent system containing oil, surfactant, cosurfactant, and probably water; therefore there may be considerable variation in size and shape depending upon the overall composition. The packing constraints which dictate size and shape of normal micelles (Section 1) should be relaxed in microemulsions because of the presence of cosurfactant and oil. However, it is possible to draw analogies between the behavior of micelles and microemulsion droplets, at least in the more aqueous media.

Reactions of hydrophilic anions in microemulsions can be treated in terms of the ion-exchange formalism. The extent of ion binding to microemulsion droplets has been estimated conductimetrically, and the rates of reactions of OH^- and F^- with *p*-nitrophenyl diphenyl phosphate are consistent with the extent of ion binding and competition between reactive and inert anions. Alternatively the data could be accommodated to variations in the estimated surface potential of the droplet (Mackay and Hermansky, 1981; Mackay, 1982).

Another approach is to measure the fractional ionization (α) of cationic head groups at a microemulsion droplet using a specific ion electrode. This technique has been applied to microemulsions or swollen micelles of CTABr and the ion exchange equation (7) has then been used to estimate the amount of a reactive ion, e.g. OH^-, at the droplet surface.[3]

The fractional ionization, α, of ionic micelles is increased by hydrophobic non-ionic solutes which decrease the charge density at the micellar surface and the binding of counterions (Larsen and Tepley, 1974; Zana, 1980; Bunton and de Buzzaccarini, 1982). Consistently, microemulsion droplets are less effective at binding counterions than otherwise similar micelles.

There is also a problem in defining the volume element of reaction in a microemulsion droplet, but despite these uncertainties second-order rate constants in the droplet are similar to those in cationic micelles for reactions of anionic nucleophiles in alcohol-swollen droplets (Bunton *et al.*, 1983b). Thus, the rate enhancements seem to be due to concentration of reactants in the droplet.

Microemulsions can be generated using n-alkylamines as cosurfactant. These amines react with a substrate, e.g. 2,4-dinitrochlorobenzene, bound to the droplet, and the rate data can be analysed satisfactorily in terms of the concentration of amine in the droplet (Bunton and de Buzzaccarini, 1981b).

All the evidence suggests that in solvents of high water content the microemulsion droplets are very similar to micelles in their ability to promote bimolecular reactions, and other probes support this view.

The rate of decarboxylation of 5-nitrobenzisoxazole carboxylate ion (24):

$$\text{5-}O_2N\text{-benzisoxazole-3-}CO_2^- \longrightarrow \text{2-}(C{\equiv}N)\text{-4-}O_2N\text{-}C_6H_3O^- + CO_2 \qquad (24)$$

is similar in micelles and microemulsions. The rate of this reaction is a sensitive probe of medium effects in bulk solvent and at submicroscopic surfaces (Section 4). The rate of the water-catalysed hydrolysis of bis(4-nitrophenyl) carbonate (25) is also similar in micelles and cationic o/w microemulsions as is the polarity of the surface as indicated by Z-values. These results

$$(O_2N\text{—}C_6H_4\text{—}O)_2C{=}O \xrightarrow{H_2O} 2O_2N\text{—}C_6H_4\text{—}O^- + CO_2 \qquad (25)$$

suggest that water is present at the surfaces of o/w microemulsion droplets and that its reactivity is not very different from that of bulk water (Bunton and de Buzzaccarini, 1981a).

[3] The cosurfactant in this approach cannot be a primary or secondary alcohol; otherwise alkoxide ion reactions will dominate at high pH.

Overall rate enhancements are generally smaller in microemulsions than in micelles, and non-ionic microemulsions often give overall inhibition because a nonionic substrate will bind to the droplet and the ionic reagent will remain in the water.

An additional point is that relatively high concentrations of surfactant, oil and cosurfactant are often used in microemulsions. Thus the volume of the microemulsion pseudophase is large and droplet-bound reactants are therefore diluted. Generally speaking, rate enhancements increase in the sequence microemulsions < micelles < vesicles simply because of a decrease in the volume of the micellar or droplet pseudophase.

However, microemulsions have a significant advantage as reaction media over micelles and vesicles in one key respect. They are excellent solubilizing media and will tolerate high concentrations of apolar solutes; they are much superior to micelles and vesicles in this respect (Mackay, 1981).

SUBMICELLAR SELF-ASSEMBLY AGGREGATES

Hydrophobic ammonium ions which are phase transfer catalysts such as tri-n-octylalkylammonium ions $(C_8H_{17})_3NR^+X^-$ (R = Me, Et, CH_2CH_2OH; X = Cl, Br, $MeSO_3$) are surface active but appear to form small nonmicellar aggregates (Okahata *et al.*, 1977; Kunitake *et al.*, 1980). The salts of these ions are only sparingly soluble in water, but they are very effective at speeding reactions of hydrophobic nucleophilic anions.

The first example of this type of reaction was provided by Kunitake and coworkers who found that deacylation of *p*-nitrophenyl acetate by functional surfactants containing oximate or hydroxamate groups was strongly accelerated by tri-n-octylmethylammonium chloride (Okahata *et al.*, 1977).

Dephosphorylation of *p*-nitrophenyl diphenyl phosphate by arenimidazolide ions is very strongly accelerated by the tri-n-octylethylammonium bromide or mesylate (TEABr or TEAMs respectively) (Scheme 9). The

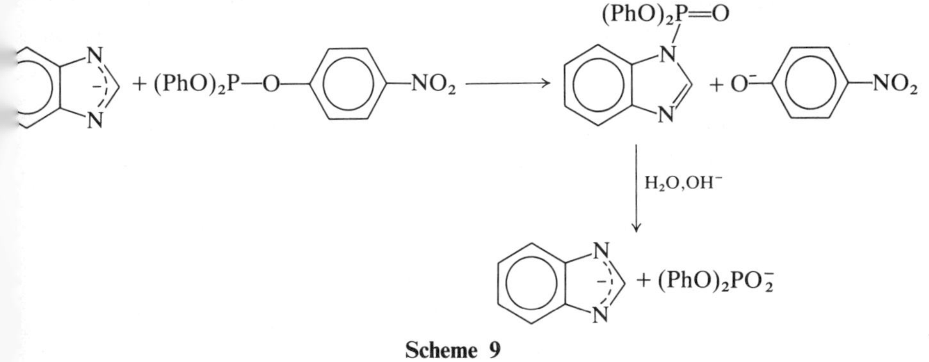

$$\text{Benzimidazolide}^- + (PhO)_2P{-}O{-}C_6H_4{-}NO_2 \longrightarrow \text{Benzimidazole-N-}P(=O)(OPh)_2 + {}^-O{-}C_6H_4{-}NO_2$$

$$\xrightarrow{H_2O,\,OH^-} \text{Benzimidazolide}^- + (PhO)_2PO_2^-$$

Scheme 9

first-order rate constants go through maxima with increasing [TEAMs], and the rate maximum is at very low [TEAMs] for the very hydrophobic naphthalene imidazolide ion (Fig. 6). These rate maxima are qualitatively similar to those observed with micelles of CTABr (Fig. 1) and have a similar origin, *viz*. the colloidal aggregate brings the two reactants together in a small

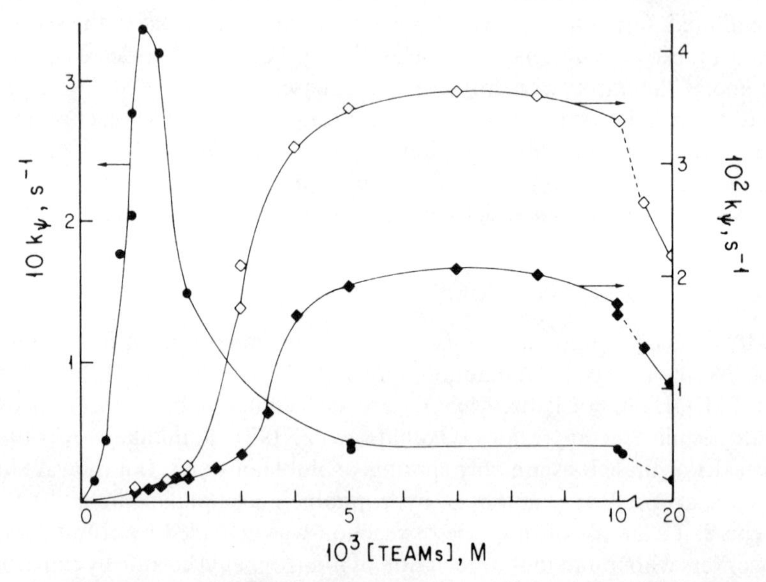

FIG. 6 Reaction of *p*-nitrophenyl diphenyl phosphate in non-micellar aggregates of tri-n-octyl ethylammonium mesylate (TEAMs) at pH 10.7: ●, 10^{-4} M naphth-2,3-imidazole; ◆ and ◇, 10^{-4} and 2×10^{-4} M benzimidazole, respectively. (Reprinted with permission of the American Chemical Society)

volume (Bunton *et al.*, 1981b). The concentrations at the surface of the aggregate can be estimated and the second-order rate constants calculated in terms of moles of reactants per associated quaternary ammonium ion and cationic micelles are very similar (Table 11). As for the reaction in micelles, the concentration of amphiphilic ammonium ion at the rate maximum decreases markedly with increasing hydrophobicity of the nucleophile.

$$(C_8H_{17})_3N^+CH_2CH_2OH \underset{}{\overset{OH^-}{\rightleftharpoons}} (C_8H_{17})_3N^+CH_2CH_2\overline{O} \qquad (26)$$

These tri-n-octylammonium ions can be functionalized giving the possibility of reactions such as (26). The alkoxide zwitterions are effective nucleophiles and the quantitative treatments described for functional micelles

TABLE 11

Comparison of rate constants of reactions in micelles and in non-micellar assemblies

Reaction system	k_M/s^{-1}	Ref.
$(PhO)_2PO.OC_6H_4NO_2$ + benzimidazole anion; CTABr	7	Bunton *et al.*, 1981b
$(PhO)_2PO.OC_6H_4NO_2$ + benzimidazole anion; $(C_8H_{17})_3\overset{+}{N}Et$	*ca* 20	Bunton *et al.*, 1981b
$(PhO)_2PO.OC_6H_4NO_2 + C_{16}H_{33}\overset{+}{N}Me_2CH_2CH_2O^-$	5.6	Biresaw *et al.*, 1984
$(PhO)_2PO.OC_6H_4NO_2 + (C_8H_{17})_3\overset{+}{N}CH_2CH_2O^-$	3	Biresaw *et al.*, 1984
$PhCO.OC_6H_4NO_2 + C_{16}H_{33}\overset{+}{N}Me_2CH_2CH_2O^-$	264	Biresaw *et al.*, 1984
$PhCO.OC_6H_4NO_2 + (C_8H_{17})_3\overset{+}{N}CH_2CH_2O^-$	113	Biresaw *et al.*, 1984

(Section 6) can be applied to reactions of the non-micellizing ions. The second-order rate constants at the surface of a functionalized micelle and tri-n-octylammonium ion are very similar (Biresaw *et al.*, 1984; Bunton and Quan, 1984). Some examples are given in Table 11. In some reactions mixtures of functional and nonfunctional amphiphile were used and allowance was made for the dilution. The second-order rate constants in Table 11 are calculated using nucleophile concentrations written as mole ratios of nucleophile to quaternary ammonium ion because it is not obvious how a molar volume element of reaction can be estimated for the tri-n-octylammonium ions.

Variations of first-order rate constants by the hydroxyethyl amphiphile, $(C_8H_{17})_3\,N^+CH_2CH_2OH\ MsO^- + NaOH$ are shown in Fig. 7 for variation of [amphiphile]. The curves are calculated in terms of the distribution of substrate between water and the aggregate. Similar observations were made for reaction with 2,4-dinitrochlorobenzene. The reaction rates depend also on the extent of deprotonation of the hydroxyethyl group (26) and the variation of rate constant with [NaOH] is shown in Fig. 8. The curves are calculated in terms of the binding of the substrate and deprotonation of the hydroxyl group (Biresaw *et al.*, 1984).

These hydrophobic ammonium ions exert a medium effect on spontaneous, unimolecular reactions. Tri-n-octylmethylammonium chloride effectively speeds decarboxylation of 5-nitrobenzisoxazole carboxylate ion (24) (Kunitake *et al.*, 1980), and tri-n-octyl ethylammonium mesylate or bromide

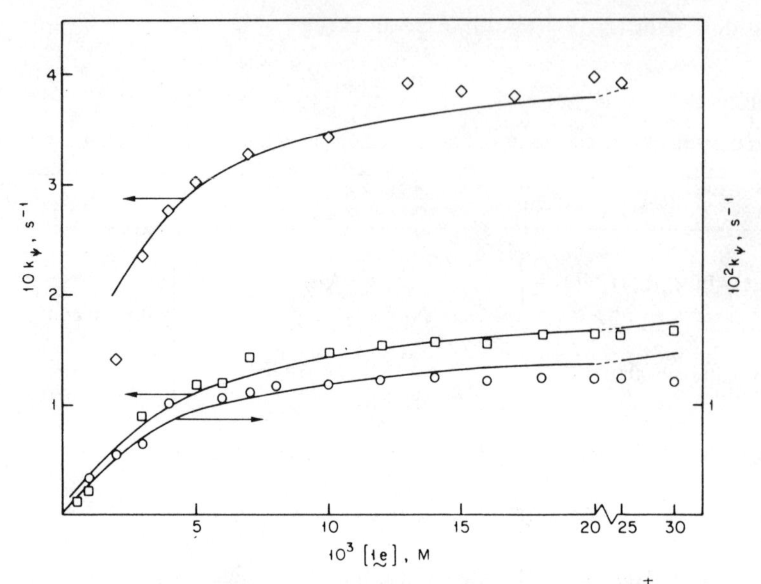

FIG. 7 Reaction of *p*-nitrophenyl diphenyl phosphate with $(C_8H_{17})_3\overset{+}{N}CH_2CH_2OH$-$OMs^-$(1e): ○, □ and ◇, 0.001, 0.01, 0.02 M NaOH, in H_2O:MeCN 90 : 10 v/v. The lines are theoretical. (Reprinted with permission of the American Chemical Society)

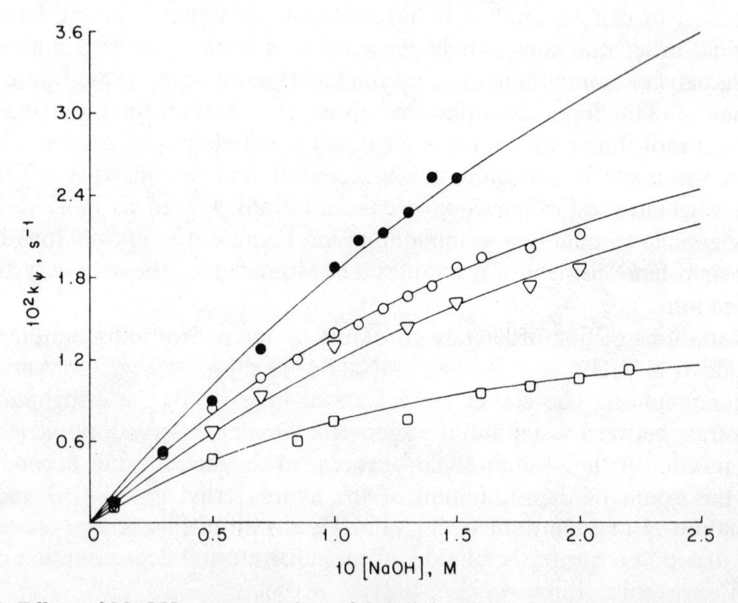

FIG. 8 Effect of NaOH upon reaction of 2,4-dinitrochlorobenzene with the hydrophobic ammonium salts, $(C_8H_{17})_3NCH_2CH_2OH.X$. Solid points in 80 vol% H_2O; open points in 70 vol% H_2O. Ammonium salt concentrations; ●, 0.008 M, X = Br: ○, 0.05 M, X = Br; □, 0.05 M, X = OMs; ▽, 0.1 M, X = OMs. The lines are theoretical. (Reprinted with permission of the American Chemical Society)

speeds hydrolysis of 2,4-dinitrophenyl phosphate dianion (Bunton and Quan, 1985). In this reaction there was a rate plateau at a (relatively) high concentration of ammonium ion. This rate plateau is also observed with micelle-assisted unimolecular reactions, and shows that a complex is formed between substrate and amphiphilic cation and that this complex is considerably more reactive than free substrate.

Little is known about the structures of these kinetically effective complexes, or even about the aggregates of the amphiphile. Both hydrophobic and coulombic interactions are important because these aggregates are much less effective than micelles at assisting reactions of hydrophilic nucleophilic anions. These observations are consistent with the view that the aggregates are much smaller than micelles. It is probable that the structures and aggregation numbers of these aggregates depend on the nature of the solutes which bind to them and Piszkiewicz (1977) has suggested that such interactions play a role in micellar kinetics.

9 Stereochemical effects

Moss and coworkers provided an early example of the way in which micellization can control the stereochemical course of a reaction. Deamination of chiral primary aliphatic amines in water proceeds with net inversion and extensive racemization, and the extent of racemization depends upon the lifetime of the carbocation-like intermediate. The situation changes dramatically if the salts of the primary amine can self-micellize, because now the nucleophile, typically water, is directed in from the front-side so that there is extensive retention of configuration (Moss *et al.*, 1973).

Micellar control of stereochemistry has also been realized in S_N1 hydrolyses of chiral sulfonic esters (Okamoto *et al.*, 1975; Sukenik and Bergman, 1976).

The changes of stereochemistry in these systems are due to changes in the preferred direction of nucleophilic attack upon a carbocationic intermediate or a related ion pair. These results therefore indicate that the substrates or reaction intermediates have preferred orientations at the micellar surface.

Stereoselectivity is small, or non-existent when rates of reaction of enantiomers are compared in micelles of chiral, but chemically inert surfactants (Moss and Sunshine, 1974). However, reactions of enantiomeric substrates in chiral functionalized micelles are stereoselective as are the corresponding reactions with chiral reagents in chemically inert micelles.

Generally speaking imidazole is the nucleophilic functional group, and many of the chiral surfactants are histidine derivatives (Brown and Bunton, 1974; Brown *et al.*, 1981). In other cases mixed systems have been used, e.g. an inert cationic surfactant plus a chiral amphiphilic histidine or hydroxamate derivative (Ihara, 1978; H. Ihara *et al.*, 1980; Y. Ihara *et al.*, 1980; Ohkubo *et al.*, 1980).

The degree of stereoselectivity is usually not large in these reactions and appears to be due to transition-state rather than initial-state interactions. In other words the diastereomeric transition states derived from the enantiomeric substrates have different free energies in the micelle. To this extent the situation is essentially no different to the stereoselectivity which is often observed in non-micellar reactions involving reactions of enantiomeric substrates with a chiral reagent. In some cases it is possible to identify the noncovalent interactions which are responsible for the stereoselectivity (Brown *et al.*, 1981).

Moss and coworkers have obtained stereoselectivity using diastereomers, and here they have observed selective interactions in both initial and transition states. Most of this work has been done using di- and tripeptides and it has been possible to identify the interactions which are responsible for selectivity. A striking exception to the relatively low enantioselectivities which are generally observed is the cleavage of L- or D-N-dodecanoylphenylalanine *p*-nitrophenyl esters by aggregates of the histidine tripeptide *Z*-L-Phe-L-His-L-Leu plus CTABr and the double chain surfactant, $(C_{14}H_{29})_2N^+Me_2Br^-$ ($2C_{14}$). The enantioselectivity increases sharply as $2C_{14}$ is added to CTABr, up to a value of *ca* 70 at 30 mole% $2C_{14}$. The solution then becomes heterogeneous, but solutions become clear at *ca* 60 mole% of $2C_{14}$ and enantioselectivity then decreases (Ueoka *et al.*, 1984). This sharp increase in enantioselectivity is associated with a sharp decrease in particle diffusivity as determined by dynamic light scattering (Ueoka *et al.*, 1985). It seems that the aggregates take up some extended alignment prior to phase separation which enforces an ordered arrangement of the reactants and hence increased stereoselectivity. This hypothesis also explains why normal micelles which have a loose, mobile structure are not very effective at inducing stereoselectivity in reactions, although Moss and coworkers have found considerable diastereoselectivity in cleavage of LL and DL-2-Trp-Pro *p*-nitrophenyl esters which is also related to a decrease in particle diffusivity (Moss *et al.*, 1984a). The assumption that formation of particles with a degree of order is related to increased stereoselectivity is consistent with the relatively high degree of enantioselectivity of the acylation of β-cyclodextrin by chiral esters which contain a ferrocenyl moiety (Trainor and Breslow, 1981). The ferrocenyl group ensures that the substrate is located in the asymmetric environment provided by the cyclodextrin.

Isocyanide polymers functionalized with amino acid groups, typically di- or tripeptides containing histidine or serine, give enantioselective deacylation and rate enhancements. Their activity is increased by addition of cationic surfactants (Visser *et al.*, 1985).

10 Applications in synthetic and analytical chemistry

Micelles exert large rate effects upon organic reactions and can in principle discriminate between different reactions, depending upon their charge type or molecularity. There are a number of examples of this type of discrimination in the literature, and they are easily explained in terms of the generally accepted models of kinetic micellar effects.

Bimolecular E2 reactions involving OH^- in aqueous solution are speeded by cationic and inhibited by anionic micelles (Minch *et al.*, 1975) whereas spontaneous S_N reactions are generally inhibited strongly by cationic micelles and less strongly by anionic micelles; it is therefore relatively easy to observe micellar control of product formation.

In much the same way it should be possible to discriminate between attack by anionic and non-ionic nucleophiles. Micelles, regardless of charge, generally speed attack by non-ionic nucleophiles, but the enhancements are typically small, whereas large inhibition or enhancement is observed for attack of nucleophilic anions, depending upon micellar charge.

Most studies of micellar effects upon rates or products of organic reactions have been made with very low concentrations of reactants, and this small scale of work is not very encouraging for the synthetic organic chemist. An additional disadvantage is that surfactants complicate product separation by extraction or distillation, and to date most studies in this general area have been exploratory and have been aimed at solving these problems.

The problem of separation can sometimes be solved by precipitating an ionic surfactant as an insoluble salt, for example quaternary ammonium perchlorates are sparingly soluble in water. Jaeger and his coworkers have developed a more general approach which is to synthesize chemically labile surfactants. For example, if a cationic surfactant contains an acid labile ketal group, the micellized surfactant can be used to assist reaction of the substrate with an anionic reagent, and the surfactant can then be destroyed on acid work up (Jaeger and Frey, 1982). This approach was applied to permanganate oxidation of piperonal and alkaline hydrolysis of α,α,α-trichlorotoluene. These systems had been studied earlier using non-destructible surfactants (Menger *et al.*, 1975).

A limited degree of regioselectivity of aromatic chlorination in surfactant solutions has been observed (Jaeger and Robertson, 1977). Somewhat similar observations were made by Onyivruka *et al.* (1983) who also used n.m.r. spectroscopy to probe the substrate orientation, and they concluded that it controlled regioselectivity.

Sukenik and coworkers have used surfactants to change the relative extents of 1,2 and 1,4 reduction of enones by BH_4^-. Cationic micelles in water favor 1,4-addition as does an alcohol of low polarity, e.g. 2-propanol, so that the micelle-mediated specificity may be simply a polarity effect (Nickles and Subenik, 1982). Specificity in diene mercuration is also improved by the use of anionic surfactants (Link *et al.*, 1980; Sutter and Sukenik, 1982).

Microemulsions are potentially useful reaction media because of their solubilizing ability (Mackay, 1981). Reductions of ketones and enones by borohydride go readily in microemulsions of CTABr, hexane and 1-butanol, and there is more 1,4-reduction than in 2-propanol (Jaeger *et al.*, 1984).

An especially interesting example of micelle-mediated specificity is the observation that formation of *trans*-dibromide or *trans*-bromohydrin from cyclohexene and Br_2 (27) is affected not only by CTABr, but also by the

$$\text{cyclohexene} + Br_2 \xrightarrow{H_2O} \textit{trans}\text{-1,2-dibromocyclohexane} + \textit{trans}\text{-2-bromocyclohexanol} \quad (27)$$

order of addition of the reagents (Bianchi *et al.*, 1984). The dibromide is a significant product of reaction in water or when Br_2 is added to cyclohexene in aqueous CTABr. But if cyclohexene is added to Br_2 in CTABr the product is almost exclusively bromohydrin. Thus, despite the high concentration of Br^- at the micellar surface, the intermediate bromonium ion is trapped almost exclusively by water. In this system selectivity probably depends upon the reaction steps being faster than solute reorganization at the micellar surface. This reaction is synthetically useful because the products are easily removed and the surfactant is reusable.

There is no reason to believe that these surfactant effects upon regioselectivity are related to differences in molecularity. In this respect they differ from micellar control of elimination to substitution ratios. The controlling factors may be the orientation or conformation of reactants or intermediates in micelles or similar aggregates and changes in the lifetimes of intermediates (cf. van der Langkruis and Engberts, 1984).

It is not clear that regioselectivity in these systems is always due to the presence of micelles or whether the surfactants and organic substrates are giving an emulsion or microemulsion, because in some systems the surfactant is not in large excess over the substrate. But in some systems they are doing more than providing a homogeneous reaction medium because Menger *et al.* (1975) obtained better yields in alkaline hydrolysis of α,α,α-trichlorotoluene in the presence of cationic or non-ionic surfactants than in aqueous organic solvent mixtures. Similar observations were made on oxidation by $KMnO_4$.

In some respects these results are somewhat similar to those obtained with phase-transfer catalysts, although here emphasis is placed on the rate of transfer of an ionic reagent from an aqueous to an apolar organic phase (Weber and Gokel, 1977). However, some workers have concluded that reactions at the interface are important.

An ionic surfactant or phase-transfer catalyst can also be immobilized by binding it to an insoluble resin. The binding is generally covalent (Brown and Jenkins, 1976; Brown and Lynn, 1980) but it can be coulombic (Brown *et al.*, 1980); the catalysts are re-usable and product separations are also simplified.

A particularly ingenious approach is that of triphase catalysis which was developed by Regen (1979). The catalyst is a quaternary ammonium residue which is covalently bound to an insoluble polystyrene resin, and reactions of anionic reagents are carried out in a two-phase water-organic solvent mixture.

The uses of micelles in chemical analysis are rapidly increasing (Hinze, 1979). Analytical reactions are carried out typically on a small scale and are based on spectrophotometry. At the same time, undesired side reactions can cause major problems, especially when the analytical procedure depends on reactions which are relatively slow and require high temperatures, exotic solvents or high reagent concentrations for completion. Micelles can suppress undesired reactions as well as speed desired ones and they also solubilize reagents which are sparingly soluble in water. In addition it is often possible to make phosphorescence measurements at room temperature in the presence of surfactants which enormously increases the utility of this very sensitive method of detection.

The use of micelles or similar aggregates in isotopic separations on the basis of nuclear spins is an especially interesting example of the way in which colloidal aggregates can keep reactive intermediates in close proximity (Turro and Kraeutler, 1980; Gould *et al.*, 1984b; Herve *et al.*, 1984).

Micelles and vesicles can also be used in the preparation of very small submicroscopic particles which may be useful as heterogeneous catalysts or because of their properties as semiconductors (Lufimpadio *et al.*, 1984; Tricot and Fendler, 1984).

Acknowledgements

Our work in this area has depended upon the help and enthusiasm of our coworkers, most of whom are cited in the references. The work has been supported by the National Science Foundation (Chemical Dynamics Program), the U.S. Army Office of Research, CNR (Rome) and the Ministero della Pubblica Istruzione.

APPENDIX

A survey of micellar effects on chemical and photochemical reactions

This section gives tabulated examples of recent work on micellar effects upon chemical and photochemical reactions. In general the examples given in this section do not duplicate material covered elsewhere in the chapter; for example micellar effects on some photochemical reactions and reactivity in reversed micelles are listed here although they are neglected in the body of the text. For many ionic reactions in aqueous micelles only overall rate effects have been reported, in many cases because the evidence did not permit estimation of the parameters which describe distribution of reactants between aqueous and micellar pseudophases. These reactions are, nevertheless, of considerable chemical importance, and they are briefly described here.

REACTIONS IN MICROEMULSIONS OR ALCOHOL-SWOLLEN MICELLES

Reaction/substrate	Conditions/conclusions	Reference
$PhCH_2Cl + Br^-$	Hexane, aq.KBr, CTABr, 1-butanol. Rate decreases with increasing hexane. Products: $PhCH_2Br$ with small amounts of alcohol and ether	Martin *et al.*, 1982
$4\text{-}O_2NC_6H_4OPO(OR)_2$ (R = Et, n-hexyl, Ph)	Hexadecane, CTABr, 1-butanol + aq.OH^- or F^-. Rates in microemulsions compared with those in H_2O, aq.dioxan and aq.CTABr	Mackay and Hermansky, 1981
$4\text{-}O_2NC_6H_4OPO(OPh)_2$	$C_{16}H_{33}\overset{+}{N}Me_2CH_2CH_2OH\ Br^-$, aq.NaOH, octane, t-amyl alcohol. Reaction involves phosphorylation of the functional surfactant	Bunton *et al.*, 1983c
1-Cl-2-NO_2-4-substituted benzene + $C_6H_{13}NH_2$	CTABr, n-$C_6H_{13}NH_2$, octane. Second-order rate constants in the microemulsion droplets calculated	Bunton and de Buzzaccarini, 1981

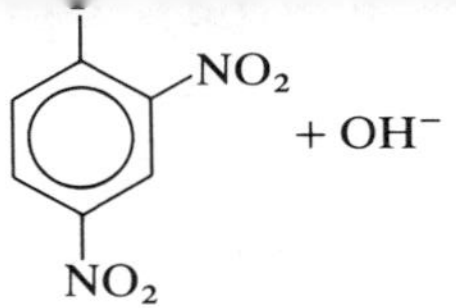 1-X-2,4-dinitrobenzene $(X = \text{[structure cut off at page edge]}) + OH^-$	CTABr, t-amyl alcohol, aq.NaOH. Second-order rate constants in the microemulsion droplets calculated.	Athanassakis *et al.*, 1982
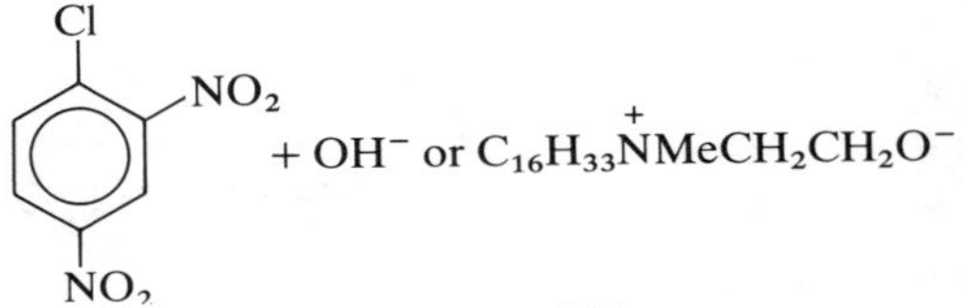 1-chloro-2,4-dinitrobenzene $+ OH^-$ or $C_{16}H_{33}\overset{+}{N}MeCH_2CH_2O^-$	CTABr or $C_{16}H_{33}\overset{+}{N}Me_2CH_2CH_2OH$ Br^-, t-amyl alcohol or 1-butanol, aq.NaOH. Overall reaction rates and products examined	Bunton *et al.*, 1983b
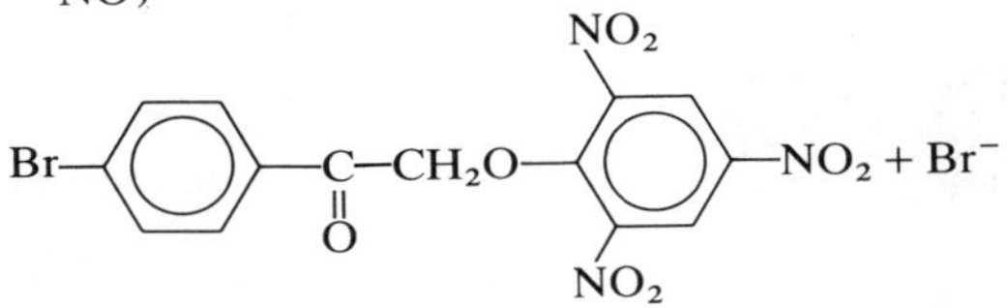 $p\text{-}BrC_6H_4\text{-}C(=O)\text{-}CH_2O\text{-}C_6H_2(NO_2)_3$ (2,4,6-trinitrophenyl) $\rightarrow\ \ldots\text{-}NO_2 + Br^-$	CTABr + C_4H_9OH, $C_5H_{11}OH$ or $C_6H_{13}OH$, + octane. Second-order rate constants in aq. micelles and microemulsion droplets compared	Bunton and de Buzzaccarini, 1982
$Fe(Xphen)_3^{2+} + H^+, OH^- + O_2, CN^-$ [X = H, 5-NO_2; phen = phenanthroline]	2-Butoxyethanol or 2-propanol + decane or hexane. Rate constants of aquation and attack of OH^- or CN^- examined	Blandamer *et al.*, 1983
Triarylmethyl cations or $C_6H_3(NO_2)_2Cl + ROH$	Rate constants in water and microemulsions compared	Blandamer *et al.*, 1984
Diels–Alder reaction:  cyclopentadiene $+ CH_2{=}C(Me)CO_2Me$	*Endo*/*exo* ratios compared in water, micelles and microemulsions	Gonzales and Holt, 1982

Reaction/substrate	Conditions/conclusions	Reference
Deacylation of *p*-nitrophenyl alkanoates	Alkylammonium alkanoates C_6H_6, H_2O. Effects of surfactant structure and H_2O examined	O'Connor and Ramage, 1980
Ester aminolysis and hydrolysis	Alkylammonium alkanoates. The amine can react nucleophilically or as a general base	M. I. El Seoud *et al.*, 1982
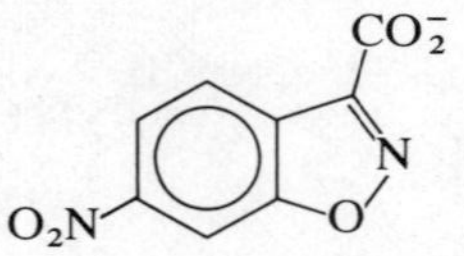 CO_2^- N O O_2N	Rates of decarboxylation compared in various solvents and in aq. micelles, reversed micelles and vesicles	Sunamoto *et al.*, 1983a
Indicator equilibria	Aerosol-OT, heptane. Apparent values of pK_a and pK_{R^+} analysed in terms of ion exchange	O. A. El Seoud *et al.*, 1982; O. A. El Seoud and Chinelatto, 1983
Indicator protonation	CTAX, $CHCl_3$, H_2O. Apparent values of pK_a measured in various mixtures.	El Seoud *et al.*, 1984
Cyclization of histidine -pyridoxal Schiff base	Reverse micelles affect relative rates of the first and second steps of cyclization	Sunamoto *et al.*, 1983b
Fe^{2+}/O_2	Air oxidation much faster in Aerosol-OT reverse micelles in cyclohexane and n-heptane than in H_2O	Inouye *et al.*, 1982
Photo-induced proton transfer	Aerosol-OT. The efficiency of proton transfer depends on H_2O. Pyranine used as a fluorescent probe.	Bardez *et al.*, 1984

Reaction/substrate	Conditions/conclusions	Reference
Thiolysis of *p*-nitrophenyl acetate	Synthetic vesicles prepared by injection. Rates compared with those in aq.CTABr	Cuccovia *et al.*, 1979
Thiolysis of *p*-nitrophenyl octanoate	Synthetic vesicles of DODACl. Very large rate enhancements due to increased reactant concentration in the vesicles	Cuccovia *et al.*, 1982
Ester hydrolysis	Lecithin vesicles and CTABr micelles. Vesicles inhibited reaction	Fatah and Loew, 1983
Ellman's reagent[a] + dithionite	Synthetic vesicles of dicetyldimethyl-ammonium bromide. Effects of vesicles and fast and slow reaction components separ-ated	Moss and Schreck, 1983
$(O_2NC_6H_4O)_2PO.OEt$ + hydroxamates	Dephosphorylations by mixed micelles, vesicles and hydrophobic hydroxamic acid compared	Okahata *et al.*, 1981
Pt-catalysed reduction of methylene blue and 10-methyl-5-deazoisoalloxazine -3-propanesulfonic acid	Polymerized surfactant vesicles. Catalytic efficiencies high in these systems	Kurihara and Fendler, 1983
Deacylation of *p*-nitrophenyl esters of amino acids	Histidine-functionalized micelles and syn-thetic vesicles. Rate enhancements and enantioselectivity observed	Murakami *et al.*, 1981b
Thiolysis of peptides	Thiol-functionalized synthetic vesicles. Diastereoselectivities lower in vesicles than in micelles	Moss *et al.*, 1982c
Deacylation of *p*-nitrophenyl esters of amino acids	Synthetic vesicles + dipeptide nucleophiles. High stereoselectivity observed	Ohkubo *et al.*, 1982
Deacylation of *p*-nitrophenyl esters of amino acids	Synthetic vesicles + histidine-functiona-lized surfactant. Cholesterol increased rates and enantioselectivity	Ueoka and Matsumoto, 1984

[a] 5,5′-Dithiobis(2-nitrobenzoic acid)

Reaction/substrate	Conditions/conclusions	Reference
Deacylation of *p*-nitrophenyl alkanoates	CTACl and hydroxy- and imidazole-functionalized micelles. Rates compared in micellar systems	Moss *et al.*, 1975
Deacylation of *p*-nitrophenyl acetate	Comicelles of hydroxy-, phenol-, thiophenoxide-, thiol- and imidazole-functionalized surfactants. Rate comparisons	Moss and Dix, 1981
Deacylation of *p*-nitrophenyl acetate	CTABr, SDS and non-functional and hydroxamate-functionalized micelles. Rate comparisons	Kunitake *et al.*, 1976
Deacylation of *p*-nitrophenyl alkanoates	Hydroxamate-, imidazole- and thiol-functionalized micelles and chymotrypsin. Rate comparisons	Anoardi *et al.*, 1978
Deacylation of *p*-nitrophenyl alkanoates	Hydroxy and imidazole functionalized micelles. Rate comparisons	Tonellato, 1977; Fornasier and Tonellato, 1980
Deacylation of *p*-nitrophenyl alkanoates	CTABr, oxime-, imidazole- and hydroxamate-functionalized micelles. Rate comparisons	Anoardi *et al.*, 1981
Thiolysis of *p*-nitrophenyl alkanoates	Thiol-functionalized micelles. Rate comparisons	Fornasier and Tonellato, 1982b
Thiolysis of *p*-nitrophenyl alkanoates	Micelles of dihydrolipoyl derivatives. Rate comparisons	Fornasier and Tonellato, 1982c
Hydrolysis of $CF_3CON(n\text{-}Bu)C_6H_3(NO_2)_2$ (2,4-dinitrophenyl)	Hydroxy-, oxime-, hydroxamate- and imidazole-functionalized micelles. Rate comparisons	Fornasier and Tonellato, 1982a

Deacylation of *p*-nitrophenyl alkanoates	Micelles and comicelles functionalized with histidinamide or alaninamide moieties. Rate comparisons	Murakami *et al.*, 1981b
Deacylation of *p*-nitrophenyl acetate	Comicelles of histidine functionalized surfactants. Rate comparisons	Ihara *et al.*, 1983a
Deacylation of $CF_3CONEtPh$	CTABr and hydroxy- and imidazole-functionalized micelles. Rate comparisons	O'Connor and Porter, 1981
Acyl transfer from hydrophobic esters	Hydroxy-functionalized micelles. Rate comparisons	Pillersdorf and Katzhendler, 1979
Dephosphorylation of $(PhO)_2PO\cdot O{-}C_6H_4{-}NO_2$ and $EtOPO_2^-\cdot O{-}C_6H_3(NO_2)_2$	CTABr and hydroxy and phenol functionalized micelles. Rate comparisons	Moss and Ihara, 1983
Nucleophilic addition to Malachite Green	$n\text{-}C_{16}H_{33}\overset{+}{N}Me_2CH_2CH_2OH$ comparison with reactivity of $Me_3\overset{+}{N}CH_2CH_2OH$	Bunton and Paik, 1976
Deacylation of $PhCH_2OCONHCH(CH_2Ph)CO_2C_6H_4NO_2$-(4)	Chiral imidazole-functionalized micelles. Comparison of rates and enantioselectivities	Ihara, 1980, Ihara *et al.*, 1981, Ihara and Hosako, 1982
Deacylation of chiral esters	Imidazole-functionalized carboxylate ion micelles. Rate comparisons	Ihara *et al.*, 1983b
Deacylation of chiral esters by dipeptides	CTABr + histidine dipeptides. Comparison of rates and enantioselectivities	Ihara *et al.*, 1983c

REACTIONS IN AQUEOUS FUNCTIONAL MICELLES *continued*

Reaction/substrate	Conditions/conclusions	Reference
Deacylation of *p*-nitrophenyl derivatives of amino acids	Chiral histidine-derived surfactants + CTABr. Rates and enantioselectivities examined	Matsumoto and Ueoka, 1983
Deacylation of hydrophobic *p*-nitrophenyl alkanoates	Hydroxamic acid and phenyl ester derivatives had alkyl or fluoroalkyl substituents. Rate effects depend on selectivity of binding of fluoro- and hydro-carbon derivatives	Kunitake *et al.*, 1984
Deacylation of *p*-nitrophenyl derivatives of amino acids	Chiral surfactants derived from amino acids. Comparisons of rates and enantioselectivities	Ono *et al.*, 1981
Cu(II) mediated oxidation of DOPA	Cu(II) derivatives of dodecylhistidine surfactants. Comparison of rates and enantioselectivities	Yamada *et al.*, 1980
Deacylation or hydrolysis of chiral carbamates, carbonates and alkanoates	Micelles and comicelles of N-hexadecyl-N-methylephedrinium bromide or N′-myristoyl-histidine with CTABr. Rate effects and enantioselectivities examined	Fornasier and Tonellato, 1984

Reaction/substrate	Conditions/conclusions	Reference
Anilide + OH^-	CTABr + NaOH. Rate comparisons	Broxton and Duddy, 1980
Amide + OH^-	CTABr + NaOH or buffers, Rate comparisons	Broxton, 1981b
1-Cl(F)-2,4-$(NO_2)_2C_6H_3$ + NO_2^-	CTABr + $NaNO_2$. Rate comparisons	Broxton, 1981a,c
Aspirin hydrolysis	CTABr and cetylpyridinium chloride + NaOH or buffer. Enhancement at high and inhibition at low pH	Broxton, 1982
Carbamate + OH^-, Hydrolysis and decarboxylation	CTABr + OH^-. Rate comparison with variation of substrate	Broxton, 1984
Substituted benzamides + OH^-	CTABr + OH^-. Possible micellar effects on mechanism	Broxton *et al.*, 1981
$(PhO)_2PO.OC_6H_4NO_2$-(4) + OH^-, $CH_2(OH)CH_2O^-$	CTACl, CTABr + NaOH, aq.diol. Comparison of rate constants in micelle and solvent	Bunton *et al.*, 1983a
$(PhO)_2PO.OC_6H_4NO_2$-(4) + OH^-, RO_2^-	CTABr, CTACl + NaOH and peroxy-anions. Rate comparisons and quantitative rate analysis	Bunton *et al.*, 1984
$PhSO_3Me$ + Cl^-, Br^-	CTACl, CTABr. Second-order rate constants for halide attack in micelles and water are similar	Al-Lohedan *et al.*, 1983

NUCLEOPHILIC REACTIONS IN NON-FUNCTIONAL MICELLES *continued*

Reaction/substrate	Conditions/conclusions	Reference
N-Alkylhydroxamic acid hydrolysis	$C_{12}H_{25}SO_3Na + H_3O^+$, CTABr + OH^-. An attempt made to separate electronic and hydrophobic effects on the micellar reaction	Berndt *et al.*, 1984
Methyl Violet + OH^-	Anionic and cationic micelles. Effect of surfactant structure examined	Malaviya and Katiyar, 1984
$(2\text{-}ClC_6H_4)\overset{+}{C}(C_6H_4\text{-}NMe_2)_2 + CN^-$	Reaction speeded by CTABr, inhibited by SDS	Srivastava and Katiyar, 1980
Transamination of barbiturate derivatives with semicarbazide	Rate enhancement by CTABr depends on substrate structure	Reddy and Katiyar, 1981
Hydrolysis and acyl migration of corticosteroid esters	Substrate micellization inhibits hydrolysis and acyl migration	Anderson *et al.*, 1983
Penicillin derivatives + OH^-	CTABr 0.05 M NaOH. Effect of structure	Gesmantel and Page, 1982
p-Nitrophenyl acetate + oximate ions	CTABr, pH 9.56. Effect of oxime structure	Meyer and Viout, 1981
Triphenylmethane dyes + OH^-	Effect of pressure on cationic micelles	Taniguchi and Iguchi, 1983

Reaction/substrate	Conditions/conclusions	Reference
N-trifluoroacetylindole (CF_3CO on N) $+ H_3O^+$	SDS, aq.HCl. $k_2^m/k_W \approx 0.03$	Cipiciani *et al.*, 1981
$IrCl_6^{2-}$, $Os(bpy)_3^{3+}$, $Os(dmbpy)_3^{3+}$ $+ Fe^{2+}$	SDS. Rates and equilibria of outer sphere electron transfer analysed in terms of electrostatic and non-electrostatic effects	Pramauro *et al.*, 1982
$Fe(Cp)_2 + Co(phen)_3^{3+}$	CTABr. Rates of electron transfer measured	Cavasino *et al.*, 1983
Dihydroxybenzene + iridium or molybdenum complex	CTABr. Rate effects of CTABr and substrate	Pelizzetti and Pramauro, 1984
N-Methylphenothiazine + iron or molybdenum complex	Micellar effects upon rates and equilibria of electron transfer	Minero *et al.*, 1983
Acyloximes; hydrolysis and cyclization	Cationic micelles inhibit hydrolysis but not cyclization	Soto *et al.*, 1981
N-acylimidazole (RCO on N) $+ H_2O, H_3O^+$	CTABr, SDS. Micellar inhibition depends on surfactant and substrate hydrophobicity	Linda *et al.*, 1983

Reaction/substrate	Conditions/conclusions	Reference
Cyclic dienes + $Hg(OAc)_2$, $NaBH_4$	SDS, aq.THF. Regioselectivity and product composition are micelle-sensitive	Link *et al.*, 1980; Sutter and Sukenik, 1982
Cyclopentadiene + dienophiles	SDS, CTABr, EtOH. *Endo*/*exo* ratio is micelle-sensitive	Breslow *et al.*, 1983
Diels–Alder reaction: 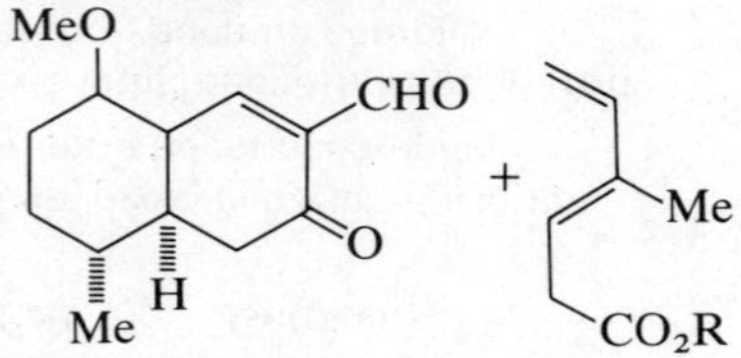	Self-micellization of the diene affects yield and product composition	Grieco *et al.*, 1983
Alkyl *p*-trimethylammonium-benzenesulfonates, hydrolysis	Substitution/elimination rate sensitive to substrate micellization	Sukenik and Bergman, 1976
Photochemical enone cycloaddition	Products and rates differ in anionic micelles and organic solvents	Berenjian *et al.*, 1982
Cyclopropene reduction by $HMn(CO)_5$	*Cis*/*trans* ratio and yields different in anionic micelles and organic solvents	Matsui and Orchin, 1983
Diazotization	Diazonium ion decomposition gives a phenol in water and reduction in SDS	Abe *et al.*, 1983
Amines + 2,4-dinitrofluorobenzene	Analytical method is improved by use of CTABr	Wong and Connors, 1982
Diazonium ions + H_2O/Br^-	Micellization of a hydrophobic diazonium ion affects its capture by H_2O or Br^-	Moss *et al.*, 1982a

SPONTANEOUS WATER-CATALYSED REACTIONS

Reaction/substrate	Conditions/conclusions	Reference
Hydrolysis of 1-methylheptyl mesylate	Dioxan-H_2O. Rate decrease with increasing [substrate] due to substrate association	Mengelsberg *et al.*, 1980
Hydrolysis of bis(*p*-nitrophenyl) carbonate and *p*-chlorodiphenylmethyl chloride	CTABr, aq.dioxan. Micellar effects consistent with an aqueous micellar surface	Menger *et al.*, 1981
6-nitrobenzisoxazole-3-carboxylate (O_2N, CO_2^-)	$C_{12}H_{25}C_5H_4\overset{+}{N}MeI^-$. Decarboxylation is assisted by micellized pyridinium ion	Rupert and Engberts, 1982
Fe(II) $(phen)_3 + H_2O$	Aquation is accelerated by dodecylpyrazinium chloride	Burrows *et al.*, 1982

ELIMINATION

Reaction/substrate	Conditions/conclusions	Reference
E2 elimination of 3-bromo-3-phenylpropionate	Cationic micelles assist base-mediated elimination	Bunton *et al.*, 1974
p-Nitrophenyl cyanoacetate (E1cB mechanism)	CTABr micelles inhibit decomposition of the conjugate base, but assist reaction if $pH < pK_a$ of substrate	Al-Lohedan and Bunton, 1981
DDT, DDD and DDM	CTABr and $C_{16}H_{33}\overset{+}{N}Me_2CH_2CH_2OH$ Br^-. Second-order rate constants estimated in the micelles	Rezende *et al.*, 1983

PHOTOCHEMISTRY

Reaction/substrate	Conditions/conclusions	Reference
Photodimerization of coumarins	Yield and configuration of dimer compared in MeOH, C_6H_6, H_2O, CTABr, SDS and Triton	Muthuramu and Ramamurthy, 1982; Muthuramu *et al.*, 1983
Photodimerization of 7-alkoxy- and 7-methyl-7-alkoxy-coumarins	Yields of *syn* head to tail dimers depend on substrate structure and on CTACl or CTABr	Ramnath and Ramamurthy, 1984
Photocycloaddition of enones to alkenes	Micelles affect regiochemistry and hydrogen-atom transfer	Berenjian *et al.*, 1982
Fluorescence quenching of amphiphilic anthracenes	Aq.CPyCl(Br,NO_3)	Miola *et al.*, 1983
Photoproton transfer from benzoin	CTABr + amphiphilic azobenzenes. Light and dark reactions are compared	Shinkai *et al.*, 1982
Excimer formation of bis(arylmethyl)ammonium chlorides	Aq. micelles affect excimer formation with change in geometry of binding	Emert *et al.*, 1981
Photo-oxidation of protoporphyrin IX and its dimethyl ester	Aq. micelles, vesicles and CH_2Cl_2 give different product distribution	Cox *et al.*, 1982
Norrish type II reaction of amphiphilic ketoacid	Quantum yields differ for t-BuOH, C_6H_6, and aq.CTACl and SDS	Winckle *et al.*, 1983
Quenching of naphthalene fluorescence by Ni^{2+}	Quenching used to estimate binding of naphthalene to CTABr	Abuin and Lissi, 1980
Photochemical proton transfer to indicators	Steady-state and time resolved fluorescence used to examine proton transfer	Politi and Fendler, 1984
Quenching with MV^{2+} and Cu^{2+} of RuL_3^{4-} [L = 4,4′-dicarboxy-2,2′-bipyridyl]	Solubilization in anionic micelles was examined in various conditions	Bonilha *et al.*, 1982

Fluorescence depolarization of 9-anthryloxy-fatty acids in CTABr	Molecular motions in CTABr micelles examined	Blatt *et al.*, 1982
Photodecarbonylation of *p*-tolyl benzyl ketone	Formation of cross products differed in CTABr from that in H_2O or C_6H_6	Turro and Cherry, 1978
Photolysis of dibenzyl ketones with $CuCl_2$ scavenger	Cage effects in CTACl depended upon substrate structure and magnetic field	Turro and Weed, 1983
Photochemistry of benzyl sulfones	Cage effects in SDS depend on sensitizer, substrate structure and magnetic field	Gould *et al.*, 1984a
Photochemistry of benzophenene	Reactions of benzophenone-cyclohexadienyl radical pair depend on surfactant, magnetic field and ^{13}C content	Scaiano *et al.*, 1982

Quantitative treatment of micelle-assisted bimolecular reactions

Effects of micelles upon reactions have generally been analysed using a pseudophase model in which concentrations of both reactants in the micellar pseudophase have been estimated. In some cases, e.g. with non-ionic nucleophiles, the amount of substrate in the micellar pseudophase has been measured directly. For reactions of some ions, e.g. di- or tri-valent metal ions, complete binding to a counterionic micelle has been assumed. Generally the binding of reactive hydrophilic counterions in the presence of inert counterions has been estimated using the ion-exchange equation, but in a few cases it has been calculated from the assumed surface potential of the micelle. In the absence of inert counterions, i.e. in reactive-ion micelles, complete binding of less hydrophilic ions, e.g. Br^- or CN^-, has been assumed, but with very hydrophilic ions, e.g. OH^-, ion binding has generally been assumed to follow a mass action-like equation.

The value of k_2^m/k_W depends upon assumptions regarding α, and the molar volume of reaction, V_M, in the micellar pseudophase. In a few systems authors have postulated a role for direct reaction of a hydrophilic ion in the aqueous pseudophase with micellar bound substrate.

The symbols, IE or MA indicate that counterion binding was calculated using the ion exchange or mass action models and ST that the micelle was assumed to be saturated with counterion.

Reaction/substrate	Conditions/conclusions	Reference
Ethyl benzoate, *p*-aminobenzoate and *p*-nitrobenzoate + OH^-	CTABr + NaOH/NaBr. $k_2^m/k_W = 0.01–0.08$; $\alpha = 0.2$; $V_M = 0.33L$; $K_{Br}^{OH} = 10$. IE	Funasaki and Murata, 1980
p-Nitrophenyl alkanoates + OH^-	CTABr + borate buffer. $k_2^m/k_W = 0.29–0.37$; $\alpha = 0.2$; $V_M = 0.37L$. Allowance made for ion-exchange of borate and OH^-. IE	Quina *et al.*, 1980
p-Nitrophenyl acetate + ArS^-	CTABr, $V_M = 0.37L$; $k_2^m/k_W = 0.63–1.08$	Cuccovia *et al.*, 1978
$(ArCO)_2O + OH^-, HCO_2^-$	CTABr + NaOH, HCO_2Na. $k_2^m/k_W = 0.06$ (OH^-, Ar=Ph), = 0.21 (HCO_2^-, Ar=Ph), = 0.70 (OH^-, Ar=$O_2NC_6H_4$); $\alpha = 0.25$; $V_M = 0.14L$. Similar results obtained in CTAOH and $CTAHCO_2$. IE	Al-Lohedan and Bunton, 1982
n-BuI + $S_2O_3^{2-}$	CTABr + $Na_2S_2O_3$. $k_2^m/k_W = 8$; $\alpha = 0.2$; $V_M = 0.37$. Similar results obtained in $(CTA)_2S_2O_3$. Competition between mono- and dianions analysed. IE or MA	Cuccovia *et al.*, 1982
DDT + OH^-	CTABr + NaOH. Analysis involved introduction of reaction with OH_W^- and micellar bound DDT at high OH^-. IE	Nome *et al.*, 1982
Malachite Green + OH^-, BH_4^-, 1-benzyldihydronicotinamide	CTABr. $k_2^m/k_W \approx 1$ for ionic reactions; $k_2^m/k_W < 1$ for reaction with the nicotinamide in CTABr and SDS. IE	Bunton *et al.*, 1978a
N-Alkyl-2-bromopyridinium ion + OH^-	CTABr, CTACl. $k_2^m/k_W = 0.06–0.18$; $\alpha = 0.25$; $K_{Cl}^{OH} = 4$; $K_{Br}^{OH} = 9$; $V_M = 0.14L$. IE	Al-Lohedan *et al.*, 1982

N-Methyl-4-cyanopyridinium ion + OH^-	CTABr or $C_{14}H_{29}Me_3Cl$ + OH^-. Reaction occurs only in the water, giving $K^{OH}_{Br} = 12.5$ and $K^{OH}_{Cl} = 7.1$. IE	Chaimovich *et al.*, 1979
2,4-Dinitrochlorobenzene and -naphthalene, $PhSO_3Me$ or $PhCO.OC_6H_3(NO_2)_2$	CTABr, CTACl, CTAOMS, + NaN_3. $\alpha \approx 0.2$; $K^{N3}_{Br} = 2$; $K^{N3}_{Cl} = 1.3$; $K^{N3}_{OMe} = 1.1$; $k_2^m/k_W \approx 1$ except for aromatic substitution where $k_2^m/k_W = 50$–400. Similar results in $CTAN_3$. IE or MA	Bunton *et al.*, 1982b
p-Nitrophenyl diphenyl phosphate + aldoximate and aryloxide ions	CTABr + oxime or phenol. $V_M = 0.14$L; $k_2^m/k_W = 0.3$–0.5 for a variety of aryloxide and oximate ions	Bunton and Sepulveda, 1979 Bunton *et al.*, 1979b
$4\text{-}O_2NC_6H_4CH(OEt)_2 + H_3O^+$ $PhNH.NHPh + 2H_3O^+$	SDS + H_3O^+. $K^H_{Na} \approx 1$; $V_M = 0.14$L; for acetal $k_2^m/k_W = 0.05$; for benzidine rearrangement (third order, $k_3^m/k_W = 0.009$). IE	Bunton *et al.*, 1978b
Dioxolan or acyl arylaldoximes + H_3O^+	SDS + HCl. In $[H^+] > 0.1$ M reaction postulated between H^+_W and micellar bound substrate	Gonsalves *et al.*, 1985
4-(Pyridine-2-azo)dimethylaniline + Ni^{2+}	SDS. $k_2^m/k_W \approx 1$ based on binding constants of amine and Ni^{2+}.	Diekmann and Frahm, 1979
$HCO_2H + I^-$	$C_{12}H_{25}NMe_3Cl$, $C_{10}H_{21}NMe_3Cl$. $k_2^m/k_W = 0.5$–1; $\alpha = 0.27$	Hayakawa *et al.*, 1983
Aromatic nucleophilic substitution and amide hydrolysis	CTABr, CTAF + OH^-. Relative rate effects of CTABr and CTAF did not fit simple pseudophase model	Broxton and Sango, 1983
Acetylsalicylic and 3-acetoxy-2-naphthoic acid + OH^-	CTACl, CTABr, CTAOH + OH^-. $k_2^m/k_W \approx 0.01$. Kinetic data analysed using either independent binding or ion exchange constants. IE and MA	Rodenas and Vera, 1985

Reaction/substrate	Conditions/conclusions	Reference
Acetals + H_3O^+	RSO_3H (R = hydrophobic residue). $k_2^m/k_W = 0.020$–0.076; $\alpha = 0.25$; $V_M = 0.14$L. $k_2^m/k_W = 0.046$ in SDS. ST	Bunton *et al.*, 1979a
$O_2NC_6H_4OP(O)(OPh)_2$ (*p*NPDPP) + OH^-, F^-. 2,4-Dinitrochloro-benzene (DNCB) and -naphthalene (DNCN) + OH^-	CTAOH + NaOH. $V_M = 0.14$L; $K'_{OH} = 55$; $K'_F = 40$. For OH^-, $k_2^m/k_W \approx 0.2$ (*p*NPDPP); ≈ 2.3 (DNCN); ≈ 5 (DNCB) and 0.6 for F^- + *p*NPDPP. MA	Bunton *et al.*, 1981a
$PhSO_3Me + Cl^-$, Br^-	CTACl + NaCl, CTABr + NaBr. $k_2^m/k_W \approx 0.8$ (Cl^-), ≈ 1.5 (Br^-). MA	Al-Lohedan *et al.*, 1983
3-$CONH_2$-pyridinium $\overset{+}{N}$—R + CN^-; R = $C_{12}H_{25}$, $C_{14}H_{29}$, $C_{16}H_{33}$	CTACN, CTABr + CN^-. $k_2^m/k_W \approx 1$; $\alpha = 0.25$; $V_M = 0.14$L. ST	Bunton *et al.*, 1980a
DDT,DDD,DDM + OH^-	CTAOH + OH^-. At high $[OH^-]$ an additional reaction postulated between OH_W^- and micellar-bound substrate	Stadler *et al.*, 1984

References

Abe, M., Suzuki, N. and Andogino, K. (1983). *J. Colloid Interface Sci.* **93**, 285

Abuin, E. and Lissi, E., (1980). *J. Phys. Chem.* **84**, 2605

Abuin, E., Lissi, E., Bianchi, N., Miola, L. and Quina, F. H. (1983a). *J. Phys. Chem.* **87**, 5166

Abuin, E., Lissi, E., Araujo, P. S., Aleixo, R. M. V., Chaimovich, H., Bianchi, N., Miola, L. and Quina, F. H. (1983b). *J. Colloid Interface Sci.* **96**, 293

Albrizzio, J. P. de and Cordes, E. H. (1979). *J. Colloid Interface Sci.* **68**, 292

Albrizzio, J., Archila, J., Rodulfo, T. and Cordes, E. H. (1972), *J. Org. Chem.* **37**, 871

Al-Lohedan, H. and Bunton, C. A. (1981). *J. Org. Chem.* **46**, 3929

Al-Lohedan, H. and Bunton, C. A. (1982). *J. Org. Chem.* **47**, 1160

Al-Lohedan. H., Bunton, C. A. and Romsted, L. S. (1981). *J. Phys. Chem.* **85**, 2123

Al-Lohedan, H., Bunton, C. A. and Romsted, L. S. (1982a). *J. Org. Chem.* **47**, 3528

Al-Lohedan, H., Bunton, C. A. and Mhala, M. M. (1982b). *J. Am. Chem. Soc.* **104**, 6654

Al-Lohedan, H., Bunton, C. A. and Moffatt, J. R. (1983). *J. Phys. Chem.* **87**, 332

Almgren, M. and Rydholm, R. (1979). *J. Phys. Chem.* **83**, 360.

Anderson, B. D., Conradi, R. A. and Johnson, K. (1983). *J. Pharm. Sci.* **72**, 448

Aniansson, G. E. A. (1978). *J. Phys. Chem.* **82**, 2805

Anoardi, L., de Buzzaccarini, F., Fornasier, R. and Tonellato, U. (1978). *Tetrahedron Lett.* 3945

Anoardi, L., Fornasier, R. and Tonellato, U. (1981). *J. Chem. Soc. Perkin Trans. 2,* 260

Armstrong, D. W. (1981). *Anal. Chem.* **53**, 11

Athanassakis, V., Bunton, C. A., de Buzzaccarini, F. (1982). *J. Phys. Chem.* **86**, 5002

Athanassakis, V., Moffatt, J. R., Bunton, C. A., Dorshow R. B., Savelli, G. and Nicoli, D. F. (1985). *Chem. Phys. Lett.* **115**, 467

Balekar, A. A. and Engberts, J. B. F. N. (1978). *J. Am. Chem. Soc.* **100**, 5914

Banthorpe, D. V. (1970). *Chem. Rev.* **70**, 295

Barden, R. E. and Holt, S. L. (1979). *In* "Solution Chemistry of Surfactants" (K. L. Mittal, ed.) Vol. 2, p. 707. Plenum Press, New York

Bardez, E., Goguillon, B. T., Keh, E. and Valuer, B. (1984). *J. Phys. Chem.* **88**, 1909

Bartet, D., Gamboa, C. and Sepulveda, L. (1980). *J. Phys. Chem.* **84**, 272

Bell, G. M. and Dunning, A. J. (1970). *Trans. Faraday Soc.* **66,** 500

Bellocq, A. M., Biais, J., Bothorel, P., Clin, B., Fourche, G., Lalanne, P., Lemaire, B., Lemanceau, B. and Roux, D. (1984). *Adv. Colloid Interface Sci.* **20**, 167

Bentley, T. W., Carter, G. E. and Harris, H. C. (1984). *J. Chem. Soc. Chem. Commun.* 388

Berenjian, N., de Mayo, P., Sturgeon, M. E., Sydnes, L. K. and Weedon, A. C. (1982). *Can. J. Chem.* **60**, 425

Berndt, D. C., Utrapiromsuk, N. and Conzan, E. (1984). *J. Org. Chem.* **49**, 106

Bianchi, M. T., Cerichelli, G., Mancini, G. and Marinelli, F. (1984). *Tetrahedron Lett.* 5205

Biresaw, G. and Bunton, C. A. (1986). *J. Org. Chem.* **51**

Biresaw, G., Bunton, C. A., Quan, C. and Yang, Z.-Y. (1984). *J. Am. Chem. Soc.* **106**, 7178

Biresaw, G., Bunton, C. A. and Savelli, G. (1985). *J. Org. Chem.* **50**, 5374

Blandamer, M. J., Burgess, J. and Clark, B. (1983). *J. Chem. Soc. Chem. Commun.* 659

Blandamer, M. J., Clark, B. and Burgess, J. (1984). *J. Chem. Soc. Faraday Trans. 1,* **80**, 1651
Blatt, E. B., Ghiggino, K. P. and Sawyer, W. H. (1982). *J. Phys. Chem.* **86**, 4461
Bonilha, J. B. S., Foreman, T. K. and Whitten, D. G. (1982). *J. Am. Chem. Soc.* **104**, 4215
Bonilha, J. B. S. Chiericato, G., Martins-Franchetti, S. M., Ribaldo, E. J. and Quina, F. H. (1984). Quoted in Romsted, L. S. (1984)
Brady, J. E., Evans, D. F., Kachar, B. and Ninham, B. W. (1984). *J. Am. Chem. Soc.* **106**, 4279
Breslow, R., Maitra, V. and Rideout, D. (1983). *Tetrahedron Lett.* 1901
Briggs, J., Dorshow, R. B., Bunton, C. A. and Nicoli, D. F. (1982). *J. Chem. Phys.* **76**, 775
Brown, J. M. (1979). In "Colloid Science, A. Specialist Periodical Report" (D. H. Everett, senior reporter) Vol. 3, p. 253. Chemical Society, London
Brown, J. M. and Bunton, C. A. (1974). *J. Chem. Soc. Chem. Commun.* 969
Brown, J. M. and Jenkins, J. A. (1976). *J. Chem. Soc. Chem. Commun.* 458
Brown, J. M. and Lynn, J. L., Jr. (1980). *Ber. Bunsenges. Phys. Chem.* **84**, 95
Brown, J. M., Bunton, C. A. and Diaz, S. (1974). *J. Chem. Soc. Chem. Commun.* 971
Brown, J. M., Bunton, C. A., Diaz, S. and Ihara, Y. (1980). *J. Org. Chem.* **45**, 4169
Brown, J. M., Elliott, R. L., Griggs, C. G., Helmchen, G. and Nill, G. (1981). *Angew. Chem. Int. Ed. Eng.* **20**, 890
Broxton, T. J. (1981a). *Aust. J. Chem.* **34**, 969
Broxton, T. J. (1981b). *Aust. J. Chem.* **34**, 1615
Broxton, T. J. (1981c). *Aust. J. Chem.* **34**, 2313
Broxton, T. J. (1982). *Aust. J. Chem.* **35**, 1357
Broxton, T. J. (1984). *Aust. J. Chem.* **37**, 47
Broxton, T. J. and Duddy, N. W. (1980). *Aust. J. Chem.* **33**, 1771
Broxton, T. J. and Sango, D. B. (1983). *Aust. J. Chem.* **36**, 711
Broxton, T. J., Fernando, D. B. and Rowe, J. E. (1981). *J. Org. Chem.* **46**, 3522
Buchwald, S. L. and Knowles, J. R. (1982). *J. Am. Chem. Soc.* **104**, 1438.
Bunton, C. A. (1973a). In "Reaction Kinetics in Micelles" (E. H. Cordes, ed.) p. 73. Plenum Press, New York
Bunton, C. A. (1973b). *Prog. Solid State Chem.* **8**, 239
Bunton, C. A. (1977). *Pure Appl. Chem.* **49**, 969
Bunton, C. A. (1979a). *Cat. Rev. Sci. Eng.* **20**, 1
Bunton, C. A. (1979b). In "Solution Chemistry of Surfactants" (K. L. Mittal, ed.) Vol. 2, p. 519.
Bunton, C. A. (1984). In "The Chemistry of Enzyme Action" (M. I. Page, ed.) Ch. 13. Elsevier, Amsterdam.
Bunton, C. A. and Cerichelli, G. (1980). *Internat. J. Chem. Kinetics,* **12**, 519
Bunton, C. A. and de Buzzaccarini, F. (1981a). *J. Phys. Chem.* **85**, 3139
Bunton, C. A. and de Buzzaccarini, F. (1981b). *J. Phys. Chem.* **85**, 3142
Bunton, C. A. and de Buzzaccarini, F. (1982). *J. Phys. Chem.* **86**, 5010
Bunton, C. A. and Diaz, S. (1976). *J. Am. Chem. Soc.* **98**, 5663
Bunton, C. A. and Huang, S. K. (1972). *J. Org. Chem.* **37**, 1790
Bunton, C. A. and Ionescu, L. G. (1973). *J. Am. Chem. Soc.* **95**, 2912
Bunton, C. A. and Ljunggren, S. (1984). *J. Chem. Soc. Perkin Trans. 2,* 355
Bunton, C. A. and McAneny, M. (1977). *J. Org. Chem.* **42**, 475
Bunton, C. A. and Paik, C. H. (1976). *J. Org. Chem.* **41**, 40
Bunton, C. A. and Quan, C. (1984). *J. Org. Chem.* **49**, 5012

Bunton, C. A. and Quan, C. (1985). *J. Org. Chem.* **50**, 3230
Bunton, C. A. and Robinson, L. (1968). *J. Am. Chem. Soc.* **90**, 5972
Bunton, C. A. and Robinson, L. (1969). *J. Org. Chem.* **34**, 773, 780
Bunton, C. A. and Romsted, L. S. (1979). In "The Chemistry of the Functional Groups. Supplement B, The Chemistry of Acid Derivatives" (S. Patai, ed.), Part 2. p. 945. Wiley-Interscience, New York
Bunton, C. A. and Romsted, L. S. (1982). In "Solution Behavior of Surfactants" (K. L. Mittal and E. J. Fendler eds) Vol. 2, p. 975. Plenum Press, New York
Bunton, C. A. and Rubin, R. J. (1976). *J. Am. Chem. Soc.* **98**, 4236
Bunton, C. A. and Supulveda, L. (1979). *Israel J. Chem.* **18**, 298
Bunton, C. A. and Wolfe, B. (1973). *J. Am. Chem. Soc.* **95**, 3742
Bunton, C. A., Fendler, E. J. and Fendler, J. H. (1967). *J. Am. Chem. Soc.* **89**, 1221
Bunton, C. A., Fendler, E. J., Sepulveda, L. and Yang, K.-U. (1968). *J. Am. Chem. Soc.* **90**, 5512
Bunton, C. A., Kamego, A. and Minch, M. J. (1972). *J. Org. Chem.* **37**, 1388
Bunton, C. A., Minch, M. J., Hidalgo, J. and Sepulveda, L. (1973). *J. Am. Chem. Soc.* **95**, 3262.
Bunton, C. A., Kamego, A. A. and Ng. P. (1974). *J. Org. Chem.* **39**, 3469
Bunton, C. A., Kamego, A. A., Minch, M. J. and Wright, J. L. (1975). *J. Org. Chem.* **40**, 1321
Bunton, C. A., Ohmenzetter, K. and Sepulveda, L. (1977). *J. Phys. Chem.* **81**, 2000
Bunton, C. A., Carrasco, N., Huang, S. K., Paik, C. H. and Romsted, L. S. (1978a). *J. Am. Chem. Soc.* **100**, 5420
Bunton, C. A., Romsted, L. S. and Smith, H. J. (1978b). *J. Org. Chem.* **43**, 4299
Bunton, C. A., Romsted, L. S. and Savelli, G. (1979a). *J. Am. Chem. Soc.* **101**, 1253
Bunton, C. A., Cerichelli, G., Ihara, Y. and Sepulveda, L. (1979b). *J. Am. Chem. Soc.* **101**, 2429
Bunton, C. A., Romsted, L. S. and Thamavit, C. (1980a). *J. Am. Chem. Soc.* **102**, 3900
Bunton, C. A., Frankson, J. and Romsted, L. S. (1980b). *J. Phys. Chem.* **84**, 2607
Bunton, C. A., Romsted, L. S. and Sepulveda, L. (1980c). *J. Phys. Chem.* **84**, 2611
Bunton, C. A., Hamed, F. H. and Romsted, L. S. (1980d). *Tetrahedron Lett.* 1217
Bunton, C. A., Gan, L.-H., Moffatt, J. R., Romsted, L. S. and Savelli, G. (1981a). *J. Phys. Chem.* **85**, 4118
Bunton, C. A., Hong, Y.-S., Romsted, L. S. and Quan, C. (1981b). *J. Am. Chem. Soc.* **103**, 5784, 5788
Bunton, C. A., Hong, Y.-S. and Romsted, L. S. (1982a). In "Solution Behavior of Surfactants" (K. L Mittal and E. J. Fendler, eds) Vol. 2, p. 1137. Plenum Press, New York
Bunton, C. A., Moffatt, J. R. and Rodenas, E. (1982b). *J. Am. Chem. Soc.* **104**, 2653
Bunton, C. A., Gan, L. H., Hamed, F. H. and Moffatt, J. R. (1983a). *J. Phys. Chem.* **87**, 336
Bunton, C. A., de Buzzaccarini, F. and Hamed, F. H. (1983b). *J. Org. Chem.* **48**, 2461
Bunton, C. A., de Buzzaccarini, F. and Hamed, F. H. (1983c). *J. Org. Chem.* **48**, 2457
Bunton, C. A., Mhala, M. M. and Moffatt, J. R. (1984a). Internat. Symp. Surfactants, Bordeaux
Bunton, C. A., Mhala, M. M., Moffatt, J. R., Monarres, D. and Savelli, G. (1984b) *J. Org. Chem.* **49**, 426
Bunton, C. A., Mhala, M. M. and Moffatt, J. R. (1985). *J. Org. Chem.* **50**, 4921

Burrows, H. D., Ige, J. and Umoh, S. A. (1982). *J. Chem. Soc. Faraday Trans. 1,* **78**, 947
Butcher, J. A. and Lamb, G. W. (1984). *J. Am. Chem. Soc.* **106**, 1217
Cabane, B. (1981). *J. Phys. Paris,* **42**, 847
Cabane, B., Duplessix, R., and Zemb, T. (1984). In "Surfactants in Solution" (K. L. Mittal and B. Lindman, eds) Vol. 1, p. 373. Plenum Press. New York
Cavasino, F. P., Sbriziolo, C. and Pelizzetti, E. (1983). *Ber. Bunsenges Phys. Chem.* **87**, 843
Chaimovich, H., Bonilha, J. B. S., Politi, M. J. and Quina, F. H. (1979). *J. Phys. Chem.* **83**, 1851
Chaimovich, H., Cuccovia, I. M., Bunton, C. A. and Moffatt, J. R. (1983). *J. Phys. Chem.* **87**, 3584
Chaimovich, H., Bonilha, J. B. S., Zanette, D. and Cuccovia, I. M. (1984). *In* "Surfactants in Solution" (K. L. Mittal and B. Lindman, eds) Vol. 2, p. 1121. Plenum Press, New York
Cipiciani, A. and Savelli, G. (1985). Unpublished results
Cipiciani, A., Linda, P. and Savelli, G. (1979). *J. Heterocyclic Chem.* **16**, 679
Cipiciani, A., Linda, P., Savelli, G. and Bunton, C. A. (1981). *J. Org. Chem.* **46**, 911
Cipiciani, A., Linda, P., Savelli, G. and Bunton, C. A. (1983a). *J. Phys. Chem.* **87**, 5259
Cipiciani, A., Germani, R., Savelli G. and Bunton, C. A. (1983b). *J. Phys. Chem.* **87**, 5262
Cipiciani, A., Germani, R., Savelli, G. and Bunton, C. A. (1984). *Tetrahedron Lett.* 3765
Cipiciani, A., Germani, R., Savelli, G. and Bunton, C. A. (1985). *J. Chem. Soc. Perkin Trans. 2,* 527
Cordes, E. H. and Gitler, C. (1973). *Progr. Bioorg. Chem.* **2**, 1
Corti, M. and Degiorgio, V. (1981). *J. Phys. Chem.,* **85**, 711
Cox, G. S., Krieg, M. and Whitten, D. G. (1982). *J. Am. Chem. Soc.* **104**, 6930
Cuccovia, I. M., Schroter, E. M., Montiero, P. M. and Chaimovich, H. (1978). *J. Org. Chem.* **43**, 2248.
Cuccovia, I. M., Aleixo, R. M. V., Mortara, R. A., Filho, P. B., Bonilha, J. B. S., Quina, F. H. and Chaimovich, H. (1979). *Tetrahedron Lett.* 3065
Cuccovia, I. M., Aleixo, R. M. V., Erismann, N. E., van der Zee, N. T. E., Schreir, S. and Chaimovich, H. (1982a). *J. Am. Chem. Soc.* **104**, 4554
Cuccovia, I. M., Quina, F. H. and Chaimovich, H. (1982b). *Tetrahedron,* **38**, 917
Cuenca, A. (1985). Thesis, University of California, Santa Barbara
Danielsson, I. and Lindmann, B. (1981). *Colloids Surf.* **3**, 391
Debye, P. (1949). *J. Phys. Colloid Chem.* **53**, 1
Debye, P. and Anacker, E. W. (1951). *J. Phys. Colloid Chem.* **55**, 644
Diekmann, S. and Frahm, J. (1979), *J. Chem. Soc. Faraday Trans 1,* **75**, 2199
Dill, K. A. (1984). In "Surfactants in Solution" (K. L. Mittal and B. Lindman, eds) Vol. 2, p. 307. Plenum Press, New York
Dill, K. A. and Flory, P. J. (1980). *Proc. Natl. Acad. Sci. (U.S.A.)* **77**, 3115
Dill, K. A. and Flory, P. J. (1981). *Proc. Natl. Acad. Sci. (U.S.A.)* **78**, 676
Dill, K. A., Koppel, D. E., Cantor, R. S., Dill, J. D., Bendedouch, D. and Cheng S.-H. (1984). *Nature,* **309**, 42
Dorshow, R., Briggs, J., Bunton, C. A. and Nicoli, D. F. (1982). *J. Phys. Chem.* **86**, 2388.
Dorshow, R. B., Bunton, C. A. and Nicoli, D. F. (1983). *J. Phys. Chem.* **87**, 1409

Duynstee, E. F. J. and Grunwald, E. (1959). *J. Am. Chem. Soc.* **81**, 4540, 4542
El-Seoud, M. I., Viera, R. C. and El-Seoud, O. A. (1982). *J. Org. Chem.* **47**, 5137
El-Seoud, O. A. and Chinelatto, A. M. (1983), *J. Colloid Interface Sci.* **95**, 163.
El Seoud, O. A., Martins, A., Barbur, L. P., de Silva, M. J. and Aldrigue, W. (1977), *J. Chem. Soc. Perkin Trans. 2,* 1674
El-Seoud, O. A., Chinelatto, A. M. and Shimizu, M. R. (1982). *J. Colloid Interface Sci.* **88**, 420
El-Seoud, O. A., Viera, R. C. and Chinelatto, A. M. (1984). *J. Chem. Res.* 80
Emert, J., Phalon, P., Catena, R. and Kodali, D. (1981). *J. Chem. Soc. Chem. Commun.* 759
Evans, D. F. and Wightman, P. J. (1982). *J. Colloid Interface Sci.* **86**, 515
Evans, D. F. and Ninham, B. W. (1983). *J. Phys. Chem.* **87**, 5025
Fadnavis, N. and Engberts, J. B. F. N. (1982). *J. Org. Chem.* **47**, 152
Fadnavis, N. and Engberts, J. B. F. N. (1984). *J. Am. Chem. Soc.* **106**, 2636
Fatah, A. A. and Loew, L. M. (1983). *J. Org. Chem.* **48**, 1886
Fendler, J. H. (1982). "Membrane Mimetic Chemistry". Wiley Interscience, New York
Fendler, J. H. and Fendler, E. J. (1975). "Catalysis in Micellar and Macromolecular Systems". Academic Press, New York
Fendler, J. H. and Hinze, W. (1981). *J. Am. Chem. Soc.* **103**, 5439.
Fendler, E. J., Liechti, R. R. and Fendler, J. H. (1970). *J. Org. Chem.* **35**, 1658
Fernandez, M. S. and Fromherz, P. (1977). *J. Phys. Chem.* **81**, 1755
Fornasier, R. and Tonellato, U. (1980). *J. Chem. Soc. Faraday Trans. 1,* **76**, 1301
Fornasier, R. and Tonellato, U. (1982a). *J. Chem. Soc. Perkin Trans. 2,* 899
Fornasier, R. and Tonellato, U. (1982b). *Gazz. Chim. Ital.* **112**, 261
Fornasier, R. and Tonellato, U. (1982c). *Bioorg. Chem.* **11**, 428
Fornasier, R. and Tonellato, U. (1984). *J. Chem. Soc. Perkin Trans. 2,* 1313
Fromherz, P. (1981). *Ber Bunsenges. Phys. Chem.* **85**, 891
Funasaki, N. (1979) *J. Phys. Chem.* **83**, 1998
Funasaki, N. and Murata, A. (1980). *Chem. Pharm. Bull.* **28**, 805
Fry, A. (1970). In "Isotope Effects in Chemical Reactions" (C. J. Collins and N. S. Bowman, eds) p. 402. Van Nostrand Reinhold, New York
Gamboa, C., Sepulveda, L. and Soto, R. (1981). *J. Phys. Chem.* **85**, 1429
Gan, L.-H. (1985). *Can. J. Chem.* **63**, 598
Gani, V., Lapinte, C. and Viout, P. (1973). *Tetrahedron Lett.* 4435
Garnett, C. J., Lambie, A. T., Beck, W. H. and Liler, M. (1983). *J. Chem. Soc. Faraday Trans. 1,* **79**, 953, 965
Gesmantel, N. B. and Page, M. I. (1982). *J. Chem. Soc. Perkin Trans. 2,* 147, 155
Gobbo, M., Fornasier, R. and Tonellato, U. (1984). In "Surfactants in Solution" (K. L. Mittal and B. Lindman, eds) Vol. 2, p.1169. Plenum Press, New York
Gonsalves, M., Probst, S., Rezende, M. C., Nome, F., Zucco, C. and Zanette, D. (1985). *J. Phys. Chem.* **89**, 1127
Gonzales A. and Holt, S. L. (1982). *J. Org. Chem.* **47**, 3186
Gould, I. R., Turig, C., Turro, N. J., Given, R. S. and Matuszewski, B. (1984a). *J. Am. Chem. Soc.* **106**, 1789
Gould, I. R., Turro, N. J. and Zimmit, M. B. (1984b). *Adv. Phys. Org. Chem.* **20**, 1
Grieco, P. A., Garner, P. and He, Z. (1983). *Tetrahedron Lett.* **24**, 1897
Gruen, D. W. R. (1984). In "Surfactants in Solution" (K. L. Mittal and B. Lindman, eds) Vol. 1, p. 279. Plenum Press, New York
Gunnarsson, G., Jonsson, B. and Wennerstrom, H. (1980). *J. Phys. Chem.* **84**, 3114

Haberfield, P. and Pessin, J. (1982). *J. Am. Chem. Soc.* **104**, 6191
Hartley, G. S. (1948). *Quart. Rev.* **2**, 152
Hartley, G. S. (1977). In "Micellization, Solubilization and Microemulsions" (K. L. Mittal, ed.) p. 23. Plenum Press, New York
Hashimoto, S., Thomas, J. K., Evans, D. F., Mukherjee, S. and Ninham, B. W. (1983). *J. Colloid Interface Sci.* **95**, 594
Hayakawa, K., Kanda, M. and Satake, I. (1983). *Bull. Chem. Soc. Jpn.* **56**, 75
Herve, P., Nome, F. and Fendler, J. H. (1984). *J. Am. Chem. Soc.* **106,** 8291
Hicks, J. R. and Reinsborough, V. C. (1982). *Aust. J. Chem.* **35**, 15.
Hinze, W. L. (1979). *In* "Solution Chemistry of Surfactants" (K. L. Mittal, ed.) Vol. 1, p. 79. Plenum Press, New York
Hoffman, H., Rehage, H., Schorr, W. and Thurn, H. (1984). *In* "Surfactants in Solution" (K. L. Mittal and B. Lindman, eds) Vol 1, p. 425. Plenum Press, New York
Ihara, H., Ono, S., Shosenji, H. and Yamada, K. (1980). *J. Org. Chem.* **45**, 1623
Ihara, Y. (1978). *J. Chem. Soc. Chem. Commun.* 984
Ihara, Y. (1980). *J. Chem. Soc. Perkin Trans. 2,* 1483
Ihara, Y. and Hosako, R. (1982). *Bull. Chem. Soc. Jpn.* **55**, 1979
Ihara, Y., Nango, M. and Kuroki, N. (1980). *J. Org. Chem.* **45**, 5009
Ihara, Y., Hosako, R., Nango, M. and Kuroki, N. (1981). *J. Chem. Soc. Chem. Commun.* 393
Ihara, Y., Nango, M., Kimura, Y. and Kuroki, N. (1983a). *J. Am. Chem. Soc.* **105**, 1252
Ihara, Y., Hosako, R., Nango, M. and Kuroki, N. (1983b). *J. Chem. Soc. Perkin Trans. 2,* 5
Ihara, Y., Kumikiyo, N., Kumikasa, T., Kimura, Y., Nango, M. and Kuroki, N. (1983c). *J. Chem. Soc. Perkin Trans. 2,* 1741
Ionescu, L. G. and Fung, D. S. (1981). *J. Chem. Soc. Faraday Trans. 1,* **77**, 2907
Ionescu, L. G. and Nome, F. (1984). *In* "Surfactants in Solution" (K. L. Mittal and B. Lindman, eds) Vol. 2, p. 789. Plenum Press, New York
Ionsescu, L. G., Romanesio, L. S. and Nome, F. (1984). *In* "Surfactants in Solution" (K. L. Mittal and B. Lindman, eds) Vol. 2, p. 789. Plenum Press, New York
Ingold, C. K. (1969). "Structure and Mechanism in Organic Chemistry". 2nd edn, Chap. 7. Cornell University Press, Ithaca, N.Y.
Inouye, K., Endo, R., Otsuka, Y., Miyashiro, K., Kaneko, K. and Ishitawa, T. (1982). *J. Phys. Chem.* **86,** 1465
Israelachvili, J. N., Mitchell, D. J. and Ninham, B. W. (1976). *J. Chem. Soc. Faraday Trans. 2,* **72**, 1525
Jaeger, D. A. and Robertson, R. E. (1977). *J. Org. Chem.* **42**, 3298
Jaeger, D. A. and Frey, M. R. (1982). *J. Org. Chem.* **47**, 311
Jaeger, D. A., Ward, M. D. and Martin, C. A. (1984). *Tetrahedron,* **40**, 2691
Johnson, S. L. (1967). *Adv. Phys. Org. Chem.* **5**, 237
Jonsson, B., Nilsson, P.-G., Lindman, B., Guldbrand, L. and Wennerstrom, H. (1984). *In* "Surfactants in Solution" (K. L. Mittal and B. Lindman eds) Vol. 1, p. 3. Plenum Press, New York
Kachar, B., Evans, D. F. and Ninham, B. W. (1984). *J. Colloid Interface Sci.* **99**, 593
Kachar, B., Evans, D. F. and Ninham, B. W. (1985). *J. Colloid Interface Sci.* **100**, 287
Kemp, D. S. and Paul, K. (1980). *J. Am. Chem. Soc.* **95**, 3262
Kirby, A. J. and Varvoglis, A. G. (1967). *J. Am. Chem. Soc.* **89**, 415
Kitahara, A. (1980). *Adv. Colloid. Interface Sci.* **12**, 109

Kunitake, T. and Shinkai, S. (1980). *Adv. Phys. Org. Chem.* **17**, 435
Kunitake, T., Shinkai, S. and Okohata, Y. (1976). *Bull. Chem. Soc. Jpn.* **49**, 540
Kunitake, T., Okahata, Y., Ando, R., Shinkai, S., and Hirakawa, S. (1980). *J. Am. Chem. Soc.* **102**, 7877
Kunitake, T., Ihara, H. and Hashiguchi, Y. (1984). *J. Am. Chem. Soc.* **106**, 1156
Kurihara, K. and Fendler, J. H. (1983). *J. Am. Chem. Soc.* **105**, 6152
Kurz, J. L. (1962). *J. Phys. Chem.* **66**, 2239
Lapinte, C. and Viout, P. (1973). *Tetrahedron Lett.* 1113
Lapinte, C. and Viout, P. (1979). *Tetrahedron,* **35**, 1931
Larsen, J. W. and Tepley, L. B. (1974). *Colloid Interface Sci.* **49**, 113
Lianos, P. and Zana, R. (1983). *J. Phys. Chem.* **87**, 1289
Linda, P., Stener, A., Cipiciani, A. and Savelli, G. (1983). *J. Chem. Soc. Perkin Trans. 2,* 821
Lindman, B. and Wennerstrom, H. (1980). *Top. Curr. Chem.* **87**, 1
Link, C. M., Jansen, D. K. and Sukenik, C. N. (1980). *J. Am. Chem. Soc.* **102**, 7798
Lissi, E., Abuin, E. B., Sepulveda, L. and Quina, F. H. (1984). *J. Phys. Chem.* **88**, 81
Lufimpadio, N., Nagy, J. B. and Derouane, E. (1984). *In* "Surfactants in Solution" (K. L. Mittal and B. Lindman, eds) Vol. 3, p. 1483. Plenum Press, New York
Mackay, R. A. (1981). *Adv. Colloid Interface Sci.* **15**, 131
Mackay, R. A. (1982). *J. Phys. Chem.* **86**, 4756
Mackay, R. A. and Hermansky, C. (1981). *J. Phys. Chem.* **85**, 739
Malaviya, S. and Katiyar, S. S. (1984). *Z. Phys. Chem.* **265**, 26
Martin, C. A., McCrann, P. M., Angelos, G. H. and Jaeger, D. A. (1982). *Tetrahedron Lett.* 4651
Martinek, K., Yatsimirski, A. K., Levashov A. V. and Berezin, I. V. (1977). *In* "Micellization, Solubilization and Microemulsions" (K. L. Mittal, ed.) Vol. 2, p. 489. Plenum Press, New York
Matsui, Y. and Orchin, M. (1983). *Organometallic Chem.* **244**, 369
Matsumoto, Y. and Ueoka, R. (1983). *Bull. Chem. Soc. Jpn.* **56**, 3370
Mengelsberg, I., Langhals, H. and Rüchardt, C. (1980). *Chem. Ber.* **113**, 2424
Menger, F. M. (1979a). *Acc. Chem. Res.* **12**, 111
Menger, F. M. (1979b). *Pure Appl. Chem.* **51**, 999
Menger, F. M. and Whitesell, L. G. (1985). *J. Am. Chem. Soc.* **107**, 707
Menger, F. M. and Portnoy, C. E. (1967). *J. Am. Chem. Soc.* **89**, 4698
Menger, F. M. and Donohue, J. A. (1973). *J. Am. Chem. Soc.* **95**, 432
Menger, F. M. and Jerkunica, J. M. (1979). *J. Am. Chem. Soc.* **101**, 1896
Menger, F. M. and Venkatasubban, K. S. (1976). *J. Org. Chem.* **41**, 1868
Menger, F. M., Rhee, J. V. and Rhee, H. K. (1975). *J. Org. Chem.* **40**, 3803
Menger, F. M., Yoshinaga, H., Venkatasubban, K. S. and Das, A. R. (1981). *J. Org. Chem.* **46**, 415
Meyer, C. and Viout, P. (1981). *Tetrahedron,* **37**, 2269
Mille, M. and Vanderkooi, G. (1977). *J. Colloid Interface Sci.* **59**, 211
Miller, J. (1968). "Aromatic Nucleophilic Substitution". Elsevier, Amsterdam
Minch, M. J., Giaccio, M. and Wolff, R. (1975). *J. Am. Chem. Soc.* **97**, 3766
Minero, C., Pramauro, E., Pelizzetti, E. and Misel, D. (1983) *J. Phys. Chem.* **87**, 399
Miola, L., Abarkerli, R. B., Ginani, M. F., Toscano, V. G. and Quina, F. H. (1983). *J. Phys. Chem.* **87**, 4417
Missel, P. J., Mazer, N. A., Benedek, G. B. and Young, C. Y. (1980). *J. Phys. Chem.* **84**, 1044.
Mizutani, T. and Whitten, D. G. (1985). *J. Am. Chem. Soc.* **107**, 3621

Mortimer, J. (1982). "Clinging to the Wreckage", Preface. Weidenfeld and Nicolson, London
Moss, R. A. and Bizzigotti, G. O. (1981). *J. Am. Chem. Soc.* **103**, 65
Moss, R. A. and Dix, F. M. (1981). *J. Org. Chem.* **46**, 3029
Moss, R. A. and Hui, Y. (1983). *Tetrahedron Lett.* **24**, 1961
Moss, R. A. and Ihara, Y. (1983). *J. Org. Chem.* **48**, 588
Moss, R. A. and Schreck, R. P. (1983). *J. Am. Chem. Soc.* **105**, 6767
Moss, R. A. and Shin, J.-S. (1983). *J. Chem. Soc. Chem. Commun.* 1027
Moss, R. A. and Sunshine, W. L. (1974). *J. Org. Chem.* **39**, 1083
Moss, R. A., Talkowski, C. J., Reger, D. W. and Powell, C. E. (1973). *J. Am. Chem. Soc.* **95**, 5215
Moss, R. A., Nahas, R. C., Ramaswami, S. and Sanders, W. J. (1975). *Tetrahedron Lett.* 3379
Moss, R. A., Nahas, R. C. and Ramaswami, S. (1977). *J. Am. Chem. Soc.* **99**, 627
Moss, R. A., Dix, F. M. and Romsted, L. S. (1982a). *J. Am. Chem. Soc.* **104**, 5048
Moss, R. A., Ihara, Y. and Bizzigotti, G. O. (1982b) *J. Am. Chem. Soc.* **104,** 7476
Moss, R. A., Taguchi, T. and Bizzigotti, G. O. (1982c). *Tetrahedron Lett.* **23**, 1985
Moss, R. A., Alwis, K. W. and Bizzigotti, G. O. (1983). *J. Am. Chem. Soc.* **105**, 681
Moss, R. A., Chiang, Y.-C. P. and Hui, Y. (1984a). *J. Am. Chem. Soc.* **106**, 7506
Moss, R. A., Swarup S., Hendrickson, T. F. and Hui, Y. (1984b). *Tetrahedron Lett.* 4079
Moss, R. A., Kim, K. Y. and Swarup, S. (1986). *J. Am. Chem. Soc.* **108**, 788
Motsavage, V. A. and Kostenbauder, H. B. (1963). *J. Colloid Sci.* **18**, 603
Mukerjee, P. and Mysels, K. J. (1970). "Critical Micelle Concentrations of Aqueous Surfactant Systems". National Bureau of Standards, Washington, D.C.
Mukerjee, P., Cardinal, J. R. and Desai, N. R. (1977). *In* "Micellization, Solubilization and Microemulsions" (K. L. Mittal, ed.) Vol. 1, p. 241. Plenum Press, New York.
Muller, N. and Simsohn, H. (1971). *J. Phys. Chem.* **75**, 942
Murakami, Y., Nakano, A., Yoshimatsu, A. and Fukuya, K. (1981a). *J. Am. Chem. Soc.* **103**, 728
Murakami, Y., Nakano, A., Yoshimatsu, A. and Matsumoto, K. (1981b). *J. Am. Chem. Soc.* **103**, 2750
Muthuramu, K. and Ramamurthy, V. (1982). *J. Org. Chem.* **47**, 3976
Muthuramu, K., Ramnath, N. and Ramamurthy, V. (1983). *J. Org. Chem.* **48**, 1872
Nickles, J. A. and Sukenik, C. N. (1982). *Tetrahedron Lett.* 4211
Ninham, B. W., Evans, D. F. and Wei, G. J. (1983). *J. Phys. Chem.* **87**, 4538
Nome, F., Rubira, A. F., Franco, C. and Ionescu, L. G. (1982). *J. Phys. Chem.* **86**, 1881
O'Connor, C. J. and Porter, A. J. (1981). *Aust. J. Chem.* **34**, 1603
O'Connor, C. J. and Ramage, R. E. (1980). *Aust. J. Chem.* **33**, 757, 771, 779
O'Connor, C. J., Lomax, T. D. and Ramage, R. E. (1982). *In* "Solution Behavior of Surfactants" (K. L. Mittal and E. J. Fendler eds) Vol. 2, p. 803. Plenum Press, New York.
O'Connor, C. J., Lomax, T. D. and Ramage, R. E. (1984). *Adv. Colloid Interface Sci.* **20**, 21
Ohkubo, K., Sugahara, K., Yoshinaga, K. and Ueoka, R. (1980). *J. Chem. Soc. Chem. Commun.* 637
Ohkubo, R., Matsumoto, N. and Ohta, H. (1982). *J. Chem. Soc. Chem. Commun.* 738
Okahata, Y., Ando, R. and Kunitake, T. (1977). *J. Am. Chem. Soc.* **99**, 3067

Okahata, Y., Ihara, H. and Kunitake, T. (1981). *Bull. Chem. Soc. Jpn.* **54**, 2072
Okamoto, K., Kinoshita, T. and Yoneda, H. (1975). *J. Chem. Soc. Chem. Commun.* 922
Ono, S., Shosenji, H. and Yamada, K. (1981), *Tetrahedron Lett.* 2391
Onyivruka, S. O., Suckling, C. J. and Wilson, A. A. (1983). *J. Chem. Soc. Perkin Trans. 2.* 1103
Pearson, R. G. (1966). *Science,* **151**, 172
Pearson, R. G. and Songstad, J. (1967). *J. Am. Chem. Soc.* **89,** 1827
Pelizzetti, E. and Pramauro, E. (1984). *J. Phys. Chem.* **88**, 990.
Pelizzetti, E., Pramauro, E., Meisel, D. and Borgarello, E. (1984). *In* "Surfactants in Solution" (K. L. Mittal and B. Lindman, eds) Vol. 2, p. 1159. Plenum Press, New York
Pillersdorf, A. and Katzhendler, J. (1979). *Israel J. Chem.* **18**, 330
Piszkiewicz, D. (1977). *J. Am. Chem. Soc.* **99**, 1550, 7695
Politi, M. J. and Fendler, J. H. (1984). *J. Am. Chem. Soc.* **106**, 265
Pramauro, E., Pelizzetti, E., Diekmann, S. and Frahm, J. (1982). *Inorg. Chem.* **21**, 2432
Quina, F. H. and Chaimovich, H. (1979). *J. Phys. Chem.* **83**, 1844
Quina, F. H., Politi, M. J., Cuccovia, I. M., Baumgarten, E., Martins-Franchetti, S. M. and Chaimovich, H. (1980). *J. Phys. Chem.* **84**, 361
Quina, F. H., Politi, M. J., Cuccovia, I. M., Martins-Franchetti, S. M. and Chaimovich, H. (1982). *In* "Solution Behavior of Surfactants" (K. L. Mittal and E. J. Fendler, eds) Vol. 2, p. 1125. Plenum Press, New York.
Ramnath, N. and Ramamurthy, V. (1984). *J. Org. Chem.* **49**, 2827
Ray, A. (1971). *Nature,* **231**, 313
Ray, A. and Nemethy, G. (1971). *J. Phys. Chem.* **75**, 809
Ramachandran, C., Pyter, R. A. and Mukerjee P. (1982). *J. Phys. Chem.* **86**, 3198
Reddy, I. A. K. and Katiyar, S. S. (1981). *Tetrahedron Lett.* 585
Regen, S. L. (1979). *Angew. Chem. Int. Ed. Eng.* **18**, 421
Ritchie, C. D. and Sawada, M. (1977). *J. Am. Chem. Soc.* **99**, 3754
Rezende, M. C., Rubira, A. F., Franco, C. and Nome, F. (1983). *J. Chem. Soc. Perkin Trans 2,* 1075
Robinson, B. H., Steytler, D. C. and Tack, R. D. (1979). *J. Chem. Soc. Faraday Trans. 1,* **75**, 481
Rodenas, E. and Vera, S. (1985). *J. Phys. Chem.* **89**, 513
Romsted, L. S. (1977). *In* "Micellization, Solubilization and Microemulsions" (K. L. Mittal, ed.) Vol. 2, p. 509. Plenum Press, New York
Romsted, L. S. (1984). *In* "Surfactants in Solution" (K. L. Mittal and B. Lindman, eds) Vol. 2, p. 1015. Plenum Press, New York
Romsted, L. S. (1985). *J. Phys. Chem.* **89**, 5107, 5113
Rupert, L. A. M. and Engberts, J. B. F. N. (1982). *J. Org. Chem.* **47**, 5015
Russell, J. C. and Whitten, D. G. (1982). *J. Am. Chem. Soc.* **104**, 593
Russell, J. C., Whitten, D. G. and Braun, A. M. (1981). *J. Am. Chem. Soc.* **103**, 3219
Savelli, G. and Si, V. (1985). Unpublished results
Scaiano, J. C., Abuin, E. B. and Stewart, L. C. (1982). *J. Am. Chem. Soc.* **104**, 5673
Schiffman, R., Chevion, M., Katzhendler, J., Rav-Acha, Ch. and Sarel, S. (1977a). *J. Org. Chem.* **42,** 856
Schiffman, R., Rav-Acha, Ch., Chevion, M., Katzhendler, J. and Sarel, S. (1977b). *J. Org. Chem.* **42**, 3279
Sepulveda, L. (1974). *J. Colloid Interface Sci.* **46**, 372

Shine, H. J. (1967). "Aromatic Rearrangements", Chap. 3. Elsevier, Amsterdam
Shine, H. J., Zmuda, H., Park, K. H., Kwart, H., Morgan, A. G. and Brechbiel, M. (1982). *J. Am. Chem. Soc.* **104**, 2501
Shinkai, S., Matsuo, K., Harada, A. and Manabe, O. (1982). *J. Chem. Soc. Perkin Trans 2,* 1261
Soto, R., Meyer, G. and Viout, P. (1981). *Tetrahedron,* **37**, 2977
Srivastava, S. K. and Katiyar, S. S. (1980). *Ber. Bunsenges. Phys. Chem.* **84**, 1214
Stadler, E., Zanette, D., Rezende, M. C. and Nome, F. (1984). *J. Phys. Chem.* **88**, 1892
Stigter, D. (1964). *J. Phys. Chem.* **68**, 3603
Stigter, D. (1978). *Prog. Colloid Polymer Sci.* **65**, 45
Sudhölter, E. J. R., van der Langkruis, G. B. and Engberts, J. B. F. N. (1979). *Recl. Trav. Chim. Pay-Bas Belg.* **99**, 73
Sukenik, C. N. and Bergman, R. G. (1976). *J. Am. Chem. Soc.* **98**, 6613
Sunamoto, J., Iwamoto, K., Nagamatsu, S. and Kondo, H. (1983a). *Bull. Chem. Soc. Jpn.* **56**, 2469
Sunamoto, J., Kondo, H., Kibuchi, J., Yoshinaga, H. and Takei, S. (1983b) *J. Org. Chem.* **48**, 2423
Sutter, J. K. and Sukenik, C. N. (1982). *J. Org. Chem.* **47**, 4174
Talmon, Y., Evans, D. F. and Ninham, B. W. (1983). *Science.* **221**, 1047
Tanford, C. (1978). *Science* **200**, 1012
Tanford, C. (1980). "The Hydrophobic Effect". 2nd edn. Wiley-Interscience, New York
Taniguchi, Y. and Iguchi, A. (1983), *J. Am. Chem. Soc.* **105**, 6782
Thomas, J. K. (1984). "Chemistry of Excitation at Interfaces". A.C.S. Monograph 184. Washington, D.C.
Thomson, A. (1970). *J. Chem. Soc. B.* 1198
Tonellato, U. (1977). *J. Chem. Soc. Perkin Trans. 2,* 822
Tonellato, U. (1979). *In* "Solution Chemistry of Surfactants" (K. L. Mittal, ed.) Vol. 2, p. 541. Plenum Press, New York
Trainor, G. L. and Breslow, R. (1981). *J. Am. Chem. Soc.* **103,** 154
Tricot, Y.-M. and Fendler, J. H. (1984). *J. Am. Chem. Soc.* **106**, 7359
Turro, N. J. and Cherry, W. R. (1978). *J. Am. Chem. Soc.* **100**, 7431
Turro, N. J. and Kraeutler, B. (1980). *Acc. Chem. Res.* **13**, 369
Turro, N. J. and Weed, G. C. (1983). *J. Am. Chem. Soc.* **105,** 186
Ueoka, R. and Matsumoto, Y. (1984). *J. Org. Chem.* **49**, 3774
Ueoka, R. and Murakami, Y. (1983). *J. Chem. Soc. Perkin Trans. 2,* 219
Ueoka, R., Matsumoto, Y. and Ihara, Y. (1984). *Chem. Lett.* 1807
Ueoka, R., Moss, R. A., Swarup, S., Matsumoto, Y., Strauss, G. and Murakami, Y. (1985). *J. Am. Chem. Soc.* **107**, 2185
van der Langkruis, G. B. and Engberts, J. B. F. N. (1984). *J. Org. Chem.* **49**, 4152
Visser, H. G. J., Nolte, R. J. M. and Drenth, W. (1983). *Recl. Trav. Chim. Pays-Bas Belg.* **102**, 417
Visser, H. G. J., Nolte, R. J. M. and Drenth, W. (1985) *Macromolecules,* in press
Weber, W. P. and Gokel, G. W. (1977). "Phase Transfer Catalysis in Organic Synthesis". Springer, Heidelberg
Wennerström, H. and Lindman, B. (1979). *J. Phys. Chem.* **83,** 2931
Whitten, D. G., Bonilha, J. B. S., Schauze, K. S. and Winkle, J. R. (1984). *In* "Surfactants in Solution" (K. L. Mittal and B. Lindman eds) Vol. 1, p. 585. Plenum Press, New York

Winkle, J. R., Worsham, P. R., Schanze, K. S. and Whitten, D. G. (1983). *J. Am. Chem. Soc.* **105**, 3951
Wong, M. P. and Connors, K. A. (1982). *J. Pharm. Sci.* **72**, 146
Yamada, K., Shosenji, H., Otsubo, Y. and Ono, S. (1980). *Tetrahedron Lett.* **21**, 2649
Zana, R. (1980). *J. Colloid Interface Sci.* **78**, 330
Zana, R., Yiv, S., Strazielle, C. and Lianos, P. (1981). *J. Colloid Interface Sci.* **80**, 208

Notes Added in Proof

1. Rate constants of bimolecular, micelle-assisted, reactions typically go through maxima with increasing concentration of inert surfactant (Section 3). But a second rate maximum is observed in very dilute cationic surfactant for aromatic nucleophilic substitution on hydrophobic substrates. This maximum seems to be related to interactions between planar aromatic molecules and monomeric surfactant or submicellar aggregates. These second maxima are not observed with nonplanar substrates, even such hydrophobic compounds as *p*-nitrophenyl diphenyl phosphate (Bacaloglu, R. 1986, unpublished results).

2. Interactions between cationic micelles and uni- and divalent anions have been treated quantitatively by solving the Poisson–Boltzmann equation in spherical symmetry and considering both Coulombic and specific attractive forces. Predicted rate-surfactant profiles are similar to those based on the ion-exchange and mass-action models (Section 3), but fit the data better for reactions in solutions containing divalent anions (Bunton, C. A. and Moffatt, J. R. (1985) *J. Phys. Chem.* 1985, **89**, 4166; 1986, **90**, 538).

Structure and Reactivity of Carbenes having Aryl Substituents

GARY B. SCHUSTER

Department of Chemistry, University of Illinois, Urbana, Illinois U.S.A.

1 Introduction 311
2 Orbitals and energetics of carbenes 312
3 The menagerie of aromatic carbenes 316
4 Experimental investigation of the chemical and physical properties of carbenes 320
 Low temperature epr and optical spectroscopy 321
 Room temperature spectroscopy on a short timescale 324
 Reactivity as a probe of spin state 326
5 The structure and reactivity of aromatic carbenes 331
 Mesitylbora-anthrylidene (BA) 331
 9-Xanthylidene (XA) 338
 Fluorenylidene (FL) 341
 3,6-Dimethoxyfluorenylidene (DMFL) 344
 2,7-Dichlorofluorenylidene (DCFL) 346
 2,3-Benzofluorenylidene (BFL) 347
 Anthronylidene (AN) 348
 Diphenylmethylene (DPM), and its derivatives, dimesitylcarbene (DMC), dibenzocycloheptadienylidene (DCH), and sila-anthrylidene (SA) 349
 Phenylmethylene (PM), α-naphthylmethylene (ANM), and β-naphthylmethylene (BNM) 351
6 Understanding the properties of aryl-substituted carbenes 352
7 Conclusions 356
Acknowledgements 357
References 357

1 Introduction

One of the central objectives of the study of chemistry is sufficient knowledge to permit the forecast of the chemical and physical properties of a substance directly and reliably from its structure. This objective has been achieved to a remarkable degree for the host of isolable compounds that have been prepared and investigated. These findings have been codified in generalizing

ADVANCES IN PHYSICAL ORGANIC CHEMISTRY
VOLUME 22 ISBN 0 12 033522 0

theories which, in turn, permit prediction of the properties of substances not yet prepared or impossible to isolate. Reactive intermediates, by their very definition, fall into the category of unisolable substances. The ephemeral existence of these materials generally permits only indirect investigation of their chemical and physical properties. However, it is the properties of these reactive intermediates which often control the outcome of chemical reactions. It is for this reason that, even though difficult and often frustrating, the study of reactive intermediates has occupied a central position in physical organic chemistry.

The subject of this chapter is carbenes with aryl substituents (aromatic carbenes). These materials are short-lived reactive intermediates in which the normal tetravalency of carbon is reduced by two. Carbenes have been the object of speculation and investigation for more than 80 years. Nevertheless, there still is considerable uncertainty about their chemical and physical properties. In the last five years the pace of research in carbene chemistry has quickened. This is a consequence of the development of high-speed pulsed lasers that permit, for the first time, direct observation of carbenes under the conditions in which they react. This research has provided new information on the effect of structure on the chemical and physical properties of carbenes.

The presentation of carbene chemistry in this chapter is by no means complete or exhaustive. The subject is simply too large to be covered in detail within the allotted space. However, there are available comprehensive monographs and other more sharply focused reviews covering features of carbene chemistry neglected here (Closs, 1968; Bethell, 1969; Kirmse, 1971; Moss and Jones, Jr., 1973; Jones, Jr. and Moss, 1975, 1978, 1981, 1985; Abramovitch, 1980; Griller *et al.*, 1984a; Eisenthal *et al.*, 1984; Wentrup, 1984).

This report begins with a brief review of the electronic and structural features that underlie all of carbene chemistry. Next, we introduce the set of related aromatic carbenes that are the basis for our dissection of the effects of structure on carbene properties. The chemical and spectroscopic techniques and procedures used to probe these carbenes are described and explained briefly in the succeeding section. Then, the results of the application of these probes to the chosen carbenes are presented. Finally, the revealed relation of a carbene's structure to its chemical and physical properties is placed within the predictive framework of molecular orbital theory. Our objective in this report is to present sufficient information to permit us to forecast the properties of an aromatic carbene directly and reliably from its structure.

2 Orbitals and energetics of carbenes

The simplest carbene is methylene (H_2C:). Despite its simplicity, there has been significant difficulty in defining the properties of this compound.

However, there seems to be general current agreement about its most important attributes. Our attempt to understand the properties of aromatic carbenes begins with an analysis of methylene.

A defining feature for carbenes is the existence of two non-bonding orbitals on one carbon atom. There are two electrons to distribute among these two orbitals and their placement defines the electronic state of the molecule. A simple representation showing the electron occupancy of the non-bonding orbitals is displayed in Fig. 1. The orbital perpendicular to the

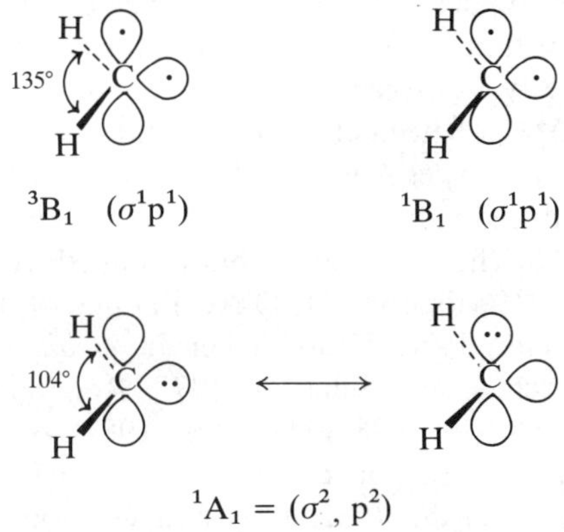

$^3B_1$ (σ^1p^1) $^1B_1$ (σ^1p^1)

$^1A_1 = (\sigma^2, p^2)$

FIG. 1 Electronic configuration of the lowest states of methylene

plane defined by the three atoms is designated as "p" while that parallel to this plane is call "σ" (Hoffman *et al.*, 1968). The four lowest energy configurations of methylene have electronic occupancies described as σ^1p^1, σ^2, or p^2. The electron spins in the σ^1p^1 configuration may be paired, a singlet, or parallel to form a triplet. Of course, the σ^2 and p^2 configurations must be electron-paired singlets.

Methylene has C_{2v} symmetry and the states resulting from these configurations are therein designated as $^3B_1$ (σ^1p^1 triplet), $^1B_1$ (σ^1p^1 singlet) and $^1A_1$ (both σ^2 and p^2). It is important to recognize that the $^1A_1$ state is best thought of as a hybrid formed from the mixing of the σ^2 and p^2 configurations (Harrison, 1971). This state thus responds to changes in carbene structure that involve either the p or σ non-bonded orbitals.

Early spectroscopic experiments on the structure of methylene were interpreted to show that the $^3B_1$ state was linear or nearly linear (Herzberg, 1961). However, theoretical calculations and eventually observation of the epr spectrum of triplet methylene (Bernheim *et al.*, 1970; Wasserman *et al.*, 1970) led to a reinterpretation of the earlier experiments (Herzberg and Johns, 1971). Now there is universal agreement that the $^3B_1$ state of methylene is bent with a bond angle (θ) of approximately 135° (see Fig. 1.)

The structure of $^1A_1$ methylene has a more acute bond angle than does the $^3B_1$ triplet. In this case, both experiment (Herzberg, 1961) and theory have been in consistent agreement. Both approaches place this bond angle at 104° (see Fig. 1).

The relative energies of the lowest electronic states of methylene has been a subject of intense interest for many years. All theoretical treatments have predicted that the ground-state of this molecule is the $^3B_1$ triplet. This has been confirmed experimentally. Also, theory has consistently identified the lowest singlet as the $^1A_1$ state. The energy of the $^1B_1$ singlet is calculated to be much higher than that of $^1A_1$ methylene (Bauschlicher Jr., 1980). Until quite recently, there was no general agreement on the magnitude of the energy difference (ΔG_{ST}) beteen the ground state triplet ($^3B_1$) and lowest singlet ($^1A_1$) state. Early experiments (Halberstadt and McNesby, 1967; Carr Jr. *et al.*, 1970) produced values in the range of 1–2 kcal mol^{-1}. Computational estimates of ΔG_{ST} have wandered, but recent efforts predict this gap to be *ca* 10 kcal mol^{-1} (Harrison, 1971; O'Neill *et al.*, 1971; Hay *et al.*, 1972; Meadows and Schaefer, 1976; Bauschlicher Jr. *et al.*, 1977; Harding and Goddard, 1977; Lucchese and Schaefer, 1977; Roos and Sieghahn, 1977; Bauschlicher Jr. and Shavitt, 1978; Feller *et al.*, 1982). All recent experiments have also converged on a value of *ca* 10 kcal mol^{-1} for ΔG_{ST} of methylene in the gas phase (Hayden *et al.*, 1982; Frey and Gordon, 1975; Simons and Curry, 1976; Feldmann *et al.*, 1978; Lengel and Zare, 1978; Leopold *et al.*, 1984).

The magnitude of ΔG_{ST} is expected to be sensitive to the carbene–carbon bond angle. This is clearly revealed by calculation and is displayed on Fig. 2. Contraction of this angle to values below *ca* 130° is expected to reduce the magnitude of ΔG_{ST}. Indeed, the calculations predict that the energy of $^1A_1$ methylene will drop below that of the $^3B_1$ state for carbenes with bond angles less than about 90°. These carbenes will thus have singlet ground states ($\Delta G_{ST} < 0$). This phenomenon may have been observed experimentally. Cyclopropenylidene prepared photochemically in an argon matrix at 10 K has a singlet ground state (Reisenauer *et al.*, 1984). This is no doubt a consequence, in part, of the small bond angle for this compound. However, the bond angle effect cannot be easily separated from electronic effects which, in this case, also tend to lower the energy of the single state relative to that of the triplet state of the carbene.

The properties of carbenes are also expected to depend very greatly on the electronic characteristics of substituents bound to the divalent carbon. For example, many carbenes with heteroatomic elements attached directly to the central carbon are calculated to have single ground states (Mueller *et al.*, 1981). The early, pioneering work on the stereochemistry of the reaction of carbenes with olefins was done with dibromocarbene (Skell and Garner,

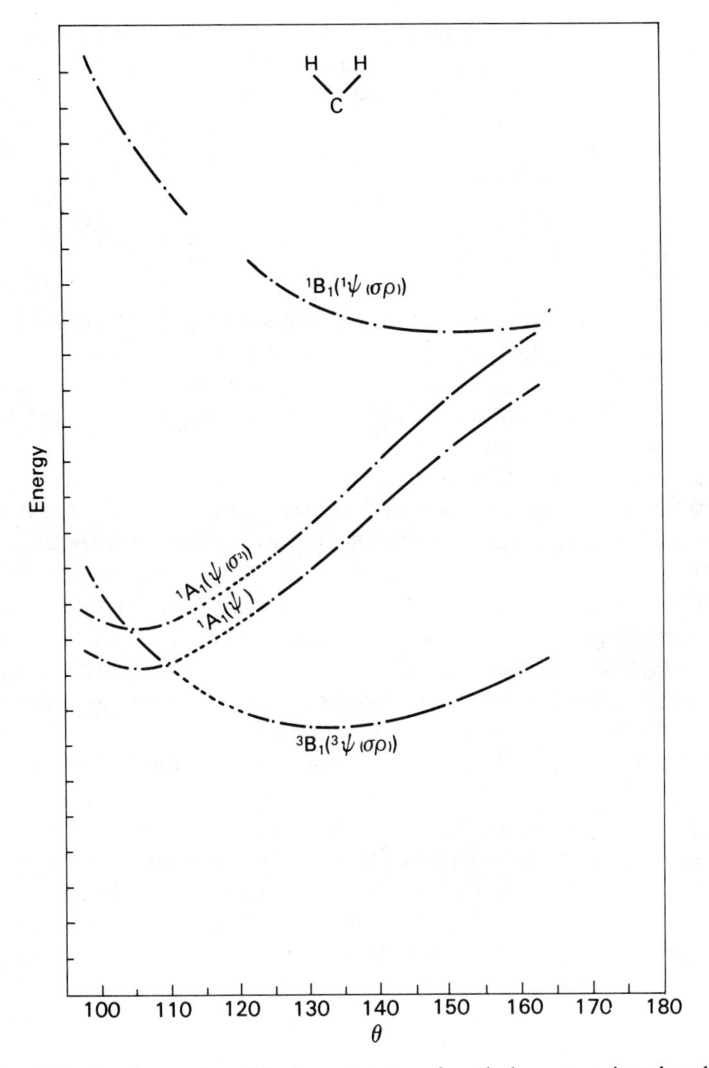

FIG. 2 Calculated energies of the lowest states of methylene at various bond angles. (Harrison, 1971, reproduced with permission)

1956). These findings, and extensive subsequent work, are consistent with a singlet ground state for this species. The chemical reactivity of these carbenes towards olefins can be related empirically but quantitatively to the electronic properties of the substituents (Moss *et al.*, 1977; Moss, 1980). An extreme example is dimethoxycarbene which does not exhibit at all the electrophilic properties normally associated with the vacant non-bonding orbital of a singlet carbene (Lemal *et al.*, 1966). These findings are easily understood by consideration of resonance structures where the singlet carbene is portrayed as an ylid (1), and the octet rule is satisfied for all atoms. From the viewpoint of molecular orbital theory, π-donation by the heteroatom changes the

$$\ddot{X}-\ddot{C}(-Y) \longleftrightarrow \overset{+}{X}=\overset{-}{C}(-Y) \qquad (1)$$

carbene ylid

nature of the lowest unoccupied orbital from one essentially localized on the carbene–carbon to a π* orbital akin to that of a carbonyl group, for example (Mueller *et al.*, 1981). In this circumstance, the energy difference between the highest occupied and lowest unoccupied orbitals is great enough to fashion a singlet ground state.

Alkyl, alkenyl and aryl substituents on the carbene–carbon have a more subtle influence on the chemical and physical properties of these intermediates. The possible range of these effects was outlined from a theoretical viewpoint some time ago (Hoffman *et al.*, 1968). Fundamentally, the conclusion from this analysis is that these substituents are capable of swaying the relative energies of the lowest carbene states. The effects of these groups is readily dissected into steric, primarily bond angle, and electronic components. This interpretation predicts that for some of these carbenes the normal ordering of triplet state below singlet can be reversed. This prediction has been extraordinarily difficult to examine experimentally. Typically, the chemical properties of carbenes are categorized by placing reactions within spin-specific groups. Thus, there are a number of reactions normally associated with singlet or with triplet carbenes. Many aromatic carbenes exhibit reactions commonly associated with both the singlet and triplet spin states. It is seldom possible to judge the characteristics of these carbenes simply from the nature of the chemical reactions they undergo. This circumstance requires a more careful examination of the physical and chemical properties of the individual carbenes if these are to be reliably linked with their structures. This examination has been underway for quite some time, but only recently has an adequate range of compounds been probed in sufficient detail to reveal clearly a relationship between structure and reactivity.

3 The menagerie of aromatic carbenes

Carbenes are commonly generated by irradiation or pyrolysis of an appropriate diazo-compound (2). Apart from differences readily traced to the change in temperature, the chemical properties of the carbenes formed from photolysis and from thermolysis are usually quite similar. These observations show that the irradiation of these diazo-compounds leads eventually to the same reactive intermediate that is formed from their thermolysis. Thus, although possible in principle, there is little evidence for the participation of the electronically excited diazo-compound in this chemistry. Similarly, although detected under special circumstances (Ahmed and Shevlin, 1983; Sitzmann *et al.*, 1983), chemical reactions of carbenes in other than the lowest state of a given multiplicity (i.e. the $^1A_1$ and $^3B_1$ states) also do not normally contribute. With these remarks in mind, the photochemistry of the diazo-compounds described below is interpreted simply on the basis of the denitrogenation reaction shown in (2).

$$\mathrm{Ar_2C{=}N_2} \xrightarrow[\text{or }\Delta]{h\nu} \mathrm{Ar_2C{:}} + \mathrm{N_2} \tag{2}$$

The diazo-compounds and corresponding aromatic carbenes that form the basis for our dissection of structure and reactivity are shown in Table 1. The carbenes in this group are carefully chosen so that the variation in structure is systematic; the theory identifies the carbene bond angle and certain electronic factors as controlling chemical and physical properties, and as far as possible, these two features are varied independently of each other for these carbenes. Table 2 lists some other aromatic carbenes that have been studied. In general, the structures of these carbenes are not simply related to each other. Nevertheless, the principles uncovered by analysis of the compounds of Table 1 can be readily extended to those of Table 2.

The carbenes listed in Table 1 are all cyclic aromatic structures which incorporate the carbene–carbon in a presumably planar ring of five or six atoms. The C_{2v} symmetry of methylene is assumed to be maintained (not strictly for BFL), and the lowest singlet and triplet states of these carbenes are designated $^1A_1$ and $^3B_1$. The carbene bond angle of those examples which contain a five-membered ring (FL, DCLF, BFL, DMFL) is expected to be constant at a value of *ca* 110° for both the triplet and singlet spin multiplicities. This notion is confirmed, in part, by the epr spectra observed for the triplet states of these compounds. For this series, the diversity of chemical behaviour, which is quite substantial, can thus be associated exclusively with changes in electronic circumstance. Similarly, the remaining carbenes in Table 1 (BA, AN, XA) contain the divalent carbon in a six-membered ring with an assumed bond angle of *ca* 120°. In part, the variation in chemical and physical properties between the five- and six-membered ring carbenes may be due to this change in bond angle. But for a given ring size, this diversity must be associated primarily with a change in electronic factors.

TABLE 1
Primary list of carbenes and their precursors

Diazo-compound (abbreviation)	Carbene (abbreviation)
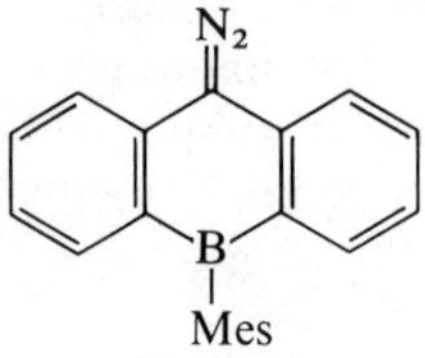 N_2 B Mes (DABA)	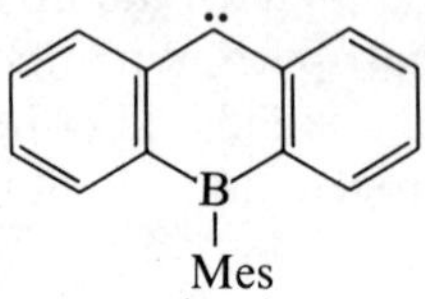 B Mes (BA)
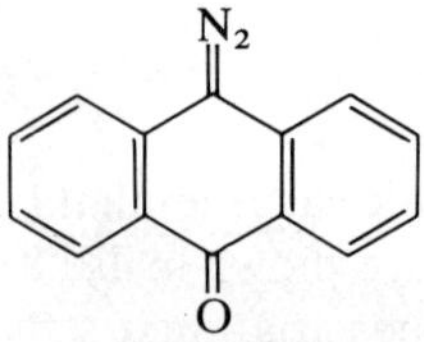 N_2 O (DAAN)	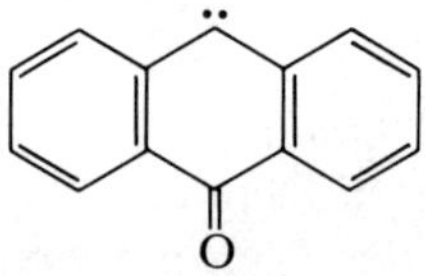 O (AN)
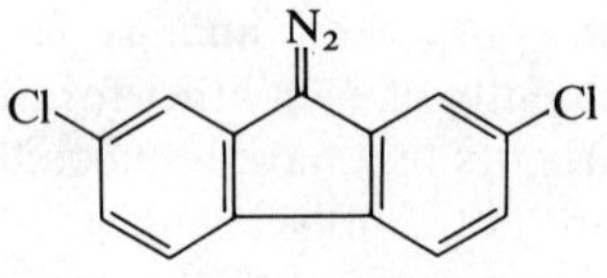 N_2 Cl Cl (DCDAF)	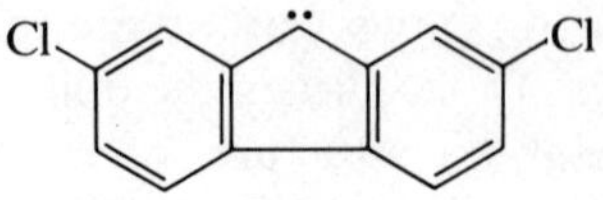 Cl Cl (DCFL)
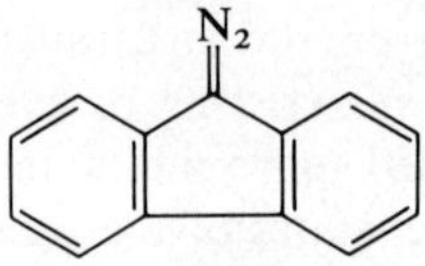 N_2 (DAF)	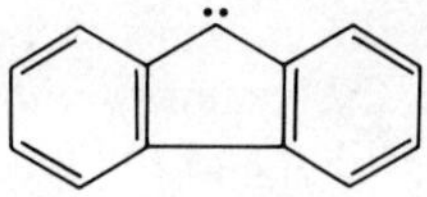 (FL)
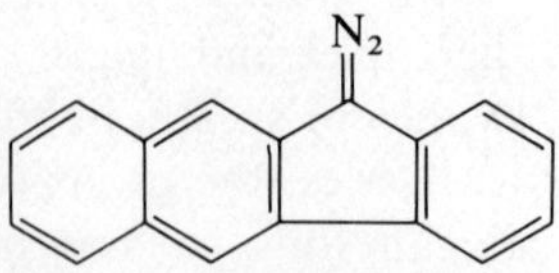 N_2 (DABF)	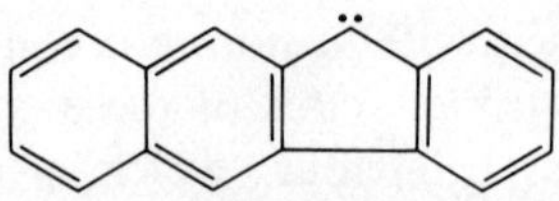 (BFL)
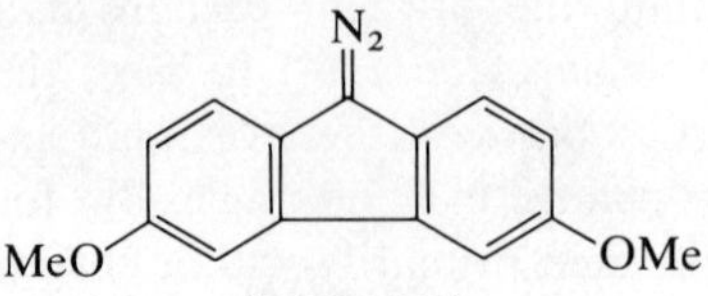 N_2 MeO OMe (DMDAF)	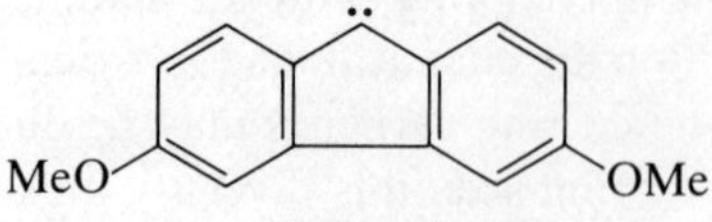 MeO OMe (DMFL)

TABLE 1 *Continued*

Diazo-compound (abbreviation)	Carbene (abbreviation)
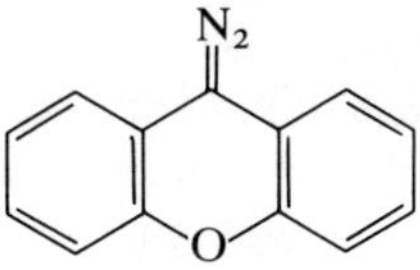 (DAX)	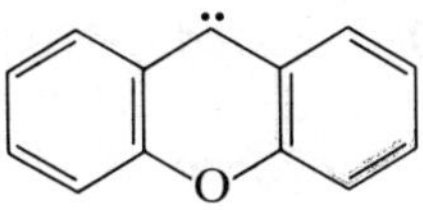 (XA)

TABLE 2

Supplementary list of carbenes

Carbene (abbreviation)	Carbene (abbreviation)
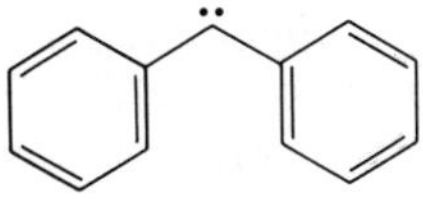 (DPM)	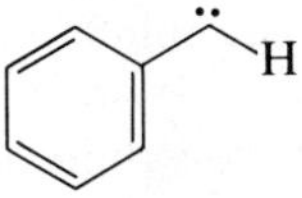 (PM)
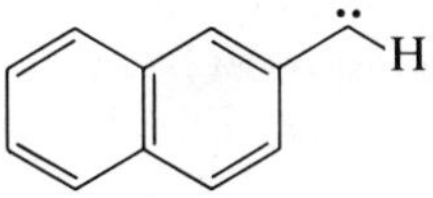 (BNM)	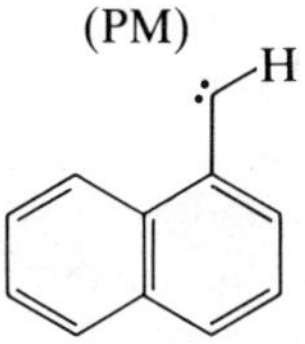 (ANM)
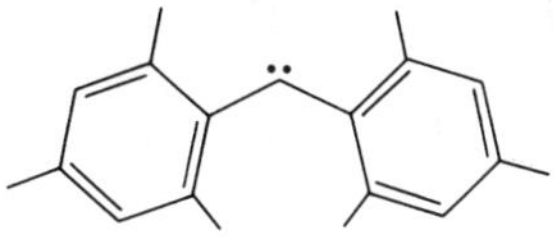 (DMC)	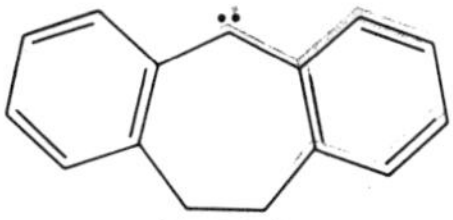 (DCH)
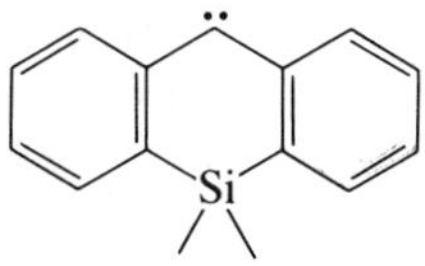 (SA)	

The electronic diversity of the carbenes listed in Table 1 is easily recognized. First, the number of electrons contained within the aromatic π-system changes from $(4n)$ to $(4n+2)$ within this group. This feature will undoubtedly affect the relative energies of the p^2 configuration of the $^1A_1$ states. For example, BA in this configuration contains 6π electrons in the ring containing the carbene–carbon and should experience stabilization. On the other hand, the analogous p^2 configuration of XA contains 8π electrons and should be destabilized. An equivalent way of stating this observation is to recognize that the symmetry of the highest occupied molecular orbital changes on going from BA to XA. The alternation in symmetry affects the energy of the two configurations that contribute importantly to the $^1A_1$ state. A second electronic factor affecting the carbenes listed in Table 1 is readily traced to the effect of substituents. For example, if FL is viewed as the midpoint, then DMFL is perturbed by electron-donor substitution and DCFL by (inductive) electron withdrawal. The consequences of these effects may be understood simply by consideration of the predictable response of the frontier molecular orbitals to these substituents. Similarly, BFL can be viewed as a modified FL and the perturbation of this system also falls within a recognizable pattern.

Of course the structural changes represented in Table 1 are much more complex than the simple bond angle plus electronic effect analysis reveals. For example, solvation of the carbene may depend on the details of its structure, and solvation undoubtedly influences chemical and physical properties (Langan *et al.*, 1984). Nonetheless, it is possible to develop a good grasp of the most important properties of aromatic carbenes from the simple considerations described above. Before we proceed to examine these relationships in more detail, the carbene properties of interest must be identified, and the experimental procedures available for measurement of these properties must be critically considered.

4 Experimental investigation of the chemical and physical properties of carbenes

An assortment of spectroscopic, kinetic, and chemical techniques and procedures have been developed to permit assessment of the properties of carbenes. In general, each of these reveals a different aspect of the carbene. A complete picture, accurately portraying these intermediates, results only when several of these experiments can be evaluated together. In this section a brief description of the most useful of these analytical procedures is presented. The objective is to provide a basis for the interpretation of the findings that is presented later.

LOW TEMPERATURE EPR AND OPTICAL SPECTROSCOPY

Carbenes are such highly reactive intermediates that their direct observation requires extraordinary efforts. One set of conditions that has proved quite valuable is low temperature isolation. Carbenes can be generated by irradiation of an appropriate precursor within a glass or more ordered inert matrix at very low temperatures. The low temperature of the experiment stops or slows reactions of the carbene with the matrix material. Also, the rigidity of the medium prevents diffusion and the dimerization of the carbene is stopped. Many carbenes can be stabilized at the boiling point of nitrogen (77 K); others require liquid helium temperatures (4 K).

Carbenes with triplet ground states are, of course, paramagnetic. Thus they exhibit characteristic epr spectra which can be used to confirm their presence, to provide some data on their structure, and, at least in principle, to reveal ΔG_{ST} in favorable cases. The first epr spectrum of a randomly oriented triplet carbene was observed for DPM (Murray *et al.*, 1962). In subsequent years spectra of many carbenes have been recorded (Table 3). These spectra are commonly interpreted by fitting the results to a spin Hamiltonian **H** (3), to give the zero field parameters D and E (Stevens, 1952). In the usual

TABLE 3

Epr spectra of some aromatic carbenes

Carbene	$\lvert D\rvert/\mathrm{cm}^{-1}$	$\lvert E\rvert/\mathrm{cm}^{-1}$	$\lvert D/E\rvert$	Reference
BA	0.360	0.0156	23	*a*
AN	0.365	0.0177	21	*b*
DCFL	0.4246	0.0268	16	*c*
FL	0.408	0.0283	14	*d*
BFL	0.39	0.030	13	*e*
DMFL	none observed		—	*f*
XA	none observed		—	*g*
DPM	0.401	0.018	22	*h*
PM	0.510	0.0249	20	*d*
BNM	0.456 (*anti*)	0.0202 (*anti*)	22	*d*
ANM	0.471 (*anti*)	0.0209 (*anti*)	23	*d*
DMC	0.356	0.0125	28	*i*
DCH	0.393	0.0170	23	*j*
SA	0.391	0.0215	18	*k*

[a] Lapin *et al.*, 1984
[b] Devolder *et al.*, 1972
[c] Rak *et al.*, 1985
[d] Trozzolo and Wasserman, 1975
[e] Grasse *et al.*, 1985
[f] Chuang *et al.*, 1985
[g] Lapin and Schuster, 1985
[h] Trozzolo *et al.*, 1962
[i] Nazran *et al.*, 1983
[j] Moritani *et al.*, 1967
[k] Sekiguchi *et al.*, 1982

interpretation, the absolute value of D is taken as a measure of the average distance between the interacting spins of the triplet, and that of E is related to the structural deviation from cylindrical symmetry (Trozzolo, 1968). The ratio $|D|/|E|$ is thought to provide an estimate of the magnitude of the carbene–carbon bond angle (Higuchi, 1963). This analysis has been criticized (Hoffman *et al.*, 1968) and special assumptions sometimes are required to interpret the ratio (Wasserman *et al.*, 1964). However, there does appear to be a general relation between this parameter and the structure of the carbene.

$$\mathbf{H} = g\beta H \cdot \mathbf{S} + D\mathbf{S}_z^2 + E(\mathbf{S}_x^2 - \mathbf{S}_y^2) \tag{3}$$

The temperature dependence of the intensity of the epr signal for a triplet carbene can provide some information about the ordering and energetics of the lowest singlet and triplet spin states. Four general cases need to be considered. These are: (*i*) singlet far below triplet, (*ii*) singlet slightly below triplet, (*iii*) triplet slightly below singlet, and (*iv*) triplet far below singlet. For the first case, no epr signal from the triplet carbene will be observed because this state is not populated significantly at the temperatures that can be achieved before irreversible chemical consumption of the carbene occurs. When the energy difference between the singlet and triplet spin state is small (cases (*ii*) and (*iii*)) then an epr spectrum can be observed, and the intensity of magnetization (I) should deviate from the behavior predicted by the Curie Law (I inversely proportional to temperature, T) according to (4) (Poole, Jr., 1967). In fact, execution of this experiment is often ambiguous because the temperatures available are restricted to a narrow range. Thus unless the two spin states are within a few calories per mole of each other, behavior characteristics of case (*iv*) (Curie Law obeyed) is obtained. This is the plight of all of the triplet aromatic carbenes that have been studied.

$$I = \frac{C}{T}[3 + \exp(-\Delta E_{\mathrm{ST}}/RT)]^{-1} \tag{4}$$

The optical spectrum of a triplet aromatic carbene can be recorded at low temperature and generally consists of two or more absorption bands. The position of these bands is often similar to the closely related transitions found in the correpsonding radical (Trozzolo and Gibbons, 1967). For example, DPM has absorptions with maxima near 300 and 465 nm and the diphenylmethyl radical exhibits maxima at 336 and 515 nm (Porter and Strachan, 1958).

Measurement of the optical absorption spectrum of a carbene can be quite complicated. Unlike the epr spectrum, the optical spectrum reveals virtually no structural information. Thus, irradiation of a carbene precursor at low temperature often creates new absorption features, but it is difficult to assign these bands to a particular structure with certainty. To assist in the assignment, experiments can be carried out that narrow the range of possible choices. For example, if the new absorption feature does indeed belong to the carbene, then warming the matrix will cause the disappearance of the feature and the simultaneous appearance of the expected chemical products. Of course, this behavior does not ensure that the spectrum obtained is that of the carbene. It does verify, however, that the carbene was formed, and that a species stable at low temperature and unstable to warming did exist. Similarly, comparison of the optical and epr spectroscopic results obtained under identical conditions can corroborate the simultaneous existence of both the carbene and the absorbing species. This result is particularly meaningful if there are no other odd-electron species (radicals) apparent in the epr spectrum. Nevertheless, assignments made on this basis must be considered less than certain since the epr measurement is completely insensitive to all but paramagnetic species and the optical spectroscopy reports simply the strongest absorber. The more or less certainly assigned absorption maxima for a number of aromatic carbenes are listed in Table 4. These data play an important part in the interpretation of the transient spectroscopic results described below. In turn, these short timescale experiments help to

TABLE 4

Optical spectra of some aromatic carbenes

Carbene	λ_{max}/nm	Reference
BA	359,460,530	*a*
AN	353,520,565	*b*
DCFL	470	*c*
FL	440,470	*d*
BFL	434,485	*e*
DMFL	409,428	*f*
XA	376,385,393	*g*
DPM	300,465	*h*
DMC	330	*i*
DCH	350,500	*j*
SA	343,450,520	*k*

[a] Lapin *et al.*, 1984
[b] Bourlet *et al.*, 1972; Devolder *et al.*, 1974
[c] Rak *et al.*, 1985
[d] Grasse *et al.*, 1983
[e] Grasse *et al.*, 1985
[f] Chuang *et al.*, 1985
[g] Lapin and Schuster, 1985
[h] Gibbons and Trozzolo, 1966
[i] Nazran and Griller, 1984
[j] Moritani *et al.*, 1967
Sugawara *et al.*, 1983a

confirm the spectral assignments made on the basis of the low temperature measurements.

ROOM TEMPERATURE SPECTROSCOPY ON A SHORT TIMESCALE

A second way to overcome the high reactivity of carbenes and so permit their direct observation is to conduct an experiment on a very short timescale. In the past five years this approach has been applied to a number of aromatic carbenes. These experiments rely on the rapid photochemical generation of the carbene with a short pulse of light (the pump beam), and the detection of the optical absorption (or emission) of the carbene with a probe beam. These pump-probe experiments can be performed on timescales ranging from picoseconds to milliseconds. They provide an important opportunity absent from the low temperature experiments, namely, the capability of studying chemical reactions of the carbene under normal conditions. Before proceeding to discuss the application of these techniques to aromatic carbenes, a few details illuminating the nature of the data obtained and the limitations of the experiment need to be introduced.

The pump beam comes from a laser. The necessity of high light intensity in a short time demands this. Exceptions are possible for relatively unreactive intermediates; a flash lamp was used in the first direct detection of a carbene (Closs and Rabinow, 1976), but the availability of modern high-power, pulsed uv-lasers has made this approach obsolete. One requirement then is that the precursor to be irradiated absorb at an available laser frequency. For aromatic carbenes, this is not a restrictive requirement.

The nature of the probe beam and the way the intermediate is detected vary with the timescale of the measurement. For experiments lasting more than a few nanoseconds the probe beam can be generated with a xenon lamp synchronized to the firing of the pump laser. In this circumstance, probe light with frequencies ranging from the ultraviolet through the visible into the infrared are easily available, and the intermediate is detected by its optical absorption. A limitation to this experiment comes from the power of the pump beam. The power of the pump lasers commonly used is sufficient to generate concentrations of intermediates *ca* 1×10^{-4}M in the irradiated zone. Experience has shown that the signal-to-noise level achievable in these experiments will permit accurate measurements of absorbance changes greater than *ca* 0.05. This translates, through Beer's Law, to detectability of intermediates with extinction coefficients greater than *ca* $500\mathrm{M}^{-1}\mathrm{cm}^{-1}$ in spectral regions where the irradiated precursor does not absorb strongly. This is a somewhat restrictive requirement.

The probe beam for picosecond timescale experiments must originate from the same laser as does the pump beam. The time period between creation of the intermediate (arrival of the pump beam at the sample) and its detection (arrival of the probe beam at the sample) is controlled by sending the probe over a slightly longer route to the sample than the pump (1 ps ≡ 0.3 mm). The common origin of the two beams puts severe limitations on the character of the probe. It may be restricted to a single frequency corresponding to the fundamental, or an harmonic, of the laser. On the other hand, it can be a pseudo-continuum created by focusing the beam into an appropriate medium. In the latter case, generally, only visible light can be created, and intermediates that absorb in the ultraviolet cannot be detected. Detection of intermediates on a picosecond timescale usually relies on absorption, and the same extinction-coefficient limitation applies as in the nanosecond experiment. In some favorable cases, intermediates can be detected on a picosecond timescale by their fluorescence. This analysis is sometimes more sensitive, but it is often more difficult to interpret.

The absorption spectrum measured in the typical pump-probe experiment is the difference between the spectrum of the remaining irradiated precursor and the created intermediate(s). Assignment of the transient absorption spectrum typically is done by reference to the low-temperature spectra described above, and (sometimes more certainly) by analysing the chemical behavior of the intermediate. For example, many carbenes are known to react with alcohols to give ethers (see below). If the detected intermediate can be observed to react with an alcohol, then this is taken as additional evidence for its assignment as a carbene.

A powerful consequence of the capacity to detect these intermediates under normal reaction conditions is the ability to measure their rates of reaction. For example, increasing the concentration of alcohol in the presence of a carbene can shorten the carbene lifetime (i.e. increase its rate of reaction). These results can be interpreted to yield the specific rate constant for the reaction of the carbene and alcohol. Of course, carbenes react with reagents other than alcohols, and these rate constants for reaction can be similarly obtained. The kinetic and spectroscopic results can be combined with product analyses to confirm the assignment of the intermediate. For example, many carbenes react with olefins to give cyclopropanes (see below). Thus, formation of a carbene in a solution containing both an alcohol and an olefin will give both ether and cyclopropane. The yields of these products can be measured experimentally as well as predicted from the calculated rate constants for the carbene reaction. If the calculation and the experiment agree, this is additional evidence that the intermediate is properly assigned and that the rates of its consumption correspond with those for the formation of the products. In this way the low temperature and short timescale measurements support each other in the identification of the detected intermediates.

REACTIVITY AS A PROBE OF SPIN STATE

An impressive array of reactions characteristic of carbenes has been disscovered that define the chemical properties of these intermediates. Many of the reactions are believed to originate specifically with the carbene either in its singlet or triplet spin state. In some cases, assignment of spin-specific reactivity is muddled because a carbene shown by epr spectroscopy to be a ground-state triplet may react predominantly from its first-formed singlet state, i.e. reactions of the singlet (inter- or intramolecularly) are faster than is intersystem crossing to the lower energy triplet. One way to resolve this apparent dilemma is by photochemical triplet sensitization.

Direct irradiation or thermolysis of a diazo-compound (2) is believed to generate the carbene initially in its singlet spin state. Triplet sensitization (5) is presumed to give the triplet carbene directly without first forming its singlet state via the triplet diazo-compound. In some cases, careful comparison of the results of direct irradiation experiments with those from triplet sensitization can provide useful information to identify the spin state initiating a reaction.

$$\text{Sens} \xrightarrow{h\nu} \text{Sens}^{*1} \xrightarrow[\text{fast}]{\text{isc}} \text{Sens}^{*3} \xrightarrow{R_2C{=}N_2} {}^{3}R_2C\!: + N_2 \qquad (5)$$

Another source of complexity in the assignment of spin-specific properties of carbenes arises when the energy difference between the lowest states is small. In this circumstance, thermal reactivation and reaction from the upper state can compete with irreversible direct reaction of the ground state. Thus, even though a carbene is formed in the triplet ground state, it may react predominantly from the singlet state which has closely similar but higher energy. One approach, sometimes found to be helpful for identifying this complication, is an examination of the temperature dependence of the reaction (Moss and Dolling, 1971). This often introduces further complexity, however, because the change in temperature affects all aspects of the reaction. For example, at sufficiently low temperature, hydrogen-atom tunneling plays an important role in the reaction of triplet carbenes (Senthilnathan and Platz, 1980; Platz *et al.*, 1982). The introduction of this process masks the effect of temperature on reaction from the upper state.

Despite the complications, and with a few reservations (see DPM below), the chemical properties of singlet and triplet carbenes are generally distinct and separable. For the most part, triplet carbenes behave as biradicals, and the singlets as electrophiles. Of course there are exceptions, but this generalization appears to be true for the aromatic carbenes which are the subject of this report. In the remaining parts of this section we discuss some of the particular reactions used to characterize the carbenes listed in Tables 1 and 2.

Alcohols

The reaction of carbenes with alcohols to form ethers (6) occupies a central and critical position in the network of observations that define the properties

$$\mathrm{Ar\ddot{C}Ar} + \mathrm{ROH} \longrightarrow \mathrm{Ar_2C(H)OR} \qquad (6)$$

of these intermediates. It is generally agreed that this reaction is characteristic of carbenes in their singlet state (Kirmse, 1963; Bethell *et al.*, 1970; Tomioka and Izawa, 1977). Recently, it was proposed that a spin-forbidden, single-step reaction of the triplet with alcohols may also occur (Griller *et al.*, 1984b). However, there are alternative interpretations of the data supporting this provocative suggestion (Zupancic *et al.*, 1985), and this important detail is treated more fully in the discussion of the chemistry of DPM.

The mechanism for the addition of singlet carbenes to alcohols has been studied in some detail (Bethell *et al.*, 1971; Kirmse *et al.*, 1981). By and large, the evidence supports two routes. The first, more common, sequence features initial formation of an ylid. Under some circumstances this reaction is reversible (Zupancic *et al.*, 1985; Liu and Subramanian, 1984; Warner and Chu, 1984). Next, proton transfer, either intramolecularly, which may be slowed by symmetry constraints, or by a pair of intermolecular protonation and deprotonation steps, gives the ether. These reactions are outlined in (7).

$$\mathrm{Ar_2C}{\uparrow\downarrow} + \mathrm{ROH} \rightleftharpoons \mathrm{Ar_2\overset{-}{C}\text{—}\overset{+}{O}(R)H} \xrightarrow{\sim(\mathrm{H^+})} \mathrm{Ar_2C(OR)H} \qquad (7)$$

The second route for ether formation is initiated by protonation of the carbene followed by capture of the cation with alcohol (8). Finally a concerted insertion into the oxygen–hydrogen bond of the alcohol (8) has been considered, but there is no experimental support for this path. The

$$\mathrm{Ar_2C}{\uparrow\downarrow} \xrightarrow[(+\mathrm{H^+})]{\mathrm{ROH}} \underset{\text{cation}}{\mathrm{Ar_2\overset{+}{C}\text{—}H}} \xrightarrow[(-\mathrm{H^+})]{\mathrm{ROH}} \mathrm{Ar_2C(OR)H} \qquad (8)$$

$$\mathrm{Ar_2C}{\uparrow\downarrow} + \mathrm{ROH} \longrightarrow \left[\mathrm{Ar_2C}\!:\!\cdots\mathrm{H}\cdots\mathrm{OR}\right]^{\ddagger} \longrightarrow \mathrm{Ar_2C(OR)H} \qquad (9)$$

choice of either the route shown in (7) or (8) appears to depend primarily on the stability of the cation formed by protonation of the carbene (Kirmse *et al.*, 1981; Ono and Ware, 1983).

Whichever mechanism operates, it appears to be generally true that singlet aromatic carbenes react with the lower alcohols to form ethers at rates approaching the diffusion limit. On the other hand, aromatic carbenes that are clearly triplets do not give any ether at all from reaction with alcohols. Instead, these triplets behave as is expected of "biradicals" and abstract a hydrogen atom from the oxygen bearing carbon of the alcohol. The stable products of this reaction are those due to the combination and disproportionation (10) of the pair of radicals (Lapin *et al.*, 1984). The more complicated cases, where one carbene shows properties of both singlet and triplet spin-states, require more discussion. This is postponed until after some of the experimental evidence has been reviewed.

$$Ar_2C{\uparrow\uparrow} + HCR_2OH \longrightarrow Ar_2\dot{C}\text{—}H + {}^{\bullet}CR_2OH \longrightarrow Ar_2CH_2 + (Ar_2CH)_2, \text{ etc.} \quad (10)$$

Hydrocarbons

A second process that has a central position in the analysis of the chemical properties of carbenes is their reaction with hydrocarbons. As is the case for alcohols, singlet and triplet carbenes react with hydrocarbons in distinctive ways. It has long been held that very electrophilic singlet carbenes can insert directly into carbon–hydrogen bonds (11) (Kirmse, 1971). On the other hand, triplet carbenes are believed to abstract hydrogen atoms to generate radicals that go on to combine and disproportionate in subsequent steps (12) (Closs and Trifunac, 1970; Baldwin and Andrist, 1971; Lepley and Closs, 1972; Bethell and McDonald, 1977). The formation of free radicals from aromatic carbenes is often easily detected by the fast laser spectroscopic techniques discussed earlier. The radicals generally have characteristic absorption spectra and reactivity patterns that make their identification certain. The direct insertion reaction of singlet carbenes is not expected to generate free radicals.

$$Ar_2C{\uparrow\downarrow} + RH \longrightarrow Ar_2C(R)H \quad (11)$$

$$Ar_2C{\uparrow\uparrow} + RH \longrightarrow Ar_2\dot{C}\text{—}H + R^{\bullet} \longrightarrow Ar_2CH_2 + (Ar_2CH)_2 + Ar_2CHR, \text{ etc.} \quad (12)$$

A simple and straightforward way to distinguish between a direct insertion process and one going through free radicals is by a crossover experiment.

This is illustrated for a generalized hydrocarbon R_2CH_2 and its deuteriated counterpart R_2CD_2. If a mixture of these hydrocarbons reacts with a carbene exclusively by a direct insertion route, then the addition product will consist entirely of undeuteriated and dideuteriated compounds (13).

$$Ar_2C{\uparrow\downarrow} + (R_2CH_2/R_2CD_2) \longrightarrow Ar_2C(H){-}CR_2(H) + Ar_2C(D){-}CR_2(D) \quad (13)$$

Alternatively, if some of the reaction occurs by the hydrogen abstraction-recombination route proceeding through free radicals, then some mono-deuteriated (crossed) products will be formed (14). The former result is taken to be characteristic of the reaction of a singlet carbene, the latter that of a triplet.

$$Ar_2C{\uparrow\uparrow} + (R_2CH_2/R_2CD_2) \longrightarrow Ar_2C((D)H){-}CR_2(D(H)) + Ar_2C((H)D){-}CR_2(H(D)) \quad (14)$$

Some care must be exercised in setting up the crossover equipment to account for kinetic isotope effects associated with abstraction or insertion into a carbon–hydrogen (deuterium) bond. In general, abstraction is expected to exhibit a larger isotope effect than insertion, and this appears to be the case (see below). To accommodate this, and to increase the sensitivity of the experiment, it is often necessary to employ a smaller amount of the hydrocarbon than of its deuterated analog.

Cyclohexane and cyclohexane-d_{12} have been used as the probe for crossover and, hence, the reactive multiplicity of the subject carbenes. By combining direct and triplet-sensitized generation of the carbene with kinetic analysis from laser spectroscopy and the results of the crossover experiments, a rather complete picture of the reaction of aromatic carbenes with hydrocarbons emerges.

Olefins

Perhaps the most important and characteristic reaction of a carbene is its addition to an olefin to form a cyclopropane. Apart from its utility in the diagnosis of the multiplicity of the reacting carbene, this reaction has useful synthetic applications. These two features have combined to encourage the study of this process in some detail.

From an historical point of view, the earliest indication of spin-selective reactivity of carbenes was exhibited by the stereochemistry of the cyclopropanation reaction. The Skell Hypothesis (Skell and Woodworth, 1956) suggests that a spin-prohibition requires the addition of a triplet carbene to an olefin to occur in at least two steps. In turn, the obligatory formation of an intermediate provides an opportunity for the loss of stereochemistry in the reaction. Over the years this hypothesis has been expanded and amplified so that it is usually taken to imply that singlet carbenes add to olefins stereospecifically and triplets with loss of stereochemistry. It will be seen below that for the aromatic carbenes this expansion is well-justified.

A complicating factor associated with experimental application of the Skell Hypothesis is that triplet carbenes abstract hydrogen atoms from many olefins more rapidly than they add to them. Also, in general, the two cyclopropanes that can be formed are diastereomers, and thus there is no reason to expect that they will be formed from an intermediate with equal efficiency. To allay these problems, stereospecifically deuteriated α-methylstyrene has been employed as a probe for the multiplicity of the reacting carbene. In this case, one bond formation from the triplet carbene is expected to be rapid since it generates a particularly well-stabilized 1,3-biradical. Also, the two cyclopropane isomers differ only in isotopic substitution and this is anticipated to have only a small effect on the efficiencies of their formation. The expected non-stereospecific reaction of the triplet carbene is shown in (15) and its stereospecific counterpart in (16).

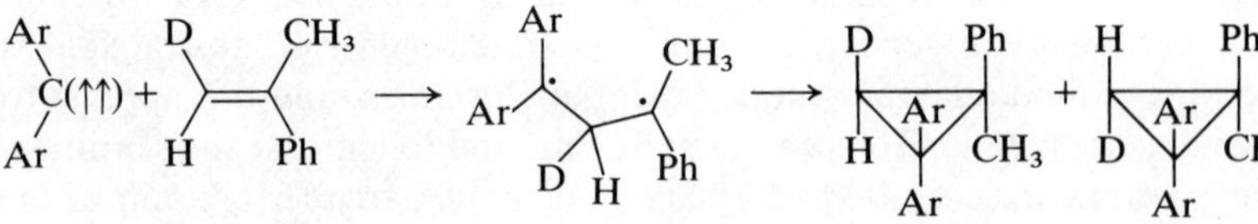

$$Ar_2C(\uparrow\uparrow) + (D)(H)C{=}C(CH_3)(Ph) \longrightarrow Ar_2\dot{C}{-}C(D)(H){-}\dot{C}(CH_3)(Ph) \longrightarrow \text{cyclo-}[C(D)(H)\,C(Ph)(CH_3)\,CAr_2] + \text{cyclo-}[C(H)(D)\,C(Ph)(CH_3)\,CAr_2] \qquad (15)$$

$$Ar_2C(\uparrow\downarrow) + (D)(H)C{=}C(CH_3)(Ph) \longrightarrow \text{cyclo-}[C(H)(D)\,C(Ph)(CH_3)\,CAr_2] \qquad (16)$$

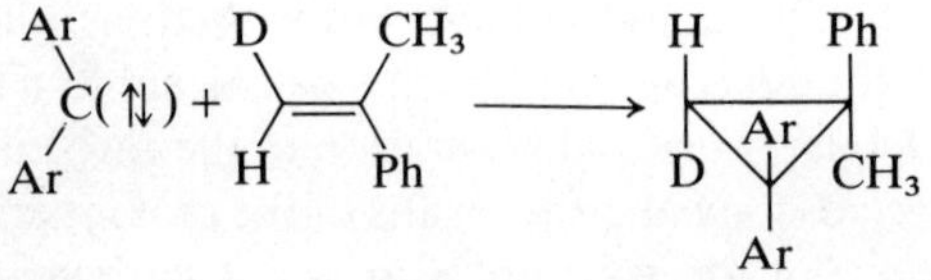

Oxygen

Some of the earliest studies of triplet carbenes in frozen media by epr spectroscopy revealed that these intermediates react rapidly with molecular oxygen (Trozzolo and Gibbons, 1967). This should not come as a surprise since the combination of a triplet carbene with triplet oxygen is a spin-allowed process. Indeed, recent measurements show that this reaction proceeds with a rate that is approximately at the diffusion limit. The product of this reaction (17) is the expected carbonyl oxide (Werstiuk *et al.*, 1984;

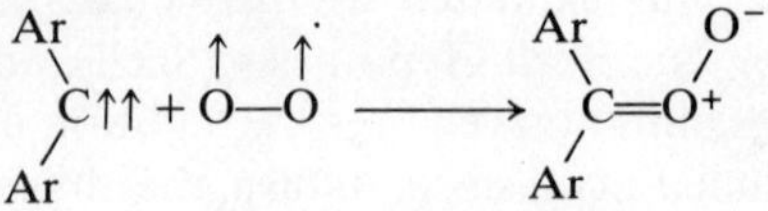

$$Ar_2C\uparrow\uparrow + \overset{\uparrow}{O}{-}\overset{\uparrow}{O} \longrightarrow Ar_2C{=}O^{+}{-}O^{-} \qquad (17)$$

Casal *et al.*, 1984). On the other hand, reaction of a singlet carbene with oxygen should be quite slow. Not only do spin restrictions demand formation of a high-energy biradical intermediate, but oxygen does not normally react rapidly with electrophiles.

The susceptibility of a carbene to reaction with oxygen can be used as a probe of its spin multiplicity. If a carbene is insensitive to oxygen, this is taken to be an indication that it is a ground-state singlet and that the singlet does not readily form the triplet carbene.

5 The structure and reactivity of aromatic carbenes

Genuine exposition of the chemical properties of an aromatic carbene comes from the fusion of all of the types of experiment described above. The significance of each type becomes clear when considered within the context of the whole array of theoretical and experimental findings. The chemical properties of a particular carbene, in turn, become categorizable only with respect to other related examples. Finally, a pattern connecting structure to reactivity emerges when an entire host of clearly understood cases are compared. A pattern of structure and reactivity for the carbenes listed in Table 1 will be developed. The analysis begins at one extreme with BA, and then jumps, for contrast, to another extreme with an account of XA. With the boundaries defined, the other examples fall clearly into place.

MESITYLBORA-ANTHRYLIDENE (BA)

The chemical properties of BA have been studied in detail (Lapin *et al.*, 1984). Low temperature epr spectroscopy shows clearly that the ground state of BA is the triplet (^{3}BA). The zero field parameters (Table 3) reveal some details of this structure. When the irradiation is performed at 4.6 K in a 2-methyltetrahydrofuran glass no epr signals from radical species are apparent. The optical spectrum under these conditions shows absorptions (Table 4) which disappear when the glass is warmed. From these findings the absorption bands are assigned tentatively to ^{3}BA. This conclusion is strongly supported by results from laser flash photolysis experiments.

Irradiation of a benzene solution of DABA at room temperature with a nitrogen laser (Horn and Schuster, 1982) gives the transient absorption spectrum shown in Fig. 3. This spectrum was recorded 50 ns after irradiation of the diazo-compound and decays over a period of *ca* 250 μs by a path exhibiting complex kinetic behavior. This transient spectrum is essentially identical with the low temperature optical spectrum described above, and thus is similarly assigned to ^{3}BA.

Irradiation of DABA at room temperature with a mode-locked frequency

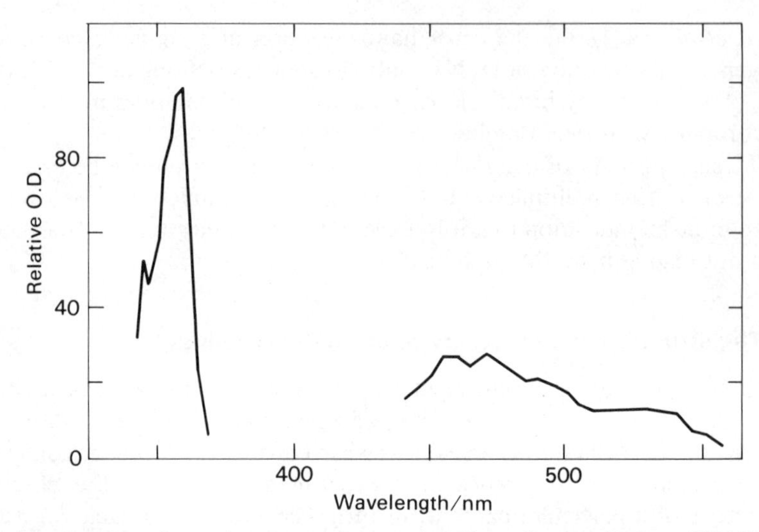

FIG. 3 Transient absorption spectrum recorded 50 ns after irradiation of DABA in benzene solution. The region between 370 and 440 nm is obscured by the absorption of the diazo compound

tripled Nd-YAG laser (Grasse *et al.*, 1983) permits estimation of the rise-time of ^{3}BA. The strong ultraviolet band (359 nm) cannot be detected with this apparatus for the reasons outlined earlier. The low intensity and breadth of the visible transitions (460, 530 nm) lead to noisy spectra. Nevertheless, these data show that ^{3}BA is formed in less than 100 ps from a precursor presumed to be the singlet carbene (^{1}BA). These results are summarized in (18).

N_2 — B — Mes (DABA) $\xrightarrow[C_6H_6,\ -N_2]{h\nu}$ ⇅ B — Mes (^{1}BA) $\xrightarrow{<100\text{ ps}}$ ↑↑ B — Mes (^{3}BA)

(18)

Examination of DABA photolysis in cyclohexane instead of benzene solution leads to predictably different results. The laser spectroscopy shows that ^{3}BA is formed but, in this solvent, the triplet carbene undergoes an additional rapid reaction to generate the mesitylbora-anthryl radical (BAH$^{\bullet}$). This radical is identified by comparison of its spectrum with that of an authentic sample prepared from dihydrobora-anthracene. The half-life of

TABLE 5

Reaction of aromatic carbenes with cyclohexane at room temperature

Carbene[a]	$k(C_6H_{12})/M^{-1}s^{-1}$	$k(C_6D_{12})/M^{-1}s^{-1}$	Cross-over
^{3}BA	$8.1 \pm 0.8 \times 10^5$	$9.9 \pm 1 \times 10^4$	yes
^{1}XA	$< 1.7 \times 10^2$	—	no
FL	$7.7 \pm 3 \times 10^7$	$4 \pm 2 \times 10^7$	(yes)[b]
^{1}DMFL	$7.5 \pm 1.2 \times 10^6$	$4.4 \pm 0.4 \times 10^6$	no
BFL	$4.6 \pm 2 \times 10^8$	—	no

[a] A word is needed about the assignment of rate constants to specific carbene spin states. Where a measured rate constant can be attributed with some confidence to a particular spin multiplicity, that multiplicity is indicated (i.e. ^{1}XA and ^{3}BA). Where the multiplicity is uncertain, the experimentally determined rate constant is reported and no spin state is indicated (i.e. FL). In the latter cases, the reported rate constant can often be viewed as the product of the actual bimolecular rate constant and the equilibrium constant (Table 8) connecting the carbene spin states

[b] Griller *et al.*, 1984c. This conclusion is reached solely from the analysis of products in C_6H_{12}

^{3}BA in neat cyclohexane at room temperature is 75 ns. Rate constants for reaction of cyclohexane with ^{3}BA and other carbenes are reported in Table 5.

The products obtained from photolysis of DABA in cylcohexane support the assignments made above. The major product is the dimer formed from the coupling of BAH$^\bullet$. This is accompanied by the other expected radical products, dihydrobora-anthracene and cyclohexybora-anthracene, as is shown in (19).

N_2 ... B—Mes $\xrightarrow{C_6H_{12},\ h\nu}$ (H, C_6H_{11})–B–Mes + Mes—B (H)(H) B—Mes + (H H)–B–Mes (19)

Irradiation of DABA in C_6D_{12} also generates the bora-anthryl radical. Laser spectroscopy reveals a kinetic isotope effect for the reaction of ^{3}BA with cyclohexane of *ca.* 7 (Table 5). The magnitude of this isotope effect is consistent with rate-limiting hydrogen atom abstraction by the carbene (Collins and Bowman, 1971). When DABA is irradiated in an 8 : 1 mixture of C_6D_{12} and C_6H_{12} the mixed coupling product, cyclohexylbora-anthracene, is shown by mass spectroscopy to contain 22% of the d_{11} compound. This crossed-product confirms the formation and subsequent coupling of free radicals. There may be some direct insertion of the singlet carbene into the carbon–hydrogen bonds of the solvent, but for BA this can amount to no more than *ca* 5% of its total reaction.

The transient triplet carbene formed from irradiation of DABA in benzene can be observed to react with styrene. For this process laser spectroscopy reveals a bimolecular rate constant (k_{sty}) equal to $1.2 \times 10^7 M^{-1} s^{-1}$. The product of the reaction is the expected cyclopropane. This observation clearly supports the spectral assignment of the carbene made above. When deuteriated α-methylstyrene is substituted for styrene as a trap for the carbene, the cyclopropane that results is a 1 : 1 mixture of stereoisomers, (20) (Table 6). This finding indicates that BA is reacting in this sequence exclusively from its ground triplet state.

TABLE 6

Reaction of aromatic carbenes with styrenes

Carbene	Solvent	$k_{sty}/M^{-1}s^{-1}$ [a]	$k_{mst}/M^{-1}s^{-1}$	Stereospecificity (%)
^{3}BA	C_6H_6	1.2×10^7	—	0
^{1}XA	C_6H_{12}	6×10^5	5.1×10^6	>95
FL	CH_3CN	9.4×10^8	1.6×10^9	49
^{1}DMFL	C_6H_6	4.3×10^7	1.3×10^8	>95
^{3}DCFL	CH_3CN		$\sim 3.6 \times 10^9$	(0–79)
BFL	CH_3CN	2.8×10^8	5.2×10^8	80
^{3}AN	C_6H_6	6.4×10^7	7.5×10^7	0

[a] k_{sty} is the bimolecular rate constant for reaction of the carbene with styrene, k_{mst}, with α-methylstyrene both at room temperature

The reaction of BA with isopropyl alcohol is particularly revealing. Direct irradiation of DABA in neat isopropyl alcohol gives the ether (17%) expected from the reaction of the singlet carbene, and the radical coupling products characteristic of the triplet spin state (21). The ratio of the yields of radical-derived products to ether depends inversely on the concentration of isopropyl alcohol in a benzene solution. This behavior is particularly meaningful when compared with that of FL, XA and DPM. For these

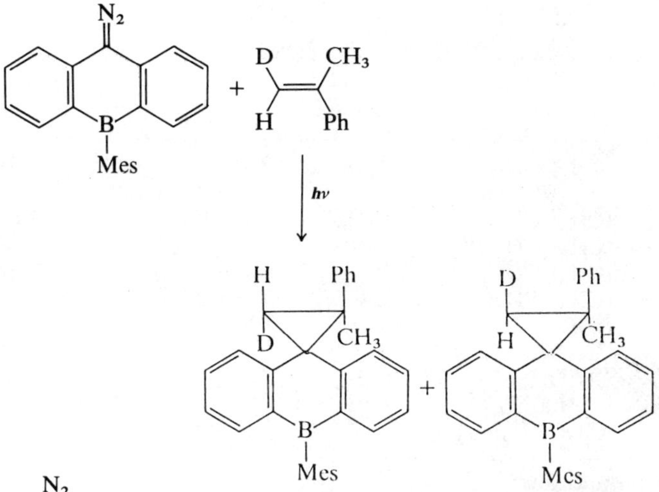

(20)

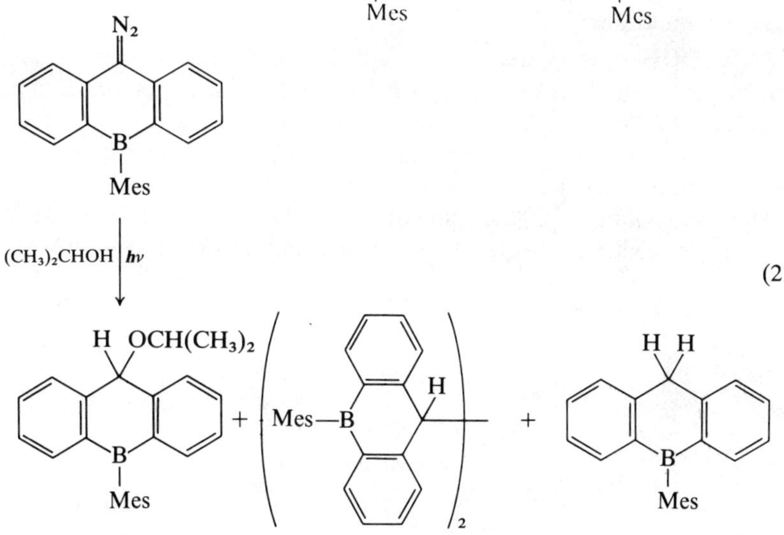

(21)

carbenes the yield of ether from isopropyl alcohol is very high, and the amount of this product formed is almost insensitive to the alcohol concentration (Bethell *et al.*, 1970).

The triplet-sensitized irradiation of DABA in neat isopropyl alcohol does not give a detectable amount of the ether. For comparison, triplet sensitization of DAF or DAX (see below) gives the appropriate ethers in a high yield which is essentially unchanged from the direct irradiation.

The reaction of ^{3}BA with isopropyl alcohol can be monitored by laser spectroscopy. This gives a rate constant for the bimolecular process of $3.4 \times 10^6 M^{-1} s^{-1}$ (Table 7). When this reaction is examined using

TABLE 7

Reaction of aromatic carbenes with nucleophiles

Carbene	Reagent	Solvent	$k_{reagent}/M^{-1}s^{-1}$[a]	Product
^3BA	$(CH_3)_2CHOH$	C_6H_6	$3.4 \pm 0.9 \times 10^6$	Radicals
^3BA	$(CH_3)_2CDOH$	C_6H_6	$6.6 \pm 0.8 \times 10^5$	Radicals
^3AN	$(CH_3)_2CHOH$	C_6H_6	$1.7 \pm 0.1 \times 10^7$	Radicals
^1BA	$(CH_3)_2CHOH$	neat	$6 \pm 2 \times 10^8$	Ether
^1XA	$(CH_3)_3COH$	n-C_5H_{12}	$3.4 \pm 0.3 \times 10^9$	Ether
^1XA	$(CH_3)_3COH$	cyclo-C_6H_{12}	$1.8 \pm 0.2 \times 10^9$	Ether
^1XA	$(C_2H_5)_3N$	n-C_5H_{12}	1.4×10^6	—
FL	CH_3OH	CH_3CN	$8.6 \pm 0.8 \times 10^8$	Ether
^1DMFL	CH_3OH	C_6H_6	$5.6 \pm 0.3 \times 10^8$	Ether
^1DMFL	$(C_2H_5)_3N$	C_6H_6	$\leq 1 \times 10^7$	—
BFL	CH_3OH	CH_3CN	$3.6 \pm 0.7 \times 10^9$	Ether

[a] At room temperature

$(CH_3)_2CDOH$ it is slowed by a factor of *ca* 5. This confirms that the primary reaction of ^3BA with isopropyl alcohol is abstraction of the methine hydrogen to give $BAH^\bullet$, which is observed spectroscopically, and the 2-hydroxy-2-propyl radical. These radicals combine and disproportionate to form the stable products.

The results obtained in the study of BA can be explained by consideration of a simple reaction scheme (shown in Scheme 1) in combination with the

$$\text{DABA} \xrightarrow[-N_2]{h\nu} {}^1\text{BA} \underset{k_{TS}}{\overset{k_{ST}}{\rightleftharpoons}} {}^3\text{BA} \xleftarrow[-N_2]{\text{Sens}^{*3}} \text{DABA}$$

(diazo compound: N_2 ... B–Mes)

Bimolecular Reactions

"Singlet" products "Triplet" products

Scheme 1

notions of spin-specific reactivity detailed earlier. For example, direct irradiation of DABA in benzene containing α-methylstyrene gives first ^1BA, but its lifetime is too short (< 100 ps) to react with a low concentration of the styrene. Instead, ^1BA intersystem crosses to form ground-state ^3BA. This species lives long enough to undergo bimolecular reaction with the styrene, and non-stereospecific cyclopropanation is the consequence.

Similarly, direct irradiation of DABA in neat isopropyl alcohol generates

^{1}BA. Reaction of this spin-state of the carbene with the alcohol gives the ether. This reaction occurs in competition with intersystem crossing to ^{3}BA. Decreasing the alcohol concentration slows the rate of the ether-forming reaction, and consequently the yield of ether declines. Triplet sensitization reveals that ^{3}BA does not give any ether in neat isopropyl alcohol. This observation shows that ^{3}BA does not intersystem cross to ^{1}BA before it is irreversibly consumed. It is already known that there must be an activation barrier for this reaction (k_{TS}) since ^{1}BA is of higher energy than ^{3}BA.

The rate constants k_{TS} and k_{ST} define an equilibrium constant (K_{eq}) connecting the singlet and triplet carbenes. An estimate of K_{eq}, and hence ΔG_{ST}, for BA can be obtained from the experiments described above. The time resolved spectroscopic measurements indicate that ^{1}BA reacts with isopropyl alcohol with a rate constant some five times slower than the diffusion limit (Table 7). This, in conjunction with the picosecond timescale measurements, gives a value for k_{ST}. The absence of ether formation from the sensitized irradiation, when combined with the measured rate constant for reaction of ^{3}BA with isopropyl alcohol, gives an upper limit for k_{TS}. These values give K_{eq} and thus $\Delta G_{ST} \geqq 5.2\,\text{kcal mol}^{-1}$ (Table 8).

TABLE 8

Physical and chemical properties of aromatic carbenes

Carbene	Solvent	k_{ST}/s^{-1}	k_{TS}/s^{-1}	K_{eq}	$\Delta G_{ST}{}^a$ kcal mol^{-1}	Characteristic reactivity
BA	C_6H_6	$\geqq 1 \times 10^{10}$	$\leqq 2 \times 10^6$	$\geqq 4000$	$\geqq 5.2$	triplet
DCFL	CH_3CN	—	—	1000	4.1	triplet/singlet
DPM[b]	CH_3CN	3.2×10^9	$\sim 10^7$	200	3.2	triplet/singlet
FL	CH_3CN	2.5×10^9	1.1×10^8	20	1.9	singlet/triplet
BFL	CH_3CN	$\geqq 2 \times 10^{10}$	$< 4 \times 10^9$	5	1.0	singlet/triplet
DMFL	C_6H_6	—	—	~ 0.03	~ -2	singlet/?
XA	C_6H_{12}	—	—	$\leqq 0.0002$	$\leqq -5$	singlet

[a] ΔG_{ST} for BA, XA and DMFL are limits based on unobserved reactions; for the other carbenes ΔG_{ST} is calculated by assuming a diffusion controlled reaction rate for the singlet carbene with methyl alcohol. All kinetic parameters refer to room temperature
[b] Sitzmann and Eisenthal, 1983

The significance of a value of $\Delta G_{ST} \geqq 5.2\,\text{kcal mol}^{-1}$ is that most of the bimolecular reactions of the triplet carbene will be faster than the formation of the singlet from the triplet. Thus intersystem crossing from ^{1}BA to ^{3}BA is irreversible, and if ^{3}BA is formed first (triplet sensitization), then there will be no reaction at all of ^{1}BA. However, in the direct irradiation of DABA, some reaction of ^{1}BA does occur when the rate of this process is fast (high concentration of reagent) in comparison with that of intersystem crossing (k_{ST}).

Finally, recall that ΔG_{ST} is *ca* 10 kcal mol^{-1} for methylene. The corresponding value is smaller for BA. This can be viewed as the major consequence of the structural change. It will be a point of special interest to show how this parameter is affected in the other carbenes that are discussed.

9-XANTHYLIDENE (XA)

The results from a series of early studies of the products formed from irradiation of DAX appeared to show that XA does not react with hydrocarbons or simple olefins (Reverdy, 1976a,b,c.). However, later reports (G. W. Jones *et al.*, 1978, 1979) seemed to contradict many of these claims. Our more recent investigation of XA resolves the apparently conflicting results and provides further information on the forces that relate structure to reactivity for aromatic carbenes (Lapin and Schuster, 1985).

Photolysis of DAX in a methylcyclohexane glass at 77 K creates a metastable species detected by its optical absorption spectrum (Table 4). This solution does not exhibit an epr spectrum characteristic of a triplet carbene, nor is one observed when DAX is irradiated in Fluorolube (where triplet carbenes are generally more stable). Warming the frozen solution causes the reaction of the metastable transient and the formation of dimeric xanthone azine.

It is difficult to reach a firm conclusion about the structure of the transient detected optically at low temperature. However, the epr results rule out its being ^{3}XA. Other reasonable candidates are ^{1}XA, the xanthyl radical (XAH$^\cdot$), or the xanthylium cation (XAH$^+$). The last two possibilities can be excluded by examination of the chemical behavior of the transient.

Irradiation of DAX in pentane with a nitrogen laser creates a transient species whose spectrum is quite similar to that of the species formed by the low temperature irradiation. The half-life of this intermediate in very dry solvent is *ca* 50 μs. Its reaction under these conditions follows a complex kinetic path that gives ultimately dimeric azine and bi-xanthylidene (22). These findings help in the assignment of a structure to the transient. Neither XAH$^\cdot$ or XAH$^+$ are expected to give these products, but ^{1}XA easily accounts for both.

The detected transient reacts with alcohols to give the expected ethers in very high yield (23). The time-dependence of this reaction can be accurately monitored using laser spectroscopic techniques. In pentane solution it reacts with t-butyl alcohol with a bimolecular rate constant of $3.4 \times 10^9 M^{-1} s^{-1}$. When the solvent is changed to cyclohexane, the rate constant decreases by a

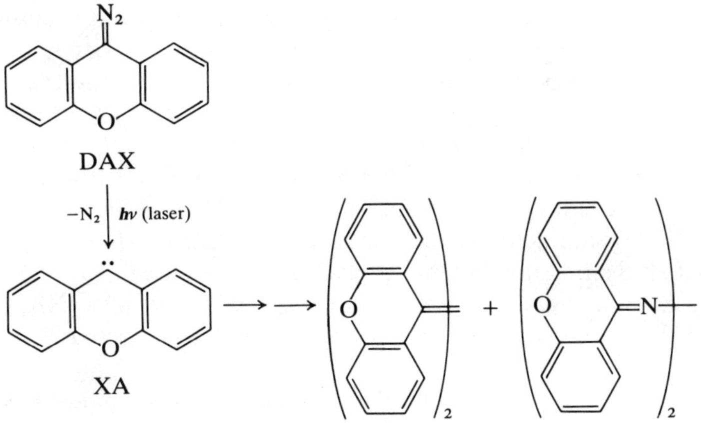

(22)

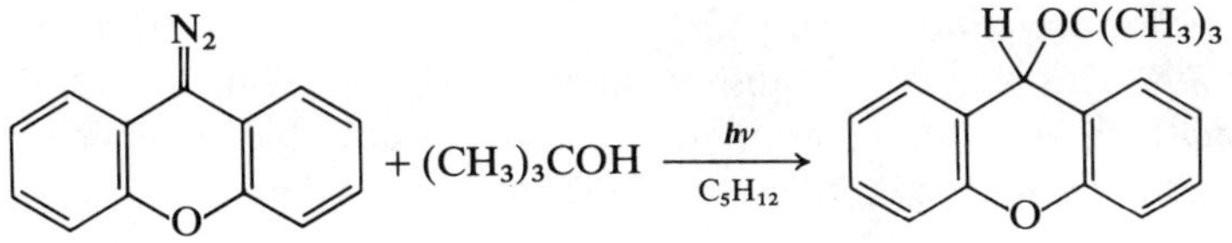

(23)

factor of about two. This is viewed as a direct consequence of the difference in viscosity between pentane and cyclohexane and points to a reaction rate for the alcohol with the transient limited in part by diffusion. The rate constant for reaction of the transient with triethylamine is at least 2000 times less than with t-butyl alcohol. This observation helps to eliminate finally the possibility that the transient is XAH^+ (Sudjak *et al.*, 1976) and points to its assignment as ^1XA. These data are summarized in Table 7.

In contrast to the behavior of BA, triplet sensitized formation of XA in the presence of alcohol gives exactly the same result as does the direct irradiation; ether is formed in very high yield. Thus, even when ^3XA is formed first, the chemical behavior observed is that of ^1XA. This finding is consistent with a carbene having a singlet ground state; a view at least supported by the absence of an epr spectrum for the triplet carbene as noted above.

The species identified as ^1XA reacts with styrene to give the expected cyclopropane. The rate constant for this reaction is *ca* 200 times less than the corresponding rate constant for ^3BA (Table 6). Also, use of the deuterium-labeled α-methylstyrene reveals that the cyclopropanation occurs with essentially total retention of stereochemistry. Moreover, precisely the same result is obtained when this carbene is formed by triplet sensitization rather than by direct irradiation. These findings also point to a reaction originating from a singlet carbene.

Xanthylidene does not react measurably with cyclohexane at room temperature (Table 5). Thermolysis of DAX at high temperature does, however, give some of the expected coupling product 9-cyclohexylxanthene. The crossover experiment (1 : 1 C_6H_{12}, C_6D_{12}) reveals that this product is not formed by the abstraction-recombination sequence. This observation is consistent with the direct insertion characteristic of a singlet carbene.

Xanthylidene also does not react measurably with O_2. The lifetime of ^1XA is the same in O_2-saturated cyclohexane as it is in solutions which have been deoxygenated. Bearing in mind that triplet carbenes react with O_2 at nearly the diffusion limited rate, if ^3XA were in rapid equilibrium with ^1XA, then O_2 should shorten the apparent lifetime of the singlet by reacting with the triplet.

The simple reaction scheme outlined in Scheme 2 easily accommodates the results obtained in the investigation of XA. The basic scheme is just the same as it is for BA except that the relative energies of the singlet and triplet states of the carbene are reversed. All of the evidence points to a singlet ground state for XA. This includes the absence of an epr signal, the nearly diffusion-controlled formation of ether in direct and sensitized experiments, the stereospecific cyclopropanation, and the absence of a rapid reaction of the carbene with O_2.

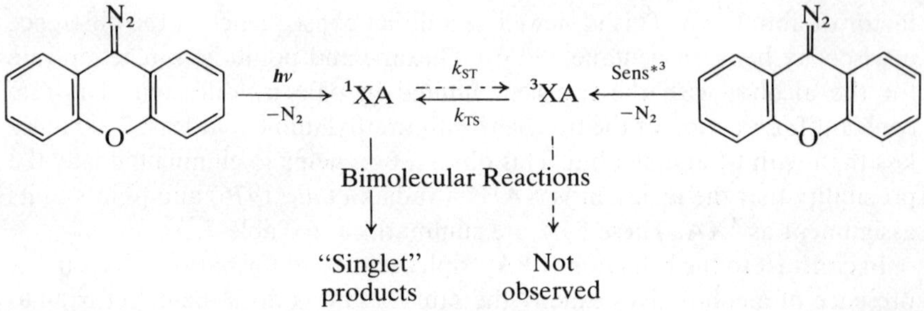

N_2 N_2

$h\nu$, $-N_2$ → ^1XA ⇄ (k_{ST}, k_{TS}) ^3XA ← Sens*[3], $-N_2$

Bimolecular Reactions

"Singlet" products   Not observed

Scheme 2

Consideration of the reactions outlined in Scheme 2 permits an estimate of ΔG_{ST} for XA to be obtained. With the knowledge that triplet carbenes react irreversibly with O_2 with a rate constant close to the diffusion limit (k_{diff}), K_{eq} can be expressed as the quotient shown in (24) where $k_{obs}^{O_2}$ is the rate constant for reaction of ^1XA in oxygen saturated solution, and $k_{obs}^{N_2}$ is the rate constant in the absence of O_2. Since no significant difference in these rate constants can be detected, the estimate of K_{eq} is a limit set in part by the experimental uncertainty in k_{obs} (Table 8). This value, $K_{eq} \geqq 1 \times 10^{-4}$ corresponds to $\Delta G_{ST} \leqq -5$ kcal mol^{-1}, the negative sign indicating that the singlet carbene is lower in energy than the triplet.

Quite evidently, changing the structure of the aromatic carbene from BA to XA has a profound effect on ΔG_{ST}. There is a difference of more than 10 kcal mol^{-1} in this physical property for these two structures. This difference, in turn, appears to control and determine the chemical properties of the carbenes. With the assumptions outlined earlier, this effect can be wholly attributed to a perturbation of the electronic character engendered by replacement of the boron in BA with the oxygen of XA. It will be seen shortly that these two carbenes represent extremes of a nearly continuously tunable range of carbene properties.

$$K_{eq} = \frac{k_{diff}[O_2]}{(k_{obs}^{O_2} - k_{obs}^{N_2})} \quad (24)$$

FLUORENYLIDENE (FL)

The chemical and physical properties of FL have been the subject of investigation for more than 30 years (Horner and Lingnau, 1955). The examination has followed a tortuous path led by the enigmatic nature of this carbene. In the last few years, almost in spite of the application of the fast laser techniques, general agreement has been reached on the physical and chemical properties properly assigned to FL (Grasse *et al.*, 1983; Griller *et al.*, 1984c).

The epr spectrum of ^{3}FL in a frozen matrix shows clearly that the triplet is the ground state of this carbene (Trozzolo *et al.*, 1962). The optical spectrum of ^{3}FL at low temperature was first misassigned (Zupancic and Schuster, 1980) but now has been almost certainly identified as consisting of at least two bands, a relatively strong absorption at 470 nm, and a weaker one at 440 nm (Griller *et al.*, 1982; Brauer *et al.*, 1982). The misassignment led to some incorrect conclusions (Zupancic and Schuster, 1980; Zupancic *et al.*, 1981; Wong *et al.*, 1981), but these appear to have all been resolved (Grasse *et al.*, 1983; Griller *et al.*, 1984c). Here we present only the currently accepted interpretations of what have always been reliable and reproducible experiments.

Irradiation of DAF in acetonitrile solution at room temperature with a *ca* 30 ps pulse results in the appearance of the absorption band at 470 nm assigned to ^{3}FL. This band does not develop immediately, but grows in with a rise-time of 280 ps from an unseen precursor presumed to be ^{1}FL. Fluorenylidene does not live very long in acetonitrile: it reacts with the solvent to form primarily an ylid which absorbs strongly at 400 nm. There is also a smaller amount of hydrogen-atom abstraction from the solvent to generate the fluorenyl radical (FLH$^\bullet$) which absorbs both at 470 and 495 nm. These results are summarized in (25) which illustrates the dichotomous character of this carbene. Instead of the characteristic ground-state triplet behavior observed for BA and the "pure" singlet of XA, FL abstracts hydrogen as does a triplet, and reacts with nucleophiles as does a singlet. This dual nature appears in nearly every reaction of this unusual carbene.

$$\text{DAF} \xrightarrow[-N_2]{h\nu} {}^1\text{FL} \xrightarrow{280\text{ ps}} {}^3\text{FL} \xrightarrow{17\text{ ns}} \text{ylide } (CH_3C{\equiv}N^+\text{–}FL^-) \quad (25)$$

The time-resolved, chemical behavior of FL depends on the solvent. Irradiation of DAF in cyclohexane gives FLH˙. The lifetime of FL in cyclohexane is 1.4 ns, and the ratio of products obtained (26) indicates that both direct insertion and abstraction-recombination mechanisms are operating (Griller *et al.*, 1984b). Replacement of the cyclohexane by its deuterated counterpart reveals a kinetic isotope effect of *ca* 2 (Table 5).

$$\text{DAF} \xrightarrow[C_6H_{12}]{h\nu} \text{9-}C_6H_{11}\text{-fluorene} + \text{fluorene} + \{\text{bicyclohexyl}\} \quad (26)$$

One of the attractive features about the chemical properties of FL is that it gives a good yield of cyclopropanes with many olefins (M. Jones Jr. *et al.*, 1970; Gaspar *et al.*, 1984). Irradiation of DAF in acetonitrile containing styrene gives the expected cyclopropane in high yield. When deuteriated α-methylstyrene is used, cyclopropanation occurs with partial retention of configuration (Table 6). Similar results are observed for other olefins.

The reaction of FL with methyl alcohol gives the ether (92%). This process plays a pivotal role in the analysis of the properties of this carbene. The results are analysed within the spin-specific reaction framework where the ether is taken to be the product of the singlet carbene and this reaction rate is approximately diffusion limited (as it is for ^{1}XA). It is further assumed that ^{3}FL will react with methyl alcohol as does ^{3}BA, i.e., by hydrogen-atom abstraction from carbon, a relatively slow process in comparison with reaction of the singlet carbene (see Table 7).

Laser spectroscopic study of the reaction of FL with alcohols in either acetonitrile or hydrocarbon solution appears to show that ^{3}FL is consumed rapidly by the alcohol; the absorptions associated with ^{3}FL decay faster in the presence of alcohol than in its absence. However, the product of this reaction is the ether expected from the singlet carbene (Table 7). Moreover, triplet-sensitized irradiation of DAF in acetonitrile solution containing methyl alcohol gives an undiminished yield of the ether even though the ground-state triplet is expected to be the first formed carbene (27).

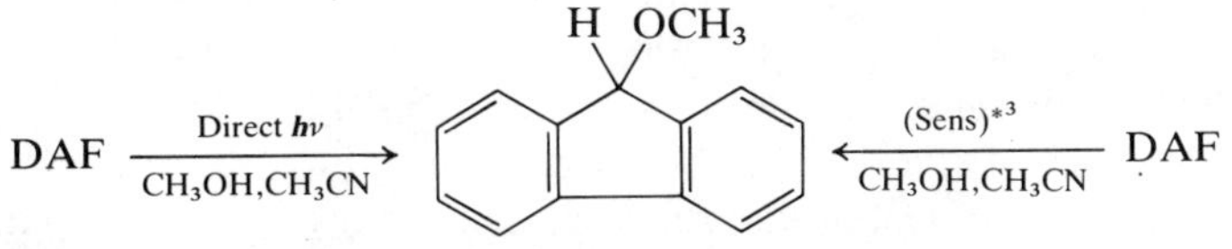

$$\text{DAF} \xrightarrow[\text{CH}_3\text{OH,CH}_3\text{CN}]{\text{Direct } h\nu} \text{9-methoxyfluorene (H, OCH}_3\text{)} \xleftarrow[\text{CH}_3\text{OH,CH}_3\text{CN}]{(\text{Sens})^{*3}} \text{DAF} \tag{27}$$

These findings can be neatly explained if ^{1}FL and ^{3}FL are in rapid equilibrium under these reaction conditions. This result is contrary to the earlier conclusion that intersystem crossing of ^{1}FL in hexafluorobenzene is irreversible (Jones Jr. and Rettig, 1965a,b). Equilibration of singlet and triplet aromatic carbenes has been invoked previously to explain the behavior of DPM (Bethell *et al.*, 1965; Closs and Rabinow, 1976). In this circumstance, triplet-like behavior is obtained when a reagent capable of rapid reaction with ^{3}FL is used to trap the carbene. Reactions characteristic of ^{1}FL are observed when a reagent reacts rapidly with the singlet. In cases where either spin state can react rapidly with a single substance, they both do. These conclusions are summarized in the reactions shown in Scheme 3.

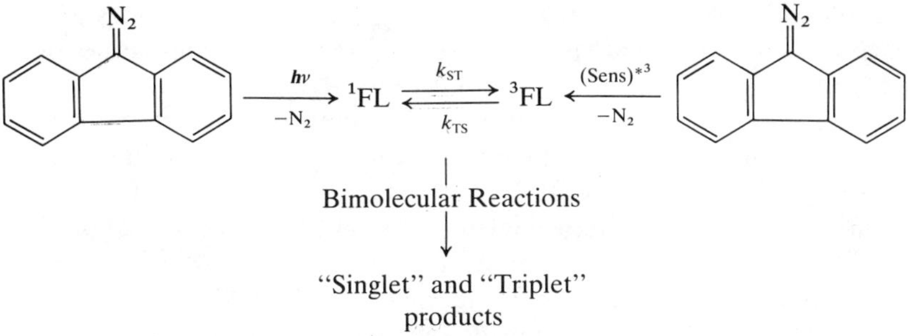

$$\text{DAF } (N_2) \xrightarrow[-N_2]{h\nu} {}^1\text{FL} \underset{k_{TS}}{\overset{k_{ST}}{\rightleftharpoons}} {}^3\text{FL} \xleftarrow[-N_2]{(\text{Sens})^{*3}} \text{DAF } (N_2)$$

Bimolecular Reactions

↓

"Singlet" and "Triplet" products

Scheme 3

A value for ΔG_{ST} for FL can be obtained from analysis of the rate of reaction of the carbene with methyl alcohol within the spin-specific reaction framework identified above. Basically, the observed rate of reaction of ^{3}FL with the alcohol is a measure of the amount of ^{1}FL in the equilibrium mixture. This gives (28) which links K_{eq} with the measured rate constant. The equilibrium constant in turn gives ΔG_{ST} and, when combined with the picosecond spectroscopic results, k_{ST} and k_{TS} (Table 8).

$$K_{eq} = (k_{diff}/k_{MeOH}) - 1 \tag{28}$$

This approach for FL gives $\Delta G_{ST} = 1.9\ \text{kcal mol}^{-1}$ in acetonitrile solution. An identical analysis reveals this value to be *ca* $1.0\ \text{kcal mol}^{-1}$ in hexafluorobenzene, a solvent of at least historical importance. The solvent dependence of ΔG_{ST} points to a potential general problem in the use of reaction kinetics to measure K_{eq}. The reagent added to examine the rate of the carbene reaction changes the reaction medium. This may be particularly important when the medium is a non-polar hydrocarbon and the reagent is an alcohol (Griller *et al.*, 1984b; Langan *et al.*, 1984).

Consideration of the three carbenes discussed thus far reveals a simple pattern. The specific structure of the carbene controls ΔG_{ST}. The magnitude of ΔG_{ST} largely determines the chemical behavior expressed by the carbene. The next carbenes presented will show that this pattern is general and will expose additional details of how structure effects ΔG_{ST}.

3,6-DIMETHOXYFLUORENYLIDENE (DMFL)

Optical spectroscopic analysis reveals that irradiation of frozen solutions of DMDAF creates an absorbing intermediate (Table 4) that is unstable to warming (Chuang *et al.*, 1985). Unlike that formed on irradiation of DAF, this intermediate does not have an epr spectrum characteristic of a triplet carbene. In principle, it could be the singlet carbene (^{1}DMFL), the dimethoxyfluorenyl radical (DMFLH$^\bullet$), or the cation (DMFLH$^+$) formed by protonation of the singlet carbene; other, less likely structures are also conceivable. However, the laser spectroscopy and chemical properties of this intermediate are most consistent with it being ^{1}DMFL.

Irradiation of DMDAF in benzene solution at room temperature generates a transient intermediate that appears within the rise-time of the laser and decays by a pseudo first-order kinetic path with a half-life of 51 ns. The optical spectrum of the intermediate is essentially the same as that observed in the low temperature irradiation experiment.

Photolysis of DMDAF in benzene containing methyl alcohol gives the ether expected from the reaction of the singlet carbene. Monitoring this reaction by laser spectroscopy reveals that the detected transient reacts with the alcohol with a bimolecular rate constant very near the diffusion limits. In contrast, the transient reacts with triethylamine at least 100 times more slowly than it does with alcohol (Table 7). This behavior is inconsistent with identification of the transient as the cation or radical and points to its assignment as the singlet carbene.

Irradiation of DMDAF in cyclohexane solution gives mainly insertion and only very minor amounts of the coupling and disproportionation products expected if hydrogen-atom abstraction were a major process. The lifetime of the carbene in cyclohexane is *ca* 11 ns and increases to 19 ns in C_6D_{12}. When the irradiation is performed in a 1 : 1 mixture of C_6H_{12} and C_6D_{12} the insertion product shows no crossover (Table 5). These chemical properties are those normally associated with a singlet carbene.

As expected, DMFL reacts with styrene to give the appropriate cyclopropane. Irradiation of the diazo-compound in benzene containing the labeled α-methylstyrene gives cyclopropane with essentially complete retention of stereochemistry (Table 6).

When the reaction of DMFL with alcohols, cyclohexane, or α-methylstyrene is initiated by triplet senitization, the outcome is virtually the same as it is for the direct irradiation. Thus ethers are formed in high yield with the alcohols, direct insertion accounts for the major product in cyclohexane, and the olefin cyclopropanation is stereospecific.

The chemical and physical properties of DMFL contrast most sharply with those of FL. The geometrical features, in particular the carbene–carbon bond angle, of these two carbenes are expected to be identical. The most important difference between DMFL and FL is that the spin multiplicities of the lowest electronic states appear to have been inverted. The experiments indicate that the ground-state of DMFL is the singlet. This conclusion is outlined in the reactions shown in Scheme 4.

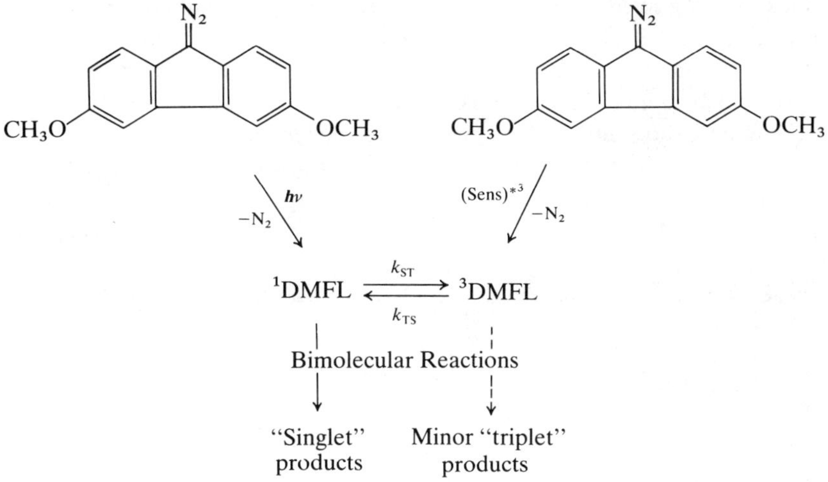

N_2 CH_3O OCH_3 N_2 CH_3O OCH_3

$h\nu$, $-N_2$ (Sens)^{*3}, $-N_2$

$^1DMFL \underset{k_{TS}}{\overset{k_{ST}}{\rightleftharpoons}} {}^3DMFL$

Bimolecular Reactions

"Singlet" products Minor "triplet" products

Scheme 4

It is of some importance to attempt to assign a value to ΔG_{ST} for DMFL. Unfortunately, with the available data this cannot be done except in a qualitative way. The results indicate that ^{3}DMFL hardly contributes to the chemistry of this carbene. On this basis, and with the assumption that the kinetic behaviour of ^{3}DMFL will be similar to that of ^{3}BA, ΔG_{ST} is estimated to be at most -2 kcal mol^{-1}.

An important conclusion, independent of the actual magnitude of ΔG_{ST}, is that the electron donating methoxy-substituents on the fluorenylidene nucleus lower the energy of the singlet state of the carbene more than that of the corresponding triplet. This makes ΔG_{ST} smaller for DMFL than for FL. One might guess that electron withdrawing substituents will have the opposite effect. There is, however, no need to guess.

2,7-DICHLOROFLUORENYLIDENE (DCFL)

Low temperature epr spectroscopy shows, not unexpectedly, that DCFL has a triplet ground state (Table 3). Pulsed irradiation of the diazo-compound in acetonitrile reveals transient spectra assignable to the triplet carbene, the radical (DCFLH$^{\bullet}$) and the ylid that are nearly identical with those recorded from irradiation of DAF under similar conditions (Rak *et al.*, 1986).

As expected, irradiation of DCDAF in acetonitrile containing methyl alcohol gives the appropriate ether. However, the laser spectroscopy reveals that the reaction of DCFL with methyl alcohol is much less efficient than is the corresponding reaction of FL; it takes an alcohol concentration *ca* 10 times higher to trap DCFL than it does to trap FL. It may be recalled (Scheme 3) that, because of rapid equilibration of carbene spin states, ^{3}FL which is long-lived compared with ^{1}FL, can give characteristic singlet carbene products. If reformation of the singlet carbene from the triplet is slow compared with irreversible reaction of the triplet (as it is for ^{3}BA), then the singlet trapping reaction must occur in the brief time before the singlet carbene undergoes intersystem crossing to the triplet.

DCFL reacts with α-methylstyrene to give the now familiar cyclopropane. The stereochemical outcome of this reaction depends on the details of the

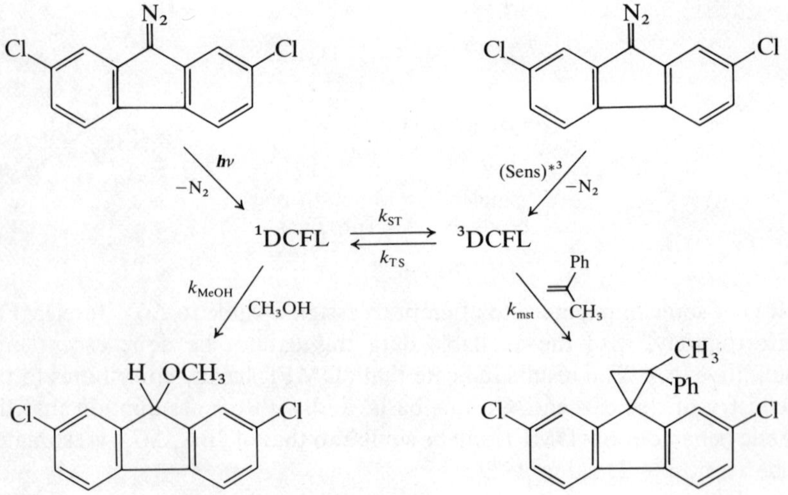

Scheme 5

experiment. Direct irradiation in an acetonitrile solution give 79% retention. Triplet-sensitized reactions of DCDAF with α-methylstyrene are practically non-stereospecific. Laser spectroscopy shows that the rate constant for reaction of the α-methylstyrene with DCFL is slightly greater than it is for FL (Table 6).

It is particularly informative to compare the results of direct and triplet-sensitized irradiation of DCDAF in acetonitrile containing both methyl alcohol and α-methylstyrene. As the reactions shown in Scheme 5 clearly demonstrate, both ether and cyclopropane products are to be expected. The relative yields of these products, however, could depend on whether the reaction is direct or sensitized, and on the rate of reaction with methyl alcohol and α-methylstyrene in comparison with intersystem crossing (k_{ST} and k_{TS}). The ratio of ether to cyclopropane for the triplet-sensitized reaction of DCDAF should vary according to (29). This prediction is verified by experiments where the alcohol concentration is held constant and the α-methylstyrene concentration is varied. Combination of this result with the rate constant for the reaction of the triplet carbene with α-methylstyrene and assuming that ^{1}DCFL reacts with methyl alcohol at approximately the diffusion-limited rate (cf. ^{1}XA and ^{1}DMFL) gives $K_{eq} = 1000$ which corresponds to $\Delta G_{ST} = 4\ \text{kcal mol}^{-1}$.

$$\frac{\text{Cyclopropane}}{\text{Ether}} = \frac{K_{eq}\, k_{mst}\, [\alpha\text{-methylstyrene}]}{k_{MeOH}\ [\text{MeOH}]} \tag{29}$$

The chlorine substituents on the 2- and 7-positions of the fluorenylidene nucleus have a "meta" relationship to the carbene–carbon atom. Halogen substituents in these positions are electron withdrawing inductively. Since these substituents increase ΔG_{ST}, it is clear that electron donating and electron withdrawing groups have opposite influence on the magnitude of ΔG_{ST} and thereby, in a predictable way, the substituents control the chemical properties of these carbenes.

2,3-BENZOFLUORENYLIDENE (BFL)

The usual analysis of frontier-orbital effects on organic chemical reactions divides substituents into three groups: electrons donating (X), electron withdrawing (Z) and conjugating (C) (Fleming, 1976). The fluorenylidene derivatives DMFL and DCFL represent the first two categories; BFL is an example of the third (Grasse *et al.*, 1985).

The low temperature spectroscopy (Tables 3 and 4) and laser spectroscopy of BFL unveil a ground state triplet carbene. This carbene is consumed in cyclohexane solution with a half-life of 260 ps (Table 5). The major product of this reaction is that expected from the direct insertion of the carbene into a carbon–hydrogen bond. However, a measurable amount of abstraction-recombination products are also evident.

Direct irradiation of BDAF in benzene or acetonitrile containing methyl alcohol gives the expected ether in high yield. The triplet-sensitized reactions give the same result. Similarly, BFL reacts with styrene and with α-methylstyrene to give cyclopropanes. In the latter case, 80% retention of configuration is observed for both the direct and triplet-sensitized irradiations.

The chemical properties of BFL are very similar to those of FL. The greatest difference is that under similar conditions there is more "singlet" reaction from BFL than from FL. This observation is reflected in the estimate of ΔG_{ST} obtained, exactly as it was for FL, from the observed rate of reaction with methyl alcohol in acetonitrile. For BFL, use of (28) gives $\Delta G_{ST} = 1.0\ \mathrm{kcal\ mol^{-1}}$. This value implies that there is a significant amount of ^{1}BFL in the equilibrium mixture, and that the effect of the conjugating benzo-substituent is to stabilize the singlet carbene more than the triplet.

ANTHRONYLIDENE (AN)

The epr spectrum of AN clearly shows that it is a ground-state triplet carbene (Devolder *et al.*, 1972). The optical absorptions of this species were assigned at low temperature (Bourlet *et al.*, 1972) and confirmed recently by laser spectroscopy (Tables 3 and 4) (Field and Schuster, 1985). The chemical properties of AN are now readily recognized as those characteristic of a ground-state triplet carbene where intersystem crossing to the singlet is slow (Cauquis and Reverdy, 1975a,b).

Irradiation of DAAN in benzene gives ^{3}AN. This carbene reacts with oxygen very rapidly to give an intermediate believed to be the carbonyl oxide. The triplet carbene reacts with labeled α-methylstyrene to give the cyclopropane with total loss of stereochemistry (Table 6). Direct irradiation in neat isopropyl alcohol gives the ether in low yield (relative to the yields from XA, DMFL, FL, and BFL). The other products are those expected to result from hydrogen-atom abstraction. Triplet-sensitized irradiation of DAAN in the alcohol does not give a detectable amount of the ether.

The results of the study of AN resemble very closely those from BA. The geometry at the carbene–carbon atom for these two species appears from their epr spectra to be similar. Moreover, the structure of AN can be viewed as isoelectronic with BA (30). With these results in mind, it seems safe to

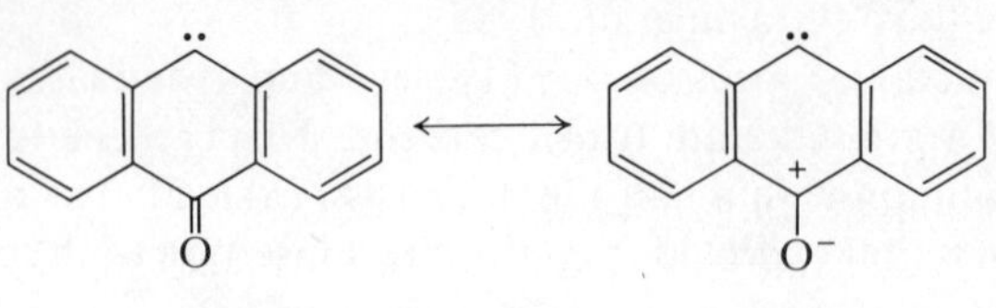

(30)

predict that ΔG_{ST} for AN is greater than *ca* 4 kcal mol^{-1}. However, there are insufficient data currently available to allow a more precise estimate of this value.

DIPHENYLMETHYLENE (DPM) AND ITS DERIVATIVES, DIMESITYL-CARBENE (DMC), DIBENZOCYCLOHEPTADIENYLIDENE (DCH) AND SILA-ANTHRYLIDENE (SA)

Diphenylmethylene is certainly the most exhaustively studied of the aromatic carbenes. Low temperature epr spectroscopy (Trozzolo *et al.*, 1962) clearly established the ground state of this carbene as the triplet. The optical spectrum of the triplet was recorded first in a 1,1-diphenylethylene host crystal (Closs *et al.*, 1966) and later in frozen solvents (Trozzolo and Gibbons, 1967).

The chemical properties of DPM have been probed with each of the procedures identified earlier. This carbene is known to react with alcohols to give ethers (Kirmse, 1963; Bethell *et al.*, 1965), it adds to olefins non-stereospecifically to form cyclopropanes (Skell, 1959; Baron *et al.*, 1973; Gaspar *et al.*, 1980; Tomioka *et al.*, 1984), and it is rapidly converted to a carbonyl oxide with oxygen (Werstiuk *et al.*, 1984; Casal *et al.*, 1984).

Diphenylmethylene was the first carbene to be studied using fast, time-resolved spectroscopic methods (Closs and Rabinow, 1976). Since then both nanosecond and picosecond laser techniques have been used to probe this intermediate (Eisenthal *et al.*, 1980, 1984; Hadel *et al.*, 1984a,b; Griller *et al.*, 1984b; Langan *et al.*, 1984; Sitzmann *et al.*, 1984). The results of these experiments are essentially undisputed, but the interpretation of them still remains somewhat controversial.

The early product and spectroscopic studies were interpreted to show that ^{3}DPM and ^{1}DPM are in equilibrium with each other under many reaction conditions. These experiments were supported by the later application of time-resolved spectroscopic measurements, and ΔG_{ST} was calculated to be *ca* 3–5 kcal mol^{-1}. This estimate relies in part on the assumption that reaction of the carbene with methyl alcohol to form ether requires thermal reactivation of ^{3}DPM to ^{1}DPM. However, measurement of the temperature dependence of this reaction revealed an activation energy smaller than could be easily accommodated by this requirement (Griller *et al.*, 1984b). This finding led to the suggestion that ^{3}DPM does not act simply as a reservoir for ^{1}DPM, but that the triplet carbene can react directly with the alcohol to give the ether. There is no certain precedent for such a spin-forbidden reaction, and interpretation of the temperature-dependent behavior requires detailed knowledge of the reaction mechanism (Zupancic *et al.*, 1985). Also, it is certainly reasonable to expect ΔG_{ST} to be sensitive to the polarity of the reaction medium (Langan *et al.*, 1984) which, in turn, may be temperature dependent. With these thoughts in mind, and with BA and XA as relatively clear examples of spin-specific reactivity of carbenes with alcohols, it seems inappropriate to abandon the rapid equilibrium model for DPM.

The chemical properties of DPM are intermediate between those of FL (which is more electrophilic—"singlet-like") and DCFL (more "triplet-like"). The estimates of ΔG_{ST} for FL and DCFL, *ca* 1.9 and 4.0 kcal mol^{-1} respectively, straddle that for DPM. These values all rely in part on the twin assumptions of the equilibrium model and spin-specific reactivity. It is clear, however, that the chemical behavior of these carbenes can be forecast simply from the estimated magnitude of ΔG_{ST}. The equilibrium model, if nothing else, does provide a convenient way to organize and categorize these properties.

There have been attempts to reverse the normal ordering of triplet below single state of DPM with substituents (Ono and Ware, 1983; Humphreys and Arnold, 1977). By comparing DCFL and DMFL with their diphenylmethyl counterparts, it is clear that substituents have a greater effect on FL than on DPM. This is a reasonable consequence of the co-operative operation of two factors. First, the measurements of ΔG_{ST} for FL and DPM indicate that the former is smaller by about a factor of two. Thus, a small substituent effect may tip the energy balance for FL but not for DPM. Second, the planar structure of FL is expected to transmit the effect of substitution from the aromatic ring more effectively than will the canted DPM structure.

Dimesitylcarbene (DMC) can be viewed as a substituted DPM in which both the electronic and geometrical variables are changed simultaneously. Steric interaction of the *o*-methyl substituents are expected to force the carbene center to adopt a greater bond-angle, and the methyl groups are electron-donating in comparison with hydrogen. This carbene has been studied in some detail (Zimmerman and Paskovich, 1964; Nazran and Griller, 1984; Nazran *et al.*, 1984).

The epr spectrum of DMC confirms the opening of the carbene–carbon bond angle. Consistent with expectations from theory, this change appears to lower the energy of the triplet with respect to the singlet carbene. The effect of the bond-angle change is apparently greater than that caused by electron-donation which should preferentially lower the energy of the singlet state.

The chemical properties of DMC are precisely those now recognized as characteristic of a ground-state triplet carbene with large ΔG_{ST} (i.e. BA, AN). The ground-state reacts rapidly with oxygen but does not give ethers with alcohols. The short-lived singlet state formed on direct irradiation has only one chance to react with alcohol and it does so with a large rate constant. Cyclopropanation of olefins, when it occurs, is non-stereospecific.

The properties of 10,11-dihydro-5H-dibenzo[*a,d*]cycloheptenylidene

(DCH) can also be understood by viewing this carbene as a substituted version of DPM. Low temperature epr spectroscopy established ^{3}DCH as the ground-state (Murahashi *et al.*, 1971). The chemical properties of DCH in methyl alcohol at low temperature are consistent with the greater bond angle increasing the magnitude of ΔG_{ST} (Wright and Platz, 1984). Flash-photolysis studies reveal that DCH reacts with hydrocarbons primarily by hydrogen-atom abstraction and behaves in many other respects similarly to DPM (Hadel *et al.*, 1984a). A related conclusion comes from measurement of intersytem crossing rates (Langan *et al.*, 1984). As with DMC, the anticipated bond-angle effect overwhelms the electronic perturbation by the substituent.

The final carbene in this group is dimethylsila-anthrylidene, SA. Low temperature spectroscopy identifies the triplet as the ground state (Sekiguchi *et al.*, 1982). This carbene appears to react rapidly with methanol and with cyclohexane (Sugawara *et al.*, 1983b). This pattern of reactivity and the reported rate constants fits quite well within the equilibrium model. However, the products of these reactions have not yet been identified. Nonetheless, it is possible to conclude tentatively that ΔG_{ST} is smaller for SA than for DPM. If true, this means that the electronic effects of the silicon substituent outweigh any increase in the carbene-centre bond angle.

PHENYLMETHYLENE (PM), α-NAPHTHYLMETHYLENE (ANM) AND β-NAPHTHYLMETHYLENE (BNM)

Investigation of the chemical and physical properties of PM, ANM and BNM is incomplete. One reason for this is that the absorption spectra of some of these carbenes appear to fall mainly under that of their diazo-compound precursors. This means the time-resolved spectroscopic study of these species is difficult or impossible to accomplish. Nonetheless, other probes of the properties of these carbenes permits some conclusions to be reached.

The triplet is the ground state for each of these carbenes. The epr spectroscopy of ANM and BNM reveals two stable conformations which have been designated *syn* and *anti* (Trozzolo *et al.*, 1965; Senthilnathan and Platz, 1981). This observation provides unambiguous evidence that triplet aromatic carbenes are not linear.

The chemical behavior of PM is characteristic of a carbene for which equilibration of spin-states is faster than irreversible reaction. In particular, direct and triplet-sensitized irradiations of 2-n-butylphenyldiazomethane give exactly the same products (Baer and Gutsche, 1971). A similar conclusion was reached in the study of ANM (Hadel *et al.*, 1983).

It is informative to compare the behavior of ANM and BNM in cyclohexane solution (Chateauneuf and Horn, private communication; Griffin and Horn, private communication). Product and isotope-effect studies indicate that ANM reacts primarily as a singlet carbene and inserts into carbon–hydrogen bonds. On the other hand, these same probes indicate that BNM behaves more like a triplet carbene in that one of its major modes of reaction is hydrogen-atom abstraction.

It is tempting to conclude from their epr spectra that the bond angles for ANM and BNM are the same in both *syn* and *anti* isomers and that the different behavior of these carbenes is due solely to electronic effects. The frontier-orbital electron population at the α-position of naphthalene is nearly twice that at the β-position (Fleming, 1976). We will detail below how this difference can influence ΔG_{ST} and thereby the properties of these carbenes.

6 Understanding the properties of aryl-substituted carbenes

The characteristic reactivity of aromatic carbenes ranges from that typically associated with triplets (BA) to that of singlets (XA) (Table 8). The equilibrium model provides a simple explanation of this behavior. The energy of ^{3}BA is well below that of ^{1}BA. Thus, intersystem crossing from ^{3}BA to ^{1}BA (k_{TS}) is too slow to compete with the irreversible, bimolecular reactions of the triplet. At the opposite extreme, conversion of ^{1}XA to its higher energy triplet does not occur to a significant extent before the singlet carbene is consumed. Other aromatic carbenes fall into intermediate positions in this range; for these cases, reactions characteristic of the two spin states are observed since significant concentrations of both can be present simultaneously in the equilibrium mixture.

Listed in Table 8 are the values of K_{eq} estimated from the kinetic measurements described earlier. Calculation of K_{eq} in this way requires presumption of a reaction mechanism. The mechanism we have chosen demands equilibration of the singlet and triplet carbene and the formation of ethers exclusively by reaction of the singlet with alcohols. The same mechanism is applied to all of the carbenes. Thus, at the very worst, the reported values of K_{eq} can be viewed simply as quantitative descriptors of observed chemical behavior. However, we feel confident that the equilibrium model employed is a satisfactory, but perhaps incomplete (Langan *et al.*, 1984), description of the mechanism. In this case, the reported values of K_{eq} provide a measure of ΔG_{ST}. A brief inspection of Table 8 reveals that the magnitude of ΔG_{ST} varies regularly among the carbenes listed. The change in structure is thus "felt" by the carbene as a change in ΔG_{ST}. It is a simple matter to apply the basic concepts of molecular orbital theory formulated by Gleiter and Hoffmann (1968) to recognize how this ensues.

The magnitude of ΔG_{ST} is directly related to the energy difference between the p- and σ-non-bonding orbitals of the carbene. The maximum value of ΔG_{ST} occurs when these orbitals are degenerate, as they would be for linear methylene. In this circumstance, ΔG_{ST} can be associated with the increased electron–electron repulsion caused by confining two electrons to one orbital in the singlet state. Bending methylene removes the orbital degeneracy and reduces ΔG_{ST}. This is easily pictured since forming the triplet state from the singlet for the bent carbene requires promotion of an electron from a lower to a higher energy orbital. The magnitude of ΔG_{ST} thus becomes the electron–electron repulsion term minus the energy required to promote an electron from the σ to the p-non-bonding orbital. As the carbene–carbon bond angle is further contracted, the σ-orbital picks up more s-character and consequently moves even lower in energy. The smaller the bond angle, the more energy it takes to promote an electron from the σ- to the p-orbital and the smaller ΔG_{ST} becomes.

Of course, the bond angle is not the only structural feature that can affect the σ- and p-orbitals. In particular, the substituents bound to the carbene centre can interact with these orbitals to change their energies. This appears to be the major factor controlling the properties of the aromatic carbenes listed in Table 8.

A simple way to analyse the effect of this perturbation by substitution is to superimpose the orbitals of a prototypical carbene on those of the π-system of the substituent, and, with suitable weighting factors, add the two together. This is shown schematically on Fig. 4. The left side of Fig. 4 shows the σ- and p-orbitals of the prototype carbene in a six-membered ring. The energy difference between these orbitals is set by the bond angle which defines ΔG_{ST}. On the right of Fig. 4A, the π-system of the "bora-anthrylidene substituent" is represented as a single unoccupied orbital. Adding these two components gives the molecular orbitals of bora-anthrylidene shown in the center of Fig. 4A. Note in particular that the energy difference between the singly occupied orbitals of BA is less than it is for the prototype carbene. The σ-orbital of the carbene cannot interact with the π-orbital of the substituent, so that its energy is unchanged in this approximation. The π-orbital of the substituent does mix with the p-orbital of the carbene and this interaction lowers the energy of this non-bonding orbital. This simple analysis leads to the prediction that the non-bonding orbitals of BA should be closer in energy than they are in the prototype. Thus, ΔG_{ST} of BA is expected to be greater than in the prototype.

An analogous picture is drawn for analysis of XA in Fig. 4B. It should be remembered that the highest occupied molecular orbital of the "xanthylidene-substituent" is the lowest unoccupied orbital for the BA system. Interaction of this occupied aryl orbital with those of the prototype carbene gives the orbitals of XA shown in the center of Fig. 4B. The energy

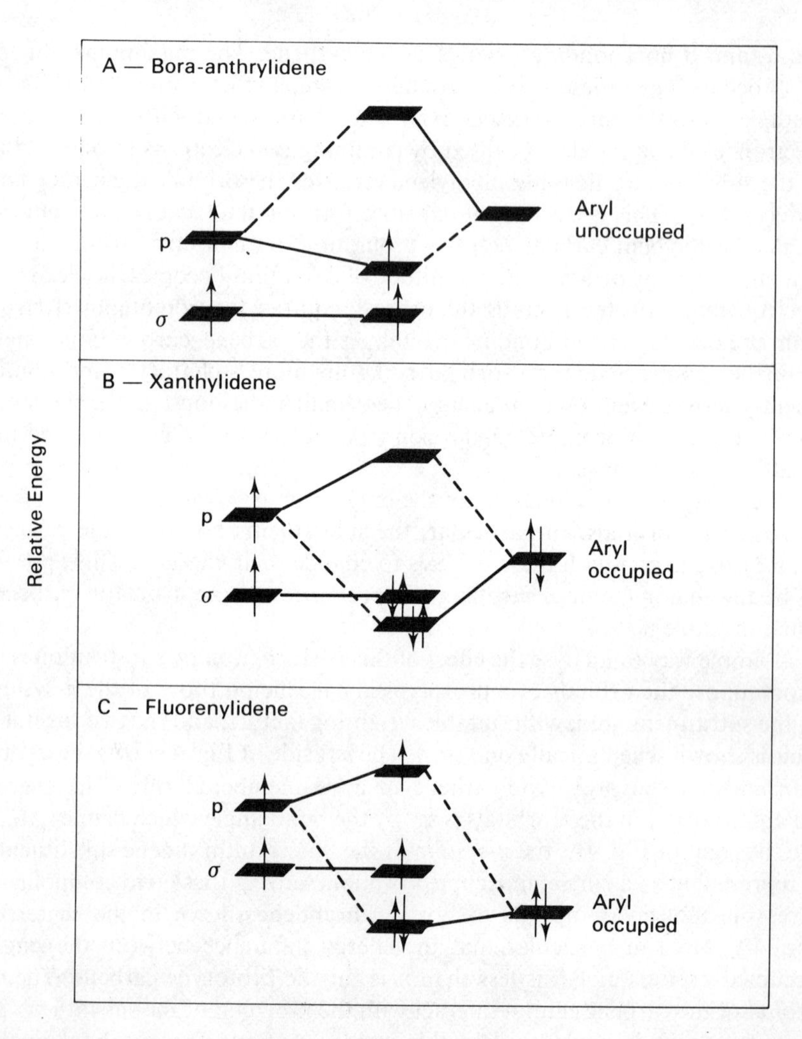

FIG. 4 Analysis of "substituent" effects on aromatic carbenes by qualitative application of Molecular Orbital Theory

difference between the frontier orbitals of XA is increased beyond that of the prototype. Hence, this simple analysis predicts that ΔG_{ST} for XA should be less than that of the prototype, just the opposite of the behavior of BA. These predictions are wholly consistent with the experimental observations.

The rate constants for reaction of ^{1}XA with cyclohexane (Table 5) and styrene (Table 6) indicate that it is less reactive than ^{1}DMFL, for example. This observation is easily understood by consideration of Fig. 4. The lowest unoccupied orbital of ^{1}XA (centre, Fig. 4B) has a higher energy than the prototype carbene. Of course, the higher the energy of this orbital, the less electrophilic and less reactive will be the singlet carbene.

The decision to add an unoccupied orbital with a large weighting factor to the prototype carbene in the case of BA and an occupied one for XA is easily made. The boron atom introduces a low energy unoccupied orbital and the oxygen a high energy occupied one into the aromatic π-system. The choice is less clear for FL and its derivatives. Providence, however, provides the same conclusion from either choice.

The orbitals of a prototypical five-membered carbene are drawn on the left side of Fig. 4C. The smaller bond angle of the five-membered ring increases the energy difference between the σ- and p-orbitals and thus ΔG_{ST} starts out smaller here than in the six-membered ring cases discussed above. The right side of Fig. 4C shows an occupied π-orbital of the "substituent". Adding these two together gives the orbitals of FL shown in the center of Fig. 4C. Just as was the case for XA, the energy difference between the frontier orbitals of FL is greater than it is in the prototype and a decrease in ΔG_{ST} is anticipated. Substituting FL with electron donor groups, as in DMFL for example, raises the energy of the occupied π-orbital added to the prototype. This causes an even greater increase in the energy separating the frontier orbitals for DMFL, and the smaller ΔG_{ST}-value obtained experimentally is just what is forecast from this model. The electron-withdrawing substituents of DCFL are expected to have just the opposite effect. These groups will lower the energy of the π-occupied orbital, making mixing with the prototype less effective, and leaving the energy difference between the frontier orbitals of DCFL more like the prototype. In this circumstance ΔG_{ST} is expected to be greater for DCFL than it is for FL. This is also consistent with the experimental results. Similarly, the occupied aryl π-orbital mixed into BFL will be higher in energy than it is for FL and the observed decrease in ΔG_{ST} is thus readily anticipated. To reach these same conclusions starting from an unoccupied π-orbital of the "fluorenylidene-substituent", it is simply necessary to recall that an electron-donating substituent raises the energy of this orbital and an electron-withdrawing group lowers it.

The value of the qualitative application of molecular orbital theory outlined above is that it provides a rationale for predicting the effect that a change in structure will have on the properties of an aromatic carbene. Since the energy of the substituent orbital added to the prototype carbene is unknown, the quantitative utility of this approach is limited to judgements involving closely related structures (BA with XA, FL with DCFL and DMFL). It is reasonable to hope that as computational methods improve, it will become possible to perform meaningful calculations on aromatic carbenes. The experimental measures of ΔG_{ST} can then serve to judge the reliability of the theory.

It is satisfying that analysis of the substituent effects on carbenes from a valence-bond viewpoint leads to exactly the same conclusions as the molecular orbital approach outlined above. Electron-donor groups, the oxygen in XA or the methoxy-substituents of DMFL, stabilize the electrophilic singlet carbene more than they do the radical-like triplet. On the other hand, electron-withdrawing groups, the boron atom in BA, the carbonyl in AN group or the *m*-chlorine atoms in DCFL, destabilize the singlet and lead to a greater ΔG_{ST}.

It is more difficult to analyse the effect structure has on the properties of the carbenes listed on Table 2. Nevertheless, direct application of the concepts detailed above can provide a guide that has some predictive value. The bond angle of DPM is greater than it is in FL and the mixing of the aromatic π-system with the carbene orbitals is less effective in the former because of its canted structure. Both of these forces push the value of ΔG_{ST} of DPM toward a larger value than in FL. Substituents can increase the bond angle (DMC) to increase ΔG_{ST} or increase the orbital mixing (SA) to diminish ΔG_{ST}. An informative, but very speculative, comparison can be made for ANM and BNM. Recall that ΔG_{ST} for BNM appears to be greater than it is in ANM. If it is assumed that this is an effect due to electronic differences between these two carbenes, then the simple orbital mixing model readily accommodates this observation. The coefficient of the occupied π-orbital of the "naphthalene-substituent" is greater at the α- than at the β-position. Thus, adding this orbital, as above, to the carbene prototype will decrease ΔG_{ST} more for AMN than for BMN.

7 Conclusions

There have been remarkable advances in the physical organic chemistry of aromatic carbenes in the past five years. The intensive study of these reactive intermediates engendered by the development of high-speed pulsed lasers has rekindled interest in this topic. The capacity to observe these species directly under actual reaction conditions permits examination of details heretofore only wished for. These studies have provided answers for old questions and raised new horizons.

The stated objective of this report was to present sufficient data about aromatic carbenes to permit the forecast of their properties directly and reliably from their structures. This has been accomplished to a reasonable degree. Coupling of the theoretical framework with the experimental measurements allows confident prediction of the outcome of many chemical reactions. The rates of the important processes controlling aromatic carbene behavior can be estimated, and thus even yields can be forecast in many cases. These conclusions form the foundation upon which further advances can be built.

It is always difficult to guess what direction an investigation will take in the future. The study of aromatic carbenes is no exception in this regard. However, with a foundation linking structure to reactivity in place, it seems likely that attempts to create carbenes of unusual reactivity by manipulation of structure will proceed. For example, although "stable" radicals have been known for some time, no carbene has ever been isolated under normal conditions. This task no longer seems impossible. At the other extreme, there is a need for extremely reactive electrophilic singlet carbenes as photoaffinity labels. The design of these species can proceed now on a rational basis from the foundation that has been created. In an entirely different vein, triplet polycarbenes have been examined as possible organic ferromagnets (Sugawara *et al.*, 1984). The list of possibilities is endless and limited only by imagination. The coming years promise to provide additional exciting and challenging opportunities for the study of aromatic carbenes.

Acknowledgements

Many of the experiments described in this report were done by my colleagues Beth Brauer, Carl Chuang, Peter Grasse, Kurt Field, Stephen Lapin, Stanton Rak, and Joseph Zupancic. I thank specially Professor Keith Horn of Tufts University for communicating the results on ANM and BNM before publication. This work has been supported by the National Science Foundation.

References

Ahmed, S. N. and Shevlin, P. B. (1983). *J. Am. Chem. Soc.* **105**, 6488
Abramovitch, R. A. (1980). "Reactive Intermediates", Vol. 1. Plenum, New York
Baer, T. A. and Gutsche, C. D. (1971). *J. Am. Chem. Soc.* **93**, 5180
Baldwin, J. E. and Andrist, A. H. (1971). *J. Chem. Soc. D.* 1512
Baron, W. J., Hendrich, M. E. and Jones M. Jr. (1973) *J. Am. Chem. Soc.* **95**, 6286
Bauschlicher, C. W. Jr., (1980). *Chem. Phys. Lett.* **74**, 273.
Bauschlicher, C. W. Jr., Schaeffer, H. F., III and Bagus, P. S. (1977). *J. Am. Chem Soc.* **99**, 7106
Bauschlicher, C. W. Jr. and Shavitt, I. (1978). *J. Am. Chem. Soc.* **100**, 739
Bernheim, R. A., Bernard, H. W., Wang, P. S., Woop, L. S., Skell, P. S. (1970). *J. Chem. Phys.* **53**, 1280
Bethell, D. (1969). *Adv. Phys. Org. Chem.* **7**, 153
Bethell, D. and McDonald, K. (1977) *J. Chem. Soc. Perkin II,* 671
Bethell, D., Whittaker, D. and Callister, J. D. (1965). *J. Chem. Soc.* 2466
Bethell, D., Stevens, G. and Tickle, P. (1970). *Chem. Commun.* 792
Bethell, D., Newall, A. R., Whittaker, D. (1971). *J. Chem. Soc. B,* 23

Bourlet, P., Devolder, P., Goodmand, P. and Laffitte, P. M. (1972). *C. R. Acad. Sci. Paris Sec. C*, **275**, 1161

Brauer, B. E., Grasse, P. B., Kaufmann, K. J. and Schuster, G. B. (1982). *J. Am. Chem. Soc.* **104**, 6814

Carr, R. W. Jr., Eder, T. W. and Topor, M. G. (1970). *J. Chem. Phys.* **53**, 4716

Casal, H. L., Sugamori, S. E. and Scaiano, J. C. (1984). *J. Am. Chem. Soc.* **106**, 7623.

Cauquis, G. and Reverdy, G. (1975a). *Bull. Soc. Chim. Fr.* 1841

Cauquis, G. and Reverdy, G. (1975b). *Bull. Soc. Chim. Fr.* 1845

Chuang, C., Lapin, S. C., Schrock, A. K. and Schuster, G. B. (1985). *J. Am. Chem. Soc.* **107**, 4238

Closs, G. L. (1968). *Top. Stereochem.* 193

Closs, G. L. and Trifunac, A. D. (1970). *J. Am. Chem. Soc.* **92**, 2186

Closs, G. L. and Rabinow, B. E. (1976). *J. Am. Chem. Soc.* **98**, 8190

Closs, G., Hutchinson, C. A. Jr. and Kohler, B. E. (1966). *J. Chem. Phys.* **44**, 413

Collins, C. J. and Bowman, N. S. (1971). "Isotope Effects in Chemical Reactions". Van Nostrand, New York

Devolder, P., Bourlet, P., Dupret, C. and Dessaux, O. (1972). *Chem. Phys. Lett.* **14**, 57

Eisenthal, K. B., Turro, N. J., Aikawa, M., Butcher, J. A. Jr., DuPuy, C., Hefferson, G., Hetherington, W., Korenowski, G. M. and McAuliffe, M. (1980). *J. Am. Chem. Soc.* **102**, 5663

Eisenthal, K. B., Moss, R. A. and Turro, N. J. (1984). *Science* **225**, 1439

Feldmann, D., Meier, K., Zacharias, H. and Welge, K. H. (1978). *Chem. Phys. Lett.* **59**, 171

Feller, D., McMurchie, L. E., Borden, W. T., Davidson, E. R. (1982). *J. Chem. Phys.* **77**, 6141.

Field, K. and Schuster, G. B. (1985). Abs. 189th National Meeting of the American Chemical Society. Miami, Florida

Fleming, I. (1976). "Frontier Orbitals and Organic Chemical Reactions". Wiley, New York

Frey, H. M. and Gordon, J. K. (1975). *J. Chem. Soc. Chem. Commun.* 233

Gaspar, P. P., Whitsel, B. L., Jones, M. Jr. and Lambert, J. B. (1980). *J. Am. Chem. Soc.* **102**, 6108

Gaspar, P. P., Lin, C.-T., Whitsel, B. L., Mack, D. P. and Balasubramanian, P. (1984). *J. Am. Chem. Soc.* **106**, 2128

Gibbons, W. A. and Trozzolo, A. M. (1966). *J. Am. Chem. Soc.* **88**, 172

Gleiter, R. and Hoffmann, R. (1968). *J. Am. Chem. Soc.* **90**, 5457

Grasse, P. B., Brauer, B.-E., Zupancic, J. J., Kaufmann, K. J. and Schuster, G. B. (1983). *J. Am. Chem. Soc.* **105**, 6833

Grasse, P. B., Zupancic, J. J., Lapin, S. C., Hendrich, M. P. and Schuster, G. B. (1985). *J. Org. Chem.* **50**, 2352

Griller, D., Montgomery, C. R., Scaiano, J. C., Platz, M. S. and Hadel, L. (1982). *J. Am. Chem. Soc.* **104**, 6813

Griller, D., Nazran, A. S. and Scaiano, J. C. (1984a). *Acc. Chem. Res.* **12**, 283

Griller, D., Nazran, A. S. and Scaiano, J. C. (1984b). *J. Am. Chem. Soc.* **106**, 198

Griller, D., Hadel, L., Nazran, A. S., Platz, M. S., Wong, P. C., Savino, T. G. and Scaiano, J. C. (1984c). *J. Am. Chem. Soc.* **106**, 2227

Hadel, L. M., Platz, M. S. and Scaiano, J. C. (1983). *Chem. Phys, Lett.* **97**, 446

Hadel, L. M., Platz, M. S., Wright, B. B. and Scaiano, J. C. (1984a). *Chem. Phys. Lett.* **105**, 539

Hadel, L. M., Platz, M. S. and Scaiano, J. C. (1984b). *J. Am. Chem. Soc.* **106**, 283
Halberstadt, M. L. and McNesby, J. R. (1967). *J. Am. Chem. Soc.* **89**, 3417
Harding, L. B. and Goddard, W. A. III (1977). *J. Chem. Phys.* **67**, 1777
Harrison, J. F. (1971). *J. Am. Chem. Soc.* **93**, 4112
Hay, P. J., Hunt, W. J. and Goddard W. A. III (1972). *Chem. Phys. Lett.* **13**, 30
Hayden, C. C., Neumark, D. M., Shoatake, K., Sparks, R. K. and Lee, Y. T. (1982). *Chem. Phys. Lett.* **77**, 6141
Herzberg, G. (1961). *Proc. Roy. Soc. London, Ser. A,* **262**, 291
Herzberg, G. and Johns, J. W. C. (1971). *J. Chem. Phys.* **54**, 2276
Higuchi, J. (1963). *J. Chem. Phys.* **39**, 1339
Hoffmann, R., Zeiss, G. D. and Van Dine, G. W. (1968). *J. Am. Chem. Soc.* **90**, 1485
Horn, K. A. and Schuster, G. B. (1982). *Tetrahedron Lett.* 1095
Horner, L. and Lingnau, E. (1955). *Ann.* **591**, 21
Humphreys, W. R. R. and Arnold, D. R. (1977). *Can. J. Chem.* **55**, 2286
Jones, G. W., Chang, K. T., Munjal, R. and Schechter, H. (1978). *J. Am. Chem. Soc.* **100**, 2922
Jones, G. W., Chang, K. T. and Schechter, H. (1979). *J. Am. Chem. Soc.* **101**, 3906
Jones, M. Jr. and Rettig, K. R. (1965a). *J. Am. Chem. Soc.* **87**, 4013
Jones, M. Jr. and Rettig, K. R. (1965b). *J. Am. Chem. Soc.* **87**, 4015
Jones, M. Jr. and Moss, R. A. (1975). "Carbene Chemistry". Wiley, New York
Jones, M. Jr. and Moss, R. A. (1978). "Reactive Intermediates", Vol. 1. Wiley, New York
Jones, M. Jr. and Moss, R. A. (1981). "Reactive Intermediates", Vol. 2. Wiley, New York
Jones, M. Jr. and Moss, R. A. (1985). "Reactive Intermediates", Vol. 3. Wiley, New York
Jones, M. Jr., Baron, W. J. and Shen, Y. H. (1970). *J. Am. Chem. Soc.* **92**, 4745
Kirmse, W. (1963). *Liebigs Ann. Chem.* **666**, 9
Kirmse, W. (1971). "Carbenes" (2nd edn). Academic Press, New York
Kirmse, W., Loosen, K. and Sluma, H.-D. (1981). *J. Am. Chem. Soc.* **103**, 5935
Langan, J., Sitzmann, E. V. and Eisenthal, K. B. (1984). *Chem. Phys. Lett.* **110**, 521
Lapin, S. C. and Schuster, G. B. (1985). *J. Am. Chem. Soc.* **107**, 4243
Lapin, S. C., Brauer, B.-E. and Schuster, G. B. (1984). *J. Am. Chem. Soc.* **106**, 2092
Lemal, D. M., Gosselink, E. P. and McGregor, S. D. (1966). *J. Am. Chem. Soc.* **88**, 582
Lengel, R. K. and Zare, R. N. (1978). *J. Am. Chem. Soc.* **100**, 7495
Leopold, D. G., Murray, K. K. and Lineberger, W. C. (1984). *J. Chem. Phys.* **81**, 1048
Lepley, A. R. and Closs, G. L. (1972). "Chemically Induced Magnetic Polarization". Wiley, New York
Liu, M. T. H. and Subramanian, R. (1984). *J. Chem. Soc. Chem. Commun.* 1062
Lucchese, R. R. and Schaefer H. F. III (1977). *J. Am. Chem. Soc.* **99**, 6765
Meadows, J. H. and Schaefer H. F. III (1976). *J. Am. Chem. Soc.* **98**, 4383
Moritani, I., Murahashi, S. I., Nishino, M., Yamamoto, Y., Itoh, K. and Mataga, N. (1967). *J. Am. Chem. Soc.* **89**, 1259
Moss, R. A. (1980). *Acc. Chem. Res.* **13**, 58
Moss, R. A. and Dolling, U.-H. (1971). *J. Am. Chem. Soc.* **93**, 954
Moss, R. A. and Jones M. Jr. (1973). "Carbenes". Wiley, New York
Moss, R. A., Mallon, C. B. and Ho, C.-T. (1977). *J. Am. Chem. Soc.* **99**, 4105
Mueller, P. H., Rondan, N. G., Houk, K. N., Harrison, J. F., Hooper, D., Willen, B. H. and Liebman, J. F. (1981). *J. Am. Chem. Soc.* **103**, 5049

Murahashi, S.-I., Moritani, I. and Nishino, M. (1971). *Tetrahedron Lett.* 5131
Murray, R. W., Trozzolo, A. M., Wasserman, E. and Yager, W. A. (1962). *J. Am. Chem. Soc.* **84**, 3213
Nazran, A. S. and Griller, D. (1984). *J. Am. Chem. Soc.* **106**, 543
Nazran, A. S., Gabe, E. J. LePage, Y., Northcott, D. J., Miller, J. M. and Griller, D. (1983). *J. Am. Chem. Soc.* **105**, 2912
Nazran, A. S., Lee, F. L., Gabe, E. J., LePage, Y., Northcott, D. J., Park, J. M. and Griller, D. (1984). *J. Phys. Chem.* **88**, 5251
O'Neill, S. V., Schaefer, H. F. III and Bender, C. F. (1971). *J. Chem. Phys.* **55**, 162
Ono, Y. and Ware, W. R. (1983). *J. Phys. Chem.* **87**, 4426.
Platz, M. S., Senthilnathan, V. P., Wright, B. B. and McCurdy, C. E. Jr. (1982). *J. Am. Chem. Soc.* **104**, 6494
Porter, G. and Strachan, E. (1958). *Trans. Faraday Soc.* **54**, 431
Poole, C. P. Jr. (1967). "Electron Spin Resonance". Wiley, New York
Rak, S., Lapin, S. C. and Schuster, G. B. (1986). Submitted to *J. Am. Chem. Soc.*
Reisenauer, H. P., Maier, G., Riemann, A. and Hoffman, R. W. (1984). *Z. Angew. Chem.* **96**, 596
Reverdy, G. (1976a). *Bull. Soc. Chem. Fr.* 1131
Reverdy, G. (1976b). *Bull. Soc. Chem. Fr.* 1136
Reverdy, G. (1976c). *Bull. Soc. Chem. Fr.* 1141
Roos, B. O. and Sieghhahn, P. M. (1977). *J. Am. Chem. Soc.* **99**, 7716
Sekiguchi, A., Ando, W., Sugawara, T., Iwamura, H. and Liu, M. T. H. (1982). *Tetrahedron Lett.* 4095
Senthilnathan, V. P. and Platz, M. S. (1980). *J. Am. Chem. Soc.* **102**, 7637
Senthilnathan, V. P. and Platz, M. S. (1981). *J. Am. Chem. Soc.* **103**, 5503
Simons, J. W. and Curry, R. (1976). *Chem. Phys. Lett.* **38**, 171
Sitzmann, E. V., Wang, Y. and Eisenthal, K. B. (1983). *J. Phys. Chem.* **87**, 2283
Sitzmann, E. V., Langan, J. and Eisenthal, K. B. (1984). *J. Am. Chem. Soc.* **106**, 1868
Sitzmann, E. V. and Eisenthal, K. B. (1983). "Applications of Picosecond Spectroscopy to Chemistry", pp. 41-63. D. Reidel Publishing, Dordrecht, Holland
Skell, P. S. and Garner, A. Y. (1956). *J. Am. Chem. Soc.* **78**, 3409
Skell, P. S. and Woodworth, R. L. (1956). *J. Am. Chem. Soc.* **78**, 4496
Skell, P. S. (1959). *J. Am. Chem. Soc.* **81**, 1008
Stevens, K. W. H. (1952). *Proc. Roy. Soc. London, Ser. A,* **214**, 237
Sudjak, R. L., Jones, R. L. and Dorfman. L. M. (1976). *J. Am. Chem. Soc.* **98**, 4875
Sugawara, T., Iwamura, H., Hayashi, H., Sekiguchi, A., Ando, W. and Liu, M. T. H. (1983a). *Tetrahedron Lett.* 4095
Sugawara, T., Iwamura, H., Hayashi, H., Sekiguchi, A., Ando, W. and Liu, M. T. H. (1983b). *Chem. Lett.* 1257
Sugawara, T., Bandon, S., Kimura, K., Iwamura, H., and Itoh, K. (1984). *J. Am. Chem. Soc.* **106**, 6449
Tomioka, H. and Izawa, Y. (1977). *J. Am. Chem. Soc.* **99**, 6128
Tomioka, H., Ohno, K., Izawa, Y., Moss, R. A. and Munjal, R. C. (1984). *Tetrahedron Lett.* 5415
Trozzolo, A. M. (1968). *Acc. Chem. Res.* **1**, 329
Trozzolo, A. M. and Gibbons, W. A. (1967). *J. Am. Chem. Soc.* **89**, 239
Trozzolo, A. M. and Wasserman, E. (1975). "Carbenes" (R. A. Moss and M. Jones eds), Vol. 2, p. 185. Wiley, New York
Trozzolo, A. M., Murray, R. W. and Wasserman, E. (1962). *J. Am. Chem. Soc.* **84**, 4990

Trozzolo, A. M., Wasserman, E. and Yager, W. A. (1965). *J. Am. Chem. Soc.* **87**, 129
Warner, P. M. and Chu, I.-S. (1984). *J. Am. Chem. Soc.* **106**, 5366
Wasserman, E., Barash, L., Trozzolo, A. M., Murray, R. W. and Yager, W. A. (1964). *J. Am. Chem. Soc.* **86**, 2304
Wasserman, E., Yager, W. A. and Kuck, V. J. (1970). *Chem. Phys. Lett.* **7**, 409
Wentrup, C. (1984). "Reactive Molecules". Wiley, New York
Werstiuk, N. H., Casal, H. L. and Scaiano, J. C. (1984). *Can. J. Chem.* **62**, 2391
Wong, P. C., Griller, D. and Scaiano, J. C. (1981). *J. Am. Chem. Soc.* **103**, 5394
Wright, B. B. and Platz, M. S. (1984). *J. Am. Chem. Soc.* **106**, 4175
Zimmerman, H. E. and Paskovich, D. H. (1964). *J. Am. Chem. Soc.* **86**, 2149
Zupancic, J. J. and Schuster, G. B. (1980). *J. Am. Chem. Soc.* **102**, 5958
Zupancic, J. J., Schuster, G. B. and Grasse, P. B. (1981). *J. Am. Chem. Soc.* **103**, 2423
Zupancic, J. J., Lapin, S. C. and Schuster, G. B. (1985). *Tetrahedron,* **41**, 147

Author Index

Numbers in italic refer to the pages on which references are listed at the end of each article

Abarkerli, R. B., 294, *305*
Abe, M., 292, *299*
Abramovitch, R. A., 312,*357*
Abuin, E. B., 228, 236, 243, 253, 294, 295, *299*, *305*, *307*
Ahmed, S. N., 317, *357*
Ahrens, M. L. 117, 118, *205*
Aikawa, M., 349, *358*
Albery, W. J., 121, 122, *205*
Albrizzio, J., 218, *299*
Albrizzio, J. P. de, 247, *299*
Alder, R. W., 135, 136, 140, 165, 166, 186, 187, *205*, *206*
Aldrigue, W., 218, *302*
Aleixo, R. M. V., 228, 236, 243, 253, 255, 285, 296, *299*, *302*
Alfred, E., 79, *111*
Allinger, N. L., 16, 17, 18, 29, 55, 85, *106*
Al-Lohedan, H., 229, 230, 232, 234, 237, 238, 239, 247, 248, 257, 289, 293, 296, *299*
Almgren, M., 224, 229, *299*
Al Salem, N. A., 89,*106*
Altman, L. J., 136, 137, 138, 141, 142, 166, *206*
Alwis, K. W., 260, *306*
Anacker, E. W., 215, *302*
Anderegg, G., 11, 99, *106*
Anderson, B. D., 290, *299*
Anderson, M. T., 150, 153, 154, 155, 157, 159, *210*
Anderson, V. K., 199, *208*
Ando, R., 222, 273, 275, *304*, *306*
Ando, W., 321, 323, 351, *360*
Andogino, K., 292, *299*
Andreassen, A. L., 131, 145, *206*
Andreevskii, D. N., 20, *109*
Andrist, A. H. 328, *357*
Angelos, G. H., 282, *305*
Aniansson, G. E. A., 242, *299*
Anoardi, L., 286, *299*
Araujo, P. S., 228, 243, 253, *299*
Archila, J., 218, *299*
Armstrong, D. W., 224, *299*
Arnold, D. R., 350, *359*
Athanassakis, V., 221, 237, 240, 242, 270, 271, 283, *299*
Aubard, J., 200, 201, 204, *206*
Awwal, A., 151, 162, 169, 172, 176, *206*, *209*

Bacon, G. E., 130, *206*
Baer, T. A., 351, *357*
Bagus, P. S., 314, *357*
Baker, W., 46, *106*
Balasubramanian, P., 342, *358*
Baldwin, J. E., 45, 96, *107*, 328, *357*
Balekar, A. A., 218, 224, *299*
Bandon, S., 357, *360*
Bannister, J. J., 124, *206*
Banthorpe, D. V. 258, *299*
Barbara, P. F., 146, 147, *206*, *212*
Barbur, L. P., 218, *302*
Barden, R. E., 271, *299*
Bardez, E., 284, *299*
Barman, P., 34, *110*
Barnett, G. H. 173, 174, 184, *206*
Barnett, R. E., 194, *206*
Baron, W. J., 342, 349, *357*, *359*
Barrow, M. J., 129, *206*
Bartet, D., 228, *299*
Basu, S., 63, *111*
Bauer, S. H., 131, 141, 145, *206*, *210*
Baughcum, S. L., 133, 141, *206*
Baumgarten, E., 228, 230, 237, 265, 296, *307*
Bauschlicher, C. W. Jr., 314, *357*
Bax, A., 138, *212*

Beck, W. H., 237, 254, *303*
Beckwith, A. L. J., 79, 97, *107*
Beens, H. 146, *206*
Begland, R. W., 50, *109*
Belen'kii, L. I., 34, *107*
Belin, C., 130, *211*
Bell, G. M., 221, 240, *299*
Bell, R. P., 120, 122, 125, 127, 142, *206*
Bellocq, A. M., 217, 271, *299*
Bendedouch, D., 219, 220, *302*
Bender, C. F., 314, *360*
Bender, M. L., 191, 192, *206, 210*
Benedek, G. B., 219, 220, *305*
Benedetti, F., 91, *107*
Benkovic, S. J., 96, *107*
Bennett, A. J., 186, *205*
Bennett, G. M., 2, 30, *107*
Bensaude, O., 195, 197, 200, 201, 202, 204, *206, 207*
Benson, S. W., 18, 20, 23, 55, 81, *109*
Bentley, T. W., 250, 251, *299*
Berenjian, N., 292, 294, *299*
Berezin, I. V., 224, 226, 227, 254, 257, *305*
Bergman, N-Å., 125, 126, 127, *206*
Bergman, R. G., 277, 292, *308*
Bernard, H. W., 313, *357*
Bernasconi, C. F., 114, 119, 149, 152, 163, 177, *206*
Bernheim, R. A., 313, *357*
Bernot, D. C., 290, *299*
Bethell, D., 312, 327, 328, 335, 343, 349, *357*
Biais, J., 217, 271, *299*
Bianchi, M. T., 280, *299*
Bianchi, N., 228, 236, 243, 253, *299*
Bianchin, B., 184, *206*
Bicerano, J., 133, *207*
Billmeyer, F. W. Jr., 4, *107*
Bird, R., 79, *107*
Biresaw, G., 222, 262, 264, 275, *299*
Birk, J. P., 151, *207*
Bischof, P., 98, *107*
Bizzigotti, G. O., 260, 269, 285, *305, 306*
Björnstad, S. L., 55, *107*
Blandamer, M. J., 283, *299*
Blatt, E. B., 295, *300*
Blow, D. M., 190, 191, *207*
Bonilha, J. B. S., 220, 229, 230, 237, 269, 285, 294, 297, *300, 302, 308*
Borden, W. T., 314, *358*
Borgarello, E., 218, *307*
Borgen, G., 55, *107*
Bothorel, P., 217, 271, *299*
Bouma, W. J., 133, *207*
Bourlet, P., 321, 323, 348, *358*
Bowman, N. S., 334, *358*
Bowman, P. S., 135, 136, 165, 166, *205*
Brady, J. E., 270, *300*
Brauer, B.-E., 321, 323, 328, 331, 332, 338, 341, *358, 359*
Brauman, J. I., 121, *211*
Braun, A. M., 220, *307*
Brechbiel, M., 258, *307*
Breslow, R., 62, *107,* 278, 292, *300, 308*
Briggs, J., 219, 220, *300*
Brooks, W., 89, *108*
Brown, H. C., 16, *107*
Brown, J. F., 10, 69, *107*
Brown, J. M., 215, 262, 263, 277, 278, 281, *300*
Brown, R. S., 134, 137, 142, *207*
Broxton, T. J., 289, 297, *300*
Brugger, W., 102, *110*
Bruice, T. C., 28, 76, 96, *107*
Brus, L. E., 147, *206, 208*
Bryce, M. R., 166, *205*
Buchwald, S. L., 245, *300*
Bull, T. E., 137, 141, *207*
Bunton, C. A., 218, 219, 220, 221, 222, 224, 225, 226, 227, 228, 229, 230, 231, 232, 233, 234, 235, 236, 237, 238, 239, 240, 242, 244, 245, 246, 247, 248, 249, 253, 254, 255, 257, 258, 260, 261, 262, 263, 264, 265, 266, 267, 270, 271, 272, 274, 275, 277, 281, 282, 283, 287, 289, 291, 293, 296, 297, 298, *299, 300, 301, 302*
Burgess, J., 283, *299*
Burrows, H. D., 293, *301*
Busch, J. H., 142, *207*
Butcher, J. A., 219, 220, *301*
Butcher, J. A. Jr., 349, *358*
Byström, K., 50, *107*

Cabane, B., 220, *301, 302*
Caldin, E., 124, *207*
Callister, J. D., 343, 349, *357*

Camerman, A., 131, *207*
Camerman, N., 131, *207*
Campbell-Crawford, A. N., 121, *205*
Cantor, R. S., 219, 220, *302*
Capon, B., 2, 9, 28, 35, 38, *107*
Cardinal, J. R., 220, *306*
Carmichael, J. B., 70, *107*
Corothers, W. H., 49, *109*
Carr, R. W. Jr., 314, *358*
Carrasco, N., 236, 296, *301*
Carré, D. J., 177, *206*
Carter, G. E., 250, 251, *299*
Casadei, M. A., 42, 57, 105, *107*
Casal, H. L., 330, 349, *358, 361*
Casson, A., 186, 187, *205*
Catena, R., 294, *303*
Catoni, G., 57, 104, *107*
Cavasino, F. P., 291, *302*
Cauquis, G., 348, *358*
Cerichelli, G., 93, 94, *107*, 224, 227, 235, 237, 261, 280, 297, *299, 300, 301*
Chaimovich, H., 221, 224, 227, 228, 229, 230, 233, 236, 237, 243, 253, 254, 255, 265, 269, 285, 296, 297, *302, 307*
Chalabi, P., 192, *210*
Chang, K-C., 144, 199, 200, *207, 210*
Chang, K. T., 338, *359*
Cheney, J., 187, 188, *207*
Cheng, S.-H., 219, 220, *302*
Cherry, W. R., 295, *308*
Chevion, M., 263, *307*
Chevrier, M., 200, 201, 204, *206*
Chiang, Y., 124, 125, 126, 127, 167, *206, 207*
Chiang, Y.-C. P., 278, *306*
Chiericato, G., 230, *300*
Chinelatto, A. M., 284, *302, 303*
Chock, P. B., 151, *207*
Chrisment, J., 184, *206*
Christiansen, G. A., 154, *209*
Chu, I.-S., 327, *361*
Chuang, C., 321, 323, 344, *358*
Cipiciani, A., 231, 235, 239, 258, 266, 267, 270, 291, *302, 305*
Clark, B., 283, *299*
Clark, J. H., 148, *212*
Clements, R., 132, *207*
Clin, B., 217, 271, *299*
Closs, G. L., 312, 324, 328, 343, 349, *358, 359*
Cohen, A. O., 121, *207*
Cohen, B., 141, *207*
Cohen, L. A., 28, *109*
Coleman, P. C., 46, *108*
Collins, C. J., 334, *358*
Connor, H. D., 57, *107*
Connors, K. A., 292, *308*
Conradi, R.A., 290, *299*
Conzan, E., 290, *299*
Coppens, P., 129, *211*
Cordes, E. H., 218, 220, 221, 222, 224 228, 247, *299, 302*
Corti, M., 219, 220, *302*
Cospito, G., 93, 94, *107*
Courtney, S. H., 146, *212*
Cousseau, J., 139, *207*
Cox, B. G., 189, 190, *207*
Cox, G. S., 294, *302*
Cox, J. D., 16, *107*
Cox, M. M., 126, 127, *207*
Cuccovia, I. M., 221, 224, 227, 228, 229, 230, 233, 236, 237, 254, 255, 265, 269, 285, 296, *302, 307*
Cuenca, A., 256, *302*
Curran, J. S., 121, *205*
Currie, M., 129, *206, 207*
Curry, R., 314, *360*

Dafforn, G. A., 76, *107*
Dale, J., 16, 28, 50, 55, *107*
Dalla Cort, A., 43, 47, 48, 65, 76, 83, 100, 101, *107*
Danforth, C., 28, *107*
Danielsson, I., 217, *302*
D'Antonio, P., 131, *210*
Das, A. R., 245, 293, *305*
Davidson, E. R., 314, *358*
Davidson, R. B., 91, *108*
Davies, A. G., 35, *107*
Davies, M., 35, *107*
Davis, A. M., 81, *107*
Dean, R. L., 132, *207*
De Buzzaccarini, F., 222, 237, 271, 272, 282, 283, 286, *299, 300, 301*
Debye, P., 215, *302*
DeCandis, F. X., 120, 126, 127, *208*
Degiorgio, V., 219, 220, *302*

De la Vega, J. R., 133, 142, *207, 208*
Delpuech, J. J., 184, *206*
De Maeyer, L., 114, 149, 154, *208*
De Mayo, P., 292, 294, *299*
Derouane, E., 281, *305*
Desai, N. R., 220, *306*
Deschamps, M. N., 184, *206*
De Silva, M. J., 218, *302*
Dessaux, O., 321, 323, 348, *358*
De Tar, D. F., 20, 25, 28, 85, 89, *108*
Devolder, P., 321, 323, 348, *358*
Diaz, S., 262, 263, 264, 281, *300*
Diekmann, S., 291, 297, *302, 307*
Dill, J. D., 219, 220, *302*
Dill, K. A., 219, 220, *302*
Di Martino, A., 43, 90, 91, *108*
Di Vona, M. L., 93, 94, 95, *108*
Dix, F. M., 269, 286, 292, *306*
Dodin, G., 195, 197, 200, 201, 202, 204, *206, 207*
Dogonadze, R. R., 121, *208*
Dolling, U.-H., 326, *359*
Dolman, D., 166, *207*
Donohue, J. A., 258, *305*
Dorfman, L. M., 339, *360*
Dorshow, R. B., 219, 220, 221, 240, 242, 270, *299, 300*
Drenth, W., 265, 278, *308*
Drewes, S. E., 46, *108*
Dreyfus, M., 195, 197, 200, 201, 202, 204, *206, 207*
D'Souza, V. T., 192, *210*
Dubois, J. E., 195, 197, 200, 201, 202, 204, *206, 207*
Duddy, N. W., 289, *300*
Duerst, R. W., 133, 141, *206*
Dunke, W. L., 131, *212*
Dunning, A. J., 221, 240, *299*
Duplessix, R., 220 *302*
Dupret, C., 321, 323, 348, *358*
DuPuy, C., 349, *358*
Duynstee, E. F. J., 217, 218, *302*
Dye, J. L., 187, 188, *211*

Eberle, H., 103, *111*
Eder, T. W., 314, *358*
Egan, W., 136, 137, 138, 141, *207, 208*
Eggers, F., 119, *207*
Ehrenson, S., 157, *212*
Eigen, M., 114, 115, 116, 149, 154, 178, *207, 208*
Einspahr, H., 166, *208*
Eisenthal, K. B., 312, 317, 320, 337, 344, 349, 350, 351, 352, *358, 359, 360*
Elliott, R. L., 277, 278, *300*
El-Seoud, M. I., 284, *302*
El-Seoud, O. A., 218, 284, *302, 303*
Emert, J., 294, *303*
Empsall, H. D., 89, *106*
Emsley, J., 131, 139, *208*
Endo, R., 284, *304*
Engberts, J. B. F. N., 214, 218, 224, 231, 247, 257, 265, 280, 293, *299, 303, 307, 308*
Engebretson, G. R., 131, *208*
Epstein, W. W., 150, 153, 154, 155, 156, 157, 159, *210*
Erismann, N. E., 236, 255, 285, 296, *302*
Erlbach, H., 57, *108*
Escabi-Perez, J. R., 148, *208*
Eustace, D., 124, 198, *208*
Evans, D. F., 138, *208*, 215, 221, 254, 270, *300, 303, 304, 306, 308*
Eyring, E. M., 150, 152, 153, 154, 155, 156, 157, 159, 178, *208, 209, 210*

Fadnavis, N., 247, 265, *303*
Fatah, A. A., 285, *303*
Feldmann, D., 314, *358*
Feller, D., 314, *358*
Fendler, E. J., 214, 215, 217, 218, 222, 223, 224, 244, 246, 265, *301, 303*
Fendler, J. H., 148, *208*, 214, 215, 217, 218, 220, 221, 222, 223, 224, 244, 246, 265, 268, 271, 281, 285, 294, *301, 303, 304, 305, 307*
Fenn, M. D., 136, 145, *208*
Fernandez, M. S., 221, 252, 265, *303*
Fernando, D. B., 289, *300*
Fersht, A. R., 3, *108*
Field, K., 348, *358*
Field, M. J., 204, *208*
Fife, T. H., 3, *108*
Filho, P. B., 285, *302*
Fischer, H., 120, 126, 127, *208*

Fleming, I., 347, 352, *358*
Flory, P. J., 10, 64, 65, 70, 71, 74, 83, *108, 109, 110,* 219, 220, *302*
Foreman, T. K., 294, *300*
Fornasier, R., 259, 261, 262, 265, 286, 288, *299, 303*
Forsén, S., 136, 137, 138, 141, 142, *206, 207, 208*
Fourche, G., 217, 271, *299*
Fox, M. F., 132, *211*
Frahm, J., 291, 297, *302, 307*
Franco, C., 244, 293, 296, *306, 307*
Frankson, J., 239, *301*
Fraser, S., 62, 74 *109*
Frenkiel, L., 34, *110*
Frey, H. M., 314, *358*
Frey, M. R., 279, *304*
Friedrich, D. M., 146, *212*
Friess, S. L., 114, *208*
Fromherz, P., 219, 220, 221, 252, 265, *303*
Fry, A., 251, *303*
Fueno, T., 157, *208*
Fuess, H., 133, *208*
Fujimoto, M., 180, *212*
Fujita, J., 89, *109*
Fujiwara, F. Y., 136, *208*
Fukuya, K., 263, 285, *306*
Funasaki, N., 229, 230, 296, *303*
Fung, D. S., 215, *304*

Gabe, E. J., 321, 350, *360*
Gadwood, R. C., 192, *210*
Galli, C., 6, 35, 37, 38, 39, 40, 42, 43, 45, 49, 55, 56, 57, 76, 79, 90, 91, 104, 105, *107, 108*
Gamboa, C., 228, *299, 303*
Gan, L.-H., 231, 232, 239, 240, 270, 289, 298, *301, 303*
Gandour, R. D., 3, 27, *108,* 192, *208*
Gani, V., 248, *303*
Gargano, P., 43, 57, 90, 91 105, *108*
Garner, A. Y., 314, *360*
Garner, P., 292, *303*
Garnett, C. J., 237, 254, *303*
Gaspar, P. P., 342, 349, *358*
Gaupset, G., 55, *107*
George, C., 131, *210*
German, E. D., 121, *208*
Germani, R., 231, 239, 266, 267, 270, *302*
Gerritzen, D., 143, *210*
Gesmantel, N. B., 290, *303*
Ghiggino, K. P., 295, *300*
Giaccio, M., 279, *305*
Gibbons, W. A., 322, 323, 330, 349, *358, 360*
Ginani, M. F., 294, *305*
Giovanelli, G., 55, 56, *108*
Gitler, C., 220, 221, 222, 224, 228, *302*
Given, R. S., 295, *303*
Gleiter, R., 352, *358*
Glick, R., 79, *111*
Glick, R. E., 157, *212*
Gobbo, M., 259, *303*
Goddard, W. A., 314, *359*
Goguillon, B. T., 284, *299*
Gokel, G. W., 280, *308*
Gold, V., 177, 185, *208*
Golic, L., 129, *208*
Gonsalves, M., 297, *303*
Gonzales, A., 283, *303*
Goode, N. C., 166, *205*
Goodman, J., 147, *208*
Goodman, N., 4, 16, 72, *109*
Goodman, P., 323, 348, *358*
Gordon, J. K., 314, *358*
Gormally, J., 124, *206*
Gosselink, E. P., 315, *359*
Gouin, L., 139, *207*
Gould, I. R., 281, 295, *303*
Graafland, T., 55, *109*
Grasse, P. B., 321, 323, 332, 341, 347, *358, 361*
Greenberg, A., 16, *109*
Grellman, K. H., 146, *206*
Grieco, P. A., 292, *303*
Griggs, C. G., 277, 278, *300*
Griller, D., 312, 321, 323, 327, 333, 341, 342, 344, 349, 350, *358, 360, 361*
Gruen, D. W. R., 219, 220, *303*
Grunwald, E., 20, 28, 74, 75, *109*, 121, 123, 124, 197, 198, 199, 200, *207, 208, 210,* 217, 218, *302*
Guest, M. E., 204, *208*
Guillerez, J., 204, 207
Guldbrand, L., 220, *304*

Gunnarsson, G., 136, 137, 138, 141, 142, 166, *206, 207, 208,* 216, 221, 240, 243, *303*
Gurr, M., 146, *206*
Guthrie, J. P., 3, *108*
Gutman, M., 146, 147, 148, *209*
Gutsche, C. D., 351, *357*

Haberfield, P., 263, *303*
Haddon, R. C., 134, 135, 137, 138, 141, 142, *207, 209*
Hadel, L. M., 333, 341, 349, 351, *358, 359*
Hadzi, D., 129, *210*
Hafner, K., 133, *208*
Haggert, B. E., 142, *207*
Hagopian, S., 148, *208*
Halberstadt, M. L., 314, *359*
Halpern, A. M., 60, 61, *108*
Halpern, J., 151, *207*
Hamed, F. H., 240, 272, 282, 283, 289, 262, *301*
Hameka, H. F., 142, *208*
Hammes, G. G., 114, 119, 149, *208*
Hammett, L. P., 28, 100, *108*
Hammond, G. S., 28, *108,* 121, *208*
Harada, A., 294, *308*
Harding, L. B., 314, *359*
Harris, H. C., 250, 251, *299*
Harrison, J. F., 313, 314, 316, *359, 360*
Hartley, G. S., 214, 217, 218, 219, 253, 265, *303*
Haselbach, E., 135, 167, *209*
Hashiguchi, Y., 288, *305*
Hashimoto, S., 270, *304*
Haslam, J. L., 152, 153, 154, *208, 209*
Hay, P. J., 314, *359*
Hayakawa, K., 297, *304*
Hayashi, H., 323, 351, *360*
Hayden, C. C., 314, *359*
He, Z., 292, *303*
Hechelhammer, W., 32, *111*
Heck, R., 79, *111*
Hefferson, G., 349, *358*
Helmchen, G., 277, 278, *300*
Hendrich, M. E., 349, *357*
Hendrich, M. P., 321, 323, 347, *358*
Hendrickson, T. F., 268, *306*
Hennig, J., 143, *209, 210*
Henriksson, A., 135, 167, *209*
Henshall, T., 28, *110*
Hermansky, C., 271, 282, *305*
Herve, P., 281, *304*
Herzberg, G., 313, 314, *359*
Hetherington, W., 349, *358*
Hibbert, F., 120, 122, 150, 151, 152, 153, 162, 165, 166, 167, 169, 172, 173, 174, 176, 178, 181, 182, 184, *205, 206, 209*
Hicks, J. R., 242, *304*
Hidalgo, J., 246, *301*
Higashimura, T., 52, 66, 67, 68, *110*
Higuchi, J., 322, *359*
Hilinski, E. F., 146, *209*
Hill, J. W., 49, *108*
Hillier, I. H., 194, 204, *208, 211*
Hinze, W., 268, *303*
Hinze, W. L., 281, *304*
Hirakawa, S., 222, 273, 275, *304*
Ho, C.-T., 315, *359*
Hoffman, H., 220, *304*
Hoffman, R. W., 314, *360*
Hoffmann, R., 91, *108*, 313, 316, 322, 352, *359*
Holl, H., 49, *111*
Hollander, F. J., 130, *209*
Holt, S. L., 271, 283, *299, 303*
Holzwarth, J. F., 124, *206, 207*
Hong, Y.-S., 226, 227, 266, 267, 274, 275, *301*
Hooper, D., 314, 316, *360*
Horn, K. A., 331, *359*
Horner, L., 341, *359*
Hosako, R., 287, *304*
Houk, K. N., 91, *110*, 314, 316, *360*
Hu, D. D., 121, *210*
Hu, S., 55, *106*
Huang, S. K., 236, 247, 296, *300, 301*
Hubert, A. J., 50, *107*
Hui, Y., 268, 269, 278, *306*
Huisgen, R., 40, *108*
Humphreys, W. R. R., 350, *359*
Hunsdiecker, H., 57, *108*
Hunt, W. J., 314, *359*
Hunte, K. P. P., 152, 166, 169, 172, *205, 209*
Huppert, D., 146, 147, 148, *209*
Hurd, C. D., 35, *108*
Hutchinson, C. A. Jr., 349, *358*

Ichikawa, K., 16, *107*
Ige, J., 293, *301*
Iguchi, A., 290, *308*
Ihara, H., 278, 285, 288, *304, 305*
Ihara, Y., 227, 235, 237, 261, 262, 263, 269, 278, 281, 287, 297, *300, 301, 304, 306, 308*
Illuminati, G., 2, 6, 9, 35, 37, 40, 42, 43, 45, 46, 49, 50, 51, 52, 53, 54, 55, 56, 57, 59, 65, 66, 72, 76, 79, 82, 83, 91, 93, 94, 95, 100, *107, 108*
Imanishi, Y., 3, 52, 66, 67, 68, 74, *109, 110*
Ingold, C. K., 27, 38, 92, *109*, 249, *304*
Inouye, K., 284, *304*
Inskeep, W. H., 178, *209*
Ionescu, L. G., 215, 219, 244, 264, 296, *300, 304, 306*
Ireland, J. F., 146, *209*
Ishitawa, T., 284, *304*
Israelachvili, J. N., 219, *304*
Ito, K., 72, 73, 82, *111*
Itoh, K., 321, 323, 357, *359, 360*
Itoh, M., 147, *209*
Iwamoto, K., 284, *308*
Iwamura, H., 321, 323, 351, 357, *360*
Izawa, Y., 327, 349, *360*

Jachimowicz, F., 135, 167, *209*
Jackman, L. M., 135, 138, 141, *209*
Jackowski, G., 63, 74, *111*
Jacobson, H., 10, 69, *109*
Jaeger, D. A., 279, 280, 282, *304, 305*
Jager, J., 55, *109*
Jaget, C. W., 154, *209*
James, J. C., 28, *107*
Jansen, D. K., 280, 292, *305*
Jencks, W. P., 3, 9, 25, 26, 27, 85, 99, 100, *109*, 120, 126, 127, 192, 193, 194, *206, 207, 208, 209, 211*
Jenkins, J. A., 281, *300*
Jensen, R. P., 154, 156, *209*
Jerkunica, J. M., 215, *305*
Joesten, M. D., 127, *209*
Johns, J. W. C., 313, *359*
Johnson, C. A., 138, *211*
Johnson, K., 290, *299*
Johnson, S. L., 245, 249, 250, *304*
Jones, D. L., 178, *209*
Jones, G. W., 338, *359*
Jones, M. Jr., 312, 342, 343, 349, *357, 358, 359*
Jones, R. D. G., 131, *209*
Jones, R. L., 339, *360*
Jönsson, B., 133, *209*, 216, 220, 221, 240, 243, *303, 304*
Jumper, C. F., 198, *208*

Kabo, G. Y., 20, *109*
Kachar, B., 270, *300, 304*
Kajimoto, O., 157, *208*
Kamego, A. A., 244, 246, 293, *301*
Kanda, M., 297, *304*
Kaneko, K., 284, *304*
Karle, J., 131, *210*
Karlström, G., 133, *209*
Kasha, M., 146, 147, *210, 211*
Katiyar, S. S., 223, 236, 290, *305, 307, 308*
Katzhendler, J., 259, 263, 287, *307*
Kaufman, K. J., 146, 147, 148, *209, 211*, 323, 332, 341, *358*
Kawakami, Y., 72, 73, 82, *111*
Kebarle, P., 79, *111*
Keh, E., 284, *299*
Kellogg, R. M., 57, *109*
Kemp, D. S., 244, *304*
Kendrick, R., 143, *210*
Kharkats, Y. I., 121, *208*
Kibuchi, J., 284, *308*
Kim, K. Y., 260, *306*
Kimura, K., 357, *360*
Kimura, Y., 57, *109*, 287, *304*
King, G. S. D., 50, *107*
King, T. A., 124, *206*
Kinoshita, T., 277, *306*
Kintzinger, J. P., 187, *207*
Kirby, A. J., 3, 9, 28, 37, 55, 90, 97, 100, 102, *108, 109*, 244, *304*
Kirmse, W., 312, 327, 328, 349, *359*
Kitahara, A., 218, *304*
Kjaer, A. M., 190, *209*
Knipe, A. C., 79, *107*
Knop, D., 189, 190, *207*
Knowles, J. R., 245, *300*
Kobelt, M., 34, *110*
Kodali, D., 294, *303*
Koeppl, G. W., 121, *209*

Kohler, B. E., 349, *358*
Konasewich, D. E., 121, *210*
Kondo, H., 284, *308*
Koppel, D. E., 219, 220, *302*
Korenowski, G. M., 349, *358*
Koshland, D. E., 27, 71, 76, *107, 110*
Kostenbauder, H. B., 217, 237, 254, *306*
Kraeutler, B., 281, *308*
Kramer, H. E. A., 133, *208*
Kreevoy, M. M., 121, 128, 130, 136, 138, 144, 145, 167, *205, 210*
Kresge, A. J., 120, 121, 122, 124, 125, 126, 127, 153, 156, 165, 167, 175, *206, 207, 209, 210*
Krieg, M., 294, *302*
Kruizinga, W. H., 57, *109*
Kruse, W., 114, 149, 154, 178, *208*
Ku, A. Y., 123, 124, *208*
Kuck, V. J., 313, 322, *361*
Kuhn, W., 6, 7, 8, 9, 64, 79, *109*
Kumikasa, T., 287, *304*
Kumikiyo, N., 287, *304*
Kuna, S., 132, *211*
Kunitake, T., 215, 218, 222, 259, 260, 273, 275, 285, 286, 288, *304, 305, 306*
Kunze, K. L., 142, *207*
Kurihara, K., 285, *305*
Kuroki, N., 278, 287, *304*
Kurz, J. L., 123, *210*, 217, 237, 254, *305*
Kurz, L. C., 123, *210*
Kuznetsov, A. M., 121, *208*
Kwart, H., 258, *307*
Kwiatkowski, J. S., 194, *210*

Laane, J., 128, *210*
Laffitte, P. M., 323, 348, *358*
Lalanne, P., 217, 271, *299*
Lamb, G. W., 219, 220, *301*
Lambert, J. B., 349, *358*
Lambie, A. T., 237, 254, *303*
Langhan, J., 320, 344, 349, 350, 351, 352, *359, 360*
Langhals, H., 293, *305*
Lanza, S., 98, 99, *110*
Lapin, S. C., 321, 323, 327, 328, 331, 338, 344, 346, 347, 349, *358, 359, 360, 361*
Lapinte, C., 248, *303, 305*
Larsen, J. W., 228, 272, *305*
Latimer, W. M., 27, *110*
Laungani, D., 136, 137, 138, 141, 142, 166, *206*
Laurino, J. P., 98, *110*
Lazaar, K. I., 141, *210*
Lee, C. K., 63, *111*,
Lee, F. L., 350, *360*
Lee, J., 192, *210*
Lee, R. A., 185, *208*
Lee, Y. T., 314, *359*
Leffler, J. E., 20, 28, 74, 75, *109*, 121, *210*
Leganza, M. W., 60, 61, *108*
Lehmann, M. S., 129, 130, *211*
Lehn, J.-M., 185, 187, 188, *207, 210, 211*
Lemaire, B., 217, 271, *299*
Lemal, D. M., 315, *359*
Lemanceau, B., 217, 271, *299*
Lengel, R. K., 314, *359*
Leopold, D. G., 314, *359*
LePage, Y., 321, 350, *360*
Lepley, A. R., 328, *359*
Levashow, A. V., 224, 226, 227, 254, 257, *305*
Levich, V. G., 121, *208*
Lewis, E. S., 114, 121, *208, 210*
Lewis, I. C., 157, *212*
Liang, T. M., 128, 130, 136, 138, 144, 145, 167, *210*
Lianos, P., 217, 240, 242, 270, *305, 309*
Liebman, J. F., 16, *109*, 314, 316, 360
Liechti, R. R., 246, *303*
Liler, M., 237, 254, *303*
Lillocci, C., 93, 94, 95, *107, 108*
Limbach, H-H., 143, *209, 210*
Lin, C.-T., 342, *358*
Linda, P., 235, 239, 258, 291, *302, 305*
Lindemann, R., 132, *210*
Lindman, B., 214, 215, 217, 219, 220, *302, 304, 305, 308*
Lindner, H. J., 133, *208*
Lineberger, W. C., 314, *359*
Lingnau, E., 341, *359*
Link, C. M., 280, 292, *305*
Lissi, E., 228, 236, 243, 253, 294, *299, 305*
Liu, M. T. H., 321, 323, 327, 351, *360*
Ljunggren, S., 248, 249, *300*

Loew, L. M., 285, *303*
Lomax, T. D., 218, 221, *306*
Loosen, K., 327, *359*
Loudon, G. M., 28, *107*
Löwdin, P-O., 194, *210*
Lowrey, A. H., 131, *210*
Lucchese, R. R., 314, *359*
Ludman, C. J., 139, *210*
Lufimpadio, N., 281, *305*
Luthra, N. P., 20, 25, 28, 85, 89, *108*
Lüttringhaus, A., 34, 46, 57, *111*
Luz, Z., 198, *210*
Lynn, J. L., 281, *300*

Maass, G., 114, 117, 118, 149, 154, *205, 208, 210*
Macdonald, A. L., 129, *210*
Maciel, G. E., 138, *212*
Mack, D. P., 342, *358*
Mackay, R. A., 217, 218, 221, 237, 271, 273, 280, 282, *305*
Maggiora, G. M., 192, *208*
Maharaj, U., 14, 63, *111*
Maier, G., 314, *360*
Maitra, V., 292, *300*
Malaviya, S., 290, *305*
Mallick, I. M., 192, *210*
Mallon, C. B., 315, *359*
Manabe, O., 294, *308*
Mancini, G., 280, *299*
Mandel, G. S., 130, *210*
Mandolini, L., 2, 6, 9, 23, 25, 35, 37, 38, 39, 40, 41, 42, 43, 45, 46, 47, 48, 49, 50, 51, 52, 53, 54, 55, 56, 57, 59, 65, 66, 72, 76, 77, 79, 82, 83, 85, 90, 91, 100, 101, 104, 105, *107, 108, 109*
Månsson, M., 50, *107*
Mar, A., 61, 62, 74, *109*
March, J., 98, *109*
Marcus, R. A., 121, *207, 210*
Marinelli, F., 280, *299*
Markham, R., 89, *106*
Marsh, R. E., 130, 140, 166, *208, 210, 211*
Marshal, T. H., 197, *210*
Martin, C. A., 280, 282, *304, 305*
Martin, J. S., 136, *208*
Martinek, K., 224, 226, 227, 254, 257, *305*
Martins, A., 218, *302*
Martins-Franchetti, S. M., 228, 230, 236, 237, 265, 296, *300, 307*
Masci, B., 2, 9, 43, 46, 47, 48, 50, 51, 52, 53, 54, 57, 59, 65, 66, 72, 76, 79, 82, 83, 100, 101, 105, *107, 108, 109*
Mason, S. C., 81, *107*
Masri, F. N., 132, *207*
Mastropaolo, D., 131, *207*
Mataga, N., 321, 323, *359*
Matsui, Y., 292, *305*
Matsumoto, K., 263, 285, 287, *306*
Matsumoto, N., 285, *306*
Matsumoto, Y., 278, 285, 288, *305, 308*
Matsuo, K., 294, *308*
Matuszewski, B., 295, *303*
Mayer, J. E., 19, *109*
Mayer, M. G., 19, *109*
Mayumi, J., 72, 73, 82, *111*
Mazer, N. A., 219, 220, *305*
McAneny, M., 246, *300*
McAuliffe, M., 349, *358*
McCrann, P. M., 282, *305*
McCurdy, C. E. Jr., 326, *360*
McDonald, K., 328, *357*
McGregor, S. D., 315, *359*
McManus, S. P., 2, 28, 38, *107*
McMorrow, D., 147, *210*
McMurchie, L. E., 314, *358*
McNesby, J. R., 314, *359*
McOmie, J. F. W., 46, *106*
Meadows, J. H., 314, *359*
Meiboom, S., 198, *208, 210*
Meier, K., 314, *358*
Meisel, D., 218, *307*
Mengelsberg, I., 293, *305*
Menger, F. M., 85, 102, *109*, 215, 219, 220, 222, 223, 236, 245, 258, 260, 279, 280, 293, *305*
Meyer, C., 290, *305*
Meyer, G., 291, *308*
Mhala, M. M., 247, 248, 251, 289, 296, *299, 301, 305*
Miles, M. H., 150, 153, 154, 155, 156, 157, 159, *209, 210*
Mille, M., 221, 240, *305*
Miller, J., 256, *305*
Miller, J. M., 321, *360*
Miller, M. A., 16, 17, 18, 85, *106*
Miller, N., 166, *205*

Miller, W. H., 133, *207*
Milstien, S., 28, *109*
Minch, M. J., 244, 246, 279, *301, 305*
Minero, C., 291, *305*
Minniti, D., 98, 99, *110*
Miola, L., 228, 236, 243, 253, 294, *299, 305*
Missel, P. J., 219, 229, *305*
Mitamura, T., 67, 68, *110*
Mitchell, D. J., 219, *304*
Miyashiro, K., 284, *304*
Mizutani, T., 270, *305*
Moffatt, J. R., 221, 231, 232, 233, 237, 238, 239, 240, 242, 247, 248, 251, 255, 270, 289, 297, 298, *299, 301, 305*
Monarres, D., 247, 248, 289, *301*
Montgomery, C. R., 341, *358*
Montiero, P. M., 224, 227, 254, 296, *302*
Morawetz, H., 3, 4, 16, 35, 64, 72, *109*
More O'Ferrall, R. A., 125, 127, 165, 167, *210*
Morgan, A. G., 258, *307*
Moritani, I., 321, 323, 351, *359, 360*
Mortara, R. A., 285, *302*
Mortimer, J., 243, *305*
Moss, R. A., 259, 260, 263, 268, 269, 277, 278, 285, 286, 287, 292, *305, 306, 308,* 312, 315, 326, 349, *358, 359, 360*
Moss, R. E., 140, 186, 187, *205*
Motsavage, V. A., 217, 237, 254, *306*
Mueller, P. H., 314, 316, *360*
Muir, K. W., 129, *206, 210*
Mukerjee, P., 215, 219, 220, 221, 251, *306, 307*
Mukherjee, S., 270, *304*
Muller, N., 220, *306*
Munjal, R., 338, *359*
Munjal, R. C., 349, *360*
Murahashi, S. I., 321, 323, 351, *359, 360*
Murakami, Y., 263, 278, 285, 287, *306, 308*
Murata, A., 229, 230, 295, *303*
Murdoch, J. R., 121, *211*
Murray, K. K., 314, *359*
Murray, R. W., 321, 341, 349, *360, 361*
Murray-Rust, J., 190, *207*
Murray-Rust, P., 190, *207*
Musso, H., 133, *208*
Muthuramu, K., 294, *306*
Mutter, M., 71, 74, *108, 109, 110*
Mysels, K. J., 215, 219, *306*

Nagamatsu, S., 284, *308*
Nagy, J. B., 281, *305*
Nahas, R. C., 259, 263, 286, *306*
Nakano, A., 263, 285, 287, *306*
Nakashima, T., 134, 137, 142, *207*
Nango, M., 278, 287, *304*
Nazran, A. S., 312, 321, 323, 327, 333, 341, 342, 344, 349, 350, *358, 360*
Nemethy, G., 215, *307*
Neumark, D. M., 314, *359*
Newall, A. R., 327, *357*
Ng, P., 293, *301*
Nicholson, A. W., 28, *107*
Nicole, D., 184, *206*
Nicoli, D. F., 219, 220, 221, 240, 242, 270, *299, 300*
Nill, G., 277, 278, *300*
Nilsson, P.-G., 220, *304*
Ninham, B. W., 215, 219, 254, 270, *300, 303, 304, 306, 308*
Nishigaki, Y., 157, *208*
Nishino, M., 321, 323, 351, *359, 360*
Nolte, R. J. M., 265, 278, *308*
Nome, F., 219, 244, 281, 293, 296, 297, 298, *303, 304, 306, 307, 308*
Northcott, D. J., 321, 350, *360*
Novak, A., 130, 132, *211*
Nukina, S., 97, *110*

Obi, K., 147, *209*
O'Connor, C. J., 218, 221, 284, 287, *306*
Ogden, S. D., 120, 126, 127, *208*
Ogino, H., 89, *109*
Oh, S., 121, *210*
Ohkubo, K., 278, *306*
Ohkubo, R., 285, *306*
Ohlinger, H., 103, *111*
Ohmenzetter, K., 253, *301*
Ohno, K., 349, *360*
Ohta, N., 285, *306*
Okada, Y., 147, *209*

Okahata, Y., 222, 273, 275, 285, *304, 306*
Okahata, Y., 286, *304*
Okamoto, K., 277, *306*
Ollis, W. D., 46, *106*
O'Neal, H. E., 18, 23, 55, 81, *109*
O'Neill, S. V., 314, *360*
Ono, S., 278, 288, *304, 307, 309*
Ono, Y., 327, 350, *360*
Onyivruka, S. O., 279, *307*
Orchin, M., 292, *305*
Orpen, A. G., 140, 187, *206*
Orville-Thomas, W. J., 127, *211*
Ostlund, R. E., 154, 155, 156, *210*
Otsubo, Y., 288, *309*
Otsuka, Y., 284, *304*
Ott, H., 40, *108*
Owen, J., 166, *205*

Page, M. I., 3, 9, 25, 26, 27, 81, 85, 99, 100, *107, 109*, 290, *303*
Paik, C. H., 236, 287, 296, *300, 301*
Pandit, U. K., 28, *107*
Pang, E. K. C., 139, *207, 210*
Paquette, L. A., 50, *109*
Park, C. H., 185, 186, *211*
Park, J. M., 350, *360*
Park, K. H., 258, *307*
Paskovich, D. H., 350, *361*
Pastro, D. J., 79, *109*
Paul, K., 244, *304*
Pawlak, Z., 132, *211*
Pearson, R. G., 256, *307*
Pedersen, C. J., 185, *211*
Pedley, J. B., 17, *109*
Pelizzetti, E., 218, 291, *302, 305, 307*
Pellerite, M. J., 121, *211*
Perkin, W. H., 2, *109*
Perlmutter-Hayman, B., 177, 178, 179, 183, *211*
Person, W. B., 15, *109*
Pessin, J., 263, *303*
Peters, F., 199, *210*
Petride, H., 93, 94, *107*
Pfeiffer, M., 102, *110*
Phalon, P., 294, *303*
Phillips, L. A., 148, *212*
Pickett, H. M., 133, *211*
Pilcher, G., 15, 16, *107, 110*
Pillersdorf, A., 259, 287, *307*
Pimentel, G. C., 15, *109*
Piszkiewicz, D., 223, 277, *307*
Pitzer, K. S., 13, 15, *109*
Pizer, R., 189, *211*
Platz, M. S., 326, 333, 341, 349, 351, *358, 359, 360, 361*
Ploss, G., 133, *208*
Politi, M. J., 228, 230, 236, 237, 265, 294, 296, 297, *307*
Poole, C. P. Jr., 322, *360*
Porter, A. J., 287, *306*
Porter, G., 322, *360*
Portnoy, C. E., 215, 222, 223, 236, *305*
Potapov, V. M., 46, *109*
Powell, C. E., 277, *306*
Powell, M. F., 175, *210*
Powell, R. E., 27, *110*
Pramauro, E., 218, 291, *305, 307*
Prelog, V., 34, *110*
Probst, S., 297, *303*
Pullman, A., 194, *211*
Pullman, B., 194, *210, 211*
Pyter, R. A., 221, 251, *307*

Quan, C., 222, 226, 227, 264, 274, 275, 277, *299, 300, 301*
Quina, F. H., 221, 227, 228, 229, 230, 233, 236, 237, 243, 253, 262, 265, 269, 285, 294, 296, 297, *299, 300, 302, 305, 307*

Rabinow, B. E., 324, 343, 349, *358*
Radom, L., 133, *207*
Rak, S., 321, 323, 346, *360*
Ralph, E. K., 123, *208*
Ramachandran, B. R., 60, 61, *108*
Ramachandran, C., 221, 251, *307*
Ramage, R. E., 218, 221, 284, *306*
Ramamurthy, V., 294, *306, 307*
Ramaswami, S., 259, 263, 285, *306*
Ramnath, N., 294, *306, 307*
Rasmussen, S. E., 27, *110*
Ratajczak, H., 127, *211*
Rav-Acha, C., 263, *307*
Ray, A., 215, *307*
Reddy, I. A. K., 290, *307*
Regen, S. L., 57, *109*, 281, *307*

Reger, D. W., 277, *306*
Rehage, H., 220, *304*
Reinsborough, V. C., 242, *304*
Reisenauer, H. P., 314, *360*
Rentzepis, P. M., 146, 147, *206, 209*
Rettig, K. R., 343, *359*
Reverdy, G., 338, 348, *358, 360*
Rezende, M. C., 244, 293, 297, 298, *303, 307, 308*
Rhee, H. K., 279, 280, *305*
Rhee, J. V., 279, 280, *305*
Ribaldo, E. J., 230, *300*
Rideout, D., 292, *300*
Ridl, B. A., 144, 145, *210*
Riemann, A., 314, *360*
Ritchie, C. D., 256, *307*
Robbins, H. J., 166, 167, 173, *205, 209*
Robert, J-B., 166, *208*
Roberts, J. D., 166, *208*
Robertson, R. E., 279, *304*
Robinson, B. H., 218, *307*
Robinson, L., 228, *300*
Rochester, C. H., 166, *211*
Rodenas, E., 223, 240, 255, 297, *301, 307*
Rodulfo, T., 218, *299*
Roelens, S., 47, *109*
Romanesio, L. S., 219, *304*
Romeo, R., 79, 98, 99, *110*
Romsted, L. S., 221, 224, 225, 226, 227, 228, 229, 230, 231, 232, 235, 236, 237, 238, 239, 240, 241, 242, 244, 252, 253, 257, 258, 261, 262, 264, 265, 266, 267, 268, 269, 270, 274, 275, 292, 295, 297, 298, *299, 301, 306, 307*
Rondan, N. G., 314, 316, *360*
Roos, B., 133, *209*
Roos, B. O., 314, *360*
Rose, M. C., 158, 178, *211*
Rosenberg, S., 193, *211*
Rossa, L., 2, 102, *110*
Rouvé, A., 4, 5, 6, 8, 31, 33, 34, *110*
Roux, D., 217, 271, *299*
Rowe, J. E., 289, *300*
Rowe, W. F., 133, 141, *206*
Rozeboom, M. D., 91, *110*
Rozière, J., 130, *211*
Rubin, R. J., 258, *301*
Rubira, A. F., 244, 293, 296, *306, 307*
Rüchardt, C., 293, *305*
Ruggli, P., 4, 103, *110*
Rumpel, H., 143, *210*
Rundle, R. E., 131, *208, 212*
Rupert, L. A. M., 293, *307*
Russell, J. C., 220, *307*
Ruzicka, L., 40, 85, 102, *110*
Rydholm, R., 224, 229, *299*
Rylance, J., 17, *109*

Salomon, G., 2, 5, 31, 75, *110*
Sanders, W. J., 286, *306*
Sandorfy, C., 127, *211*
Sango, D. B., 297, *300*
Sarel, S., 263, *307*
Sarfaty, R., 177, 178, *211*
Satake, I., 297, *304*
Saunders, D. S., 63, 74, *111*
Saunders, M., 138, *211*
Saunders, S., 138, *211*
Saunders, W. H., 35, *108*
Savelli, G., 221, 227, 231, 232, 235, 237, 239, 240, 242, 247, 248, 258, 261, 264, 266, 267, 270, 289, 291, 298, *299, 301, 302, 305, 307*
Savino, T. G., 333, 341, *358*
Sawada, M., 256, *307*
Sawyer, W. H., 295, *300*
Sayer, J. M., 120, 193, *211*
Sbriziolo, C., 291, *302*
Scaiano, J. C., 295, *307*, 312, 327, 330, 333, 341, 342, 344, 349, 351, *358, 359, 361*
Scanlan, M. J., 194, *211*
Schaad, L. J., 127, *209*
Schaefer, W. P., 140, *211*
Schaeffer, H. F., 133, *207*, 314, *357, 359, 360*
Schanze, K. S., 294, *308*
Schauble, J. H., 142, *207*
Schauze, K. S., 220, *308*
Schechter, H., 338, *359*
Schenck, H., 55, *109*
Schiffman, R., 263, *307*
Schinz, H., 102, *110*
Schneider, H., 189, 190, *207*
Schonbaum, G. R., 191, *206*
Schorr, W., 220, *304*
Schowen, R. L., 192, *208*

Schreck, R. P., 269, 285, *306*
Schreir, S., 236, 255, 285, 296, *302*
Schrock, A. K., 321, 323, 344, *358*
Schroter, E. M., 224, 227, 254, 296, *302*
Schulz, G., 133, *208*
Schuster, G. B., 321, 323, 327, 328, 331, 332, 338, 341, 344, 346, 347, 348, 349, *358, 359, 360, 361*
Schuster, P., 127, *211*
Schwab, A. P., 79, 98, *110*
Schwarzenbach, G., 11, 27, *110*
Searles, S., 91, 97, *110*
Sekiguchi, A., 321, 323, 351, *360*
Sellers, P., 13, *110*
Semlyen, J. A., 3, 69, 70, 71, *108, 110*
Sengupta, P. K., 146, *211*
Senthilnathan, V. P., 326, 351, *360*
Sepulveda, L., 222, 224, 225, 227, 228, 235, 237, 246, 253, 254, 261, 266, 297, *299, 301, 303, 305, 307*
Serratrice, G., 184, *206*
Serve, M. P., 79, *109*
Sessions, R. B., 140, 186, 187, *205, 206*
Seyedrezai, S. E., 91, *110*
Shaw, B. L., 89, *106*
Shen, Y. H., 342, *359*
Shevlin, P. B., 317, *357*
Shimada, K., 57, 58, 59, 60, *107, 110*
Shimizu, M. R., 284, *303*
Shimozato, Y., 58, 59, *110*
Shin, J.-S., 269, *306*
Shinar, R., 177, 178, 179, 183, *211*
Shine, H. J., 258, *307*
Shinkai, S., 215, 218, 222, 259, 260, 273, 275, 286, 294, *304, 308*
Shoatake, K., 314, *359*
Shosenji, H., 278, 288, *304, 307, 309*
Si, V., 270, *307*
Sicher, J., 16, 35, *110*
Sieghhahn, P. M., 314, *360*
Silfvast, W. T., 178, *209*
Silbermann, W. E., 28, *110*
Silver, S. M., 193, *211*
Simmons, H. E., 185, 186, *211*
Simons, J. W., 314, *360*
Simpson, G. R., 151, 167, 172, 178, 181, 182, *209*
Simsohn, H., 220, *306*
Singer, L. A., 148, *208*
Singh, T. R., 132, *207*
Sinke, G. C., 13, 17, 19, 22, 27, *110*
Sisido, M., 52, 59, 66, 67, 68, 74, *110*
Sitzmann, E. V., 317, 320, 337, 344, 349, 350, 351, 352, *359, 360*
Skell, P. S., 313, 314, 329, 349, *357, 360*
Skinner, H. A., 15, *110*
Skipper, P. L., 199, *208*
Sluma, H.-D., 327, *359*
Slusarczuk, G. M. J., 10, 69, *107*
Smith, H. J., 235, 258, 297, *301*
Smith, J. A. S., 139, *207, 210*
Smith, K. K., 147, *211*
Smith, P. B., 187, 188, *211*
Smith, Z., 133, 141, *206*
Songstad, J., 256, *307*
Sorensen, P. E., 190, *209*
Soto, R., 228, 291, *303, 308*
Sparks, R. K., 314, *359*
Speakman, J. C., 129, 130, *206, 208, 210*
Spinner, E., 136, 145, *208*
Sprague, J. T., 18, *106*
Srivastava, S. K., 223, 236, 290, *308*
Stadler, E., 244, 298, *308*
Steele, W. R. S., 135, 136, 165, 166, *205*
Stein, S. E., 26, *110*
Stener, A., 291, *305*
Stevens, E. D., 129, *211*
Stevens, G., 327, 335, *357*
Stevens, K. W. H., 321, *360*
Stewart, L. C., 295, *307*
Stewart, R., 166, *207*
Steytler, D. C., 218, *307*
Stigter, D., 219, 240, 242, *308*
Stirling, C. J. M., 77, 79, 89, 91, *110*
Stockmayer, W. H., 10, 69, *109*
Stoll, M., 4, 5, 6, 8, 31, 33, 34, 35, 102, *107, 110*
Stoll-Comte, G., 4, 5, 6, 8, 31, *110*
Storm, C. B., 143, *211*
Storm, D. R., 27, 71, *110*
Strachan, E., 322, *360*
Strandjord, A. J. G., 146, *212*
Strauss, G., 278, *308*
Strazielle, C., 217, *309*
Stridh, G., 13, *110*
Strohbusch, F., 132, *211*
Stuehr, J. E., 158, 178, *211*
Stull, D. R., 13, 17, 19, 22, 27, *110*
Sturgeon, M. E., 292, 294, *299*

Subramanian, R., 327, *359*
Suckling, C. J., 279, *307*
Sudhölter, E. J. R., 214, *308*
Sudjak, R. L., 339, *360*
Sugahara, K., 278, *306*
Sugamori, S. E., 330, 349, *358*
Sugawara, T., 321, 323, 351, 357, *360*
Sukenik, C. N., 277, 280, 292, *305, 308*
Sullivan, M. J., 138, *212*
Sunamoto, J., 284, *308*
Sundberg, R. J., 98, *110*
Sunner, S., 13, 14, *110*
Sunshine, W. L., 277, *306*
Sutter, J. K., 280, 292, *308*
Suter, U. W., 71, 74, *108, 109, 110*
Suzuki, N., 292, *299*
Swarup, S., 260, 268, 278, *306, 308*
Sydnes, L. K., 292, 294, *299*
Szeverenyi, N. M., 138, *212*
Szwarc, M., 57, 58, 59, 60, *107, 110*

Tack, R. D., 218, *307*
Taft, R. W., 157, *212*
Taguchi, T., 269, 285, *306*
Takagi, H., 67, 68, *110*
Takei, S. J., 284, *308*
Takeuchi, H., 147, *209*
Talkowski, C. J., 277, *306*
Talmon, Y., 270, *308*
Tamborra, P., 35, 37, 40, 79, 91, *108*
Tanaka, I., 147, *209*
Tanford, C., 215, 219, 220, *308*
Taniguchi, Y., 290, *308*
Tanimoto, Y., 147, *209*
Teklu, Y., 143, *211*
Templeton, D. H., 130, *209*
Tepley, L. B., 228, 272, *305*
Terrier, F., 152, 163, *206*
Thamavit, C., 232, 237, 238, 298, *301*
Thistlethwaite, P. J., 147, *212*
Thomas, J. K., 218, 220, 270, *304, 308*
Thomson, A., 244, *308*
Thurn, H., 220, *304*
Tickle, P., 327, 335, *357*
Tobe, M. L., 79, 98, 99, *110*
Tokumura, K., 147, *209*
Tolbert, M. A., 148, *212*
Tomioka, H., 327, 349, *360*
Tonellato, U., 259, 260, 261, 263, 265, 286, 288, *299, 303, 308*
Topor, M. G., 314, *358*
Toscano, V. G., 294, *305*
Toullec, J., 115, *212*
Trainor, G. L., 278, *308*
Trewella, J. C., 135, 138, 141, *209*
Tribble, M. T., 16, 17, 18, 85, *106*
Tricot, Y.-M., 281, *308*
Trifunac, A. D., 328, *358*
Trozzolo, A. M., 321, 322, 323, 330, 341, 349, 351, *358, 360, 361*
Trueman, R. E., 63, 74, 111
Truter, M. R., 166, *212*
Tse, A., 134, 137, 142, *207*
Turig, C., 295, *303*
Turner, A., 76, *107*
Turro, N. J., 281, 295, *303, 308,* 312, 349, *358*
Tusk, M., 132, *211*

Ueoka, R., 278, 285, 288, *305, 306, 308*
Ulstrup, J., 190, *209*
Umoh, S. A., 293, *301*
Utrapiromsuk, N., 290, *299*

Valuer, B., 284, *299*
Vanderkooi, G., 221, 240, *305*
van der Langkruis, G. B., 214, 231, 257, 280, *308*
van der Zee, N. T. E., 236, 255, 285, 296, *302*
Van Dine, G. W., 313, 316, 322, *359*
van Truong, N., 190, *207*
Varvoglis, A. G., 244, *304*
Venkatasubban, K. S., 245, 293, *305*
Vera, S., 240, 297, *307*
Vickery, B. L., 166, *212*
Viera, R. C., 284, *302, 303*
Vincent, M. A., 133, *207*
Viout, P., 248, 290, 291, *303, 305, 308*
Visser, H. G. J., 265, 278, *308*
Vögtle, F., 2, 102, *110*

Waddington, T. C., 139, *210*
Walker, C. R., 130, *206*
Walsh, W. M., 156, *209*

Wang, P. S., 313, *357*
Wang, Y., 317, *360*
Ward, M. D., 280, *304*
Ware, W. R., 327, 350, *360*
Warner, P. M., 327, *361*
Wasserman, E., 313, 321, 322, 341, 349, 351, *360, 361*
Watt, I., 81, *107*
Webb, S. P., 148, *212*
Weber, W. P., 280, *308*
Weed, G. C., 295, *308*
Weedon, A. C., 292, 294, *299*
Weeks, B., 89, *106*
Wei, G. J., 270, *306*
Weiss, S., 141, *207*
Weissberger, A., 114, *208*
Welge, K. H., 314, *358*
Weller, A. H., 146, *206*
Wennerström, H., 133, 136, 137, 138, 141, 142, 166, *206, 208, 209,* 214, 215, 216, 219, 220, 221, 240, 243, *303, 304, 305, 308*
Wentrup, C., 312, *361*
Werstiuk, N. H., 330, 349, *361*
Wertz, D. H., 16, 85, *106*
Westrum, E. F., 13, 17, 19, 22, 27, *110*
White, D. N. J., 129, *206*
Whitsel, B. L., 349, *358*
Whitesell, L. G., 260, *305*
Whittaker, D., 327, 343, 349, *357*
Whitten, D. G., 220, 270, 294, *300, 302, 305, 307, 308*
Whittington, S. G., 63, 74, *111*
Wightman, P. J., 221, *303*
Willen, B. H., 314, 316, *360*
Williams, D. E., 130, 131, *212*
Wilson, A. A., 279, *307*
Wilson, E. B., 133, 141, *206*
Winger, R., 70, *107*
Winkle, J. R., 220, 294, *308*
Winnik, M. A., 2, 3, 12, 14, 41, 42, 46, 59, 61, 62, 63, 64, 74, 75, 82, 89, *107, 109, 111*
Winstein, S., 79, *111*
Winterman, D. R., 135, 136, 165, 166, *205*
Wirz, J., 135, 167, *209*
Wires, R. A., 79, *109*
Wohlgemuth, K., 34, 46, 57, *111*
Wolfe, B., 224, *301*
Wolff, R., 279, *305*
Wong, M. P., 292, *308*
Wong, P. C., 333, 341, *358, 361*
Wood, J. L., 128, 132, *207, 212*
Woodworth, R. L., 329, *360*
Woolf, A. A., 166, *212*
Woolf, G. J., 147, *212*
Woop, L. S., 313, *357*
Worsham, P. R., 294, *308*
Wright, B. B., 326, 349, 351, *358, 360, 361*
Wright, J. L., 244, 246, *301*
Wulff, C. A., 14, *110*
Wyatt, P. A. H., 146, *209*

Yager, W. A., 313, 321, 322, 351, *360, 361*
Yamada, K., 278, 288, *304, 307, 309*
Yamaguchi, M., 192, *210*
Yamamoto, Y., 321, 323, *359*
Yamashita, Y., 72, 73, 82, *111*
Yamdagni, R., 79, *111*
Yang, K.-U., 222, 246, *301*
Yang, Z.-Y., 222, 262, 264, 275, *299*
Yannoni, C. S., 143, *210*
Yatsimirski, A. K., 224, 226, 227, 254, 257, *305*
Yeh, S. W., 148, *212*
Yiv, S., 217, *309*
Yoneda, H., 277, *306*
Yoshida, N., 180, *212*
Yoshikawa, E., 52, 66, 68, *110*
Yoshimatsu, A., 263, 285, 287, *306*
Yoshinaga, H., 245, 284, 293, *305, 308*
Yoshinaga, K., 278, *306*
Yoshioka, T., 157, *208*
Young, C. Y., 219, 220, *305*

Zacharias, H., 314, *358*
Zalkin, A., 130, *209*
Zalkow, V., 29, 85, *106*
Zana, R., 217, 240, 242, 270, 272, *305, 309*
Zanette, D., 229, 244, 269, 297, 298, *302, 303, 308*
Zare, R. N., 314, *359*
Zebelman, D., 145, *206*
Zeiss, G. D., 313, 316, 322, *359*

Zemb, T., 220, *302*
Zerner, B., 191, *206*
Ziegler, K., 2, 6, 32, 34, 47, 49, 57, 103, *111*
Zimmerman, H. E., 350, *361*
Zimmit, M. B., 281, *303*
Zmuda, H., 258, *307*
Zucco, C., 297, *303*
Zundel, G., 127, 132, *210, 211, 212*
Zupancic, J. J., 321, 323, 327, 332, 341, 347, 349, *358, 361*

Cumulative Index of Authors

Ahlberg, P., **19**, 223
Albery, W. J., **16**, 87
Allinger, N. I., **13**, 1
Anbar, M., **7**, 115
Arnett, E. M., **13**, 83
Bard, A. J., **13**, 155
Bell, R. P., **4**, 1
Bennett, J. E. **8**, 1
Bentley, T. W., **8**, 151; **14**, 1
Berger, S., **16**, 239
Bethell, D., **7**, 153; **10**, 53
Blandamer, M. J., **14**, 203
Brand, J. C. D., **1**, 365
Brändström, A., **15**, 267
Brinkman, M. R., **10**, 53
Brown, H. C., **1**, 35
Buncel, E., **14**, 133
Bunton, C. A., **22**, 213
Cabell-Whiting, P. W., **10**, 129
Cacace, F., **8**, 79
Capon, B., **21**, 37
Carter, R. E., **10**, 1
Collins, C. J., **2**, 1
Cornelisse, J., **11**, 225
Crampton, M. R., **7**, 211
Davidson, R. S., **19**, 1; **20**, 191
Desvergne, J. P., **15**, 63
de Gunst, G. P., **11**, 225
de Jong, F., **17**, 279
Dosunmu, M. I., **21**, 37
Eberson, L., **12**, 1; **18**, 79
Engdahl, C., **19**, 223
Farnum, D. G., **11**, 123
Fendler, E. J., **8**, 271
Fendler, J. H., **8**, 271; **13**, 279
Ferguson, G., **1**, 203
Fields, E. K., **6**, 1
Fife, T. H., **11**, 1
Fleischmann, M., **10**, 155
Frey, H. M., **4**, 147
Gilbert, B. C., **5**, 53
Gillespie, R. J., **9**, 1
Gold, V., **7**, 259
Goodin, J. W., **20**, 191
Gould, I. R., **20**, 1
Greenwood, H. H., **4**, 73
Hammerich, O., **20**, 55
Havinga, E., **11**, 225
Hibbert, F., **22**, 113
Hine, J., **15**, 1
Hogen-Esch, T. E., **15**, 153
Hogeveen, H., **10**, 29 129
Ireland, J. F., **12**, 131
Johnson, S. L., **5**, 237
Johnstone, R. A. W., **8**, 151
Jonsäll, G., **19**, 223
José, S. M., **21**, 197
Kemp, G., **20**, 191
Kice, J. L., **17**, 65
Kirby, A. J., **17**, 183
Kohnstam, G., **5**, 121
Kramer, G. M., **11**, 177
Kreevoy, M. M., **6**, 63; **16**, 87
Kunitake, T., **17**, 435
Ledwith, A., **13**, 155
Liler, M., **11**, 267
Long, F. A., **1**, 1
Maccoll, A., **3**, 91
Mandolini, L., **22**, 1
McWeeny, R., **4**, 73
Melander, L., **10**, 1
Mile, B., **8**, 1
Miller, S. I., **6**, 185
Modena, G., **9**, 185
More O'Ferrall, R. A., **5**, 331
Morsi, S. E., **15**, 63
Neta, P., **12**, 223
Norman, R. O. C., **5**, 33
Nyberg, K., **12**, 1
Olah, G. A., **4**, 305
Parker, A. J., **5**, 173
Parker, V. D., **19**, 131; **20**, 55
Peel, T. E., **9**, 1
Perkampus, H. H., **4**, 195
Perkins, M. J., **17**, 1
Pittman, C. U. Jr., **4**, 305
Pletcher, D., **10**, 155
Pross, A., **14**, 69; **21**, 99
Ramirez, F., **9**, 25
Rappoport, Z., **7**, 1
Reeves, L. W., **3**, 187
Reinhoudt, D. N., **17**, 279
Ridd, J. H., **16**, 1
Riveros, J. M., **21**, 197
Robertson, J. M., **1**, 203
Rosenthal, S. N., **13**, 279
Samuel, D., **3**, 123
Sanchez, M. de N. de M., **21**, 37
Savelli, G., **22**, 213
Schaleger, L. L., **1**, 1
Scheraga, H. A., **6**, 103
Schleyer, P. von R., **14**, 1
Schmidt, S. P., **18**, 187
Schuster, G. B., **18**, 187; **22**, 311
Scorrano, G., **13**, 83
Shatenshtein, A. I., **1**, 156
Shine, H. J., **13**, 155
Shinkai, S., **17**, 435

Silver, B. L., **3**, 123
Simonyi, M., **9**, 127
Stock, L. M., **1**, 35
Symons, M. C. R., **1**, 284
Takashima, K., **21**, 197
Tedder, J. M., **16**, 51
Thomas, A., **8**, 1
Thomas, J. M., **15**, 63
Tonellato, U., **9**, 185
Toullec, J., **18**, 1
Tüdös, F., **9**, 127
Turner, D. W., **4**, 31
Turro, N. J., **20**, 1
Ugi, I., **9**, 25
Walton, J. C., **16**, 51
Ward, B., **8**, 1
Westheimer, F. H., **21**, 1
Whalley, E., **2**, 93
Williams, D. L. H., **19**, 381
Williams, J. M. Jr., **6**, 63
Williams, J. O., **16**, 159
Williamson, D. G., **1**, 365
Wilson, H., **14**, 133
Wolf, A. P., **2**, 201
Wyatt, P. A. H., **12**, 131
Zimmt, M. B., **20**, 1
Zollinger, H., **2**, 163
Zuman, P., **5**, 1

Cumulative Index of Titles

Abstraction, hydrogen atom, from O–H bonds, **9**, 127
Acid solutions, strong, spectroscopic observation of alkylcarbonium ions in, **4**, 305
Acid-base properties of electronically excited states of organic molecules, **12**, 131
Acids and bases, oxygen and nitrogen in aqueous solution, mechanisms of proton transfer between, **22**, 113
Acids, reactions of aliphatic diazo compounds with, **5**, 331
Acids, strong aqueous, protonation and solvation in, **13**, 83
Activation, entropies of, and mechanisms of reactions in solution, **1**, 1
Activation, heat capacities of, and their uses in mechanistic studies, **5**, 121
Activation, volumes of, use for determining reaction mechanisms, **2**, 93
Addition reactions, gas-phase radical, directive effects in, **16**, 51
Aliphatic diazo compounds, reactions with acids, **5**, 331
Alkylcarbonium ions, spectroscopic observation in strong acid solutions, **4**, 305
Ambident conjugated systems, alternative protonation sites in, **11**, 267
Ammonia, liquid, isotope exchange reactions of organic compounds in **1**, 156
Aqueous mixtures, kinetics of organic reactions in water and, **14**, 203
Aromatic photosubstitution, nucleophilic, **11**, 225
Aromatic substitution, a quantitative treatment of directive effects in, **1**, 35
Aromatic substitution reactions, hydrogen isotope effects in, **2**, 163
Aromatic systems, planar and non-planar, **1**, 203
Aryl halides and related compounds, photochemistry of, **20**, 191
Arynes, mechanisms of formation and reactions at high temperatures, **6**, 1
A-S_E2 reactions, developments in the study of, **6**, 63

Base catalysis, general, of ester hydrolysis and related reactions, **5**, 237
Basicity of unsaturated compounds, **4**, 195
Bimolecular substitution reactions in protic and dipolar aprotic solvents, **5**, 173

^{13}C N.M.R. spectroscopy in macromolecular systems of biochemical interest, **13**, 279
Carbene chemistry, structure and mechanism in, **7**, 163
Carbenes having aryl substituents, structure and reactivity of, **22**, 311
Carbanion reactions, ion-pairing effects in **15**, 153
Carbocation rearrangements, degenerate, **19**, 223
Carbon atoms, energetic, reactions with organic compounds, **3**, 201
Carbon monoxide, reactivity of carbonium ions towards, **10**, 29
Carbonium ions (alkyl), spectroscopic observation in strong acid solutions, **4**, 305
Carbonium ions, gaseous, from the decay of tritiated molecules, **8**, 79
Carbonium ions, photochemistry of, **10**, 129
Carbonium ions, reactivity towards carbon monoxide, **10**, 29
Carbonyl compounds, reversible hydration of, **4**, 1
Carbonyl compounds, simple, enolisation and related reactions of, **18**, 1
Carboxylic acids, tetrahedral intermediates derived from, spectroscopic detection and investigation of their properties, **21**, 37

Catalysis by micelles, membranes and other aqueous aggregates as models of enzyme action, **17**, 435
Catalysis, enzymatic, physical organic model systems and the problem of, **11**, 1
Catalysis, general base and nucleophilic, of ester hydrolysis and related reactions, **5**, 237
Catalysis, micellar, in organic reactions; kinetic and mechanistic implications, **8**, 271
Catalysis, phase-transfer by quaternary ammonium salts, **15**, 267
Cation radicals in solution, formation, properties and reactions of, **13**, 155
Cation radicals, organic, in solution, kinetics and mechanisms of reaction of, **20**, 55
Cations, vinyl, **9**, 135
Chain molecules, intramolecular reactions of, **22**, 1
Charge density–N.M.R. chemical shift correlations in organic ions, **11**, 125
Chemically induced dynamic nuclear spin polarization and its applications, **10**, 53
Chemiluminescence of organic compounds, **18**, 187
CIDNP and its applications, **10**, 53
Conduction, electrical, in organic solids, **16**, 159
Configuration mixing model: a general approach to organic reactivity, **21**, 99
Conformations of polypeptides, calculations of, **6**, 103
Conjugated, molecules, reactivity indices, in, **4**, 73
Crown-ether complexes, stability and reactivity of, **17**, 279

D_2O–H_2O mixtures, protolytic processes in, **7**, 259
Degenerate carbocation rearrangements, **19**, 223
Diazo compounds, aliphatic, reactions with acids, **5**, 331
Diffusion control and pre-association in nitrosation, nitration, and halogenation, **16**, 1
Dimethyl sulphoxide, physical organic chemistry of reactions, in, **14**, 133
Dipolar aprotic and protic solvents, rates of bimolecular substitution reactions in, **5**, 173
Directive effects in aromatic substitution, a quantitative treatment of, **1**, 35
Directive effects in gas-phase radical addition reactions, **16**, 51
Discovery of the mechanisms of enzyme action, 1947–1963, **21**, 1
Displacement reactions, gas-phase nucleophilic, **21**, 197

Effective molarities of intramolecular reactions, **17**, 183
Electrical conduction in organic solids, **16**, 159
Electrochemical methods, study of reactive intermediates by, **19**, 131
Electrochemistry, organic, structure and mechanism in, **12**, 1
Electrode processes, physical parameters for the control of, **10**, 155
Electron spin resonance, identification of organic free radicals by, **1**, 284
Electron spin resonance studies of short-lived organic radicals, **5**, 23
Electron-transfer reactions in organic chemistry, **18**, 79
Electronically excited molecules, structure of, **1**, 365
Electronically excited states of organic molecules, acid-base properties of, **12**, 131
Energetic tritium and carbon atoms, reactions of, with organic compounds, **2**, 201
Enolisation of simple carbonyl compounds and related reactions, **18**, 1
Entropies of activation and mechanisms of reactions in solution, **1**, 1
Enzymatic catalysis, physical organic model systems and the problem of, **11**, 1
Enzyme action, catalysis by micelles, membranes and other aqueous aggregates as models of, **17**, 435

Enzyme action, discovery of the mechanisms of, 1947–1963, **21**, 1
Equilibrium constants, N.M.R. measurements of, as a function of temperature, **3**, 187
Ester hydrolysis, general base and nucleophilic catalysis, **5**, 237
Exchange reactions, hydrogen isotope, of organic compounds in liquid ammonia, **1**, 156
Exchange reactions, oxygen isotope, of organic compounds, **2**, 123
Excited complexes, chemistry of, **19**, 1
Excited molecules, structure of electronically, **1**, 365

Force-field methods, calculation of molecular structure and energy by, **13**, 1
Free radicals, identification by electron spin resonance, **1**, 284
Free radicals and their reactions at low temperature using a rotating cryostat, study of **8**, 1

Gaseous carbonium ions from the decay of tritiated molecules, **8**, 79
Gas-phase heterolysis, **3**, 91
Gas-phase nucleophilic displacement reactions, **21**, 197
Gas-phase pyrolysis of small-ring hydrocarbons, **4**, 147
General base and nucleophilic catalysis of ester hydrolysis and related reactions, **5**, 237

H_2O—D_2O mixtures, protolytic processes in, **7**, 259
Halogenation, nitrosation, and nitration, diffusion control and pre-association in, **16**, 1
Halides, aryl, and related compounds, photochemistry of, **20**, 191
Heat capacities of activation and their uses in mechanistic studies, **5**, 121
Heterolysis, gas-phase, **3**, 91
Hydrated electrons, reactions of, with organic compounds, **7**, 115
Hydration, reversible, of carbonyl compounds, **4**, 1
Hydrocarbons, small-ring, gas-phase pyrolysis of, **4**, 147
Hydrogen atom abstraction from O–H bonds, **9**, 127
Hydrogen isotope effects in aromatic substitution reactions, **2**, 163
Hydrogen isotope exchange reactions of organic compounds in liquid ammonia, **1**, 156
Hydrolysis, ester, and related reactions, general base and nucleophilic catalysis of, **5**, 237

Intermediates, reactive, study of, by electrochemical methods, **19**, 131
Intermediates, tetrahedral, derived from carboxylic acids, spectroscopic detection and investigation of their properties, **21**, 37
Intramolecular reactions, effective molarities for, **17**, 183
Intramolecular reactions of chain molecules, **22**, 1
Ionization potentials, **4**, 31
Ion-pairing effects in carbanion reactions, **15**, 153
Ions, organic, charge density–N.M.R. chemical shift correlations, **11**, 125
Isomerization, permutational, of pentavalent phosphorus compounds, **9**, 25
Isotope effects, hydrogen, in aromatic substitution reactions, **2**, 163
Isotope effects, magnetic, magnetic field effects and, on the products of organic reactions, **20**, 1
Isotope effects, steric, experiments on the nature of, **10**, 1

Isotope exchange reactions, hydrogen, of organic compounds in liquid ammonia, **1**, 150
Isotope exchange reactions, oxygen, of organic compounds, **3**, 123
Isotopes and organic reaction mechanisms, **2**, 1

Kinetics and mechanisms of reactions of organic cation radicals in solution, **20**, 55
Kinetics, reaction, polarography and, **5**, 1
Kinetics of organic reactions in water and aqueous mixtures, **14**, 203

Least nuclear motion, principle of, **15**, 1

Macromolecular systems of biochemical interest, ^{13}C N.M.R. spectroscopy in **13**, 279
Magnetic field and magnetic isotope effects on the products of organic reactions, **20**, 1
Mass spectrometry, mechanisms and structure in: a comparison with other chemical processes, **8**, 152
Mechanism and structure in carbene chemistry, **7**, 153
Mechanism and structure in mass spectrometry: a comparison with other chemical processes, **8**, 152
Mechanism and structure in organic electrochemistry, **12**, 1
Mechanisms and reactivity in reactions of organic oxyacids of sulphur and their anhydrides, **17**, 65
Mechanisms, nitrosation, **19**, 381
Mechanisms of proton transfer between oxygen and nitrogen acids and bases in aqueous solution, **22**, 113
Mechanisms, organic reaction, isotopes and, **2**, 1
Mechanisms of reaction in solution, entropies of activation and, **1**, 1
Mechanisms of solvolytic reactions, medium effects on the rates and, **14**, 10
Mechanistic applications of the reactivity–selectivity principle, **14**, 69
Mechanistic studies, heat capacities of activation and their use, **5**, 121
Medium effects on the rates and mechanisms of solvolytic reactions, **14**, 1
Meisenheimer complexes, **7**, 211
Methyl transfer reactions, **16**, 87
Micellar catalysis in organic reactions: kinetic and mechanistic implications, **8**, 271
Micelles, aqueous, and similar assemblies, organic reactivity in, **22**, 213
Micelles, membranes and other aqueous aggregates, catalysis by, as models of enzyme action, **17**, 435
Molecular structure and energy, calculation of, by force-field methods, **13**, 1

Nitration, nitrosation, and halogenation, diffusion control and pre-association in, **16**, 1
Nitrosation mechanisms, **19**, 381
Nitrosation, nitration, and halogenation, diffusion control and pre-association in, **16**, 1
N.M.R. chemical shift–charge density correlations, **11**, 125
N.M.R. measurements of reaction velocities and equilibrium constants as a function of temperature, **3**, 187
N.M.R. spectroscopy, ^{13}C, in macromolecular systems of biochemical interest, **13**, 279
Non-planar and planar aromatic systems, **1**, 203
Norbornyl cation: reappraisal of structure, **11**, 179

Nuclear magnetic relaxation, recent problems and progress, **16**, 239
Nuclear magnetic resonance, *see* N.M.R.
Nuclear motion, principle of least, **15**, 1
Nucleophilic aromatic photosubstitution, **11**, 225
Nucleophilic catalysis of ester hydrolysis and related reactions, **5**, 237
Nucleophilic displacement reactions, gas-phase, **21**, 197
Nucleophilic vinylic substitution, **7**, 1

OH–bonds, hydrogen atom abstraction from, **9**, 127
Oxyacids of sulphur and their anhydrides, mechanisms and reactivity in reactions of organic, **17**, 65
Oxygen isotope exchange reactions of organic compounds, **3**, 123

Permutational isomerization of pentavalent phosphorus compounds, **9**, 25
Phase-transfer catalysis by quaternary ammonium salts, **15**, 267
Phosphorus compounds, pentavalent, turnstile rearrangement and pseudorotation in permutational isomerization, **9**, 25
Photochemistry of aryl halides and related compounds, **20**, 191
Photochemistry of carbonium ions, **9**, 129
Photosubstitution, nucleophilic aromatic, **11**, 225
Planar and non-planar aromatic systems, **1**, 203
Polarizability, molecular refractivity and, **3**, 1
Polarography and reaction kinetics, **5**, 1
Polypeptides, calculations of conformations of, **6**, 103
Pre-association, diffusion control and, in nitrosation, nitration, and halogenation, **16**, 1
Products of organic reactions, magnetic field and magnetic isotope effects on, **30**, 1
Protic and dipolar aprotic solvents, rates of bimolecular substitution reactions in, **5**, 173
Protolytic processes in H_2O–D_2O mixtures, **7**, 259
Protonation and solvation in strong aqueous acids, **13**, 83
Protonation sites in ambident conjugated systems, **11**, 267
Proton transfer between oxygen and nitrogen acids and bases in aqueous solution, mechanisms of, **22**, 113
Pseudorotation in isomerization of pentavalent phosphorus compounds, **9**, 25
Pyrolysis, gas-phase, of small-ring hydrocarbons, **4**, 147

Radiation techniques, application to the study of organic radicals, **12**, 223
Radical addition reactions, gas-phase, directive effects in, **16**, 51
Radicals, cation in solution, formation, properties and reactions of, **13**, 155
Radicals, organic application of radiation techniques, **12**, 223
Radicals, organic cation, in solution kinetics and mechanisms of reaction of, **20**, 55
Radicals, organic free, identification by electron spin resonance, **1**, 284
Radicals, short-lived organic, electron spin resonance studies of, **5**, 53
Rates and mechanisms of solvolytic reactions, medium effects on, **14**, 1
Reaction kinetics, polarography and, **5**, 1
Reaction mechanisms, use of volumes of activation for determining, **2**, 93
Reaction mechanisms in solution, entropies of activation and, **1**, 1
Reaction velocities and equilibrium constants, N.M.R. measurements of, as a function of temperature, **3**, 187

Reactions of hydrated electrons with organic compounds, **7**, 115
Reactions in dimethyl sulphoxide, physical organic chemistry of, **14**, 133
Reactive intermediates, study of, by electrochemical methods, **19**, 131
Reactivity indices in conjugated molecules, **4**, 73
Reactivity, organic, a general approach to: the configuration mixing model, **21**, 99
Reactivity–selectivity principle and its mechanistic applications, **14**, 69
Rearrangements, degenerate carbocation, **19**, 223
Refractivity, molecular, and polarizability, **3**, 1
Relaxation, nuclear magnetic, recent problems and progress, **16**, 239
Short-lived organic radicals, electron spin resonance studies of, **5**, 53
Small-ring hydrocarbons, gas-phase pyrolysis of, **4**, 147
Solid-state chemistry, topochemical phenomena in, **15**, 63
Solids, organic, electrical conduction in, **16**, 159
Solutions, reactions in, entropies of activation and mechanisms, **1**, 1
Solvation and protonation in strong aqueous acids, **13**, 83
Solvents, protic and dipolar aprotic, rates of bimolecular substitution-reactions in, **5**, 173
Solvolytic reactions, medium effects on the rates and mechanisms of, **14**, 1
Spectroscopic detection of tetrahedral intermediates derived from carboxylic acids and the investigation of their properties, **21**, 37
Spectroscopic observations of alkylcarbonium ions in strong acid solutions, **4**, 305
Spectroscopy, ^{13}C N.M.R., in macromolecular systems of biochemical interest, **13**, 279
Spin trapping, **17**, 1
Stability and reactivity of crown-ether complexes, **17**, 279
Stereoselection in elementary steps of organic reactions, **6**, 185
Steric isotope effects, experiments on the nature of, **10**, 1
Structure and mechanisms in carbene chemistry, **7**, 153
Structure and mechanism in organic electrochemistry, **12**, 1
Structure and reactivity of carbenes having aryl substituents, **22**, 311
Structure of electronically excited molecules, **1**, 365
Substitution, aromatic, a quantitative treatment of directive effects in, **1**, 35
Substitution, nucleophilic vinylic, **7**, 1
Substitution reactions, aromatic, hydrogen isotope effects in, **2**, 163
Substitution reactions, bimolecular, in protic and dipolar aprotic solvents, **5**, 173
Sulphur, organic oxyacids of, and their anhydrides, mechanisms and reactivity in reactions of, **17**, 65
Superacid systems, **9**, 1

Temperature, N.M.R. measurements of reaction velocities and equilibrium constants as a function of, **3**, 187
Tetrahedral intermediates derived from carboxylic acids, spectrosopic detection and the investigation of their properties, **21**, 37
Topochemical phenomena in solid-state chemistry, **15**, 63
Tritiated molecules, gaseous carbonium ions from the decay of **8**, 79
Tritium atoms, energetic, reactions with organic compounds, **2**, 201
Turnstile rearrangements in isomerization of pentavalent phosphorus compounds, **9**, 25

Unsaturated compounds, basicity of, **4**, 195

Vinyl cations, **9**, 185
Vinylic substitution, nucleophilic, **7**, 1
Volumes of activation, use of, for determining reaction mechanisms, **2**, 93

Water and aqueous mixtures, kinetics of organic reactions in, **14**, 203